Topology at Infinity of Discrete Groups

CONTEMPORARY MATHEMATICS

812

Topology at Infinity of Discrete Groups

AMS Special Session on Ends and Boundaries of Groups
In honor of Michael Mihalik's 70th Birthday
April 15–16, 2023
University of Cincinnati, Cincinnati, OH

Ross Geoghegan
Craig R. Guilbault
Kim Ruane
Editors

AMERICAN MATHEMATICAL SOCIETY

2020 *Mathematics Subject Classification.* Primary 20F65.

Library of Congress Cataloging-in-Publication Data

Names: AMS Special Session on Ends and Boundaries of Groups in Honor of Michael Mihalik's 70th Birthday (2023 : Cincinnati, Ohio) | Geoghegan, Ross, 1963– editor. | Guilbault, Craig R., 1959– editor. | Ruane, Kim, 1969– editor. | Mihalik, Michael L., 1951– honoree.

Title: Topology at infinity of discrete groups : AMS Special Session on Ends and Boundaries of Groups in Honor of Michael Mihalik's 70th Birthday, April 15–16, 2023, University of Cincinnati, Cincinnati, Ohio / Ross Geoghegan, Craig R. Guilbault, Kim Ruane, editors.

Other titles: AMS Special Session on Ends and Boundaries of Groups in Honor of Michael Mihalik's 70th Birthday, April 15-16, 2023, University of Cincinnati, Cincinnati, Ohio

Description: Providence, Rhode Island : American Mathematical Society, [2025] | Series: Contemporary mathematics, 0271-4132 ; volume 812 | Includes bibliographical references.

Identifiers: LCCN 2024035053 | ISBN 9781470475345 (paperback) | ISBN 9781470478636 (ebook)

Subjects: LCSH: Discrete groups. | Geometric group theory. | AMS: Group theory and generalizations – Special aspects of infinite or finite groups – Geometric group theory. | LCGFT: Festschriften.

Classification: LCC QA178 .A47 2025 | DDC 512/.55–dc23/eng/20241203
LC record available at https://lccn.loc.gov/2024035053

Contemporary Mathematics ISSN: 0271-4132 (print); ISSN: 1098-3627 (online)

DOI: https://doi.org/10.1090/conm/812

Color graphic policy. Any graphics created in color will be rendered in grayscale for the printed version unless color printing is authorized by the Publisher. In general, color graphics will appear in color in the online version.

♾ The paper used in this book is acid-free and falls within the guidelines
established to ensure permanence and durability.
Visit the AMS home page at `https://www.ams.org/`

10 9 8 7 6 5 4 3 2 1 30 29 28 27 26 25

To Michael Mihalik in honor of his seventieth birthday.

Contents

Preface

This collection of papers grew out of a Special Session in honor of Michael Mihalik's 70th birthday, which took place during the AMS Sectional Meeting at the University of Cincinnati on April 15–16, 2023. In recognition of Mihalik's work, the session focused on topics related to the topology at infinity of discrete groups. The key idea underlying geometric group theory is to study a group G by identifying a space X on which G acts, then looking for geometric and topological properties of X that shed light on G. For sufficiently nice actions, those properties of X can be directly attributed to G. In the cases of interest here, G is infinite and X is a locally compact noncompact complex. The manner in which X exhibits its noncompactness can often be captured by looking toward infinity. For example, the "number of ends" is the most basic property at infinity of X (and hence G); its importance is classical and is captured in famous work by Hopf and Stallings. But that is just the beginning. For example, one can sometimes distinguish 1-ended (also called *connected at infinity*) spaces and their groups, by defining, calculating, and comparing homology and cohomology groups of that end. If, instead, we turn to homotopy properties of an end, we arrive at a fundamental problem. When studying an end of a space X, proper rays are the analogs of basepoints in ordinary homotopy theory; and the analog of path connectedness is semistability—meaning that any two proper rays in X are properly homotopic. Spaces that are connected at infinity but not semistable exist, but the following fundamental problem is open.

Semistability Question. *If X is the Cayley complex of a finitely presented group G, must X be semistable? In other words, are all finitely presented groups semistable?*

Much of what is known about this question so far—including a multitude of positive answers under various conditions—is due to Mihalik and his collaborators. Appropriately, many of the talks in the Special Session and articles in this Proceedings have direct or indirect connections to this open question.

A method for capturing the elusive "topology at infinity" of X, in the form of a (typically complicated) compact space, involves the notion of a "boundary." When X is hyperbolic or CAT(0), a boundary for X is neatly determined by the geometry of geodesic rays in X. In more general cases, tools from infinite-dimensional topology and shape theory—in particular, $\mathcal{Z}$-sets and $\mathcal{Z}$-set compactifications, can sometimes be used to identify a boundary. Shape theory can then be used to parse out the topology at infinity—and in some cases, prove semistability. Several articles in this proceedings focus on group boundaries and the topology of $\mathcal{Z}$-sets. Others take their own perspective on understanding G by analyzing the topology (at infinity) of X.

All the papers in this volume make significant and interesting additions to our knowledge of geometric group theory and surrounding fields. We thank the authors for their contributions. We also thank the other speakers at the Special Session for making the event such a success. But above all we thank Mike Mihalik, whose hard work, good mathematics, and widely known enthusiasm and support for his colleagues, made this event and this volume a long-overdue recognition.

Ross Geoghegan, Craig R. Guilbault, Kim Ruane

Contemporary Mathematics
Volume **812**, 2025
https://doi.org/10.1090/conm/812/16259

Groups of proper homotopy equivalences of graphs and Nielsen Realization

Yael Algom-Kfir and Mladen Bestvina

ABSTRACT. For a locally finite connected graph X we consider the group $Maps(X)$ of proper homotopy equivalences of X. We show that it has a natural Polish group topology, and we propose these groups as an analog of big mapping class groups. We prove the Nielsen Realization theorem: if H is a compact subgroup of $Maps(X)$ then X is proper homotopy equivalent to a graph Y so that H is realized by simplicial isomorphisms of Y.

1. Introduction

The group $Out(F_n)$ of outer automorphisms of the free group of rank n can be thought of as the group of homotopy equivalences of a finite graph X with $\pi_1(X) \cong F_n$, up to homotopy. In this paper we begin the study of the analogous group associated with a *locally* finite graph X.

DEFINITION 1.1. Let X be a locally finite connected graph. The *mapping class group* $Maps(X)$ of X is the group of proper homotopy equivalences of X, up to proper homotopy.

Recall that $f : X \to Y$ is a *proper homotopy equivalence* if it is proper and there is a proper map $g : Y \to X$ such that both fg and gf are properly homotopic to the identity. For an example of a proper map which is homotopy equivalence but not a proper homotopy equivalence see Example 4.1.

We will equip $Maps(X)$ with a natural topology which will make it a Polish group (recall that this means that the underlying topological space is separable and admits a complete metric). See Section 4. We thus propose $Maps(X)$ as the "big $Out(F_n)$" equivalent of mapping class groups of surfaces of infinite type (or "big mapping class groups"), for a survey of the subject see [**2**]. Comparison with mapping class groups has shown to be very useful in the study of $Out(F_n)$, and we expect that comparison between $Maps(X)$ and big mapping class groups will likewise prove fruitful. We remark that the group of all automorphisms $Aut(F_\infty)$ of the free group of countable rank has a natural structure of a Polish group (e.g. it is a closed subgroup of the Polish group of all permutations of F_∞), and so does $Out(F_\infty)$ since the group of inner automorphisms is discrete. However, the

2020 *Mathematics Subject Classification.* Primary 57S99, 20F65.

The second author gratefully acknowledges the support by the National Science Foundation under the grant number DMS-1905720.

groups $Maps(X)$ seem more appealing as they have a more topological flavor and come in great variety since they depend on the graph X. Even when X is a tree the group $Maps(X)$ is of interest, it coincides with the group $Homeo(\partial X)$ of homeomorphisms of the space of ends ∂X of X (see Corollary 2.3). Note that if $h : X \to Y$ is a proper homotopy equivalence with inverse $h' : Y \to X$, then $f \mapsto hfh'$ induces an isomorphism $Maps(X) \to Maps(Y)$, which will turn out to be an isomorphism of topological groups, see Corollary 4.5.

In this paper we focus on compact subgroups of $Maps(X)$ and prove the following version of the Nielsen Realization theorem.

MAIN THEOREM. *Let H be a compact subgroup of $Maps(X)$. Then there is a locally finite graph Y proper homotopy equivalent to X so that under the induced isomorphism $Maps(X) \cong Maps(Y)$ the group H is realized as a group of simplicial isomorphisms of Y.*

Recall that the original Nielsen Realization theorem for finite type surfaces with negative Euler characteristic was proved by Kerckhoff [**18**], stating that any finite subgroup of the mapping class group of a surface of negative Euler characteristic can be realized by isometries of a complete hyperbolic metric with finite area. The version for $Out(F_n)$, proved in [**5, 15, 19, 22**], states that a finite subgroup of $Out(F_n)$ can be realized as a group of simplicial isomorphisms of a finite graph with fundamental group F_n. For big mapping class groups Nielsen Realization was proved recently by Afton-Calegari-Chen-Lyman [**1**]. Among the consequences is that compact subgroups of big mapping class groups are finite. This is not the case for $Maps(X)$. For example, let X be the graph obtained from $[0, \infty)$ by attaching two loops at every integer point. The group of symmetries of this graph is the compact group $H^\infty = \prod_{i=1}^{\infty} H$, where H is the group of symmetries of order 8 of the wedge of two circles. Note that X is proper homotopy equivalent to the graph Y obtained from $[0, \infty)$ by attaching three circles at every integer point, and the group of symmetries of Y is G^∞, where G is the group of order 48 of symmetries of the wedge of three circles. The groups $Maps(X)$ and $Maps(Y)$ are isomorphic as topological groups, but the realization using different graphs displays different compact subgroups.

In addition to big mapping class groups $Mod(\Sigma)$ and groups $Maps(X)$, the groups $Homeo(Z)$ of homeomorphism groups of compact totally disconnected metrizable spaces with compact-open topology have many similar properties and are studied more classically. Excluding the cases when Σ and X have finite type and Z is finite, all these groups have underlying space homeomorphic to the irrationals, they all admit clopen subgroups forming a basis of neighborhoods of the identity, and they all satisfy the Nielsen Realization theorem (for $Homeo(Z)$ this is proved quickly in Section 6, it states that for any compact subgroup $H < Homeo(Z)$ there is a metric on Z so that H consists of isometries).

Since this paper was first circulated, Domat, Hoganson and Kwak [**6**] investigated the coarse geometry of the pure subgroup $PMaps(X)$ of $Maps(X)$. Their work points out the differences between these three classes of topological groups.

Plan of the paper. We start by recalling the Classification theorem for locally finite graphs in Section 2. We also introduce the notation and review the homotopy extension theorem in our setting and some of its consequences. In Section 3 we explore the natural homomorphism $Maps(X) \to Out(\pi_1(X))$ and in particular we look for conditions that guarantee that a proper map $X \to X$ is properly homotopic to the identity. In the case when X is a core graph (i.e. it is the union of immersed loops) the criterion is particularly simple: if f induces the identity in $Out(\pi_1(X))$, it is properly homotopic to the identity. In the other extreme, when X is a tree, f is properly homotopic to the identity whenever it fixes the end space ∂X. The general case is more complicated since rays attached to the core graph could wrap around the core, but we show that if f is identity in π_1, fixes the ends, and preserves proper lines, then f is properly homotopic to the identity.

Section 4 is devoted to defining the topology on $Maps(X)$ and establishing that it is a Polish group. As long as X is of infinite type, we show that the underlying topological space of $Maps(X)$ is homeomorphic to the set of irrationals, but of course the group structure will depend on X.

The remainder of the paper is devoted to the proof of the Main Theorem. This is also divided into cases, with the two extremes of X being a core graph and being a tree discussed first. When X is a tree, by averaging we find an H-invariant metric on the space of ends ∂X. From this we construct H-invariant finite covers by disjoint clopen sets that refine each other and with mesh going to 0. The mapping telescope of this sequence is the desired tree Y.

The heart of the argument is the case when X is a core graph. For concreteness imagine that X is the graph obtained from the ray $[0, \infty)$ by attaching a circle at every integer point. We then cover $[0, \infty)$ by large intervals $J_1, J_2, \cdots$ so that $J_i \cap J_j = \emptyset$ if $|i - j| > 1$ and so that $J_i \cap J_{i+1}$ are large as well, controlling the properness of elements of H. Each J_i and $J_i \cap J_{i+1}$ defines a subgraph of X and a free factor of $\pi_1(X)$. By intersecting the H-translates of these free factors we obtain H-invariant free factors F_i^* and $F_{i,i+1}^*$ respectively. Using Nielsen Realization in finite rank we find finite graphs $\Gamma_{i,i+1}$ where H acts by simplicial isomorphisms realizing $F_{i,i+1}^*$. We then use the Relative Nielsen Realization, due to Hensel-Kielak [**14**], to construct finite graphs Γ_i where H acts by simplicial isomorphisms realizing F_i^*, and that contain $\Gamma_{i-1,i}$ and $\Gamma_{i,i+1}$ as disjoint invariant subgraphs. Finally, we glue the Γ_i's along these subgraphs to obtain Y. In general, when X is a tree with circles attached at vertices, the above outline still works, but instead of free factors we have to consider free factor systems which makes the notation a bit more complicated.

In the general case, we first use the case of core graphs to reduce to the situation where H is already acting on the core by simplicial isomorphisms. The graph X is obtained from the core by attaching trees, and the central part of the proof in this case is to see how to attach new trees in an equivariant fashion. The new trees are going to be mapping telescopes made of partition elements (as in the tree case) subordinate to suitable clopen sets in ∂X. To accomplish this we prove a fixed point theorem (see Lemma 7.3) that provides a suitable point in the core where these telescopes are attached.

In order to verify that the action of H on the new graph Y is conjugate to the given action on X we use the machinery developed in Section 3, see Corollary 3.6.

2. The classification of locally finite connected graphs

Let X be a locally finite infinite connected graph. The fundamental group $\pi_1(X)$ is free and we denote its rank by $g(X) \in \{0, 1, 2, \cdots, \infty\}$ and think of it as the "genus" of X. Let

$$\partial X = \lim_{\longleftarrow K \subset X} \pi_0(X \smallsetminus K)$$

be the space of ends of X with its usual inverse limit topology, where the limit runs over all compact subsets $K \subset X$. Then ∂X is a totally disconnected compact metrizable space (recall that these are precisely the spaces homeomorphic to a closed subset of the Cantor set). The union $\hat{X} = X \sqcup \partial X$ has a natural topology that makes it compact; it is the *Freudenthal* (or *end*) compactification of X. The basis of open sets in $\hat{X}$ consists of open sets in X, and for every compact $K \subset X$ and every component U of $X \smallsetminus K$ the set $\hat{U}$ which is the union of U and the set of ends that map to U. We will sometimes abuse notation and talk about a neighborhood U in X of an end $\beta \in \partial X$; what we mean is the intersection of such a neighborhood $\hat{U}$ in $\hat{X}$ with X. The end compactification can also be constructed in the same way for connected, locally finite cell complexes. Every proper map $f : X \to Y$ between such complexes extends continuously to a map $\hat{X} \to \hat{Y}$ between their end compactifications. For simplicity we will usually denote this extension, as well as its restriction $\partial X \to \partial Y$, by f as well. Properly homotopic maps induce the same map between the boundaries.

Denote by $X_g \subset X$ the *core* of X, i.e. the smallest subgraph that contains all immersed loops. Thus $X_g = \emptyset$ precisely when X is a tree. Let $\partial X_g \subseteq \partial X$ be the space of ends of X_g; it is a closed subspace of ∂X (and consists of ends "accumulated by genus"). Thus $\partial X_g = \emptyset$ precisely when $g(X) < \infty$. In general, X is either a tree or is obtained from X_g by attaching trees.

DEFINITION 2.1 (Characteristic pairs). If $g(X) < \infty$, its characteristic pair is $(\partial X, g(X))$, otherwise its characteristic pair is $(\partial X, \partial X_g)$.

The following is the analog of Kerékjártó's classification theorem for surfaces and was proved by Ayala-Dominguez-Marquez-Quintero [**3**].

THEOREM 2.2. *Let X, Y be locally finite connected graphs. Then a homeomorphism of characteristic pairs extends to a proper homotopy equivalence. If X and Y are trees the extension is unique up to proper homotopy.*

In the case when the genus is finite, a homeomorphism of characteristic pairs means a homeomorphism between the spaces of ends together with the information that the genera are equal.

If $f : X \to X$ is a proper homotopy equivalence, then the extension $\partial X \to \partial X$ is a homeomorphism which preserves ∂X_g. Thus we have a well defined homomorphism

$$\sigma : Maps(X) \to Homeo(\partial X, \partial X_g)$$

to the group of homeomorphisms of the pair $(\partial X, \partial X_g)$. The following is then an immediate corollary of the classification theorem.

COROLLARY 2.3. *The homomorphism σ is always surjective. If X is a tree the map $\sigma : Maps(X) \to Homeo(\partial X)$ is an isomorphism.*

In light of this we will usually focus on the kernel of σ, which is the *pure mapping class group* of X, and we denote it by $PMaps(X)$.

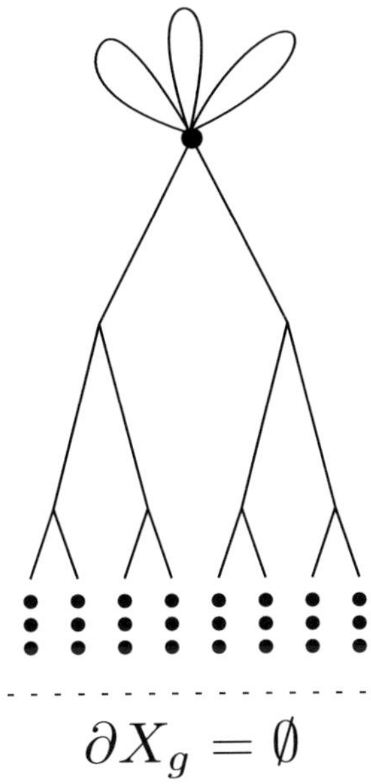

FIGURE 1. A graph with a finitely generated fundamental group. In this case $\partial X_g = \emptyset$ and the space of ends can be any closed subset of the Cantor set.

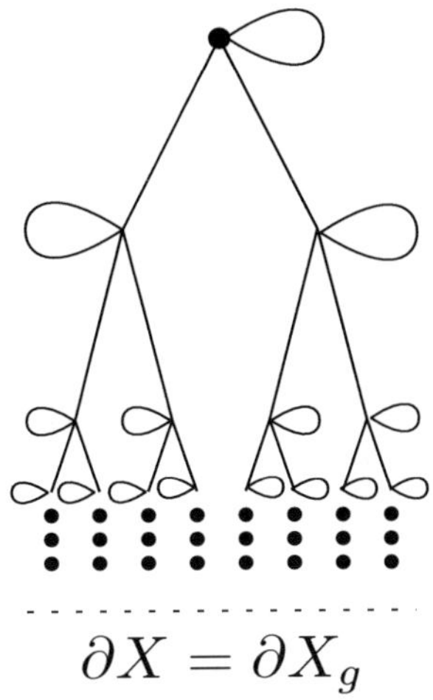

FIGURE 2. This is a core graph, $\partial X = \partial X_g$. By deleting a part of the Cantor tree we get a standard model for (A, A) where A is any closed subspace of the Cantor set.

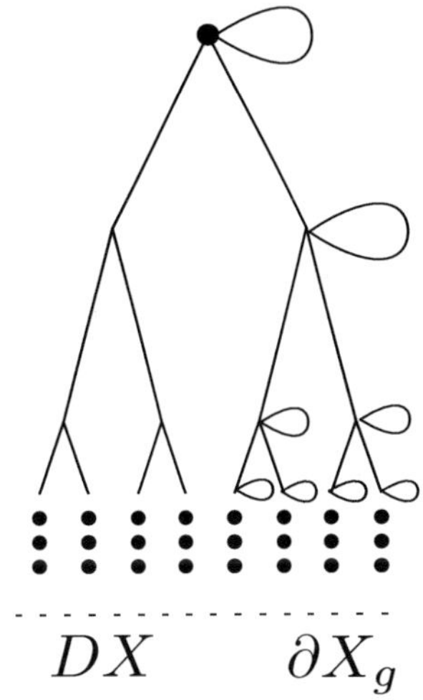

FIGURE 3. In this case both ∂X_g and its complement $DX := \partial X \setminus \partial X_g$ are non-empty. By deleting loops/subtrees from the space in Figure 2 we get a model for any pair of closed subsets $\partial X = A \supset B = \partial X_g$ of the Cantor set.

DEFINITION 2.4. The pure group $PMaps(X)$ is the subgroup of $Maps(X)$ consisting of $f \in Maps(X)$ so that $f : \partial X \to \partial X$ is the identity.

We shall make use of the following concepts and lemmas from [**3**].

DEFINITION 2.5 (Standard Models). The Cantor tree T is the rooted binary tree embedded in the plane so that its boundary is the standard trinary Cantor set in $[0, 1] \times \{0\}$. For each closed subset B of the Cantor set, let T_B be the union of the set of rays in T initiating at the root and terminating in B. For a characteristic pair (B, g) where B is a closed subset of the Cantor set and g is a natural number let $X_{(B,g)}$ be the tree T_B with g loops attached at the root. The characteristic set of $X_{(B,g)}$ is (B, g). For two closed non-empty subsets $A \subseteq B$ of the Cantor set, let $X_{(B,A)}$ be the tree T_B with a one edge loop attached at each vertex of the subtree T_A. Again, the characteristic set of $T_{(B,A)}$ is (B, A). These trees are called the Standard Models.

COROLLARY 2.6. *Every locally finite connected infinite graph is proper homotopy equivalent to a Standard Model.*

In particular, we can assume that X has no valence 1 vertices, and that $X = X_g$ if $\partial X_g = \partial X$.

The following is the standard Homotopy Extension Theorem, see [**13**]. We will use it often, usually without saying it. Given a subgraph $Y \subset X$, the frontier of Y is the set of vertices in $Y - int(Y)$ denoted $Fr(Y)$.

PROPOSITION 2.7. *Let Y be a subgraph of X, let $H\colon Y \times I \to X$ be a proper homotopy from h to f. Let $u\colon X \to X$ be a proper map so that $u|_Y = h$. Then there is a proper homotopy $H'\colon X \times I \to X$ extending H and u, i.e. $H'(x, 0) = u(x)$. Moreover, if $K \subset X$ is a subset so that $H(Fr(Y)\times I) \cap K = \emptyset$ and $u(\overline{X \smallsetminus Y}) \cap K = \emptyset$ then $H'((X \smallsetminus Y) \times I) \cap K = \emptyset$.*

PROOF. First define the extension on the vertices v of X outside of Y by $H'(v, t) = u(v)$. Now let e be an edge with endpoints a, b. If both a, b are outside of Y define H' on $e \times I$ to be stationary as well: $H'(x, t) = u(x)$ for $x \in e$. Otherwise let $\alpha : [0, 1] \to e$ be a parametrization of e. Let P be a retraction from $I \times I$ to $(\{0\} \times I) \cup (I \times \{0\}) \cup (\{1\} \times I)$. Then $e \times I$ can be identified with $I \times I$ (with $0 \times I$ and $1 \times I$ identified if e is a loop). The homotopy is already defined on $(\{0\} \times I) \cup (I \times \{0\}) \cup (\{1\} \times I)$ and we extend it to $I \times I$ by composing with P.

We leave the verification that H' is proper and the last sentence to the reader. □

We will also sometimes have a need to restrict $f \in Maps(X)$ to an "invariant" subgraph $Y \subset X$ e.g. $Y = X_g$. Since f is defined only up to proper homotopy, Y will usually not satisfy $f(Y) \subseteq Y$, but this will be true after a proper homotopy H. We will also want to know that the proper homotopy class of the restriction is independent of the choice of H. A bad example to keep in mind is the projection $\pi : \mathbb{R}^2 \to \mathbb{R}$. There are proper maps $f : \mathbb{R} \to \mathbb{R}^2$ such that πf is not proper (and there are analogous examples with Cayley graphs $Cay(\mathbb{Z}) \to Cay(\mathbb{Z}^2)$), and there are pairs of properly homotopic maps $f, g : \mathbb{R} \to \mathbb{R}^2$ such that πf and πg are proper, but not properly homotopic.

LEMMA 2.8. *Let Y be a subgraph of X such that inclusion $Y \hookrightarrow X$ induces an injection $\partial Y \hookrightarrow \partial X$. Let $\pi : X \to Y$ be a retraction with the following property: for every $\beta \in \partial Y$ and every neighborhood U of β in Y there is a neighborhood V of β in X so that $\pi(V) \subseteq U$. Let Z be a locally compact metrizable space.*

(i) *If $f : Z \to X$ is a proper map such that $f(\partial Z) \subseteq \partial Y$ then $\pi f : Z \to Y$ is proper.*
(ii) *If $f, g : Z \to Y$ are proper maps such that they are properly homotopic within X, then they are properly homotopic within Y.*

PROOF. To prove (i), note that if $\gamma \in \partial Z$ then $f(\gamma) = \beta \in \partial Y \subseteq \partial X$. If U is a neighborhood of β in Y, there is a neighborhood V of β in X so that $\pi(V) \subseteq U$. Since f is proper, there is a neighborhood W of γ in Z so that $f(W) \subseteq V$. It follows that $\pi f(W) \subseteq U$, so πf is proper. For (ii), apply (i) to a proper homotopy $H : Z \times I \to X$. The fact that $H(\partial(Z \times I)) \subseteq \partial Y$ follows from the assumption that $\partial Y \subseteq \partial X$. □

PROPOSITION 2.9. *Suppose X, Y are two locally finite connected graphs, $f : X_g \to Y$ a proper map that induces $\overline{f} : \partial X_g \to \partial Y$ and let $\overline{F} : \partial X \to \partial Y$ be an extension of $\overline{f}$. Then there is a proper map $F : X \to Y$ that extends f and induces $\overline{F}$.*

PROOF. First consider the case when X and Y are trees, and choose base vertices $x_0 \in X$, $y_0 \in Y$. We are given $\overline{F} : \partial X \to \partial Y$ and we have to construct a proper map $F : X \to Y$. If $\overline{F}$ is a homeomorphism the existence of F follows from the Classification theorem (and in fact it is unique up to proper homotopy). In general, we can construct F as follows. When v is a vertex of X let the shadow $Sh_X(v) \subseteq \partial X$ be the set of endpoints of rays that start at x_0 and pass through v. When $A \subseteq \partial X$ contains at least two points, let $\sup A$ be the vertex $v \in X$ with the largest distance $|v|$ from x_0 satisfying $Sh_X(v) \supseteq A$. If $A = \{\beta\} \subseteq \partial X$ is a single point, define $\sup A$ to be β and let $|\beta| = \infty$. Make the similar definition for subsets of ∂Y. For a vertex $v \in X$ consider $\sup \overline{F}(Sh_X(v))$. If this is a vertex w at distance $|w| \leq |v|$ from y_0 then define $F(v) = w$, and otherwise define $F(v)$ as the vertex at distance $|v|$ from y_0 along the segment (or ray) $[y_0, w]$. Extend F linearly to the edges of X.

Now consider the general case. We may assume that f sends vertices to vertices. First suppose that there is a maximal tree $T \subseteq Y$ such that $\partial T = \partial Y$. Since X is simplicial and locally finite, $X \smallsetminus X_g$ is a countable (or finite) union of trees, and let T_i for $i \in \mathbb{N}$ be the closure of a component of $X \smallsetminus X_g$. We denote by x_i the point of intersection of T_i and X_g, so x_i is the root of T_i. For each $i \in \mathbb{N}$ let $F_i \colon T_i \to T \subseteq Y$ be the map constructed in the first paragraph so that $F_i(x_i) = f(x_i)$ and $\partial F_i = \overline{F}|_{\partial T_i}$. We define F by gluing the maps f and F_i for all $i \in \mathbb{N}$.

One way to avoid constructing a special maximal tree is as follows. Let Z be a Standard Model proper homotopy equivalent to Y. Then the underlying tree in Z has the same ends. So we may apply the above paragraph to the composition $X_g \to Y \to Z$ and get an extension $X \to Z$, which we then compose with the inverse proper homotopy equivalence $Z \to Y$ to get $F : X \to Y$. The map F may not agree with f on X_g but it is properly homotopic to it, so we conclude by applying the Homotopy Extension Theorem (Proposition 2.7). □

3. Relationship with $Out(\pi_1(X))$

In this section we will investigate the relationship between $Maps(X)$ and $Out(\pi_1(X))$. First, there is a natural homomorphism

$$\Psi : Maps(X) \to Out(\pi_1(X))$$

that sends $h \in Maps(X)$ to (the outer automophism class of) $h_* : \pi_1(X) \to \pi_1(X)$. If the genus of X is infinite, Ψ is not onto since there are automorphisms not realized by proper maps. On the other hand, Ψ is onto when $g(X) < \infty$. We will show first that Ψ is injective if X is a core graph (meaning $X = X_g$), and in general we will describe the kernel of Ψ. The next theorem is analogous to the fact that a homeomorphism of a surface with nonabelian fundamental group that induces identity in π_1 is isotopic to the identity, see [**9, 16**].

THEOREM 3.1. *Suppose X is a core graph and let $f \colon X \to X$ be a proper map so that $f_* = id \in Out(\pi_1(X))$. Then f is properly homotopic to the identity on X. In other words, Ψ is injective.*

PROOF. In the proof we will use the fundamental property of graphs that disjoint nontrivial loops are not homotopic and that nullhomotopic loops can be nullhomotoped within their images. We may assume that X is a Standard Model. Note that f necessarily induces the identity on the space of ends. Indeed, if $f(\beta) \neq$

β for an end β, there will be an immersed loop α in X near β such that $f(\alpha)$ is disjoint from α, and in particular f does not fix the conjugacy class of α.

Next, we can assume, by applying a proper homotopy (using Proposition 2.7) that f fixes all vertices and moreover, by homotoping the root v around a loop, that $f_* : \pi_1(X, v) \to \pi_1(X, v)$ is the identity.

We will now construct a proper homotopy between the identity and f. If w is a vertex, let $e_1 e_2 \cdots e_k$ be the edge path in the underlying tree T from v to w and define $H : \{w\} \times I \to X$ to be the tightened path $\bar{e}_k \ldots \bar{e}_1 f(e_1) \ldots f(e_k)$. Also define H on $X \times \{0\}$ to be identity and on $X \times \{1\}$ to be f (see Figure 4). We will argue below that H defined so far is proper. If e is an edge in T then H is defined on $\partial(e \times I)$ and is nullhomotopic on this loop. Thus we may extend H to all such 2-cells keeping the image contained in the image of $\partial(e \times I)$. Finally, H extends to the cylinders $x_w \times I$, where x_w is the loop attached at w, using the fact that $f(e_1 \cdots e_k x_w \bar{e}_k \cdots \bar{e}_1) \simeq e_1 \cdots e_k x_w \bar{e}_k \cdots \bar{e}_1$. We again ensure that the image of the extension is contained in the image of the boundary of the 2-cell, so the extended H will be proper.

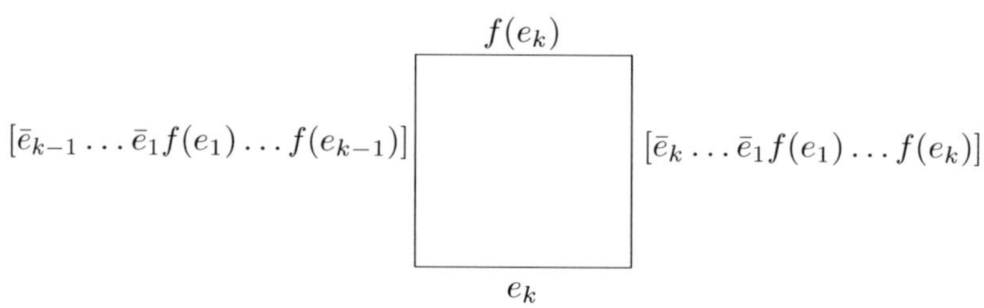

FIGURE 4. The homotopy H. The brackets signify that we take the immersed path homotopic to the given one rel endpoints.

It remains to show that H defined on the 1-skeleton is proper. This is where we will use the assumption that X is a core graph. Let $K \subset X$ be a finite subgraph. Since f is proper, there is a finite subgraph $L \subset X$ such that $f(X \smallsetminus L) \subseteq X \smallsetminus K$. Let w be a vertex outside of L, $e_1 \cdots e_k$ the edge path from v to w in T, and x_w the loop attached to w. Thus $f(x_w) \cap K = \emptyset$. From the fact that

$$f(e_1 \ldots e_k) f(x_w) f(\bar{e}_k \ldots \bar{e}_1) \simeq e_1 \ldots e_k \cdot x_w \cdot \bar{e}_k \ldots \bar{e}_1$$

we see that after tightening $f(e_1 \ldots e_k)$ does not cross any loops attached to vertices in K. For example, the fundamental group can be thought of as the free group on the attached loops, and if y is the last loop in K crossed by $[f(e_1 \cdots e_k)]$ the word $[f(e_1 \cdots e_k)] \cdot [f(x_w)] \cdot [f(\bar{e}_k \cdots \bar{e}_1)]$ could be tightened by tightening the portion between the corresponding y and $\bar{y}$, and would not yield the trivial word. Therefore $H(\{w\} \times I)$ is disjoint from K. $\square$

The fundamental group of X does not "see" the ends of X not accumulated by genus. For example, if X is a tree the mapping class group $Maps(X)$ is isomorphic to the homeomorphism group $Homeo(\partial X)$ and may be quite nontrivial, while $\pi_1(X) = 1$. It is therefore natural, when studying the kernel of Ψ, to restrict to the pure mapping class group $PMaps(X)$.

We will now describe the kernel of the restriction

$$\Psi_P : PMaps(X) \to Out(\pi_1(X))$$

of Ψ. This is well-known when X is obtained from a finite graph, say of rank r so that $\pi_1(X) = F_r$, by attaching a finite number, say n, of rays. These rays can be thought of equivalently as distinguished points in the finite graph. When $n = 1$ we have $Maps(X) \cong Aut(F_r)$ and when $n > 1$ then $PMaps(X) \cong Aut(F_r) \ltimes F_r^{n-1}$ with the natural diagonal action of $Aut(F_r)$ on F_r^{n-1}. The F_r^{n-1} factor can be thought of as measuring the marking of $(n-1)$ distinguished points with respect to the remaining distinguished point, which is considered to be the basepoint. This will be generalized in Corollary 3.9. When $r \geq 2$ the kernel of Ψ_P is isomorphic to F_r^{n-1} represented by maps that are identity on the finite graph and send each ray R to a ray of the form $w_R R$ for some loop w_R in the finite graph.

Let X be a locally finite graph and we assume $DX := \partial X - \partial X_g \neq \emptyset$ and choose $\alpha_0 \in DX$. This will be the basepoint "at infinity". Let $\pi_1(X, \alpha_0)$ be the set of proper homotopy classes of lines $\sigma \colon \mathbb{R} \to X$ so that $\lim_{t\to-\infty} \sigma(t) = \lim_{t\to\infty} \sigma(t) = \alpha_0$, with concatenation as the group operation. Notice that concatenation makes sense since $\alpha_0 \in DX$ and any two rays limiting to α_0 eventually coincide, up to a proper homotopy. Given $x_0 \in X$, there is an isomorphism

$$\pi_1(X, x_0) \to \pi_1(X, \alpha_0)$$

given by $\gamma \to \bar{\rho}_{x_0} \gamma \rho_{x_0}$ where ρ_{x_0} is a fixed ray in X from x_0 to α_0. Moreover, if $f \in Maps(X)$ fixes α_0 then f induces a map $f_0 \in Aut(\pi_1(X, \alpha_0))$.

We first consider the case of the graph $X = X_g^*$ obtained from a core graph (or a point) X_g by attaching a single ray.

LEMMA 3.2. *For $X = X_g^*$ the kernel of $\Psi : Maps(X) \to Out(\pi_1 X)$ (or $\Psi_P : PMaps(X) \to Out(\pi_1 X)$) is isomorphic to $\pi_1(X) = \pi_1(X_g)$ when this group is nonabelian, and otherwise it is trivial.*

When the genus $n = g(X)$ is finite, the lemma says that the kernel $Aut(F_n) \to Out(F_n)$ is F_n for $n > 1$ and otherwise it is trivial.

PROOF. Let $f \in Maps(X)$ induce identity in $Out(\pi_1(X))$. Using Lemma 2.8 applied to the nearest point projection $\pi : X \to X_g$ we see that after a proper homotopy we may assume that f preserves the core X_g, and thus by Theorem 3.1 we may assume f is identity on X_g. Let ρ_0 denote the geodesic ray in X that intersects X_g at one point and such that $X = X_g \cup \rho_0$. Let $c(f)$ be the homotopy class of $\bar{\rho}_0 f(\rho_0)$ in $\pi_1(X, \alpha_0)$. Then $f_0 \in Aut(\pi_1(X, \alpha_0))$ is just conjugation by $c(f)$. The map $f \mapsto c(f)$ is a homomorphism $Ker(\Psi) \to \pi_1(X, \alpha_0)$. When $\pi_1(X)$ is nonabelian it is an isomorphism. If X_g is a circle, f can be homotoped to the identity by a homotopy that rotates the circle to unwind the attached ray, so $Ker(\Psi)$ is trivial. □

We now consider the general case. Let X be a Standard Model which is not a tree, let X_g be the core subgraph of X, and we assume $DX \neq \emptyset$. Fix $\alpha_0 \in DX$ and let ρ_0 be the ray in X intersecting X_g in a point and limiting to α_0. Let $T \subset X$ be the underlying tree, and let $T_g = X_g \cap T$ be the underlying tree in the core, and likewise let $T_g^* = T_g \cup \rho_0$ be the underlying tree in $X_g^* = X_g \cup \rho_0$. Thus $X_g \subset X_g^* \subseteq X$ and both inclusions are homotopy equivalences. We note that restriction maps

$$PMaps(X) \to PMaps(X_g^*) \to PMaps(X_g)$$

are well-defined by Lemma 2.8, where for the retraction π we take the nearest point projection. In fact, we have a factorization of $PMaps(X) \to Out(\pi_1(X))$ as

$$PMaps(X) \to PMaps(X_g^*) \to PMaps(X_g) \to Out(\pi_1(X))$$

Our next goal is to describe $Ker(PMaps(X) \to PMaps(X_g^*))$. To that end, we introduce the group $\mathcal{R}$ that measures how the rays towards the ends in DX wrap around the loops in the core graph X_g.

DEFINITION 3.3. The group $\mathcal{R}$ as a set is the collection of functions $h\colon DX \to \pi_1(X_g^*, \alpha_0)$ satisfying

(R0) $h(\alpha_0) = 1$.

(R1) h is locally constant.

(R2) For all $\beta \in \partial X_g$ and every neighborhood U of β in X there exists a ray ρ_U from a point in U to α_0 which is the concatenation of a segment in T and ρ_0 so that if $(\beta_i)_{i=1}^{\infty} \subset DX$ limits to β then for large enough i, $h(\beta_i) = \bar{\rho}_U * \gamma_i * \rho_U$ where γ_i is a loop contained in U.

The group operation in $\mathcal{R}$ is pointwise multiplication in $\pi_1(X, \alpha_0)$.

DEFINITION 3.4. We start by assigning an element $\Phi_T(f) \in \mathcal{R}$ to certain proper maps $f : X \to X$. More precisely assume:

(i) $f : \partial X \to \partial X$ is the identity, and

(ii) either DX is compact or $f_* : \pi_1(X, \alpha_0) \to \pi_1(X, \alpha_0)$ is the identity.

Note that we do not assume that f is a proper homotopy equivalence, cf. Example 4.1.

If $\beta \in DX \smallsetminus \{\alpha_0\}$ let ℓ_β be the bi-infinite line in T connecting α_0 to β. Define the map

$$\Phi_T(f) = h\colon DX \to \pi_1(X, \alpha_0)$$

$$h(\beta) = \begin{cases} \ell_\beta f(\bar{\ell}_\beta) & \beta \in DX - \{\alpha_0\}, \\ 1 & \beta = \alpha_0 \end{cases}$$

We claim that indeed $h \in \mathcal{R}$. That h is locally constant follows from the observation that if $\beta \in DX$ there is a neighborhood $U \subset DX$ of β so that if ℓ is a line joining two distinct points in U then $f(\ell)$ is properly homotopic to ℓ.

Condition (R2) is vacuous if DX is compact so suppose $f_* = id$. Thus we may assume that f is identity on X_g^*. Let ℓ_β be the line in T from α_0 to β, and similarly let ℓ_{β_i} be the line in T from α_0 to β_i. Note that the lines ℓ_{β_i} converge ℓ_β. The map f fixes ℓ_β and thus for large i takes ℓ_{β_i} to a line that agrees with ℓ_{β_i} outside a given neighborhood U of β. This proves (R2).

THEOREM 3.5. *Assume $f\colon X \to X$ is proper, $\alpha_0 \in DX$, and $f_* = id\colon \pi_1(X, \alpha_0) \to \pi_1(X, \alpha_0)$. If $f\colon \partial X \to \partial X$ is identity and $\Phi_T(f) = 1$ then f is properly homotopic to the identity. Moreover,*

$$\Phi_T : Ker(PMaps(X) \to PMaps(X_g^*)) \to \mathcal{R}$$

is an isomorphism.

PROOF. Let $\mathcal{K} = Ker(PMaps(X) \to PMaps(X_g^*))$. We start by arguing that $\Phi_T\colon \mathcal{K} \to \mathcal{R}$ is a homomorphism. Let $f, g \in \mathcal{K}$, and we assume that f, g are identity on X_g^*. Let ℓ_β be the line as in the definition of Φ_T. We have

$$\Phi_T(gf)(\beta) = \ell_\beta \cdot gf(\bar{\ell}_\beta) = \ell_\beta \cdot g(\bar{\ell}_\beta) \cdot g(\ell_\beta \cdot f(\bar{\ell}_\beta)) = \Phi_T(g)(\beta) \cdot g(\Phi_T(f)(\beta))$$

which equals $\Phi_T(g)(\beta)\cdot\Phi_T(f)(\beta)$ since g acts as the identity on $\pi_1(X,\alpha_0)$.

We next show that if $\Phi_T(f)=1$, then $f\simeq id$. We may assume f is identity on X_g^*. Consider the universal cover $\tilde{X}$ of X and let $\tilde{X}_g$ be the preimage of X_g to $\tilde{X}$ (which is connected). Let $\tilde{f}$ be the lift of f that restricts to the identity on $\tilde{X}_g$. The assumption that $\Phi_T(f)(\beta)=1$ for every $\beta\in DX$ amounts to saying that $\tilde{f}$ fixes the ends of $\tilde{X}$. The straight line homotopy $\tilde{H}$ from $\tilde{f}$ to id is equivariant with respect to the deck group and descends to the homotopy $H\colon X\times I\to X$ from f to id. It remains to show that H is proper. It is useful to describe H directly. If $x\in X$ consider a ray ρ_x from x to α_0 in X. The path $H(\{x\}\times I)$ is the tightened path $f(\rho_x)\bar{\rho}_x$.

Let K be a compact set in X. As in the proof of Theorem 3.1 it is enough to show that there is a compact set S so that for each vertex v outside S, $H(\{v\}\times I)\cap K=\emptyset$.

Assume $x\in X$ is in a small neighborhood of an end $\beta\in\partial X$. If $\beta\in\partial X_g$ then $f(\rho_x)$ and ρ_x agree outside a bit larger neighborhood by the assumption that f fixes β and X_g^*, so the path $H(\{x\}\times I)$ is also close to β. If $\beta\in DX$ consider the concatenation $\bar{\rho}_x\gamma$ where γ is the ray from x to β in T. This concatenation is properly homotopic to the line ℓ_β from the definition of Φ_T, and so by our assumption that $\Phi_T(f)=1$ we have $f(\bar{\rho}_x\gamma)\simeq\bar{\rho}_x\gamma$. Since f fixes β, $f(\gamma)$ is in a bit larger neighborhood of γ, so $f(\rho_x)\bar{\rho}_x\simeq f(\gamma)\bar{\gamma}$ is also close to β.

Finally, we argue that $\Phi_T:\mathcal{K}\to\mathcal{R}$ is onto. Let $h\in\mathcal{R}$. We define a proper homotopy equivalence $f\colon X\to X$. Let $f|_{X_g^*}=id$. Let S_w be the tree attached to a vertex $w\in X_g^*$, i.e. the closure of a component of $T\smallsetminus T_g^*$. Then ∂S_w is compact and from the fact that h is locally constant we see that there is some distance C so that for every edge $e\subset S_w$ at distance C from w, h is constant on the ends of the unbounded component of $S\smallsetminus e$. Define $f|S_w$ to be identity on all edges of S other than those at distance C from w. On such an edge e define f to be the immersed path with the same endpoints as e and so that if ℓ_β is a line that crosses e then $f(\ell_\beta)\bar{\ell}_\beta\in\pi_1(X,\alpha_0)$ represents $h(\beta)$ (equivalently, $\bar{\rho}_e f(e)\bar{e}\rho_e$ represents $h(\beta)$ for a suitable ray ρ_e in T with e oriented towards β). Since $h(\alpha_0)=1$ the map f will be identity on X_g^*, as no edges e where we modify the identity are along the ray to α_0. We must show that f is a proper map. Let $\beta\in\partial X$ and assume that the edge e as above is close to β. This means that $\beta\in\partial X_g$ (ends in DX have a neighborhood not containing any edges e as above). It follows that $f(e)$ is close to β by (R2). Thus f is a proper map. Let g be constructed similarly for $h^{-1}\in\mathcal{R}$. Then $gf\in PMaps(X)$ has the property that $\Phi_T(gf)=\Phi_T(g)\Phi_T(f)=h^{-1}h=id$, so $gf\simeq id$ by the injectivity of Φ_T, and similarly $fg\simeq id$. □

We now have a useful criterion when a proper map $X\to X$ is properly homotopic to the identity, without assuming it is a proper homotopy equivalence.

COROLLARY 3.6. *Let $f:X\to X$ be proper. Then f is properly homotopic to the identity if and only if it preserves the homotopy class of every oriented closed curve and the proper homotopy class of every oriented proper line in X that in each direction converges to an end in $DX=\partial X\smallsetminus\partial X_g$.*

PROOF. If X is a core graph we are assuming that $f_*\in Out(\pi_1(X))$ preserves all conjugacy classes in $\pi_1(X)$. This implies that $f_*=id$ and the conclusion follows from Theorem 3.1. If $DX\neq\emptyset$ choose some $\alpha_0\in DX$. We now see that $f_*:\pi_1(X,\alpha_0)\to\pi_1(X,\alpha_0)$ is identity, since f preserves lines that start and end at α_0, and similarly f fixes all ends of X. Finally, we see that $\Phi_T(f) = 1$ since f preserves all lines joining α_0 with any $\beta \in DX$, so the statement follows from Theorem 3.5. □

COROLLARY 3.7. *Suppose $f : X \to Y$ is proper, induces a homeomorphism $\partial X \to \partial Y$ and the restriction $X_g \to Y_g$ is a proper homotopy equivalence. Then f is a proper homotopy equivalence.*

PROOF. Using Proposition 2.9 we have a proper map $g : Y \to X$ so that the restriction $Y_g \to X_g$ is the homotopy inverse to $f : X_g \to Y_g$ and so that fg and gf are identity on the boundaries. But then both are proper homotopy equivalences by Theorem 3.5, and thus both f and g are as well. □

Recall that $\mathcal{R} \cong Ker(PMaps(X) \to PMaps(X_g^*))$.

COROLLARY 3.8. *If $\alpha_0 \in DX$ and if $\pi_1(X, \alpha_0)$ is nonabelian then*

$$K = Ker(PMaps(X) \to PMaps(X_g))$$

fits in an exact sequence

$$1 \to \mathcal{R} \to K \to \pi_1(X, \alpha_0) \to 1$$

If $X_g = S^1$ then $Ker(PMaps(X) \to PMaps(X_g) = \mathbb{Z}/2\mathbb{Z})$ is isomorphic to $\mathcal{R}$.

PROOF. We focus on the first statement. The horizontal and vertical sequences in the commutative diagram below are exact by Theorem 3.5 and Lemma 3.2. The diagonal sequence is exact by the definition of K. The construction of the red arrows and the exactness of the sequence are a diagram chase.

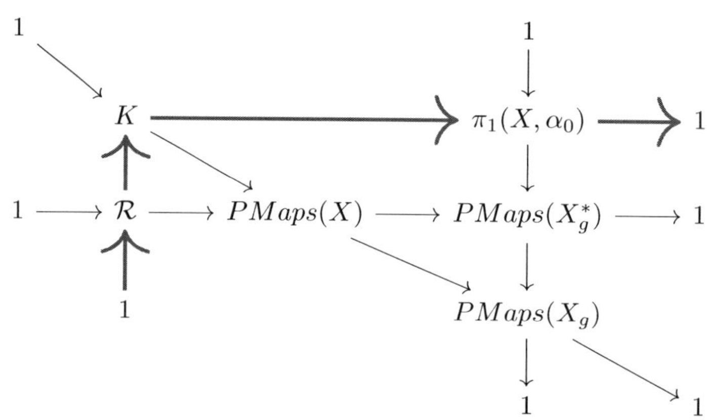

□

When DX is compact there is a more refined statement. Note that in that case condition (R2) in the definition of $\mathcal{R}$ is vacuous.

COROLLARY 3.9. *If DX is compact and nonempty then*

$$PMaps(X) \cong \mathcal{R} \rtimes PMaps(X_g^*)$$

where $PMaps(X_g^*)$ acts on $\mathcal{R}$ by $g \cdot h(\beta) = g_*(h(\beta))$, where $g_* : \pi_1(X_g^*, \alpha_0) \to \pi_1(X_g^*, \alpha_0)$ is the homomorphism induced by g.

PROOF. By Definition 3.4, we have $\Phi_T : PMaps(X) \to \mathcal{R}$, and let $R : PMaps(X) \to PMaps(X_g^*)$ be the restriction. Thus we have a function

$$(\Phi_T \times R) : PMaps(X) \to \mathcal{R} \rtimes PMaps(X_g^*)$$

That this is a homomorphism follows from the displayed calculation in the proof of Theorem 3.5. That the map is 1-1 and onto follows from Theorem 3.5 plus the observation that R is onto. □

When DX is not compact, Φ_T may not be a well-defined function to $\mathcal{R}$ since (R2) may fail.

4. Topology

It takes a bit of care to define the topology on $Maps(X)$. Let $\hat{X}$ denote the Freudenthal compactification $X \cup \partial X$ by the ends of X. If a map $f : X \to X$ is a proper homotopy equivalence then it induces an isomorphism of $\pi_1(X)$ and it extends to a continuous map $\hat{X} \to \hat{X}$ that restricts to a homeomorphism of ∂X. However, the converse of this statement is false.

EXAMPLE 4.1. Let X be the ray $[0, \infty)$ with the circle x_n attached at n. Then there is a proper map whose action on π_1 is $x_0 \mapsto x_0$ and $x_n \mapsto x_n x_{n-1}$ for $n > 0$. The inverse sends every x_n to a word that involves x_0 so it cannot be realized by a proper map.

To circumvent this pathology, we will consider the space of pairs of maps which are proper homotopy inverses of each other. This way the inverse is "built in", cf. the proof that inversion is continuous, Proposition 4.4. More precisely, let $(\hat{X} \to \hat{X})$ be the space of all continuous maps $\hat{X} \to \hat{X}$ equipped with compact open topology. If we fix a metric d on $\hat{X}$ then $(\hat{X} \to \hat{X})$ has the associated sup metric which we also denote by d. This metric is also complete, and composition is continuous, and $(\hat{X} \to \hat{X})$ is separable (see e.g. [**17**, Theorem 4.19]).

Next, we look at the space $PH(X) \subset (\hat{X} \to \hat{X})^2$ consisting of pairs $(\hat{f}, \hat{g})$ such that:
$\hat{f}, \hat{g}$ are extensions to $\hat{X}$ of proper homotopy equivalences $f, g : X \to X$ that are each other's inverses.

In particular, $\hat{f}, \hat{g}$ are homeomorphisms when restricted to ∂X and they are each other's inverses. We put the product topology on $(\hat{X} \to \hat{X})^2$ and the subspace topology on $PH(X)$.

Now define the function $\pi : PH(X) \to Maps(X)$ by

$$\pi(\hat{f}, \hat{g}) = [f]$$

where $[f]$ is the proper homotopy class of the restriction of $\hat{f}$ to X. This function is surjective and we put the quotient topology on $Maps(X)$.

PROPOSITION 4.2. *The quotient map π is an open map.*

PROOF. Let $U \subseteq PH(X)$ be open; we need to show that $\pi^{-1}\pi(U) \subseteq PH(X)$ is open. Let $(\hat{f}, \hat{g}) \in \pi^{-1}\pi(U)$. Thus we have proper homotopies H, K from f, g to h, k respectively, and $(\hat{h}, \hat{k}) \in U$. Therefore there is $\epsilon > 0$ such that if $(\hat{h}', \hat{k}') \in PH(X)$ and $d(\hat{h}, \hat{h}') < \epsilon$, $d(\hat{k}, \hat{k}') < \epsilon$ then $(\hat{h}', \hat{k}') \in U$. We now claim that there is $\delta > 0$ such that if $(\hat{f}', \hat{g}') \in PH(X)$, $d(\hat{f}, \hat{f}') < \delta$, $d(\hat{g}, \hat{g}') < \delta$, then there are homotopies of f', g' to maps h', k' as above, and this will show that $(\hat{f}', \hat{g}') \in \pi^{-1}\pi(U)$, finishing the proof.

We will prove the claim using Proposition 2.7. Note that all proper homotopies between maps on X extend continuously to ∂X and are stationary on all points of ∂X. It follows that in the complement of a sufficiently large finite subgraph the tracks (i.e. paths traversed by points) of each such homotopy are as small as we like. In addition, the edges outside a large finite subgraph are as small as we like.

Choose a large finite subgraph $L \subset X$ and choose $\delta_f > 0$ so that if $d(\hat{f}, \hat{f}') < \delta_f$ then we have a homotopy between $f'|L$ and $f|L$ whose tracks have small size, and we also have a homotopy between f and h with small tracks for points in $\overline{X \setminus L}$. Applying Proposition 2.7 we get a homotopy from f' to some map h' that agrees with h on L and whose tracks of points outside L are small. Thus h' is close to f outside L, which in turn is close to h outside L. Thus h' is close to h everywhere. In a similar way we find $\delta_g > 0$ and a homotopy from g' to k'. Then we set $\delta = \min\{\delta_f, \delta_g\}$. □

Corollary 4.3. *Let $PHE(X) \subset (\hat{X} \to \hat{X})$ be the subspace of maps $\hat{f}$ that are extensions to $\hat{X}$ of proper homotopy equivalences $f\colon X \to X$. Then the map $q\colon PHE(X) \to Maps(X)$ that to $\hat{f}$ assigns $[f]$ is open. Thus alternatively we could use this map to define the quotient topology on $Maps(X)$.*

Proof. The projection $(\hat{X} \to \hat{X})^2 \to (\hat{X} \to \hat{X})$ to the first coordinate restricts to a continuous map $PH(X) \to PHE(X)$. The composition with $q\colon PHE(X) \to Maps(X)$ is open by the proposition, so q is open. □

Proposition 4.4. *With this topology $Maps(X)$ is T_1 and it is a topological group.*

Proof. We first show $Maps(X)$ is a topological group. That composition is continuous follows from the fact that the map $PH(X) \times PH(X) \to PH(X)$ defined by $((\hat{f}, \hat{g}), (\hat{f}', \hat{g}')) \mapsto (\hat{f}\hat{f}', \hat{g}'\hat{g})$ is continuous, being the restriction of the analogous map $(\hat{X} \to \hat{X})^4 \to (\hat{X} \to \hat{X})^2$.

$$\begin{array}{ccc} PH(X) \times PH(X) & \longrightarrow & PH(X) \\ \downarrow{\scriptstyle \pi\times\pi} & & \downarrow{\scriptstyle \pi} \\ Maps(X) \times Maps(X) & \longrightarrow & Maps(X) \end{array}$$

Product of open maps is open so both vertical arrows are quotient maps. Continuity of the bottom horizontal map now follows.

For the inverse we consider the map $PH(X) \to PH(X)$, $(\hat{f}, \hat{g}) \mapsto (\hat{g}, \hat{f})$ and the argument is similar.

To prove that $Maps(X)$ is T_1 it suffices to show that the identity point is closed, since $Maps(X)$ is a topological group. This amounts to showing that the space of

$\hat{f} : \hat{X} \to \hat{X}$ so that f is properly homotopic to the identity is a closed set. This follows from Corollary 3.6 since preserving a loop or a line is a closed condition. □

COROLLARY 4.5. *Let X, Y be two connected locally finite graphs that are proper homotopy equivalent. Then $Maps(X)$ and $Maps(Y)$ are isomorphic as topological groups.*

PROOF. Let $F : X \to Y$ and $G : Y \to X$ be proper homotopy equivalences with FG and GF properly homotopic to the identity. Then F, G extend to maps $\hat{F} : \hat{X} \to \hat{Y}$ and $\hat{G} : \hat{Y} \to \hat{X}$ and we have the induced maps $(\hat{X} \to \hat{X}) \to (\hat{Y} \to \hat{Y})$, $\hat{f} \mapsto \hat{F} \circ \hat{f} \circ \hat{G}$ and $(\hat{Y} \to \hat{Y}) \to (\hat{X} \to \hat{X})$, $\hat{g} \mapsto \hat{G} \circ \hat{g} \circ \hat{F}$. These maps restrict to $PHE(X) \to PHE(Y)$ and $PHE(Y) \to PHE(X)$, which induce homomorphisms $Maps(X) \to Maps(Y)$ and $Maps(Y) \to Maps(X)$. These are each other's inverses and they are both continuous by Corollary 4.3. □

4.1. Clopen subgroups. We next show that $Maps(X)$ has a countable basis consisting of clopen sets, which are in fact cosets of subgroups. We will use this to show that $Maps(X)$ is homeomorphic to $\mathbb{Z}^\infty := \prod_1^\infty \mathbb{Z}$, i.e. the irrationals, and is hence a totally disconnected Polish group.

We define clopen subgroups of $Maps(X)$, analogs of pointwise stabilizers of compact subsurfaces in big mapping class groups. Let $K \subset X$ be a finite subgraph and define U_K to be the set of equivalence classes $[f] \in Maps(X)$ with a representative such that:

(i) $f = id$ on K,
(ii) f preserves each complementary component of K,
(iii) there is a representative g of $[f]^{-1}$ that also satisfies (i) and (ii),
(iv) there are proper homotopies $gf \simeq 1$ and $fg \simeq 1$ that are stationary on K and preserve complementary components of K.

LEMMA 4.6. *U_K is open.*

PROOF. We must show that $\pi^{-1}(U_K)$ is open. Let $\pi(h, k) \in U_K$. Let f, g be the representatives satisfying (i)-(iv). We then have proper homotopies $H, H' : X \times I \to X$ with $H(x, 0) = h(x), H(x, 1) = f(x), H'(x, 0) = k(x), H'(x, 1) = g(x)$. Choose a compact subgraph $L \supset K$ such that both H and H' at all times map $X \smallsetminus L$ into $X \smallsetminus K$. Finally, choose $\epsilon > 0$ such that if $d(h', h) < \epsilon$ and $d(k', k) < \epsilon$ then h', k' send $X \smallsetminus L$ to $X \smallsetminus K$ and h, h' (k, k') restricted to L are homotopic by a homotopy that doesn't move points in $Fr(L)$ into K.

Now use Proposition 2.7 a total of 4 times to prove that $\pi(h', k') \in U_K$. First, we have a homotopy from h' to a map h'' extending $h|L$, and second, we have a homotopy from h'' to a map f' extending $f|L$. Thus $f'|K = 1$ and preparations above show that f' maps $X \smallsetminus L$ to $X \smallsetminus K$. Two more homotopy extensions yield a similar homotopy from k' to g'. Since (f', g') satisfies (i)-(iv) we conclude that $\pi(h', k') \in U_K$. □

PROPOSITION 4.7. *U_K is a clopen subgroup. For every neighborhood U of $1 \in Maps(X)$ there is some K so that $U_K \subseteq U$.*

PROOF. To prove that U_K is a subgroup, we must prove that if $[f_1], [f_2] \in U_K$ with f_1, f_2 preferred representatives, then $[f_1][f_2] \in U_K$. Indeed, $[f_1][f_2] = [f_1 \circ f_2]$ and $f_1 \circ f_2$ is the identity on K and preserves complementary components. The same is true for its homotopy inverse $g_2 \circ g_1$ as well as for both homotopies to id_K.

We proved in Lemma 4.6 that U_K is open. Thus U_K is also closed since its complement is a union of cosets, which are also open.

Finally, we must show that for every open $U \subset Maps(X)$ containing 1, there is a compact K such that $U_K \subset U$. We have $\pi^{-1}U$ is an open set in $PH(X)$ containing $(1,1)$. Let $\epsilon > 0$ be such that if $(f,g) \in PH(X)$ and $d(1,\hat{f}) < \epsilon$, $d(1,\hat{g}) < \epsilon$ then $(\hat{f},\hat{g}) \in \pi^{-1}(U)$. Let $K \subset X$ be compact so that all complementary components of K have diameter $< \epsilon$. Then for preferred representatives (f,g) of $[f] \in U_K$ we have $(f,g) \in \pi^{-1}(U)$, so $U_K \subseteq \pi\pi^{-1}(U) = U$. □

COROLLARY 4.8. *$Maps(X)$ has a countable basis of clopen sets, it is separable, metrizable and totally disconnected.*

PROOF. $Maps(X)$ is separable since it is the continuous image of $PH(X)$ which is separable, being a subspace of a separable metric space. So in particular each open subgroup U_K has at most countably many cosets. Choose an exhaustion K_i of X; then U_{K_i} and their cosets form a countable basis of clopen sets. Next, T_1 plus a basis of clopen sets implies regular (proof: let $x \notin A$ with A closed; then there is a basis element V with $x \in V$ and $V \cap A = \emptyset$ and so V, V^c is the required separation). Then countable basis plus regular implies normal [**8**, 1.5.16], and finally countable basis plus normal implies metrizable (Urysohn metrization theorem, [**20**, Theorem 34.1]). □

Recall that X has *finite type* if $\pi_1(X)$ is finitely generated and ∂X is finite; otherwise X has *infinite type*.

LEMMA 4.9. *Suppose that X has infinite type. Then for every finite subgraph $K \subset X$ there is a finite subgraph $L \subset X$ such that $L \supset K$ and $U_L < U_K$ has infinite (countable) index.*

PROOF. First recall that an infinite totally disconnected compact metrizable space either has infinitely many isolated points or else it is the disjoint union of a Cantor set and finitely many isolated points. Consider the complementary components of K. If one of them has genus $N > 1$ then there is $L \supset K$ such that U_K contains a subgroup H isomorphic to the infinite group $Aut(F_N)$, while $H \cap U_L = 1$. The elements of H are realized by homotopy equivalences supported on a finite subgraph of genus N. Otherwise, after perhaps enlarging K, we may assume that all complementary components are trees, and in this case ∂X is infinite. According to the dichotomy above, we can find $L \supset K$ such that one of the following holds:

- There is a complementary component of K whose boundary at infinity contains an infinite set of isolated points, and this set can be written nontrivially as $A \cup B$ where A and B are boundary points of two distinct complementary components of L.
- There is a complementary component of K whose boundary at infinity contains a Cantor set as a clopen subset, and this Cantor set can be written nontrivially as $A \cup B$ with both A, B clopen and both boundary points of two distinct complementary components of L.

In the first case, U_K contains the entire group $Perm_0(A \cup B)$ of finitely supported permutations of $A \cup B$ (cf. the classification theorem), while the intersection of this group with U_L contains only permutations that preserve A and B. Since either A or B is infinite, this subgroup has infinite index.

In the second case U_K contains the entire group $H = Homeo(A \cup B)$, while the intersection $H \cap U_L$ contains only homeomorphisms that preserve A and B, and this again has infinite index. □

LEMMA 4.10. *$PH(X) \subset (\hat{X} \to \hat{X})^2$ is a G_δ-subset.*

PROOF. A pair $(\hat{f}, \hat{g}) \in (\hat{X} \to \hat{X})^2$ is in $PH(X)$ iff (see Corollary 3.6):

(1) $\hat{f}(\partial X) \subseteq \partial X$, $\hat{g}(\partial X) \subseteq \partial X$,
(2) $\hat{f}(X) \subseteq X$, $\hat{g}(X) \subseteq X$,
(3) $\hat{f}\hat{g}$ and $\hat{g}\hat{f}$ are identity on ∂X,
(4) $\hat{f}\hat{g}$ and $\hat{g}\hat{f}$ restricted to X preserve the homotopy classes of oriented loops and proper homotopy classes of oriented proper lines joining points of DX.

Conditions (1),(3) and (4) are closed conditions and (2) is G_δ: (2) can be written as countably many conditions $\hat{f}(K_n) \subset X$, $\hat{g}(K_n) \subset X$ for an exhaustion $\{K_n\}$, $n = 1, 2, \cdots$ and these are all open. □

When X has finite type, $Maps(X)$ is countable and discrete. Otherwise we have:

PROPOSITION 4.11. *Suppose X has infinite type. Then $Maps(X)$ is a Polish group with the underlying space homeomorphic to $\mathbb{Z}^\infty$, i.e. to the set of irrationals.*

PROOF. There is a theorem of Sierpinski that if $f\colon X \to Y$ is an open surjective map between separable metric spaces and X is complete, then Y is completely metrizable (see [**8**, Exercise 5.5.8.(d)]). Since $PH(X)$ is a G_δ subset of $(\hat{X} \to \hat{X})^2$, it is completely metrizable (see [**8**, 4.3.23]) and therefore $Maps(X)$ is completely metrizable. Then the fact that $Maps(X)$ is homeomorphic to the irrationals follows from a theorem of Hausdorff (see e.g. [**7**]): If Z is separable, completely metrizable, zero dimensional (i.e. has a basis of clopen sets), and every compact subset has empty interior, then Z is homeomorphic to the irrationals. To finish the proof, if $Maps(X)$ had a compact subset with nonempty interior, then some U_K would be compact. But this contradicts Lemma 4.9, since U_K is covered by the pairwise disjoint cosets of U_L and this cover doesn't have a finite subcover. □

We finish this section by considering continuity properties of homomorphisms studied in Section 3.

Recall the surjective homomorphism $\sigma\colon Maps(X) \to Homeo(\partial X, \partial X_g)$ from Corollary 2.3. The group $Homeo(\partial X, \partial X_g)$ is equipped with the compact-open topology. This means that a basis of neighborhoods of the identity is defined by clopen subgroups $V_{\mathcal{P}}$ where $\mathcal{P}$ is a finite partition of ∂X into clopen subsets and $V_{\mathcal{P}}$ consists of the elements of $Homeo(\partial X, \partial X_g)$ that leave the partition elements invariant. Refining the partition yields a smaller clopen subgroup.

COROLLARY 4.12. *The homomorphism σ is continuous and open. In particular, when X is a tree, $\sigma : Maps(X) \to Homeo(\partial X)$ is an isomorphism of topological groups.*

PROOF. We may assume that X is a Standard Model. We will consider finite subgraphs $K \subset X$ consisting of a subtree in the underlying tree together with all circles attached to it. The complementary components of K determine a partition $\mathcal{P}_K$ of ∂X. Since every partition $\mathcal{P}$ is refined by some $\mathcal{P}_K$ and $\sigma(U_K) \subseteq V_{\mathcal{P}_K}$

it follows that σ is continuous. To prove that σ is open it suffices to argue that $\sigma(U_K) = V_{\mathcal{P}_K}$. Let W be a complementary component of K. Thus ∂W is one of the partition elements A_W of $\mathcal{P}_K$ together with one point v corresponding to the vertex of intersection $\overline{W} \cap K$. Given a homeomorphism h of $(A_W, A_W \cap \partial X_g)$, extend it by $v \mapsto v$ and view it as a homeomorphism of $(\partial W, \partial W_g)$. By the Classification Theorem there is $f_h \in Maps(W)$ that induces $h : \partial W \to \partial W$. Now define $f \in Maps(X)$ as the identity on K and as f_h on W, for each complementary component W, and observe that $\sigma(f)$ is the given homeomorphism in $V_{\mathcal{P}_K}$. □

Next, recall the homomorphism $\Psi : Maps(X) \to Out(\pi_1(X))$ to the Polish group $Out(\pi_1(X))$. It is injective when X is a core graph (Theorem 3.1).

PROPOSITION 4.13. *The homomorphism Ψ is continuous. If the genus of X is infinite, the image is not a closed subgroup. If in addition X is a core graph then Ψ is injective but it is not a homeomorphism onto its image.*

PROOF. The topology on $Aut(\pi_1(X))$ is defined as a subgroup of the symmetric group S_∞ on the countable set $\pi_1(X)$, so an automorphism is close to the identity if it fixes a large finite set. The group $Out(\pi_1(X))$ is equipped with the quotient topology. If U is an open neighborhood of the identity in $Out(\pi_1(X))$, its preimage in $Aut(\pi_1(X))$ will contain all automorphisms that fix a certain finite set F. The elements of F are realized inside some compact subgraph $K \subset X$ and it follows that $\Psi(U_K) \subseteq U$, so Ψ is continuous.

Consider $f\colon X \to X$ from Example 4.1. Let $f_n\colon X \to X$ be defined by $f_n(x_0) = x_0$, $f_n(x_k) = x_k x_{k-1}$ for $k \leq n$ and $f_n(x_k) = x_k$ for $k > n$. Then $\Phi(f_n) \to f_* \in Out(\pi_1(X))$, but f_* is not in the image of Φ.

Similarly, consider $g_n\colon X \to X$ defined by $g_n(x_k) = x_k$ when $k \leq n$ or $k \geq 2n$, $g_n(x_k) = x_k x_1$ when $n < k < 2n$. Then $\Psi(g_n) \to id$ but the sequence g_n does not converge to id (or anywhere). So Ψ is not a homeomorphism onto its image.

Generalizing these examples to other graphs is left to the reader. □

Finally, we have the following statement, whose proof is left to the reader.

PROPOSITION 4.14. *The restriction epimorphisms $PMaps(X) \to PMaps(X_g)$ and $PMaps(X) \to PMaps(X_g^*)$ are continuous and open.*

5. Proof of Main Theorem for core graphs

5.1. Free factor systems. Let $\mathbb{F}$ be a free group, possibly of infinite rank. Recall that a nontrivial subgroup $A < \mathbb{F}$ is a *free factor* of $\mathbb{F}$ if there is a subgroup $B < \mathbb{F}$ such that $A * B = \mathbb{F}$. We will only consider free factors of finite rank, and only conjugacy classes $[A]$ of such free factors. To simplify notation we will usually omit the brackets. Topologically, a (conjugacy class of a) nontrivial subgroup is a free factor if there is a graph Γ with $\pi_1(\Gamma) = \mathbb{F}$ and with A represented by a subgraph. Similarly, a finite collection $\mathcal{F}$ of (conjugacy classes of) finitely generated free factors is a *free factor system* if there are representatives $A_1, A_2, \cdots, A_n$ and a subgroup $B < \mathbb{F}$ such that $A_1 * A_2 * \cdots * A_n * B = \mathbb{F}$. Topologically, there is a graph Γ with $\pi_1(\Gamma) = \mathbb{F}$ and with the A_is represented by pairwise disjoint subgraphs.

If $\mathcal{F}$ and $\mathcal{F}'$ are two free factor systems, the intersection $\mathcal{F} \cap \mathcal{F}'$ is naturally a free factor system. It consists of conjugacy classes of nontrivial subgroups obtained by intersecting a representative of a conjugacy class in $\mathcal{F}$ with a representative of a conjugacy class in $\mathcal{F}'$. Topologically, one can represent $\mathcal{F}$ and $\mathcal{F}'$ by immersions of finite graphs $\Gamma_F \to \Gamma$ and $\Gamma_{F'} \to \Gamma$, form the pull-back (see [**21**]) and discard the contractible components to get an immersion representing the intersection.

EXAMPLE 5.1. Let $\mathbb{F} = \langle a, b, c\rangle$, $A = \langle a, b\rangle$, $B = \langle a, cbc^{-1}\rangle$. Then A and B are free factors of $\mathbb{F}$, while their intersection is the free factor system consisting of two rank 1 free factors $\langle a\rangle$ and $\langle b\rangle$. The intersection of A and $\langle c\rangle$ is the empty free factor system.

To see that the intersection $\mathcal{F} \cap \mathcal{F}'$ is a free factor system, one can arrange that one of them is represented by subgraphs of Γ and then the pullback will be represented by subgraphs of the other one. It is also possible to compute finite intersections of free factor systems by a pull-back of several immersions.

Finally, we write $\mathcal{F} < \mathcal{F}'$ if every group (representing a conjugacy class) in $\mathcal{F}$ is contained in a group in $\mathcal{F}'$. For example, $\mathcal{F} \cap \mathcal{F}' < \mathcal{F}$.

5.2. Tree of groups. We now assume that X is a core graph and is a Standard Model. Thus X is a tree T with a root vertex v and with a loop attached at every vertex. We assign length 1 to each edge and let $D_0 : T \to [0, \infty)$ be the distance function from v. We extend D_0 to all of X so that it is constant on each attached loop. Our first task is to control the sizes of maps, measured in $[0, \infty)$, representing elements of H, as well as homotopies, measured by D_0. So in effect we replace properness by metric control. Recall that for a finite subgraph $K \subset X$ we have a clopen subgroup $U_K < Maps(X)$, so $H \cap U_K$ is compact and has finite index in H.

PROPOSITION 5.2. *Let $H < Maps(X)$ be a compact subgroup. There is a sequence of integers $0 = r_0 < r_1 < r_2 < \cdots$ and for every $n > 0$ and every $[h] \in H$ there is a representative h satisfying*

(*) *h maps every element of the closed cover $\mathcal{C}(r_1, r_2, \cdots, r_n)$ of X consisting of the sets*

$$D_0^{-1}[r_0, r_1], D_0^{-1}[r_1, r_2], \cdots, D_0^{-1}[r_{n-1}, r_n], D_0^{-1}[r_n, \infty)$$

to the union of the same element with the one or two adjacent elements.

PROOF. We construct the numbers inductively, starting with $r_1 = 1$. Then (*) is vacuous.

Suppose that r_n has been constructed satisfying (*). Note that by properness for every $[h] \in H$ (and every representative h that exists by induction) there is some $r_{n+1} > r_n$ so that (*) holds for the cover $C(r_1, r_2, \cdots, r_n, r_{n+1})$ and this h. Moreover, the same r_{n+1} will also work in a neighborhood of $[h]$ by choosing representatives of the form hu where $[u] \in U_{D_0^{-1}[0, r_{n+1}]}$ i.e. u fixes $D_0^{-1}[0, r_{n+1}]$ and leaves the complementary components invariant. Now by compactness of H, there is a finite cover of H by such open sets and the maximal r_{n+1} will then satisfy the requirements. □

It will be convenient to introduce the following notation. First, let $\rho : [0, \infty) \to [0, \infty)$ be a homeomorphism such that $\rho(r_n) = n$ for $n = 0, 1, \cdots$ and let $D = \rho D_0 : X \to [0, \infty)$. Thus $D^{-1}([m, n]) = D_0^{-1}([r_m, r_n])$. We think of D as a "control function". For example, (*) says that for every n every element of H has a representative h that "moves points < 2" i.e. $|D(x) - D(h(x))| < 2$ for every $x \in D^{-1}[0, n]$. The next proposition says that homotopies "move points < 3".

PROPOSITION 5.3. *Let $0 < r_1 < r_2 < \cdots$ be as in Proposition 5.2. Fix n and let h, h' be the representatives of two elements of H that are inverses of each other as in Proposition 5.2. Then there is a proper homotopy between the identity and $h'h$ that moves each element of the cover $\mathcal{C}(r_1, \cdots, r_n)$ to the union of at most 5 elements, namely the 2-neighborhood of the given element.*

PROOF. First note that there is a canonical proper homotopy between the identity and $h'h$: lift the given homotopy to the universal cover extending the identity map, and then replace it by the straight line homotopy. We now argue that this homotopy moves within 2-neighborhoods. Fix a component P of an element of the cover and let $\tilde{P}$ be the component of the 2-neighborhood that contains it. Since a loop in P cannot be mapped by $h'h$ disjointly (since otherwise $h'h$ would not be homotopic to the identity) we see that $h'h(P) \subseteq \tilde{P}$. By lifting to the covering space of X corresponding to $\pi_1(\tilde{P})$ and then retracting to the core $\tilde{P}$, we see that $h'h|P : P \to \tilde{P}$ is homotopic to inclusion $i : P \hookrightarrow \tilde{P}$ within $\tilde{P}$. Now note that any homotopy from i to $h'h|P$ has tracks that are nullhomotopic loops (they have to represent π_1-elements that commute with $\pi_1(P)$, but since $\pi_1(P)$ and $\pi_1(\tilde{P})$ are nonabelian free groups this forces these loops to be trivial). It follows that the tracks described by the straight line homotopy are homotopic to paths in $\tilde{P}$, but since they are immersed, they must be contained in $\tilde{P}$. $\square$

If $J \subset [0, \infty)$ is a closed interval with integer endpoints, write $\mathcal{F}(J)$ for the free factor system represented by $D^{-1}(J)$. Thus the number of free factors in $\mathcal{F}(J)$ is equal to the number of components of $D^{-1}(J)$. When J is a degenerate interval (a single integer point) then each factor in $\mathcal{F}(J)$ has rank 1. We denote by $|J|$ the length of the interval.

We also set

$$\mathcal{F}'(J) = \cap_{h \in H} h_*(\mathcal{F}(J))$$

where $h_* : \pi_1(X) \to \pi_1(X)$ is the automorphism induced by h (defined up to conjugation). This is really only a finite intersection since when h is close to the identity we will have $h_*(\mathcal{F}(J)) = \mathcal{F}(J)$, so it suffices to intersect over finitely many coset representatives. Thus $\mathcal{F}'(J)$ is an H-invariant free factor system.

When $J = [a, b] \subset [0, \infty)$ with integer endpoints and with $b - a \geq 4$ we set $J^+ = [a-2, b+2] \cap [0, \infty)$ to be the 2-neighborhood of J, and likewise $J^- = [a', b-2]$ where $a' = 0$ if $a = 0$ and otherwise $a' = a + 2$ (so J^- is obtained from J by subtracting the 2-neighborhood of the complement). Note that by our assumptions on the sequence r_n we have that

$$\mathcal{F}(J^-) < \mathcal{F}'(J) < \mathcal{F}(J^+)$$

We now show that each group in $\mathcal{F}'(J)$ either contains a group in $\mathcal{F}(J^-)$ or it has trivial intersection with all of them.

LEMMA 5.4. *Let A be a free factor in $\mathcal{F}'(J)$. If A contains a nontrivial element α that also belongs to a free factor B in $\mathcal{F}(J^-)$ then $B < A$ (up to conjugacy).*

PROOF. Represent different $h_*(\mathcal{F}(J))$, $h \in H$, by immersions of finite (possibly disconnected) graphs into X. The non-tree components of the pull-back then represent the free factors in $\mathcal{F}'(J)$. Since free factors (and free factor systems) are malnormal, if an immersion to X lifts to the pull-back, it does so uniquely. Since an immersion representing B lifts, it must lift to the component representing A, since this is where α lifts. $\square$

We now set $\mathcal{F}^*(J)$ to be the free factor system consisting of the free factors in $\mathcal{F}'(J)$ that contain a free factor in $\mathcal{F}(J^-)$. Thus we still have

$$\mathcal{F}(J^-) < \mathcal{F}^*(J) < \mathcal{F}(J^+)$$

and also $J \subset J'$ implies $\mathcal{F}^*(J) < \mathcal{F}^*(J')$.

LEMMA 5.5. *If* $|J| \geq 8$ *then* $\mathcal{F}^*(J)$ *is* H*-invariant.*

PROOF. Take a free factor A in $\mathcal{F}^*(J)$. It will contain an element α corresponding to a loop in $D^{-1}(t)$ for any $t \in J$ whose distance to each endpoint is ≥ 4. For any $h \in H$ we have $h_*(\alpha)$ is an element in a free factor of $\mathcal{F}([t-2, t+2]) < \mathcal{F}(J^-)$, and the free factor of the latter that contains it is contained in a free factor B of $\mathcal{F}^*(J)$ by Lemma 5.4, and $h_*(A) = B$. □

Now fix a sequence of intervals $J_1, J_2, \cdots$ that cover $[0, \infty)$ and so that $J_n \cap J_m = \emptyset$ when $|n-m| > 1$ and $J_{n,n+1} := J_n \cap J_{n+1}$ is an interval of length ≥ 22 for $i = 1, 2, \cdots$. Now construct the following tree of groups $\mathcal{T}$. The vertices of the tree are the free factors in $\mathcal{F}(J_n)$ (or equivalently the components of $D^{-1}(J_n)$), $n = 1, 2, \cdots$. The group associated to a vertex is the underlying free factor. The edges are the free factors in $\mathcal{F}(J_{n,n+1})$ (components of $D^{-1}(J_{n,n+1})$), again with the associated group the underlying free factor. Incidence relation is inclusion. The underlying graph is a tree, the nerve of the cover of X by the components of $D^{-1}(J_n)$, $n \geq 1$.

LEMMA 5.6. $\pi_1(\mathcal{T}) \cong \pi_1(X)$.

PROOF. By induction, the subtree of groups corresponding to the first n intervals has the fundamental group of the corresponding subgraph of X. □

In a similar way we construct a tree of groups $\mathcal{T}^*$, which will be H-invariant. A vertex of *height* n is a free factor in $\mathcal{F}^*(J_n)$, with this factor as the vertex group. An edge of *height* $[n, n+1]$ is a free factor in $\mathcal{F}^*(J_{n,n+1})$, with this factor as the edge group. Such a factor is contained in a unique vertex group at height n and a unique vertex group at height $n+1$ by Lemma 5.4 and this gives incidence and edge-to-vertex inclusions. Thus $\mathcal{T}^*$ is a graph of groups and it is H-invariant by construction. Below we will show that $\mathcal{T}^*$ is a tree and $\pi_1(\mathcal{T}^*) \cong \pi_1(\mathcal{T})$.

LEMMA 5.7. *If* C *is an edge group in* $\mathcal{T}^*$ *with* A, B *the incident vertex groups of heights* $n, n+1$ *resp., then* C *is one of the free factors in* $A \cap B$.

PROOF. We have that C is contained in some group in $A \cap B$ by construction. The free factor A is a free factor in the free factor system $\mathcal{F}'(J_n)$ that contains a factor in $\mathcal{F}(J_n^-)$ and similarly for B. The intersection $A \cap B$ consists of free factors in $\mathcal{F}'(J_{n,n+1})$ and one of them contains C, which is also a free factor in $\mathcal{F}'(J_{n,n+1})$, so equality holds. □

Note here that in principle the intersection of A and B can consist of several free factors, i.e. the vertices might be joined by several edges. We will rule this out in Lemma 5.9.

There is a natural morphism (vertices to vertices and edges to edges) $\pi : \mathcal{T}^* \to \mathcal{T}$ that sends a factor in $\mathcal{F}^*(J_i)$ to the factor in $\mathcal{F}(J_i)$ that contains it, and similarly for the edges. Note that we have a height function on both trees (sending factors in $\mathcal{T}^*(J_i)$, respectively in $\mathcal{F}(J_i)$ to i) that commutes with this map.

In the sequel it will be convenient to abuse the terminology and conflate a subcomplex of X and its fundamental group, and likewise a "component" and a "free factor" in a free factor system.

LEMMA 5.8. *Every vertex of $\mathcal{T}^*$ at height $n+1>0$ is connected by an edge to a vertex at height n. There is a unique vertex of $\mathcal{T}^*$ at height 0. In particular, $\mathcal{T}^*$ is connected.*

PROOF. Suppose the vertex is A, so it contains (possibly more than one) free factor B in $\mathcal{F}(J_{n+1}^-)$. Every component of $D^{-1}(J_{n+1}^-)$ contains a (unique) component of $D^{-1}(J_{n,n+1}^-)$ and this component is contained in a unique free factor of $\mathcal{F}^*(J_{n,n+1})$, which represents an edge at height $[n,n+1]$ attached to A.

Since $D^{-1}(J_1^-)$ is connected (recall that J_1^- contains $\{0\}$) and every vertex at height 0 must contain a component of it, it follows that there is only one height 0 vertex in $\mathcal{T}^*$. □

Note that a vertex at height n may not be connected to any vertices at height $n+1$ since a component of $D^{-1}(J_n^-)$ may not contain any components of $D^{-1}(J_{n,n+1}^-)$.

LEMMA 5.9. *Let e be an edge in $\mathcal{T}$ with height in $[n,n+1]$ and consider its preimage $\pi^{-1}(e)$ in $\mathcal{T}^*$. After removing isolated vertices from $\pi^{-1}(e)$, it is a tree with one vertex w at height n and all other vertices at height $n+1$, and these are all connected to w by a unique edge. In particular, $\mathcal{T}^*$ is a tree.*

PROOF. Let $J=J_{n,n+1}$. The statement that all edges in the preimage of e have the same vertex at height n follows from the following fact. If two components of $D^{-1}(J^-)$ are contained in the same component of $D^{-1}(J)$ then they are contained in the same component of $D^{-1}(J_n^-)$ (and this is not true if J_n^- is replaced by J_{n+1}^- and there may be several vertices at height $n+1$).

We now argue that the height $n+1$ vertices of all these edges in the preimage of e are distinct. Fix some integer $k\in J$ at distance ≥ 9 from the endpoints and let x,x' be two loops in $D^{-1}(J)$ that map to k. They will lift to unique components of $\mathcal{F}^*(J)$ and any two components are determined in this way. If they lift to the same component of $\mathcal{F}^*(J_{n+1})$ then there is an immersion $q:\Gamma\to D^{-1}(J_{n+1})$ of a barbell (two disjoint loops connected by an edge) sending one loop to x and the other to x' and so that hq can be homotoped into $D^{-1}(J_{n+1})$ for every $h\in H$. Thus q is kind of a "witness" that x,x' lift to the same component of $\mathcal{F}^*(J_{n+1})$. We need a similar witness that they lift to the same component of $\mathcal{F}^*(J)$. The map q may not work, since its image may contain points of $D^{-1}(J_{n+1}\smallsetminus J)$, and we will perform a kind of surgery on q to get a better map.

Fix $h\in H$. By perturbing if necessary we may assume that hq doesn't collapse any edges and is simplicial with respect to suitable subdivisions. Then the statement that hq can be homotoped into $\mathcal{F}(J_{n+1})$ is equivalent to saying that after folding and replacing hq by an immersion, the image of the core subgraph is contained in $D^{-1}(J_{n+1})$. This same q may not map into $D^{-1}(J)$ since it may map around loops in $D^{-1}(J_{n+1})\smallsetminus D^{-1}(J)$, so we will modify it to $q':\Gamma'\to D^{-1}(J)$. First we analyze q.

Recall that a *vanishing path* for hq is an immersion $\nu:I\to\Gamma$ such that $hq\nu:I\to X$ is a nullhomotopic closed path. There are only finitely many maximal vanishing paths and the folding process can be thought of as folding maximal vanishing paths one at a time. We now claim that $hq\nu$ has D-size <10 (i.e.

diam $Im(Dhq\nu) < 10$) when h is as in Proposition 5.2 (see Figure 5 for an illustration). Indeed, $h'hq\nu$ is also a closed nullhomotopic path (where h' is as in Proposition 5.3) and there is a homotopy of $h'hq\nu$ to $q\nu$ that moves the endpoints by < 3 measured by D. Thus the immersed path $q\nu$ gets closed up to a nullhomotopic loop by a path of D-size < 6, so it must itself have D-size < 6, and so $hq\nu$ has D-size < 10.

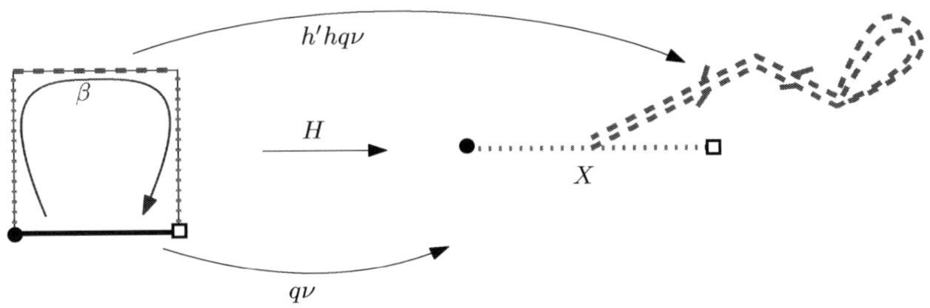

FIGURE 5. $H : I \times I \to X$ denotes the homotopy from $q\nu$ to $h'hq\nu$. The path denoted by β is mapped by H to a path homotopic to $q\nu$. Since H maps vertical segments $\{t\} \times I$ to paths whose images have D-length less than 3, the dotted subpaths are mapped by H to paths with D-length smaller than 3. The dashed part of β is null homotopic. Since $q\nu$ is immersed, its D-length is < 6.

We now observe that after folding hq the tree components of the complement of the core have D-size < 10. Indeed, choose any point $p \in \Gamma$. First fold all vanishing paths that do not contain p. After this, p is still in the core. Finally, fold the remaining vanishing paths – this operation changes only the neighborhood of p of D-size < 10.

Now consider $(Dq)^{-1}[k+1, \infty) \subset \Gamma$. It is a disjoint union of (possibly degenerate) closed intervals in the interior of the separating arc of Γ. Form a new graph $\tilde{\Gamma}$ by attaching an edge E_a to Γ for every nondegenerate arc a in this disjoint union, with $\partial E_a = \partial a$. Note that q sends both endpoints of a to the same vertex (at D-height $k+1$, i.e. distance r_{k+1} from the root vertex). Extend q to an immersion $\tilde{q} : \tilde{\Gamma} \to X$ by sending each E_a to a loop based at this vertex of combinatorial length ≤ 3 (for example, one can send it either to the attached loop based at that vertex or to the loop of the form dcd^{-1} where d is an edge that increases the distance from the root and c is the loop attached at the terminal vertex of d, see Figure 6). Let $\Gamma' \subset \tilde{\Gamma}$ be the barbell obtained by deleting the interiors of the arcs a as above, and let $q' : \Gamma' \to X$ be the restriction of $\tilde{q}$.

We now claim that $h\tilde{q}$ is homotopic into $D^{-1}(J_{n+1})$ for every $h \in H$. We can fold $h\tilde{q}$ by first folding hq, which produces a core graph with trees attached, and then adding the edges E_a. They could be attached to points in the attached trees, but all such attached trees have D-size < 10 and map to $(k+1-10, k+1+10) \subset J_{n+1}$ by D. So after removing the attached trees to which no E_a's are attached, the image is entirely contained in $D^{-1}(J_{n+1})$, which proves the claim.

In particular, $hq' : \Gamma' \to X$ is homotopic into $D^{-1}(J_{n+1})$, so q' is also a witness to the fact that x, x' lift to the same component of $\mathcal{F}^*(J_{n+1})$. By construction, the image of q' does not exceed the D-height $k+2$, so hq' does not exceed the D-height $k+4$, and we see that hq' is contained (even without homotopies) in $D^{-1}(J_n)$. It

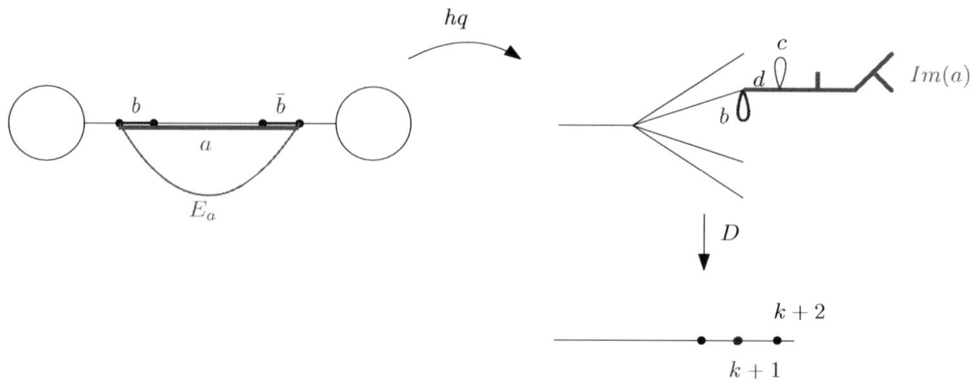

FIGURE 6. The barbell graph on the left is the graph Γ. It is mapped via hq to X. The graph $\tilde{\Gamma}$ is Γ union the edge E_a. The arc labeled a is mapped into $D^{-1}[k+1,\infty)$. Suppose the initial edge of of a is mapped to the loop b and the terminal edge is mapped to $\bar{b}$. Then in order to define $\tilde{q}$ on $\tilde{\Gamma}$ so that it will be an immersion we will let E_a map to $dc\bar{d}$ where d is the edge to the right of b, and c is the one edge loop based at the endpoint of d.

follows that hq' is homotopic into $D^{-1}(J_{n,n+1})$, and so the lifts of x, x' in $\mathcal{F}^*(J)$ are in the same component.

To see that $\mathcal{T}^*$ has no loops, we note that if there were an embedded loop then let $u \in V(\mathcal{T}^*)$ be the vertex of the loop of maximal height. Since no two vertices of the same height are connected then the loop gives us two vertices of the same height attached to u which is a contradiction. □

Our next goal is to verify that $\mathcal{T}^* \to \mathcal{T}$ induces an isomorphism between the fundamental groups of these graphs of groups. Our method is to find a sequence of folds that converts $\mathcal{T}^*$ to $\mathcal{T}$. We will do this through an intermediate tree of groups $\mathcal{T}^* \to \mathcal{T}^{**} \to \mathcal{T}$. Only $\mathcal{T}^*$ will be H-invariant.

Recall the following folding moves on simplicial G-trees T [**4**]. If e_1, e_2 are two oriented edges with the common initial vertex v such that $e_1 \cup e_2$ embeds in the quotient T/G, then we may construct a new G-tree T' by identifying e_1 and e_2 in an equivariant fashion, i.e. we identify $g(e_1)$ and $g(e_2)$ for every $g \in G$. The stabilizer of the new edge $e_1 = e_2$ is the group generated by $Stab(e_1)$ and $Stab(e_2)$, and similarly for the terminal vertices of e_1 and e_2. The effect in the quotient graph is to fold the images of e_1 and e_2. This is called Move IA in [**4**].

Similarly, suppose e_1, e_2 are two oriented edges with the common initial vertex v, each edge embeds in the quotient T/G, but they have the same images in T/G. This means that $g(e_1) = e_2$ for some $g \in Stab(v)$, so $Stab(e_2) = gStab(e_1)g^{-1}$. The equivariant folding operation has the effect that the underlying quotient graph is unchanged, but the stabilizer of $e_1 = e_2$ is now the group generated by $Stab(e_1)$ and g, and similarly for the terminal vertex. This is called Move IIA, and we think of it as pulling the element $g \in Stab(v)$ across the image edge to the terminal vertex and enlarging the stabilizers by this g. In a similar way we can pull finitely generated subgroups (or think of it as several Moves IIA performed in sequence).

Let $\mathcal{T}^{**}$ be the tree of groups obtained from $\mathcal{T}^*$ by folding each preimage of an edge to an edge, so that there is a morphism $\mathcal{T}^{**} \to \mathcal{T}$. This amounts to performing infinitely many Moves IA, but they are all independent and can be performed simultaneously. The resulting morphism $\mathcal{T}^{**} \to \mathcal{T}$ is an isomorphism of underlying trees.

It will be convenient to denote by $\mathcal{T}(e)$ the group associated to an edge e of $\mathcal{T}$, and similarly for the vertices, and for the trees $\mathcal{T}^*$ and $\mathcal{T}^{**}$.

LEMMA 5.10. *After independent Moves IIA, the morphism $\mathcal{T}^{**} \to \mathcal{T}$ becomes an isomorphism of graphs of groups.*

PROOF. The moves consist of pulling across an edge e from an endpoint w the subgroup $\mathcal{T}^{**}(w) \cap \mathcal{T}(e)$, simultaneously for all (w, e). Since $J_{n,n+1} \smallsetminus J_n^- \subset J_{n+1}^-$ then $\mathcal{F}(J_{n,n+1})$ is generated by elements of $\mathcal{F}(J_n^-)$ and $\mathcal{F}(J_{n+1}^-)$ which are contained in $\mathcal{F}^*(J_n)$ and $\mathcal{F}^*(J_{n+1})$ respectively. Therefore the group $\mathcal{T}(e)$ is generated by elements in $\mathcal{T}^*(w), \mathcal{T}^*(v)$ for the endpoints w, v of e. Thus by applying IIA moves we can promote $\mathcal{T}^{**}(e)$ to $\mathcal{T}(e)$. Similarly, $J_n \subset J_n^- \cup J_{n-1,n}^- \cup J_{n,n+1}^-$ hence $\mathcal{T}(w)$ is generated by elements in $\mathcal{T}^{**}(w)$ and $\{\mathcal{T}^{**}(e) \mid w \text{ is an endpoint of } e\}$. Therefore we can promote $\mathcal{T}^{**}(w)$ to $\mathcal{T}(w)$ using IIA moves. □

When $\mathcal{Y}$ is a locally finite graph of groups with all vertex and edge stabilizers finite rank free groups we define the *geometric realization* $GR(\mathcal{Y})$. This is the 2-complex constructed by taking a finite graph Γ_w for every vertex w so that $\pi_1(\Gamma_w) = \mathcal{Y}(w)$, and similarly taking a finite graph Γ_e for every edge e so that $\pi_1(\Gamma_e) = \mathcal{Y}(e)$, and gluing $\Gamma_e \times [0, 1]$ according to inclusion homomorphisms. Up to a proper homotopy equivalence, $GR(\mathcal{Y})$ is independent of the choices. From the lemmas above we see that the fundamental groups of graphs of groups $\mathcal{T}$, $\mathcal{T}^*$, $\mathcal{T}^{**}$ are all isomorphic to $\pi_1(X)$. We now upgrade this to proper homotopy equivalences of geometric realizations.

LEMMA 5.11. *$X, GR(\mathcal{T}), GR(\mathcal{T}^*), GR(\mathcal{T}^{**})$ are all proper homotopy equivalent.*

PROOF. $GR(\mathcal{T})$ can be built as a subspace of $X \times [0, \infty)$:

$$GR(\mathcal{T}) = \cup_{n=0}^{\infty}\big(D^{-1}(J_n) \times \{n\} \cup D^{-1}(J_n \cap J_{n+1}) \times [n, n+1]\big)$$

The map $GR(\mathcal{T}) \to X$ is the projection, and $X \to GR(\mathcal{T})$ is the map $x \mapsto (x, \phi(x))$, where ϕ equals n on $J_n \smallsetminus (J_{n-1} \cup J_{n+1})$ and is in $[n, n+1]$ on $J_n \cap J_{n+1}$. These are each other's proper homotopy inverses by homotoping along the second coordinate.

That $GR(\mathcal{T}^*) \to GR(\mathcal{T}^{**}) \to GR(\mathcal{T})$ are proper homotopy equivalences follows from the fact that Moves IA as well as IIA consisting of pulling finitely generated subgroups are proper homotopy equivalences on geometric realizations. □

To finish, we need the relative version of Nielsen Realization for graphs, proved by Hensel-Kielak.

THEOREM 5.12 ([**14**]). *Let $H < Out(F_n)$ be a finite subgroup and $\mathcal{F}$ an H-invariant free factor system. Suppose the action of H on $\mathcal{F}$ is realized as a simplicial action of H on a finite graph Γ_0 whose fundamental group is identified with $\mathcal{F}$ (so the components of Γ_0 correspond to the free factors in $\mathcal{F}$). Then there is a finite graph Γ, a simplicial action of H on Γ, an H-equivariant embedding $\Gamma_0 \hookrightarrow \Gamma$, and an identification $\pi_1(\Gamma) \cong F_n$ so that the induced $H \to Out(F_n)$ is the given embedding $H < Out(F_n)$.*

When $\mathcal{F}$ is empty, we have the (absolute) Nielsen Realization [**5**, **15**, **19**, **22**]. To apply this, we note:

LEMMA 5.13. *For every vertex w in $\mathcal{T}^*$ the incident edge groups form a free factor system in $\mathcal{T}^*(w)$.*

PROOF. This is true for the tree $\mathcal{T}$ by construction. The statement then follows from the fact that intersections of free factor systems are free factor systems. □

Now we build a graph Y. We first construct graphs associated to the edges. Note that all orbits of edges are finite. For an edge e of $\mathcal{T}^*$ choose a graph Γ_e with $\pi_1(\Gamma_e) = \mathcal{T}^*(e)$ where $Stab_H(e)$ acts inducing the given action on $\mathcal{T}^*(e)$. Of course, $Stab_H(e)$ is a compact group, but the action on $\mathcal{T}^*(e)$ factors through a finite group, so we can apply the Nielsen Realization theorem. We associate the same graph to all edges in the orbit of e, with suitable identifications on π_1, so that H now acts on the disjoint union of these graphs with the given action on π_1.

Now consider a vertex w. We have that $Stab_H(w)$ acts on $\mathcal{T}^*(w)$ and this action factors through a finite group, which also acts on the free factor system defined by the incident edges. This action is realized by the action of $Stab_H(w)$ on the disjoint union of the graphs representing the edge spaces, so the Relative Nielsen Realization provides a finite graph Γ_w that contains this disjoint union and an extension of this action. Associate such graphs to the vertices equivariantly. The union along the subgraphs associated to the edges is the desired graph Y. Thus H acts on Y simplicially. The following lemma finishes the proof of the Main Theorem in the core graph case.

LEMMA 5.14. *There is a proper homotopy equivalence $Y \to X$ that commutes with the action of H.*

PROOF. Using the same graphs to represent vertex and edge groups, the geometric realization $GR(\mathcal{T}^*)$, after collapsing the I-factors, becomes Y, and this is a proper homotopy equivalence. By composing with proper homotopy equivalences from Lemma 5.11 we have $f : X \to Y$ and $g : Y \to X$, which are each other's inverses. If $h \in H$ then by construction $h : X \to X$ and $ghf : X \to X$ induce the same element of $Out(\pi_1(X))$. It then follows from Theorem 3.1 applied to $ghf \cdot h^{-1}$ that they are properly homotopic. □

6. Proof for trees

We next prove Nielsen realization for trees.

THEOREM 6.1. *Suppose the graph X is a tree and let $H < Maps(X)$ be a compact subgroup. Then there is a tree $Y \simeq X$ where H acts by simplicial isomorphisms.*

Note that by Corollary 4.12 $Maps(X) = Homeo(\partial X)$. Fix a metric d on ∂X.

STEP 1. We replace d by an H-invariant metric d'. Let ν be a Haar measure on H and define

$$d'(p,q) = \int_H d(h(p), h(q))d\nu$$

This is an H-invariant metric. We drop the prime and assume d is H-invariant.

STEP 2. We now build equivariant finite partitions of ∂X into clopen sets. Let $\epsilon > 0$. Say $p, q \in \partial X$ are ϵ-path connected if there is a sequence $p = z_0, z_1, \cdots, z_n = q$ so that $d(z_i, z_{i+1}) < \epsilon$ for all $i = 0, \cdots, n-1$. The equivalence classes form the desired partition $\mathcal{P}_\epsilon$. Note that if $\epsilon < \epsilon'$ then $\mathcal{P}_\epsilon$ refines $\mathcal{P}_{\epsilon'}$ and if $\mathcal{P}$ is an arbitrary finite partition into clopen sets, there is $\epsilon > 0$ so that $\mathcal{P}_\epsilon$ refines $\mathcal{P}$.

STEP 3. Finally we build Y as the *mapping telescope* of a sequence of partitions from Step 2. Fix a decreasing sequence $\epsilon_n \to 0$ with $n = 1, 2, \cdots$ and let $\mathcal{P}_n := \mathcal{P}_{\epsilon_n}$. We also set $\mathcal{P}_0$ to be the trivial partition $\{\partial X\}$. Since $\mathcal{P}_{n+1}$ refines $\mathcal{P}_n$ we have a natural surjection $\mathcal{P}_{n+1} \to \mathcal{P}_n$ induced by inclusion of sets. Now let Y be the mapping telescope of this sequence. More concretely, the set of vertices is the disjoint union $\sqcup_{n=0}^{\infty} \mathcal{P}_n \times \{n\}$, and there is an edge from $P \times \{n+1\}$ to $Q \times \{n\}$ whenever $P \subseteq Q$ (here $P \in \mathcal{P}_{n+1}$ and $Q \in \mathcal{P}_n$). Then Y is a tree and ∂Y is naturally (and H-equivariantly) homeomorphic to ∂X by the homeomorphism that sends a branch $(P_n)_n$ of Y to the point $\cap_n P_n$ in ∂X. The theorem is now proved since we have natural identifications

$$Maps(X) = Homeo(\partial X) = Homeo(\partial Y) = Maps(Y)$$

and H acts simplicially on Y.

7. Proof in general

Let X be a locally finite graph which is not a tree and assume that a compact group H is acting on X by proper homotopy equivalences. The action then restricts to the core X_g (see Lemma 2.8) and by the special case of core graphs there is a core graph Y_g, an action of H by simplicial isomorphisms on Y_g, and an H-equivariant proper homotopy equivalence $f : X_g \to Y_g$.

LEMMA 7.1. *There is a locally finite graph $Y \supseteq Y_g$ and a proper homotopy equivalence $X \to Y$ that extends f.*

PROOF. Form the mapping cylinder $M = X_g \times I \sqcup Y_g / x \sim f(x)$ of f. Since f is a proper homotopy equivalence, both 0 and 1-levels of M (which can be identified with X_g and Y_g) are proper strong deformation retracts of M. For Y_g this can be seen by deforming along the mapping cylinder lines. For X_g, without the word "proper", this is a theorem of Ralph Fox [**11**], see also [**12**], but their proofs work just as well in the proper category. The statement can also be deduced from the Whitehead theorem, see [**13**], and [**10**] for the proper version. Now X is obtained from X_g by attaching trees T_v along vertices $v \in X_g$. Attach products $T_v \times I$ to M along the natural copies of $\{v\} \times I$ to obtain a space Z and note that both X and the space Y (obtained from Y_g by attaching trees T_v along $f(v)$) are proper strong deformation retracts of Z and this gives the desired proper homotopy equivalence $X \to Y$. □

We will now revert to the original notation and simply assume that H is acting by simplicial isomorphisms on X_g.

By the *convex hull* of a nonempty subset of a simplicial tree we mean the smallest simplicial subtree that contains the set. The following fixed point fact is well known.

LEMMA 7.2. *Suppose a compact group H acts continuously on a simplicial tree. Then H fixes a point in the convex hull of any orbit.*

PROOF. The convex hull is H-invariant and it is a tree of finite diameter. Iteratively remove all edges that contain a valence 1 vertex until the tree that's left is either a single vertex or a single edge. This vertex or the midpoint of the edge is then fixed by H. □

We will now use this fact to prove the following fixed point theorem, which is really the heart of the argument in this case.

LEMMA 7.3. *Suppose H fixes a point β in $DX = \partial X \setminus \partial X_g$. Then H fixes a point $\rho(\beta)$ in X_g and there is a ray (called the Nielsen ray) r from $\rho(\beta)$ to β such that $h(r)$ and r are properly homotopic rel $\rho(\beta)$ for every $h \in H$.*

PROOF. Let $\tilde{X}$ be the universal cover of X. Let r be a ray in X converging to β. The deck group acts simply transitively on the set of lifts of r and distinct lifts are not asymptotic and hence not properly homotopic. Choose one such lift $\tilde{r}$. Every $h \in H$ has a unique lift to $\tilde{X}$ that fixes the asymptotic class of rays $[\tilde{r}]$ and the set of these lifts defines an action of H on $\tilde{X}$ by proper homotopy equivalences. We will prove that the action is continuous in the next paragraph. The lifted group H preserves the preimage $\tilde{X}_g$ of X_g, which is a tree, and this defines an action of H on $\tilde{X}_g$. By Lemma 7.2 it fixes a point z. The image of z in X_g is the desired fixed point and the image of the ray that starts at z and is asymptotic to $\tilde{r}$ is the Nielsen ray.

The action is continuous: if $h \in H$ is close to the identity, we can choose a representative in its proper homotopy class that fixes a large compact set $K \subset X$ as well as the ray r, and preserves the complementary components of K. We can also arrange that $K \cup r$ is connected. Then the lift of h to $\tilde{X}$ will fix the preimage $\tilde{K}$ and will preserve its complementary components. Since K can be chosen so that $\tilde{K}$ contains any given compact set, the lift of h will be close to the identity. □

Let d be an H-invariant metric on ∂X (see Step 1 in Section 6) and let $\mathcal{P}_\epsilon$ be the partition of ∂X as in Step 2 in Section 6. Again fix a decreasing sequence $\epsilon_n \to 0$ and set $\mathcal{P}_n := \mathcal{P}_{\epsilon_n}$. Let $\pi' : X \cup DX \to X_g$ denote the nearest point projection (this is not equivariant).

Fix an H-equivariant exhaustion $\emptyset = K_0 \subset K_1 \subset K_2 \subset \cdots$ of X_g by finite connected subgraphs so that if $\beta \in X \cup DX$ and $\pi'(\beta) \notin K_{i+1}$ then $\pi'(h(\beta)) \notin K_i$ for every $h \in H$.

Call an element $P \in \mathcal{P}_n$ *good* if the following holds:

- $P \subset DX$,
- $\pi'(P)$ is a point,
- $Stab_H(P)$ fixes a point $\rho(P)$ in X_g; moreover, if $\pi'(P)$ is disjoint from K_{i+1} then $\rho(P)$ and $\pi'(P)$ are in the same component of $X_g \setminus K_i$,
- for every $x \in P$ there is a ray r_x from $\rho(P)$ to x so that all these rays (for all $x \in P$) agree along X_g and further they are permuted up to proper homotopy by $Stab_H(P)$.

So in particular r_x is a Nielsen ray with respect to $Stab_H(x) < Stab_H(P)$. We will also call the rays r_x *Nielsen rays*.

LEMMA 7.4. *For every $\beta \in DX$ there is n_0 so that for every $n \geq n_0$ the element $P \in \mathcal{P}_n$ containing β is good.*

PROOF. We first observe that $Stab_H(\beta)$ fixes a point in X_g by applying Lemma 7.3 to the induced action of $Stab_H(\beta)$ on the graph $X_g^* = X_g \cup \rho_\beta$ (see Lemma 2.8). By our assumption on the exhaustion, if $\pi'(\beta)$ misses K_{i+1} then the action restricts to the complementary component of K_i that contains $\pi'(\beta)$, so in this case the fixed point $\rho(\beta)$ can be found there. Now notice that the stabilizer of a point in X_g is a clopen subgroup of H, so when n is large the stabilizer of $P_n \in \mathcal{P}_n$ that contains β will also fix the same point. (Since H permutes the partition elements in $\mathcal{P}_n$, $Stab_H(\beta)$ will leave P_n invariant and we see that $Stab_H(\beta) = \cap_n Stab_H(P_n)$ is the intersection of clopen subgroups. By compactness we have $Stab_H(P_n) \subseteq Stab_H(\rho(\beta))$ for large n.) We will of course also have $P \subset DX$, $\pi'(P)$ is a point, and $h(\ell) \cap X_g = \emptyset$ for every line ℓ joining two points of P and every $h \in H$. □

Now we construct an H-equivariant cover $\mathcal{N}$ by pairwise disjoint good partition elements. Say an H-orbit in $\mathcal{P}_n$ (which is finite) is *good* if every (any) element in it is good. Then let $\mathcal{N}$ consist of good orbits in $\mathcal{P}_1$ as well as those good orbits in $\mathcal{P}_n$, $n = 2, 3, \cdots$ whose union is not contained in the union of any good orbit in $\mathcal{P}_{n-1}$. Define an equivariant map $\rho : \mathcal{N} \to X_g$ by letting ρ be as in the definition of a good partition element on a representative of the orbit, and then extend it equivariantly. Thus we still have the Nielsen rays for all elements of $\mathcal{N}$.

We now construct a graph Y by attaching trees to $Y_g = X_g$. For every $N \in \mathcal{N}$ we build a tree T_N as in Step 3 of Section 6 for $Stab_H(N)$, namely the mapping telescope with base vertex N and the other vertices all the partition elements contained in N. We identify ∂T_N with N. We then attach T_N to X_g by identifying the base vertex N with the point $\rho(N) \in X_g$. Doing this for all $N \in \mathcal{N}$ produces the desired graph Y. By construction H acts on Y by simplicial isomorphisms.

LEMMA 7.5. *There is a proper homotopy equivalence $F : Y \to X$ such that*

(a) *F is identity on X_g and on $DX = DY$,*
(b) *F sends the rays in T_N based at N to the Nielsen rays r_x from $\rho(N)$ to ∂N preserving the endpoints,*
(c) *F is H-equivariant.*

PROOF. The map F is uniquely defined on each T_N by (a)-(c). That this map is proper as a map $Y \to X$ follows from the fact that if $N_i \in \mathcal{N}$ converge to $\beta \in \partial X_g$, then $\rho(N_i) \to \beta$. Thus F is a proper homotopy equivalence by Corollary 3.7.

Finally we argue H-equivariance. Denote by F' the proper homotopy inverse of F which is identity on X_g. If $h \in H$ consider $F'hF \cdot h^{-1} : Y \to Y$. This is identity on X_g and on ∂X. By Corollary 3.6 it suffices to argue that this map preserves oriented loops and lines connecting points of DX. For loops this is clear since the map is identity on X_g. It also preserves lines joining points of DX since such lines can be written as a concatenation $r^{-1}sr'$ where r, r' are Nielsen rays and s is a segment in X_g. Finally, it preserves lines that connect distinct points of some $N \in \mathcal{N}$. □

This finishes the proof of the Main Theorem.

References

[1] Santana Afton, Danny Calegari, Lvzhou Chen, and Rylee Alanza Lyman, *Nielsen realization for infinite-type surfaces*, Proc. Amer. Math. Soc. **149** (2021), no. 4, 1791–1799, DOI 10.1090/proc/15316. MR4242332

[2] Javier Aramayona and Nicholas G. Vlamis, *Big mapping class groups: an overview*, In the tradition of Thurston—geometry and topology, Springer, Cham, [2020] ©2020, pp. 459–496, DOI 10.1007/978-3-030-55928-1_12. MR4264585

[3] R. Ayala, E. Dominguez, A. Márquez, and A. Quintero, *Proper homotopy classification of graphs*, Bull. London Math. Soc. **22** (1990), no. 5, 417–421, DOI 10.1112/blms/22.5.417. MR1082009

[4] Mladen Bestvina and Mark Feighn, *Bounding the complexity of simplicial group actions on trees*, Invent. Math. **103** (1991), no. 3, 449–469, DOI 10.1007/BF01239522. MR1091614

[5] Marc Culler, *Finite groups of outer automorphisms of a free group*, Contributions to group theory, Contemp. Math., vol. 33, Amer. Math. Soc., Providence, RI, 1984, pp. 197–207, DOI 10.1090/conm/033/767107. MR767107

[6] George Domat, Hannah Hoganson, and Sanghoon Kwak, *Coarse geometry of pure mapping class groups of infinite graphs*, Adv. Math. **413** (2023), Paper No. 108836, 57, DOI 10.1016/j.aim.2022.108836. MR4530621

[7] Carl Eberhart, *Some remarks on the irrational and rational numbers*, Amer. Math. Monthly **84** (1977), no. 1, 32–35, DOI 10.2307/2318303. MR644644

[8] Ryszard Engelking, *General topology*, 2nd ed., Sigma Series in Pure Mathematics, vol. 6, Heldermann Verlag, Berlin, 1989. Translated from the Polish by the author. MR1039321

[9] D. B. A. Epstein, *Curves on 2-manifolds and isotopies*, Acta Math. **115** (1966), 83–107, DOI 10.1007/BF02392203. MR214087

[10] F. T. Farrell, L. R. Taylor, and J. B. Wagoner, *The Whitehead theorem in the proper category*, Compositio Math. **27** (1973), 1–23. MR334226

[11] R. H. Fox, *On homotopy type and deformation retracts*, Ann. of Math. (2) **44** (1943), 40–50, DOI 10.2307/1969063. MR7980

[12] Martin Fuchs, *A note on mapping cylinders*, Michigan Math. J. **18** (1971), 289–290. MR287537

[13] Allen Hatcher, *Algebraic topology*, Cambridge University Press, Cambridge, 2002. MR1867354

[14] Sebastian Hensel and Dawid Kielak, *Nielsen realization by gluing: limit groups and free products*, Michigan Math. J. **67** (2018), no. 1, 199–223, DOI 10.1307/mmj/1519095620. MR3770860

[15] Sebastian Hensel, Damian Osajda, and Piotr Przytycki, *Realisation and dismantlability*, Geom. Topol. **18** (2014), no. 4, 2079–2126, DOI 10.2140/gt.2014.18.2079. MR3268774

[16] Jesús Hernández Hernández, Israel Morales, and Ferrán Valdez, *The Alexander method for infinite-type surfaces*, Michigan Math. J. **68** (2019), no. 4, 743–753, DOI 10.1307/mmj/1561773633. MR4029627

[17] Alexander S. Kechris, *Classical descriptive set theory*, Graduate Texts in Mathematics, vol. 156, Springer-Verlag, New York, 1995, DOI 10.1007/978-1-4612-4190-4. MR1321597

[18] Steven P. Kerckhoff, *The Nielsen realization problem*, Ann. of Math. (2) **117** (1983), no. 2, 235–265, DOI 10.2307/2007076. MR690845

[19] D. G. Khramtsov, *Finite groups of automorphisms of free groups* (Russian), Mat. Zametki **38** (1985), no. 3, 386–392, 476. MR811572

[20] James R. Munkres, *Topology*, Prentice Hall, Inc., Upper Saddle River, NJ, 2000. Second edition of [MR0464128]. MR3728284

[21] John R. Stallings, *Topology of finite graphs*, Invent. Math. **71** (1983), no. 3, 551–565, DOI 10.1007/BF02095993. MR695906

[22] Bruno Zimmermann, *Über Homöomorphismen n-dimensionaler Henkelkörper und endliche Erweiterungen von Schottky-Gruppen* (German), Comment. Math. Helv. **56** (1981), no. 3, 474–486, DOI 10.1007/BF02566224. MR639363

Department of Mathematics, University of Haifa, 199 Abba Khoushy Avenue Mount Carmel, Haifa, 3498838 Israel
Email address: yalgom@univ.haifa.ac.il

Department of Mathematics, University of Utah, 155 S 1400 E, RM 233 Salt Lake City, Utah 84112
Email address: mladen.bestvina@utah.edu

Contemporary Mathematics
Volume **812**, 2025
https://doi.org/10.1090/conm/812/16260

Z-sets in ANRs

Fredric D. Ancel

ABSTRACT. This paper is an exposition of the concept of a *Z-set* which played a central role in the development of the topology of infinite dimensional manifolds in the 1960's and 1970's and which reemerged in the 1990's as a key concept in the study of group boundaries in geometric group theory. The principle result of this paper, Theorem 4.1, describes the logical relations between 10 different potential candidates for the definition of *Z-set* that have appeared in the literature. We call this theorem *Torunczyk's Equivalence Theorem* in recognition of the contribution of H. Torunczyk's article [*Concerning locally homotopy negligible sets and characterization of l_2-manifolds*, Fund. Math. **101** (1978), no. 2, 93–110], which introduced several of these conditions and proved their equivalence without assuming the ambient space to be an ANR. Indeed, Theorem 4.1 establishes that five of these conditions are equivalent with no additional hypotheses, eight become equivalent if the ambient space is assumed to be an ANR, and all ten becomes equivalent if, in addition, the subset under consideration is taken to be a closed set. Theorem 4.1 is preceded by a survey of the theory of ANRs which includes background material necessary for understanding the proof of the theorem. With the goal of fostering an appreciation of the usefulness of the concept of a *Z-set*, this survey also includes statements of the main results of infinite dimensional manifold theory in which *Z-sets* played an indispensable role. The statement of Theorem 4.1 is followed by a relatively self-contained proof and two examples illustrating the necessity of certain hypotheses in the statement of the theorem. Next, the related notion of a *strong Z-set* is explored. Finally, a short alternative proof discovered by C. Guilbault and M. Moran of a result of S. Ferry is presented; this result gives a useful criterion for detecting when being a *Z-subset* is preserved by a map between spaces.

1. Introduction

The notion of a *Z-set* originated in the pioneering studies of infinite dimensional topology by R. D. Anderson in the late 1960's [**2**] and played a crucial role in the classification of Hilbert cube manifolds by T. Chapman [**6**] and of Hilbert space manifolds by H. Torunczyk [**30**] and [**31**]. This notion arose again in the study of group boundaries [**4**] where it also had a significant impact. In the literature on Z-sets, various definitions have been proposed, properties have been observed and equivalences have been proved. In particular, the results of [**29**] clarify the relations between many different fundamental properties of Z-sets in various contexts. The

2020 *Mathematics Subject Classification.* Primary 54C55; Secondary 20F65, 55M15, 57N20, 58B05.

goal of this article is to provide a comprehensive exposition of these relations that can be consulted and referenced by mathematicians whose research employs Z-sets.

The remainder of this paper is divided into 9 sections. Section 2 is a survey of important background material about absolute neighborhood retracts, with references but mostly without proofs. Some, but not all of this survey is needed to understand the material about Z-sets which follows it. Also included in this survey is information about the crucial role that absolute neighborhood retracts play in our current understanding of the theory of infinite dimensional manifolds as revealed by the theorems of West, Henderson, Chapman, Edwards and Torunczyk. An acquaintance with these results is requisite to an appreciation of the position occupied by absolute neighborhood retracts in modern topology. Section 3 defines various notions of Z-sets. Section 4 states the principal result of this paper relating different properties of Z-sets. In recognition of the contribution of [**29**], we are calling this result *Torunczyk's Equivalence Theorem*. This theorem lists ten conditions that a Z-set might satisfy and states three equivalences among these conditions that hold in different contexts. Sections 5, 6 and 7 provide proofs of these three equivalences. Section 8 describes two examples that illustrate why some of the hypotheses in Torunczyk's Equivalence Theorem can't be omitted. Section 9 investigates the relation between Z-sets and *strong Z-sets*, a concept also found in the literature on this subject. Section 10 presents an exposition of a recent alternative proof by C. Guilbault and M. Moran [**13**] of a result by S. Ferry [**12**] that we are calling *Ferry's Lemma*. Ferry's Lemma, which has been useful in the study of group boundaries, provides conditions under which a space A that is embedded in two different spaces X and Y and is a Z-set in Y will necessarily be a Z-set in X.

2. Absolute neighborhood retracts

The notion of *absolute neighborhood retract* was originated by K. Borsuk in the early 1930s [**5**]. The book [**18**] is a thorough treatment of this subject up through the 1960s. Absolute neighborhood retracts played a crucial in the explosive development of infinite dimensional topology in the late 1960s and 1970s. The 1988 book [**26**] documents the part of this development that deals with locally compact spaces. The extremely detailed 2020 book [**25**] together with its 2013 precursor [**24**] present a more comprehensive view by treating both the locally compact and non-locally compact parts of the subject. These books together with the original research articles on which they are based are our primary sources of information.

TERMINOLOGY. In this article, a *function* is not assumed to be continuous, whereas a *map* is always assumed to be a continuous function.

DEFINITION. For a subset X of a topological space Y: we call a map $r : Y \to X$ such that $r|X = id_X$ a *retraction of* Y *onto* X; and if a retraction of Y onto X exists, we call X a *retract of* Y.

NOTATION. Let $\mathcal{M}$ denote the class of all metrizable spaces.

DEFINITION. A topological space Y is an *absolute neighborhood retract* (abbreviated *ANR*) if $Y \in \mathcal{M}$ and if, whenever Y is embedded as a closed subset of a space $X \in \mathcal{M}$, an open neighborhood of Y in X retracts onto Y. Y is an *absolute retract* (abbreviated *AR*) if $Y \in \mathcal{M}$ and if, whenever Y is embedded as a closed subset of a space $X \in \mathcal{M}$, X retracts onto Y.

We remark that this definition includes the restriction "$Y \in \mathcal{M}$" to exclude the undesirable and otherwise logically unavoidable consequence of making every non-metrizable space an ANR.

We will now state a sequence of facts about absolute neighborhood retracts using footnotes to cite theorems in authoritative sources (primarily [**18**]) that justify these facts.

It is clear from the above definition that:

FACT 2.1. Every AR is an ANR.[1]

DEFINITION. A topological space Y is an *absolute neighborhood extensor* (abbreviated *ANE*) if every map $f : A \to Y$ from a closed subset A of a space $X \in \mathcal{M}$ into Y extends to a map $g : N \to Y$ from an open neighborhood N of A in X into Y. Y is an *absolute extensor* (abbreviated *AE*) if every map $f : A \to Y$ from a closed subset A of a space $X \in \mathcal{M}$ into Y extends to a map $g : X \to Y$.

Note that an ANE is not required to be metrizable; however:

FACT 2.2. Let X be a topological space. Then X is an ANR if and only if $X \in \mathcal{M}$ and X is an ANE. Similarly, X is an AR if and only if $X \in \mathcal{M}$ and X is an AE.[2]

We remark that, in the definitions of ANR and ANE, one can replace the class $\mathcal{M}$ of all metrizable spaces by the class $\mathcal{N}$ of all normal spaces. One can then develop a theory of ANRs and ANEs for the class $\mathcal{N}$ which is parallel in many respects to the theory we are about to describe for $\mathcal{M}$. In particular, one can prove an analogue of Fact 2.2 for normal spaces: for normal spaces, being an ANR for $\mathcal{N}$ is equivalent to being an ANE for $\mathcal{N}$. Furthermore, the proof of this analogue of Fact 2.2 for $\mathcal{N}$ is more elementary than the proof of Fact 2.2 for $\mathcal{M}$.[3] Observe that since $\mathcal{M}$ is a proper subclass of $\mathcal{N}$, then, for a metrizable space, being an ANE for $\mathcal{N}$ is potentially a more restrictive condition than being an ANE for $\mathcal{M}$. Indeed, a metrizable space which is an ANE for $\mathcal{N}$ must be topologically complete and have a countable basis;[4] while there is no such limitation on a metrizable space that is an ANE for $\mathcal{M}$. Beyond this point, we will not concern ourselves with spaces that are ANRs and ANEs for the class $\mathcal{N}$; we will focus exclusively on ANRs and ANEs for the class $\mathcal{M}$ and we will suppress explicit mention of "$\mathcal{M}$".

The Tietze Extension Theorem[5] and the following more general result are sources of ARs:

THEOREM 2.3 (Dugundji's Extension Theorem). *Every convex subset of a locally convex topological vector space is an AE.*[6]

Hence:

COROLLARY 2.4. $[0, 1], \mathbb{R}, [0, \infty)$, *all Euclidean spaces and all normed vector spaces are ARs. Therefore, these spaces are also ANRs.*

---

[1] [**18**], Proposition III.1.1, page 81.
[2] [**18**], Theorems III.3.1 and III.3.2, pages 83-86.
[3] [**18**], pages 83-86.
[4] [**18**], Theorem II.16.1.(d), pages 63-67.
[5] [**18**], Theorem II.3.1, pages 37-39.
[6] [**18**], Theorem II.14.1, pages 57-59.

The following two facts follow directly from the definitions:

FACT 2.5. Every retract of an AR is an AR, and every retract of an ANR is an ANR.[7]

FACT 2.6. Every open subset of an ANR is an ANR.[8]

Hence, each open subset of every space mentioned in Corollary 2.4 is an ANR.

The following facts tell us that the property of being an AR or an ANR is preserved under the formation of appropriate sorts of Cartesian products.

FACT 2.7. The Cartesian product of every countable collection of ARs is an AR.[9]

FACT 2.8. The Cartesian product of every finite collection of ANRs is an ANR.[10]

Consequently:

COROLLARY 2.9. *The Cartesian product of every countable collection of ANRs is an ANR provided that all but finitely many elements of the collection are ARs.*

To appreciate the necessity of the hypothesis of Corollary 2.9, observe that $\{0, 1\}$ is an ANR, but $\{0, 1\}^\infty$ is not.

Homotopies play a fundamental role in the theory of ANRs. We recall some terminology and notation related to homotopies.

DEFINITION. Let X and Y be topological spaces. A *homotopy* is a map of the form $h : X \times [0, 1] \to Y$. Given a homotopy $h : X \times [0, 1] \to Y$ and an element $t \in [0, 1]$, define the map $h_t : X \to Y$ by $h_t(x) = h(x, t)$. If $A \subset X$, then a homotopy $h : X \times [0, 1] \to Y$ is *stationary on* A if $h(\{x\} \times [0, 1]) = \{h_0(x)\}$ for each $x \in A$.

DEFINITION. Let X be a topological space. A subset A of X *contracts to a point in* X if there is a homotopy $h : A \times [0, 1] \to X$ such that $h_0 = id_A$ and $h_1(A)$ is a one-point set. X is *contractible* if X contracts to a point in itself. X is *locally contractible* if for every $x \in X$ and every open neighborhood U of x in X, there is an open neighborhood V of x in X such that V contracts to a point in U.

FACT 2.10. Every AR is contractible and every ANR is locally contractible.[11]

DEFINITION. Let X and Y be topological spaces and let $\mathcal{U}$ be a collection of subsets of Y. Two maps $f : X \to Y$ and $g : X \to Y$ are *$\mathcal{U}$-close* if $\{f(x), g(x)\}$ is a subset of an element of $\mathcal{U}$ for each $x \in X$. A homotopy $h : X \times [0, 1] \to Y$ is a *$\mathcal{U}$-homotopy* if $h(\{x\} \times [0, 1])$ is a subset of an element of $\mathcal{U}$ for each $x \in X$. Two maps $f : X \to Y$ and $g : X \to Y$ are *$\mathcal{U}$-homotopic* if there is a $\mathcal{U}$-homotopy $h : X \times [0, 1] \to Y$ such that $h_0 = f$ and $h_1 = g$. A map $f : X \to Y$ is a *$\mathcal{U}$-homotopy equivalence* if there is a map $g : Y \to X$ such that $g \circ f$ is $f^{-1}(\mathcal{U})$-homotopic to id_X (where $f^{-1}(\mathcal{U}) = \{f^{-1}(U) : U \in \mathcal{U}\}$) and $f \circ g$ is $\mathcal{U}$-homotopic to id_Y. Suppose, in addition, that Y is a metrizable space with assigned metric σ. For $\epsilon > 0$, let $\mathcal{U}(\epsilon)$

7 [**18**], Proposition III.7.7, page 97.
8 [**18**], Proposition III.7.9, page 97.
9 [**18**], Proposition III.7.5, page 97.
10 [**18**], Proposition III.7.6, page 97.
11 [**18**], Theorem III.7.1, page 96.

denote the collections of all subsets of Y of σ-diameter $< \epsilon$. Then two maps from X to Y are said to be *ϵ-close* if and only if they are $\mathcal{U}(\epsilon)$-close; a homotopy from $X \times [0,1]$ to Y is said to be an *ϵ-homotopy* if and only if it is a $\mathcal{U}(\epsilon)$-homotopy; two maps from X to Y are said to be *ϵ-homotopic* if and only if they are $\mathcal{U}(\epsilon)$-homotopic; and a map from X to Y is called an *ϵ-homotopy equivalence* if and only if it is a $\mathcal{U}(\epsilon)$-homotopy equivalence.

DEFINITION. Let X be a topological space, let Y be a metrizable space with assigned metric σ, and let $\alpha : X \to [0,\infty)$ be a map. Two maps $f : X \to Y$ and $g : X \to Y$ are *α-near* if $\sigma(f(x), g(x)) \leq \alpha(x)$ for each $x \in X$. A homotopy $h : X \times [0,1] \to Y$ is an *α-near homotopy* if $diam_\sigma(h(x \times [0,1])) \leq \alpha(x)$ for each $x \in X$. Two maps $f : X \to Y$ and $g : X \to Y$ are *α-nearly homotopic* if there is an α-near homotopy $h : X \times [0,1] \to Y$ such that $h_0 = f$ and $h_1 = g$. Observe the distinction between two maps $f : X \to Y$ and $g : X \to Y$ being "α-near" where α is a map from X to $[0,\infty)$ and f and g being "$\mathcal{U}$-close" where $\mathcal{U}$ is an open cover of Y. The condition that condition "g is α-near to f" is potentially more restrictive than the condition that "g is $\mathcal{U}$-close to f". More precisely, given a map $f : X \to Y$, one can prove that for every open cover $\mathcal{U}$ of Y, there is a map $\alpha : X \to [0,\infty)$ such that every map from X to Y which is α-near to f is also $\mathcal{U}$-close to f; but in general, it is false that for every map $\alpha : X \to [0,\infty)$, there is an open cover $\mathcal{U}$ of Y such that every map from X to Y which is $\mathcal{U}$-close of f is also α-near to f.

The following proposition, which has a simple statement and an equally simple proof, turns out to be an essential tool in the study of ANRs.

THE BORSUK HOMOTOPY EXTENSION PRINCIPLE 2.11. *Let Y be an ANR. If A is a closed subset of a metrizable space X, then every map $f : (X \times \{0\}) \cup (A \times [0,1]) \to Y$ extends to a homotopy $g : X \times [0,1] \to Y$.*[12] *Furthermore, if $\mathcal{U}$ is an open cover of Y and $f|A \times [0,1]$ is a $\mathcal{U}$-homotopy, then g can chosen to be a $\mathcal{U}$-homotopy.*[13]

The Borsuk Homotopy Extension Principle is instrumental in yielding elementary proofs of the following two facts.

FACT 2.12. Every contractible ANR is an AR.[14]

For a finite-dimensional[15] metrizable space, local contractibility is sufficient to ensure that the space is an ANR.

FACT 2.13. Every finite-dimensional locally contractible metric space is an ANR.[16]

Since topological manifolds and finite-dimensional simplicial complexes are locally contractible, it follows that:

COROLLARY 2.14. *Every topological manifold and every finite-dimensional metrizable simplicial complex is an ANR.*

---

[12] [**18**], Theorem IV.2.2, page 117.

[13] [**26**], Theorem 5.1.3, pages 192-193.

[14] [**18**], Proposition III.7..2, page 96.

[15] Definitions of *finite-dimensional* and *infinite-dimensional* metric spaces are found on page 150 of [**18**].

[16] [**18**], Theorem V.7.1, pages 168-169.

However, without the finite-dimensional hypothesis, local contractibility alone does not guarantee that a space is an ANR:

FACT 2.15. There is an infinite-dimensional locally contractible compact metric space that is not an ANR.[17]

Next we state one of the most useful homotopy properties of ANRs.

FACT 2.16. Let Y be an ANR. Then for every open cover $\mathcal{U}$ of Y, there is an open cover $\mathcal{V}$ of Y with the following property. If two maps $f : X \to Y$ and $g : X \to Y$ from a topological space X into Y are $\mathcal{V}$-close, then they are $\mathcal{U}$-homotopic. Furthermore, if $A \subset X$ and $f|A = g|A$, then the $\mathcal{U}$-homotopy from f to g can be chosen to be stationary on A. [18]

As previously noted, finite-dimensional metric spaces that are "locally nice" such as manifolds and finite-dimensional simplicial complexes are ANRs. Below we will state a fact that relaxes the finite-dimensionality restriction if the "locally nice" restriction is strengthened appropriately. It will then follow that certain "locally nice" infinite-dimensional metric spaces such as Hilbert cube manifolds, Hilbert space manifolds and infinite-dimensional simplicial complexes (topologized appropriately) are ANRs. In fact, the motivation for developing a theory of ANRs is that the properties that such "locally nice" spaces have because they are ANRs hold for reasons that are more clearly understood when viewed from the perspective of a general theory ANRs.

DEFINITION. A topological space X is a local ANR if X is metrizable and if every point of X has an open neighborhood that is an ANR.

FACT 2.17. Every local ANR is an ANR.[19]

Since every Hilbert cube is an AR by Fact 2.7, then every open subset of a Hilbert cube is an ANR by Fact 2.6. Therefore, every Hilbert cube manifold is a local ANR. Since every Hilbert space is an AR by Corollary 2.4, then every Hilbert space manifold is a local ANR. Consequently:

COROLLARY 2.18. *Every Hilbert cube manifold and every Hilbert space manifold is an ANR.*

Polyhedra underlying abstract simplicial complexes play an important role in the theory of ANRs. To make polyhedra useful, we must first equip them with a natural topology. We now describe two potentially distinct approaches to this task, each of which has its uses.

DEFINITION. K is an *abstract simplicial complex* if K is a set of non-empty finite subsets of a set V_K, called the *vertex set* of K, satisfying the following two properties:

(1) $v \in V_K \Rightarrow \{v\} \in K$.
(2) $\sigma \in K$ and $\emptyset \neq \tau \subset \sigma \Rightarrow \tau \in K$.

DEFINITION. Let K be an abstract simplicial complex and let $\sigma \in K$. σ is called a *simplex of* K. If $\emptyset \neq \tau \subset \sigma$, then τ is called a *face of* σ. Let $\bar{\sigma} = \{\tau : \emptyset \neq \tau \subset \sigma\}$.

---

[17] **[18]**, Section V.8, pages 169-172.
[18] **[18]**, Theorem IV.1.2, pages 112-114.
[19] **[18]**, Theorem III.8.1, page 98.

The *boundary of* σ, denoted $\partial\sigma$, is $\{\tau : \emptyset \neq \tau \subset \sigma \text{and } \tau \neq \sigma\}$. The dimension of an element σ of K, denoted $dim(\sigma)$ is n if σ has exactly $n+1$ elements. The dimension of K, denoted $dim(K)$, is $sup\{dim(\sigma) : \sigma \in K\} \in \{0, 1, 2, \cdots \infty\}$.

DEFINITION. Let K be an abstract simplicial complex. A subset L of K is a *subcomplex of* K if L is itself an abstract simplicial complex. Equivalently, L is a subcomplex of K if and only if: $\sigma \in L$ and $\emptyset \neq \tau \subset \sigma \Rightarrow \tau \in L$. In this case, the vertex set V_L of L is $\{v \in V_K : \{v\} \in L\}$. If L is a subcomplex of K, then the ordered pair (K, L) is called a *simplicial pair.* Note that if $\sigma \in K$, then $\bar{\sigma}$ and $\partial\sigma$ are subcomplexes of K.

DEFINITION. Let K be an abstract simplicial complex. The *polyhedron underlying* K, which is also called the *geometric realization of* K, is the set

$$|K| = \{\phi \in \mathbb{R}^{V_K} : \phi(V_K) \subset [0, \infty), \phi^{-1}(0, \infty) \in K \text{ and } \sum_{v \in V_K} \phi(v) = 1\}.$$

(Here, $\mathbb{R}^{V_K}$ denotes the set of all functions from V_K to $\mathbb{R}$.) For each $\sigma \in K$, let

$$|\sigma| = \{\phi \in |K| : \phi^{-1}(0, \infty) \subset \sigma\};$$

and for each $L \subset K$, let $|L| = \bigcup\{|\sigma| : \sigma \in L\}$. (Thus, for $\sigma \in K$, $|\partial\sigma| = \{\phi \in |K| : \phi^{-1}(0, \infty) \subset \sigma \text{ and } \phi^{-1}(0, \infty) \neq \sigma\}$.)

DEFINITION. Let K be an abstract simplicial complex. Define the *taxicab metric* ρ on $|K|$ by the formula

$$\rho(\phi, \psi) = \sum_{v \in V_K} |\phi(v) - \psi(v)|.$$

This metric defines a topology on $|K|$ called the *metric topology.* Let $|K|_m$ denote $|K|$ endowed with the metric topology. For each $L \subset K$, let $|L|_m$ denote $|L|$ with the topology determined by the restriction of the taxicab metric to $|L|$. Then define the *Whitehead topology on* $|K|$ to be

$$\{U \subset |K| : U \cap |\sigma| \text{ is an open subset of } |\sigma|_m \text{ for every } \sigma \in K\}.$$

Let $|K|_W$ denote $|K|$ endowed with the Whitehead topology. Since the Whitehead topology on $|K|$ clearly contains the metric topology on $|K|$, then $id_{|K|} : |K|_W \to |K|_m$ is continuous. In the situation that the metric and Whitehead topologies on $|K|$ coincide, we denote $|K|$ with either topology simply by $|K|$.

Suppose that K is an abstract simplicial complex and L is a subcomplex of K. Then $V_L \subset V_K$, $|K| \subset \mathbb{R}^{V_K}$ and $|L| \subset \mathbb{R}^{V_L}$. Define $i : \mathbb{R}^{V_L} \to \mathbb{R}^{V_K}$ by $i(\phi)(v) = \phi(v)$ for $v \in V_L$ and $i(\phi)(v) = 0$ for $v \in V_K - V_L$. Then $i||L| : |L|_m \to |K|_m$ and $i||L| : |L|_W \to |K|_W$ are embeddings (onto closed subsets[20]). From this point on, we will identify $|L|$ with $i(|L|)$. This will allow us to regard $|L|_m$ as a (closed) subspace of $|K|_m$ and $|L|_W$ as a (closed) subspace of $|K|_W$.

Let $t = m$ or W. If K is an abstract simplicial complex and L is a subcomplex of K, then there is a simplicial complex M such that $|M|_t \cong$[21] $(|K|_t \times \{0\}) \cup (|L|_t \times$

[20]To see that $|K|_m - |L|_m$ and $|K|_W - |L|_W$ are open sets, let $\phi \in |K| - |L|$. Then $\phi^{-1}(0, \infty) \notin L$. Let $\epsilon = \min\phi(\phi^{-1}(0, \infty))$. Then $\psi \in |K|$ and $\rho(\phi, \psi) < \epsilon \Rightarrow \phi^{-1}(0, \infty) \subset \psi^{-1}(0, \infty) \Rightarrow \psi^{-1}(0, \infty) \notin L$. Hence, the ϵ-neighborhood of ϕ in $|K|_m$ is disjoint from $|L|_m$. This proves $|K|_m - |L|_m$ and, hence, $|K|_W - |L|_W$ are open.

[21]$\cong$ means *is homeomorphic to.*

$[0,1]$). There is no unique choice for M, but one way to obtain M is to put a linear order $<$ on V_L and define M by

$$M = \{\sigma \times \{0\} : \sigma \in K\} \cup$$
$$\{(\tau \times \{0\}) \cup (\tau' \times \{1\}) : \tau \cup \tau' \in L \text{ and } v \leq v' \text{ for all } v \in \tau \text{ and all } v' \in \tau'\}.$$

Then $V_M = (V_K \times \{0\}) \cup (V_L \times \{1\})$. Also an embedding $|K|_t \to |M|_t$ is determined by $|\sigma| \to |\sigma \times \{0\})|$. Furthermore, two different embeddings $|L|_t \to |M|_t$ are determined by $|\tau| \to |\tau \times \{0\}|$ and $|\tau| \to |\tau \times \{1\}|$ and these embeddings are the "ends" of an embedding $|L|_t \times [0,1] \to |M|_t$.

Let $t = m$ or W. Let K be an abstract simplicial complex. A *subdivision of* K is a pair (K', h) where K' is an abstract simplicial complex and $h : |K'|_t \to |K|_t$ is a homeomorphism such that for every $\sigma' \in K'$, there is a $\sigma \in K$ such that $h(|\sigma'|) \subset |\sigma|$ and $h\big||\sigma'| : |\sigma'| \to |\sigma|$ is an affine map (i.e., for all $\phi, \psi \in |\sigma'|$ and all $t \in [0,1]$, $h((1-t)\phi + t\psi) = (1-t)h(\phi) + th(\psi)$). If $\mathcal{U}$ is an open cover of $|K|_t$, then a subdivision (K', h) of K *refines* $\mathcal{U}$ if for every $\sigma' \in K'$, $h(|\sigma'|)$ is a subset of an element of $\mathcal{U}$. Observe that if L is a subcomplex of K, (K', h) is a subdivision of K and $L' = \{\sigma' \in K' : h(|\sigma'|) \subset |L|\}$, then $(L', h\big||L'|)$ is a subdivision of L. Moreover, if (K', h) refines an open cover $\mathcal{U}$ of $|K|_t$, then $(L', h\big||L'|)$ refines $\{U \cap |L| : U \in \mathcal{U}\}$. From this point on, if (K', h) is a subdivision of K, we identify $|K'|$ with $|K|$ via h. Thus, we regard $|K'| = |K|$; and for each $\sigma' \in K'$, there is a $\sigma \in K$ such that $|\sigma'|$ is an affinely embedded subset of $|\sigma|$.

THEOREM 2.19 (The Subdivision Theorem). *For $t = m$ or W, for every abstract simplicial complex K, for every open cover $\mathcal{U}$ of $|K|_t$, there is a subdivision of K that refines $\mathcal{U}$.*[22].

A polyhedron which is assigned the Whitehead topology enjoys a property expressed in Fact 2.20 below which may give it an advantage over the same polyhedron equipped with the metric topology. However, in some circumstances, it may be desirable to restrict attention to polyhedra that lie in the class of metrizable spaces, making it preferable to assign polyhedra the metric topology. (Fact 2.21 below reveals that assigning these two topologies to a polyhedron may result in non-homeomorphic objects.) Fortunately, the validity of the important theorems about polyhedra in the theory of ANRs is not altered by switching between these two topologies. We will emphasize this invariance in the statements of subsequent facts when appropriate.

For a polyhedron with the Whitehead topology, there is a simple criterion for determining the continuity of a function from the polyhedron to a topological space:

FACT 2.20. If K is an abstract simplicial complex, then a function $f : |K|_W \to Y$ from $|K|$ to a topological space Y is continuous if and only if $f\big||\sigma|_m : |\sigma|_m \to Y$ is continuous for every $\sigma \in K$.

Note that this criterion does not necessarily ensure the continuity of $f : |K|_m \to Y$.

To compare the metric and Whitehead topologies on a given polyhedron, we require the following notion.

[22]The $t = W$ case of this theorem is the content of [**33**], Theorem 35, pages 317-320. The $t = m$ case follows from [**16**], Lemma V.7, pages 211-212 and [**23**]

DEFINITION. An abstract simplicial complex K is *locally finite* if for every $v \in V_K, \{\sigma \in K : v \in \sigma\}$ is a finite set.

FACT 2.21. Let K be an abstract simplicial complex. Then:
K is locally finite $\Leftrightarrow |K|_W$ is metrizable $\Leftrightarrow$
$id_{|K|} : |K|_W \to |K|_m$ is a homeomorphism $\Leftrightarrow$
$|K|_W$ and/or $|K|_m$ is locally compact.[23]

The usual way that an abstract simplicial complex arises in the theory of ANRs is as the nerve of an open cover. We now describe this object.

DEFINITION. If $\mathcal{U}$ is an open cover of a topological space X, then the *nerve of* $\mathcal{U}$, denoted $n(\mathcal{U})$, is the abstract simplicial complex with vertex set $\mathcal{U}$ such that $\sigma \in n(\mathcal{U})$ if and only if σ is non-empty finite subset of $\mathcal{U}$ and $\bigcap \sigma \neq \emptyset$.

DEFINITION. An open cover $\mathcal{U}$ of a topological space X is *star-finite* if for every $U \in \mathcal{U}$, $\{V \in \mathcal{U} : U \cap V \neq \emptyset\}$ is a finite set.

It is easily seen that the nerve $n(\mathcal{U})$ of an open cover $\mathcal{U}$ is locally finite (making $|n(\mathcal{U})|_W$ homeomorphic to $|n(\mathcal{U})|_m$) if and only if the open cover $\mathcal{U}$ is star-finite. If X is a separable metrizable space, then star-finite open covers of X are abundant; however, if X is a non-separable metrizable space, then star-finite open covers of X may be unavailable when sought after.[24] Thus, if we wish to study ANRs without restricting ourselves to separable metrizable spaces, then we must be willing to analyze polyhedra with the Whitehead topology and polyhedra with the metric topology separately.

The following two facts reveal that a polyhedron with either the metric topology or the Whitehead topology is as close to being an ANR as it possibly can be.

FACT 2.22. If K is an abstract simplicial complex, then $|K|_m$ is an ANR.[25]

If a polyhedron with the Whitehead topology is non-metrizable, then by definition it fails to be an ANR. However, non-metrizability does not exclude it from being an ANE:

FACT 2.23. If K is an abstract simplicial complex, then $|K|_W$ is an ANE.[26]

DEFINITION. Let X and Y be topological spaces and let $\mathcal{U}$ be an open cover of X. X is *$\mathcal{U}$-dominated by Y* if there are maps $\phi : X \to Y$ and $\psi : Y \to X$ such that $\psi \circ \phi : X \to X$ is $\mathcal{U}$-homotopic to id_X. In this situation, we will say that the maps $\phi : X \to Y$ and $\psi : Y \to X$ *witness the $\mathcal{U}$-domination of X by Y*. We say X is *dominated by* Y in the case that X is $\mathcal{U}$-dominated by Y where $\mathcal{U} = \{X\}$.

---

[23] [**18**], Section III.9, pages 99-101 and Lemma III.11.6, pages 108-109.

[24] It is well known that all separable metrizable spaces are regular and Lindelöf. Moreover, every regular Lindelöf space has the property that every open cover of it is refined by a star-finite open cover. (This is the content of [**11**], Corollary, 5.3.11, page 327.) For connected spaces, this result has a converse: every connected regular space with the property that every open cover is refined by a star-finite open cover is a Lindelöf space. ([**9**], Problem XI,7.4, page 255; or [**11**], Exercise 6.1.E, page 359.) If K is any abstract simplicial complex with uncountably many vertices such that $|K|_m$ is connected, then there must be a vertex of K which is an element of uncountably many 1-simplices of K. It follows that $|K|_m$ is not Lindelöf; and, consequently, $|K|_m$ has an open cover which is not refined by any star-finite open cover.

[25] [**18**], Theorem III.11.3, page 106.

[26] [**18**], Theorem III.10.4, page 105.

FACT 2.24_m. If $\mathcal{U}$ is an open cover of an ANR X, then there is an abstract simplicial complex K such that X is $\mathcal{U}$-dominated by $|K|_m$. Furthermore, there are maps $\phi : X \to |K|_m$ and $\psi : |K|_m \to X$ that witness the $\mathcal{U}$-domination of X by $|K|_m$ with the additional property that $\{\psi(|\sigma|) : \sigma \in K\}$ refines $\mathcal{U}$.[27]

FACT 2.24_W. If $\mathcal{U}$ is an open cover of an ANR X, then there is an abstract simplicial complex K such that X is $\mathcal{U}$-dominated by $|K|_W$. Furthermore, there are maps $\phi : X \to |K|_W$ and $\psi : |K|_W \to X$ that witness the $\mathcal{U}$-domination of X by $|K|_W$ with the additional property that $\{\psi(|\sigma|) : \sigma \in K\}$ refines $\mathcal{U}$.[28]

The proofs of Facts 2.24_m and 2.24_W cited in footnote 27 and 28 are different. However, there is an alternative approach which shows that Fact 2.24_W can be derived from Fact 2.24_m and vice versa. This approach depends on:

THEOREM 2.25 (Dowker's Theorem). *For every abstract simplicial complex K, $id_{|K|} : |K|_W \to |K|_m$ is an $\mathcal{S}$-homotopy equivalence where $\mathcal{S} = \{|\sigma| : \sigma \in K\}$.*[29]

We sketch an argument that shows how to use Dowker's Theorem to derive Fact 2.24_W from Fact 2.24_m. Assume Fact 2.24_m. Let $\mathcal{U}$ be an open cover of an ANR X. Let $\mathcal{V}$ be an open cover of X with the property that $\{V \cup V' : V, V' \in \mathcal{V}$ and $V \cap V' \neq \emptyset\}$ refines $\mathcal{U}$. Invoke Fact 2.24_m to obtain an abstract simplicial complex K and maps $\phi : X \to |K|_m$ and $\psi : |K|_m \to X$ that witness the $\mathcal{V}$-domination of X by $|K|_m$, with the additional property that $\{\psi(|\sigma|) : \sigma \in K\}$ refines $\mathcal{V}$. Let $\chi = id_{|K|} : |K|_W \to |K|_m$. Then Dowker's Theorem provides a map $\omega : |K|_m \to |K|_W$ such that $\chi \circ \omega : |K|_m \to |K|_m$ is $\mathcal{S}$-homotopic to $id_{|K|_m}$ where $\mathcal{S} = \{|\sigma| : \sigma \in K\}$. Then $\omega \circ \phi : X \to |K|_W$ and $\psi \circ \chi : |K|_W \to X$ are maps such that $(\psi \circ \chi) \circ (\omega \circ \phi)$ is $\mathcal{V}$-homotopic to $\psi \circ \phi$ which is $\mathcal{V}$-homotopic to id_X. Hence, $(\psi \circ \chi) \circ (\omega \circ \phi)$ is $\mathcal{U}$-homotopic to id_X. Also $\{\psi \circ \chi(|\sigma|) : \sigma \in K\} = \{\psi(|\sigma|) : \sigma \in K\}$ refines $\mathcal{V}$ and, hence, $\mathcal{U}$. □

There is a very useful characterization of ANRs in terms of domination.

HANNER'S CHARACTERIZATION OF ANRS 2.26. *A metrizable space X is an ANR if and only if for every open cover $\mathcal{U}$ of X, X is $\mathcal{U}$-dominated by an ANR.*[30]

C. H. Dowker proved a stronger characterization of ANRs in terms of domination which also has interesting ramifications.

DOWKER'S CHARACTERIZATION OF ANRS 2.27. *A metrizable space X is an ANR if and only if for every $\epsilon > 0$, X is ϵ-dominated by an ANR.*[31]

One consequence of Dowker's characterization is the following theorem of H. Torunczyk. Recall that a metrizable space X is *topologically complete* if the topology on X is determined by a complete metric. Also recall that a metrizable space

---

[27] [**1**], Theorem B.3, page 36.

[28] [**18**], Theorem IV.6.1, pages 138-139. The proofs of Facts 2.24_m and 2.24_W have the same outline. In the Whitehead topology case, given the open cover $\mathcal{U}$ of Y, one finds an open cover $\mathcal{V}_W$ of Y with the property that if $\mathcal{W}$ is any open cover of Y that refines $\mathcal{V}_W$, then $|n(\mathcal{W})|_W$ $\mathcal{U}$-dominates Y. In the metric topology case, given $\mathcal{U}$, one finds a possibly different open cover $\mathcal{V}_m$ of Y with the property that if $\mathcal{W}$ is any open cover of Y that refines $\mathcal{V}_m$, then $|n(\mathcal{W})|_m$ $\mathcal{U}$-dominates Y. Consequently, if $\mathcal{W}$ is an open cover of Y that refines both $\mathcal{V}_W$ and $\mathcal{V}_m$, then Y is simultaneously $\mathcal{U}$-dominated by both $|n(\mathcal{W})|_W$ and $|n(\mathcal{W})|_m$.

[29] [**7**], Section 16, Theorem 1, page 575.

[30] [**18**], Theorem IV.6.3, pages 139-140.

[31] [**8**], Section 1, Proposition 1.4, pages 105-106.

X is topologically complete $\Leftrightarrow$ X can be embedded as a G_δ subset of a topologically complete metrizable space $\Leftrightarrow$ every embedded image of X in a metrizable space Y is a G_δ subset of Y.[32]

DEFINITION. A subset Y of a topological space X is a *homotopy dense subset of* X if there is a homotopy $h : X \times [0,1] \to X$ such that $h_0 = id_X$ and $h(X \times (0,1]) \subset Y$.

Observe that every homotopy dense subset of a topological space X is a dense subset of X.

THEOREM 2.28 (Torunczyk's Completion Theorem). *Every ANR can be embedded as a homotopy dense subset of a topologically complete ANR.*[33]

Next we state a theorem attributed to Whitehead and Milnor which allows us to conclude that every ANR is homotopy equivalent to a polyhedron with either the Whitehead topology or the metric topology.

THEOREM 2.29 (The Whitehead-Milnor Theorem). *Every topological space that is dominated by a CW complex*[34] *is homotopy equivalent to a CW complex.*[35]

COROLLARY 2.30. *Every ANR is homotopy equivalent to the polyhedron underlying a simplicial complex with either the Whitehead or the metric topology.*

PROOF. Let X be an ANR. Then Fact 2.24_W provides an abstract simplicial complex K such that X is dominated by $|K|_W$. Polyhedra with the Whitehead topology are easily seen to be CW complexes. Hence, $|K|_W$ is a CW complex. Therefore, the Whitehead-Milnor Theorem implies X is homotopy equivalent to a CW complex Y. There is an abstract simplicial complex L such that Y is homotopy equivalent to $|L|_W$.[36] Dowker's Theorem implies $|L|_W$ is homotopy equivalent to $|L|_m$. □

This observation raises the question of whether every compact ANR is homotopy equivalent to a finite CW complex. The answer is provided by:

THEOREM 2.31 (West's Theorem). *Every compact ANR is homotopy equivalent to a finite CW complex*[37] *and, hence, to a compact polyhedron.*[38]

Although the statement of West's Theorem doesn't overtly suggest the relevance of infinite dimensional manifolds, his proof exploited deep properties of Z-sets in Hilbert cube manifolds and is regarded as a major contribution of infinite dimensional topology to the theory of ANRs.

West's Theorem only begins the story of the profound interplay between infinite dimensional topology and the theory of ANRs. We mention some other notable moments in this saga. First we need:

---

[32] **[9]**, Theorem XIV.8.3, page 308 or **[11]**, Theorems 4.3.23 and 4.3.24, page 838.

[33] **[29]**, Proposition 4.1, page 103. Also see **[22]**, Corollary 1, page 838.

[34] See **[14]**, page 519, for the definition of a CW complex.

[35] This result traces back to **[33]** and **[20]**. A remarkably short and ingenious proof is given in **[14]**, Proposition A.11, page 528. This proof was inspired by ideas found in **[19]**.

[36] **[14]**, Theorem 2C.5, pages 182-184.

[37] **[32]**

[38] Theorem 2C.5 on pages 182-184 of **[14]** tells us that a finite CW complex is homotopy equivalent to a compact polyhedron. Since the Whitehead and metric topologies on a compact polyhedron coincide, the topology in this situation need not be specified.

NOTATION. Let $\mathbb{Q} = [0,1]^\infty$, the *Hilbert cube*. For a set A, $\ell_2(A)$ is the Hilbert space of all square summable functions from A to $\mathbb{R}$.

DEFINITION. For $E = \mathbb{Q}$ or $\ell_2(A)$, a topological space X is an *E-manifold* if X is metrizable and if every point of X has an open neighborhood that is homeomorphic to an open subset of E.

DEFINITION. The *weight* of a metrizable space X, denoted $w(X)$, is the minimum cardinality of a dense subset of X.

THEOREM 2.32 (Edwards' Product Theorem). *If X is a locally compact ANR, then $X \times \mathbb{Q}$ is a $\mathbb{Q}$-manifold.*[39]

THEOREM 2.33 (Chapman's Triangulation Theorem). *If X is a separable $\mathbb{Q}$-manifold, then there is an abstract simplicial complex K such that X is homeomorphic to $|K|_m \times \mathbb{Q}$.*[40]

Observe that West's Theorem follows from Edwards' Product Theorem and Chapman's Triangulation Theorem.

THEOREM 2.34 (Torunczyk's Product Theorem). *If X is a topologically complete ANR and A is a set of cardinality $w(X)$, then $X \times \ell_2(A)$ is an $\ell_2(A)$-manifold.*[41]

THEOREM 2.35 (Henderson's Triangulation Theorem). *If X is an $\ell_2(A)$-manifold where A is a set of cardinality $w(X)$, then there is an abstract simplicial complex K such that X is homeomorphic to $|K|_m \times \ell_2(A)$.*[42]

The preceding four theorems of Edwards, Chapman, Torunczyk and Henderson can be derived as corollaries of Torunczyk's theorems characterizing Hilbert cube and Hilbert space manifolds. We will state these theorems for the record.[43] First, we need to define a few concepts.

DEFINITION. For topological spaces X and Y, let $C(X,Y)$ denote the set of all continuous functions from X to Y. If $f \in C(X,Y)$ and $\mathcal{U}$ is an open cover of Y, let

$$\mathcal{N}(f,\mathcal{U}) = \{g \in C(X,Y) : g \text{ is } \mathcal{U}\text{-close to } f\}.$$

Then

$$\{\mathcal{N}(f,\mathcal{U}) : f \in C(X,Y) \text{ and } \mathcal{U} \text{ is an open cover of } Y\}$$

is a basis for a topology on $C(X,Y)$ which we call the *fine topology on $C(X,Y)$*. In the following three definitions, all spaces of the form $C(X,Y)$ are assumed to have the fine topology.

DEFINITION. A metrizable space X has the *disjoint cubes property* if

$$\{f \in C(\{1,2\} \times \mathbb{Q}, X) : f(\{1\} \times \mathbb{Q}) \cap f(\{2\} \times \mathbb{Q}) = \emptyset\}$$

is a dense subset of $C(\{1,2\} \times \mathbb{Q}, X)$, where $\{1,2\}$ has the discrete topology.

[39] [**6**], Theorem 44.1, page 106. Also see [**25**], Theorem 3.8.1, pages 277-284.

[40] [**6**], Theorem 37.2, pages 83-84. Also see [**25**], Theorem 4.7.1, pages 346-351.

[41] [**28**], Theorem 4.2, page 60. Also see [**25**], Theorem 2.2.14, page 107.

[42] When X is separable, this follows from [**15**]. The non-separable case follows from [**17**] and [**28**]. Also see [**25**], Theorem 2.6.6, page 153.

[43] A beautiful exposition of Torunczyk's characterization theorems is presented in [**10**].

DEFINITION. A metrizable space X has the *discrete cubes property* if

$$\{f \in C(A \times \mathbb{Q}, X) : \{f(\{a\} \times \mathbb{Q}) : a \in A\} \text{ is a discrete collection of subsets of } X\}$$

is a dense subset of $C(A \times \mathbb{Q}, X)$, where A is a topological space with the discrete topology and cardinality equal to the weight of X.

DEFINITION. A metrizable space X has the *locally finite polyhedra property* if for every sequence $\{K_n : n \geq 1\}$ of finite-dimensional abstract simplicial complexes of cardinality no larger that the weight of X,

$$\{f \in C(\bigsqcup_{n \geq 1} |K_n|_m, X) :$$
$$\{f(|K_n|) : n \geq 1\} \text{ is a locally finite collection of subsets of } X\}$$

is a dense subset of $C(\bigsqcup_{n\geq 1} |K_n|_m, X)$:, where $\bigsqcup_{n\geq 1} |K_n|_m$ has the disjoint union topology.

THEOREM 2.36 (Torunczyk's First Characterization Theorem). *A locally compact separable ANR is a $\mathbb{Q}$-manifold if and only if it has the disjoint cubes property.*[44]

THEOREM 2.37 (Torunczyk's Second Characterization Theorem). *A separable topologically complete ANR is an $\ell_2(\mathbb{N})$-manifold (where $\mathbb{N} = \{1, 2, 3, \cdots\}$) if and only if it has the discrete cubes property.*[45]

THEOREM 2.38 (Torunczyk's Third Characterization Theorem). *If X is a non-separable topologically complete connected ANR and A is a set of cardinality equal to the weight of X, then X is an $\ell_2(A)$-manifold if and only if X has the discrete cells property and the locally finite polyhedra property.*[46]

To close this section, we briefly discuss the concept of a *Euclidean neighborhood retract* and its relation to absolute neighborhood retracts.

DEFINITION. For $n \geq 1, \mathbb{R}^n = \{(x_1, x_2, \cdots, x_n) : x_i \in \mathbb{R} \text{ for } 1 \leq i \leq n\}$. $\mathbb{R}^n$ is regarded as a vector space endowed with coordinate-wise addition and scalar multiplication. The *Euclidean norm* $\| \ \| : \mathbb{R}^n \to [0, \infty)$ is defined by $\|(x_1, x_2, \cdots, x_n)\| = (\sum_{1 \leq i \leq n} x_i^2)^{\frac{1}{2}}$. The *Euclidean metric* $d : \mathbb{R}^n \times \mathbb{R}^n \to [0, \infty)$ is defined by $d(\mathbf{x}, \mathbf{y}) = \|\mathbf{x} - \mathbf{y}\|$. The Euclidean metric on $\mathbb{R}^n$ determines the *Euclidean topology* on $\mathbb{R}^n$. Let $\mathbb{B}^n = \{\mathbf{x} \in \mathbb{R}^n : \|\mathbf{x}\| \leq 1\}$, $\partial\mathbb{B}^n = \mathbb{S}^{n-1} = \{\mathbf{x} \in \mathbb{R}^n : \|\mathbf{x}\| = 1\}$ and $int(\mathbb{B}^n) = \{\mathbf{x} \in \mathbb{R}^n : \|\mathbf{x}\| < 1\}$. Let $\mathbb{B}^0 = \{0\}$ and $\partial\mathbb{B}^0 = \emptyset$.

DEFINITION. A space is a *Euclidean neighborhood retract* (abbreviated *ENR*) if it is a retract of an open subset of $\mathbb{R}^n$ for some $n \geq 1$.

The question of which ANRs are ENRs is resolved by:

FACT 2.39. A topological space is an ENR if and only if it is a locally compact separable finite-dimensional ANR.

---

[44] **[30]**, statement I on page 432. Also see **[25]**, Theorem 3.8.6, page 284.

[45] **[30]**, statement II″ on page 432. Also see **[25]**, Theorem 3.4.7, page 248.

[46] **[30]**, statement II on page 432. Also see **[25]**, Theorem 3.4.2, pages 244-245. Torunczyk has told the author that at present (October, 2023) it is an open question as to whether "the locally finite polyhedron property" can be omitted from the statement of the Third Characterization Theorem.

PROOF. Every ENR is clearly separable and finite dimensional. A retract of a Hausdorff space must be a closed subset of the space, and a closed subset of a locally compact space must be locally compact. Hence, an ENR must be locally compact.

Conversely, suppose Y is a locally compact separable finite dimensional ANR. Then its one-point compactification $Y \cup \{\infty\}$ is compact and finite dimensional. Hence, $Y \cup \{\infty\}$ embeds in an n-sphere $\mathbb{S}^n$ as a proper subset for some sufficiently large $n \geq 1$. Therefore, Y embeds as a closed subset of $\mathbb{S}^n - \{\infty\} \cong \mathbb{R}^n$. Since Y is an ANR, then an open subset of $\mathbb{R}^n$ retracts onto Y. □

Here is a simple example of a non-locally compact separable finite-dimensional ANR that is not an ENR.

EXAMPLE 2.40. Let Z be a countable dense subset of $\partial\mathbb{B}^2$. Then $int(\mathbb{B}^2) \cup Z$ is a convex subset of $\mathbb{R}^2$ which is clearly separable and finite-dimensional. Therefore, Dugundji's Extension Theorem 2.3 tells us that $int(\mathbb{B}^2) \cup Z$ is an AR. However, $int(\mathbb{B}^2) \cup Z$ is not an ENR because it is not locally compact.

3. Various notions of Z-sets

DEFINITION. Let Z be a subset of a metrizable space X.

i) Recall $\mathbb{Q} = [0,1]^\infty$, the *Hilbert cube.* Z is a *Z-set for* $\mathbb{Q}$ if for every map $f : \mathbb{Q} \to X$ and every open cover $\mathcal{U}$ of X, f is $\mathcal{U}$-close to a map $g : \mathbb{Q} \to X - Z$.

ii) Z is a *Z-set in* X if for every open cover $\mathcal{U}$ of X, id_X is $\mathcal{U}$-close to a map $g : X \to X - Z$.

iii) Z is an *instant Z-set (in X)* if there is a homotopy $h : X \times [0,1] \to X$ such that $h_0 = id_X$ and $h(X \times (0,1]) \cap Z = \emptyset$. (Note that Z is an instant Z-set in $X \Leftrightarrow X - Z$ is homotopy dense in X.)

iv) Z is a *controlled instant Z-set (in X)* if for every open cover $\mathcal{U}$ of X, there is a $\mathcal{U}$-homotopy $h : X \times [0,1] \to X$ such that $h_0 = id_X$ and $h(X \times (0,1]) \cap Z = \emptyset$.

v) Z is a *Z-set for ANRs* if for every ANR M, every map $f : M \to X$, every metric ρ on X and every map $\alpha : M \to [0,\infty)$, f is α-near to a map $g : M \to X$ such that $g(\alpha^{-1}(0,\infty)) \cap Z = \emptyset$. (Observe that $g|\alpha^{-1}(\{0\}) = f|\alpha^{-1}(\{0\})$.)

vi) Z is a *locally homotopically negligible* (abbreviated *lhn*) if for every $x \in Z$ and every open neighborhood U of x in X, there is an open neighborhood V of x in X such that for every $k \geq 0$ and every map $f : (\mathbb{B}^k, \partial\mathbb{B}^k) \to (V, V - Z)$, there is a homotopy $h : (\mathbb{B}^k \times [0,1], \partial\mathbb{B}^k \times [0,1]) \to (U, U - Z)$ such that $h_0 = f$ and $h_1(\mathbb{B}^k) \cap Z = \emptyset$.

Notions similar to *local homotopic negligibility* and *Z-set in X* originate in the pioneer studies of the topology of infinite dimensional objects by R. D. Anderson from the late 1960's (See [**2**]). The concept of *Z-set in X* was taken as the definition of *Z-set* in the very elegant and influential mid-1970's survey *Lectures on Hilbert Cube Manifolds* by T. Chapman [**6**]. The concepts of *instant Z-set* and *Z-set for ANRs* (without these names) are taken up by Henryk Torunczyk's important 1978 paper [**29**]. In this paper, Torunczyk probed the relations among the various notions of *Z-set* and identified additional hypotheses under which they become equivalent.

4. Torunczyk's Equivalence Theorem

THEOREM 4.1 (Torunczyk's Equivalence Theorem). *Let Z be a subset of a metrizable space X and consider the following ten conditions.*

a) *Z is a Z-set for $\mathbb{Q}$.*

b) *Z is a Z-set in X.*

c) *Z is locally homotopically negligible.*

d) *Suppose (K, L) is a finite simplicial pair, $f : (|K|, |L|) \to (X, X - Z)$ is a map, and $\mathcal{U}$ is an open cover of X. Then there is a $\mathcal{U}$-homotopy $h : |K| \times [0,1] \to X$ such that $h_0 = f$, $h_1(|K|) \cap Z = \emptyset$ and h is stationary on $|L|$. (Note that because K is a finite simplicial complex, then the metric and Whitehead topologies on $|K|$ coincide and need not be specified.)*

e) *Suppose ρ is a metric on X, (K, L) is a simplicial pair, and $f : (|K|_W \times \{0\}) \cup (|L|_W \times [0,1]) \to X$ and $\alpha : |K|_W \to (0, \infty)$ are maps such that $f(|L| \times \{1\}) \cap Z = \emptyset$ and $f||L| \times [0,1]$ is an α-near homotopy. Then f extends to an α-near homotopy $g : |K|_W \times [0,1] \to X$ such that $g_1(|K|) \cap Z = \emptyset$.*

f) *Z is a Z-set for ANRs.*

g) *Z is an instant Z-set in X.*

h) *Z is a controlled instant Z-set in X.*

i) *For every open subset U of X, the inclusion $U - Z \to U$ is a homotopy equivalence.*

j) *For every open subset U of X, $\pi_k(U, U - Z) = 0$ for all $k \geq 0$.*

Then the following relations hold among these conditions.

1) *Conditions **c)**, **d)**, **e)**, **f)** and **j)** are equivalent. Furthermore,* **g)** $\Leftrightarrow$ **h)**, **g)** $\Rightarrow$ **i)**, **h)** $\Rightarrow$ **j)** *and* **h)** $\Rightarrow$ **b)** $\Rightarrow$ **a)**.

2) *If X is an ANR, then conditions **c)** through **j)** are equivalent.*

3) *If X is an ANR and Z is a closed subset of X, then all ten conditions are equivalent.*

REMARK. The contribution of Torunczyk's 1978 paper [**29**] to this topic was to add conditions **d)** through **h)** to the list of properties relevant to the study of Z-sets, to prove the equivalence of conditions **c)**, **d)**, **e)**, **f)** and **j)**, and to recognize that these proofs do not require either the assumption that X is an ANR or the assumption that Z is a closed subset of X. The paper also demonstrates that if X is assumed to be an ANR, then conditions **c)** through **j)** are equivalent.

5. The proof of 1)

The equivalence of conditions **c)**, **d)**, **e)**, **f)** and **j)** will be proved in "cyclic alphabetical order": **c)** $\Rightarrow$ **d)** $\Rightarrow$ **e)** $\Rightarrow$ **f)** $\Rightarrow$ **j)** $\Rightarrow$ **c)**.

THE PROOF THAT **c)**$\Rightarrow$**d)** . Assume **c)**. Suppose (K, L) is a finite simplicial pair, $f : (|K|, |L|) \to (X, X - Z)$ is a map, and $\mathcal{U}$ is an open cover of X. We induct on $m = max\{dim(\sigma) : \sigma \in K - L\}$.

By hypothesis, any map of $\mathbb{B}^0$ into Z can be moved off Z by an arbitrarily small homotopy. Consequently, the $m = 0$ case of **d)** is easily proved.

Let p be a positive integer and assume **d)** is true when $m = p - 1$. Let $\mathcal{V}$ be a finite cover of $f(|K|)$ by open subsets of X such that for every $V \in \mathcal{V}$, there is $U \in \mathcal{U}$ such that: for every $k \geq 0$ and for every map $\phi : (\mathbb{B}^k, \partial\mathbb{B}^k) \to (V, V - Z)$, there is a homotopy $\psi : (\mathbb{B}^k \times [0,1], \partial\mathbb{B}^k \times [0,1]) \to (U, U - Z)$ such that $\psi_0 = \phi$ and $\psi_1(\mathbb{B}^k) \cap Z = \emptyset$. Let $\mathcal{W}$ be an open cover of X that star refines $\mathcal{V} \cup \{X - f(|K|)\}$.[47] Let (K', L') be a subdivision of (K, L) so fine that $\{f(|\sigma|) : \sigma \in K'\}$ refines $\mathcal{W}$. Let $M = (K')^{p-1} \cup L'$ where $(K')^{p-1}$ is the $(p-1)$-skeleton of K'. By inductive hypothesis, there is a $\mathcal{W}$-homotopy $g : |M| \times [0,1] \to X$ such that $g_0 = f\big||M|$, $g_1(|M|) \cap Z = \emptyset$ and g is stationary on $|L'|$. Clearly, g can be extended to a map $g' : (|K'| \times 0) \cup (|M| \times [0,1]) \to X$ that satisfies $g'(y, 0) = f(y)$ for each $y \in |K'|$.

Let $\sigma \in K' - M$. Then $dim(\sigma) = p$ and $|\partial\sigma| \subset |M|$. Hence, $g'(|\partial\sigma| \times \{1\}) \cap Z = \emptyset$. Let $T_\sigma = (|\sigma| \times \{0\}) \cup (|\partial\sigma| \times [0,1])$. Then $g'(T_\sigma)$ is contained in an element V_σ of $\mathcal{V}$. There is a homeomorphism $\beta_\sigma : \mathbb{B}^p \times [0,1] \to |\sigma| \times [0,1]$ such that $\beta_\sigma(\mathbb{B}^p \times \{0\}) = T_\sigma$, $\beta_\sigma(\partial\mathbb{B}^p \times \{0\}) = |\partial\sigma| \times \{1\}$ and $\beta_\sigma((\partial\mathbb{B}^p \times [0,1]) \cup (\mathbb{B}^p \times \{1\})) = |\sigma| \times \{1\}$. Hence, $g' \circ \beta_\sigma\big|\mathbb{B}^p \times \{0\} : (\mathbb{B}^p \times \{0\}, \partial\mathbb{B}^p \times \{0\}) \to (V_\sigma, V_\sigma - Z)$. Consequently, there is a $U_\sigma \in \mathcal{U}$ and a homotopy $h_\sigma : (\mathbb{B}^p \times [0,1], \partial\mathbb{B}^p \times [0,1]) \to (U_\sigma, U_\sigma - Z)$ such that $h_\sigma\big|\mathbb{B}^p \times \{0\} = g' \circ \beta_\sigma\big|\mathbb{B}^p \times \{0\}$ and $h_\sigma(\mathbb{B}^p \times \{1\}) \cap Z = \emptyset$. Therefore, $h_\sigma \circ (\beta_\sigma^{-1}) : (|\sigma| \times [0,1], |\sigma| \times \{1\}) \to (U_\sigma, U_\sigma - Z)$ and $h_\sigma \circ (\beta_\sigma^{-1})\big|T_\sigma = g'\big|T_\sigma$.

It follows that the "union" of g' and the $h_\sigma \circ (\beta_\sigma^{-1})$'s (for $\sigma \in K' - M$) is a $\mathcal{U}$-homotopy $h : |K'| \times [0,1] \to X$ such that $h_0 = f$, $h_1(|K'|) \cap Z = \emptyset$ and h is stationary on $|L'|$. □

THE PROOF THAT **d)⇒e)**. Assume **d)**. Let ρ be a metric on X. Suppose (K, L) is a simplicial pair, and $f : (|K|_W \times \{0\}) \cup (|L|_W \times [0,1]) \to X$ and $\alpha : |K|_W \to (0, \infty)$ are maps such that $f(|L| \times \{1\}) \cup Z = \emptyset$ and $f\big|(|L| \times [0,1])$ is an α-near homotopy. Consider the collection $\mathcal{G}$ of all maps $g : (|K|_W \times \{0\}) \cup (|M|_W \times [0,1]) \to X$ where M is a subcomplex of K containing L and g is an extension of f such that $g_1(|M|) \cap Z = \emptyset$ and $g\big||M|_W \times [0,1]$ is an α-near homotopy. We partially order $\mathcal{G}$ by inclusion[48] and apply Zorn's Lemma to obtain a maximal element $g : (|K|_W \times \{0\}) \cup (|M|_W \times [0,1]) \to X$. Assume $M \neq K$. Then there is a $\sigma \in K - M$ such that $\partial\sigma \subset M$ and there is no extension of $g\big|(|\sigma| \times \{0\}) \cup (|\partial\sigma| \times [0,1])$ to an α-near homotopy $\gamma : |\sigma| \times [0,1] \to X$ such that $\gamma(|\sigma| \times \{1\}) \cap Z = \emptyset$.

Let $T = (|\sigma| \times \{0\}) \cup (|\partial\sigma| \times [0,1])$. Then $g\big|T : (T, |\partial\sigma| \times \{1\}) \to (X, X - Z)$ and $diam_\rho(g(\{\phi\} \times [0,1])) < \alpha(\phi)$ for every $\phi \in |\partial\sigma|$. Since $\alpha\big||\sigma| : |\sigma| \to (0, \infty)$ and $\phi \mapsto diam_\rho(g(\{\phi\} \times [0,1])) : |\partial\sigma| \to ([0, \infty)$ are continuous and $|\sigma|$ and $|\partial\sigma|$ are compact, then there is an $\epsilon > 0$ such that $\epsilon < min\{\alpha(\phi) : \phi \in |\sigma|\}$ and $\epsilon < min\{\alpha(\phi) - diam_\rho(g(\{\phi\} \times [0,1])) : \phi \in |\partial\sigma|\}$. Let $\mathcal{U}$ be an open cover of X such that $diam_\rho(U) < \frac{\epsilon}{2}$ for every $U \in \mathcal{U}$. Then, by hypothesis, there is a $\mathcal{U}$-homotopy $h : T \times [0,1] \to X$ such that $h_0 = g\big|T$, $h(T \times \{1\}) \cap Z = \emptyset$ and h is

[47] A collection $\mathcal{B}$ of subsets of a set X *star refines* a collection $\mathcal{A}$ of subsets of X if for every $B \in \mathcal{B}$, there is an $A \in \mathcal{A}$ such that $\bigcup\{B' \in \mathcal{B} : B \cap B' \neq \emptyset\} \subset A$. Every open cover of a metrizable space X is star refined by another open cover of X. See [**9**], VIII.3.3, VIII.3.4 and Theorem VIII.3.5, pages 167-169; or see [**11**], Theorem 5.1.12, pages 303-304 or Exercise 5.1.A(c), page 314.

[48] We regard a function f as a set of ordered pairs satisfying: (x, y) and $(x, y') \in f \Rightarrow y = y'$. Then for functions f and f': $f \subset f'$ if and only if f' is an extension of f.

stationary on $|\partial\sigma| \times \{1\}$. It follows that $h(((|\partial\sigma| \times \{1\}) \times [0,1]) \cup (T \times \{1\})) \cap Z = \emptyset$, and that $diam_\rho(h((\{\phi\} \times [0,1]) \times [0,1])) < \alpha(\phi)$ for each $\phi \in |\partial\sigma|$.

Next we need a homeomorphism $\theta : |\sigma| \times [0,1] \to T \times [0,1]$ such that $\theta(\tau) = (\tau, 0)$ for $\tau \in T$, and for each $\psi \in |\sigma|$, either $\theta(\{\psi\} \times [0,1]) = \{(\psi, 0)\} \times [0,1]$ or there is a $\phi \in |\partial\sigma|$ such that ψ is so close to ϕ and $\theta(\{\psi\} \times [0,1])$ is so close to $(\{\phi\} \times [0,1]) \times [0,1]$ that $diam_\rho(h(\theta(\{\psi\} \times [0,1]))) < \alpha(\psi)$. Notice that θ will then satisfy $\theta(T) = T \times \{0\}$ and $\theta(|\sigma| \times \{1\}) = ((|\partial\sigma| \times \{1\}) \times [0,1]) \cup (T \times \{1\})$. Once we have the homeomorphism θ, we consider the homotopy $h \circ \theta : |\sigma| \times [0,1] \to X$. It will follow that $h\circ\theta|T = g|T$, $h\circ\theta(|\sigma| \times \{1\}) \cap Z = \emptyset$ and $h\circ\theta$ is an α-near homotopy. This will contradict our earlier assumption that $M \neq K$. Hence, $g : |K|_W \times [0,1] \to X$ is an α-near homotopy that extends f such that $g_1(|K|) \cap Z = \emptyset$, proving **e)**.

We will construct the homeomorphism θ by first constructing a homeomorphism $\eta : |\partial\sigma| \times [0,1] \times [0,1] \to |\partial\sigma| \times [0,2] \times [0,1]$ and "quotient maps" $q_\sigma : |\partial\sigma| \times [0,1] \times [0,1] \to |\sigma| \times [0,1]$ and $q_T : |\partial\sigma| \times [0,2] \times [0,1] \to T \times [0,1]$ so that defining $\theta = q_T \circ \eta \circ q_\sigma^{-1}$ produces a homeomorphism satisfying the conditions stated in the first sentence of the previous paragraph. To construct the homeomorphism η and the quotient maps q_σ and q_T, let β be the barycenter of $|\sigma|$. In other words, β is the element of $|\sigma|$ determined by the equations $\beta(v) = \frac{1}{n+1}$ for every $v \in \sigma$ where $n = dim(\sigma)$ and $\beta(v) = 0$ for every $v \in V_K - \sigma$. Define q_σ and q_T by the following formulas:

$$q_\sigma(\phi, s, t) = ((1-s)\beta + s\phi, t) \text{ for } (\phi, s, t) \in |\partial\sigma| \times [0,1] \times [0,1],$$
$$q_T(\phi, s, t) = (((1-s)\beta + s\phi, 0), t) \text{ for } (\phi, s, t) \in |\partial\sigma| \times [0,1] \times [0,1] \text{ and}$$
$$q_T(\phi, s, t) = ((\phi, s-1), t) \text{ for } (\phi, s, t) \in |\partial\sigma| \times [1,2] \times [0,1].$$

Observe that q_σ maps $|\partial\sigma| \times (0,1] \times [0,1]$ homeomorphically onto $(|\sigma| - \{\beta\}) \times [0,1]$ and $q_\sigma(|\partial\sigma| \times \{0\} \times \{t\}) = \{(\beta, t)\}$ for $t \in [0,1]$. Similarly, q_T maps $|\partial\sigma| \times (0,2] \times [0,1]$ homeomorphically onto $(T - \{(\beta, 0)\}) \times [0,1]$ and $q_T(|\partial\sigma| \times \{0\} \times \{t\}) = \{((\beta, 0), t)\}$. To construct η, we must first prove the existence of an $r \in (0,1)$ with the property for every $\phi \in |\partial\sigma|$ and every $s \in [r,1]$, $diam_\rho(h(q_T(\{\phi\} \times [r,2] \times [0,1]))) < \alpha((1-s)\beta + s\phi)$. Observe that a continuous function $\delta : |\partial\sigma| \times [0,1] \times [0,1] \to \mathbb{R}$ is defined by

$$\delta(\phi, s, s') = \alpha((1-s)\beta + s\phi) - diam_\rho(h(q_T(\{\phi\} \times [s', 2] \times [0,1]))).$$

For each $\phi \in |\partial\sigma|$, since $h(q_T(\{\phi\} \times [1,2] \times [0,1])) = h((\{\phi\} \times [0,1]) \times [0,1])$ and $diam_\rho(h((\{\phi\} \times [0,1]) \times [0,1])) < \alpha(\phi)$, then $\delta(\phi, 1, 1) > 0$. Hence, $\{(\phi, s, s') \in |\partial\sigma| \times [0,1] \times [0,1] : \delta(\phi, s, s') > 0\}$ is an open subset of $|\partial\sigma| \times [0,1] \times [0,1]$ that contains $|\partial\sigma| \times \{1\} \times \{1\}$. Since $|\partial\sigma|$ is compact, there is an $r \in (0,1)$ such that $\delta(\phi, s, s') > 0$ for all $(\phi, s, s') \in |\partial\sigma| \times [r,1] \times [r,1]$. Thus, $diam_\rho(h(q_T(\{\phi\} \times [r,2] \times [0,1]))) < \alpha((1-s)\beta + s\phi)$ for all $(\phi, s) \in |\partial\sigma| \times [r,1]$. Let $\zeta : [0,1] \times [0,1] \to [0,2] \times [0,1]$ be a homeomorphism such that ζ restricts to the identity on $([0,r] \times [0,1]) \cup ([r,1] \times \{0\})$ and $\zeta(1,t) = (1+t, 0)$ for $t \in [0,1]$. Consequently, $\zeta([r,1] \times [0,1]) = [r,2] \times [0,1]$. Finally define $\eta : |\partial\sigma| \times [0,1] \times [0,1] \to |\partial\sigma| \times [0,2] \times [0,1]$ by $\eta(\phi, s, t) = (\phi, \zeta(s,t))$. Since η restricts to the identity on $|\partial\sigma| \times \{0\} \times [0,1]$, then a homeomorphism $\theta : |\sigma| \times [0,1] \to T \times [0,1]$ is defined by $\theta = q_T \circ \eta \circ q_\sigma^{-1}$.

$$\begin{array}{ccc} |\partial\sigma| \times [0,1] \times [0,1] & \xrightarrow{\ \eta\ } & |\partial\sigma| \times [0,2] \times [0,1] \\ {\scriptstyle q_\sigma}\big\downarrow & & \big\downarrow{\scriptstyle q_T} \\ |\sigma| \times [0,1] & \xrightarrow[\theta]{\qquad} & T \times [0,1] \end{array}$$

It is straightforward to verify that for $(\phi, s) \in |\partial\sigma| \times [0,1]$, $q_\sigma(\phi, s, 0) = ((1-s)\beta + s\phi, 0)$ and $q_T \circ \eta(\phi, s, 0) = (((1-s)\beta + s\phi, 0), 0)$. Similarly, for $(\phi, t) \in |\partial\sigma| \times [0,1]$, $q_\sigma(\phi, 1, t) = (\phi, t)$ and $q_T \circ \eta(\phi, 1, t) = ((\phi, t), 0)$. Thus, $\theta(\tau) = (\tau, 0)$ for $\tau \in T$. To complete the verification that θ has the necessary properties, let $\psi \in |\sigma|$. Then there is a $\phi \in |\partial\sigma|$ and an $s \in [0,1]$ such that $\psi = (1-s)\beta + s\phi$. In the case that $s \in [0, r]$, $q_\sigma(\{\phi\} \times \{s\} \times [0,1]) = \{\psi\} \times [0,1]$ and $q_T \circ \eta(\{\phi\} \times \{s\} \times [0,1]) = \{(\psi, 0)\} \times [0,1]$. Hence, $\theta(\{\psi\} \times [0,1]) = \{(\psi,0)\} \times [0,1]$. Thus, $diam_\rho(h \circ \theta(\{\psi\} \times [0,1])) = diam_\rho(h(\{(\psi,0)\} \times [0,1])) < \epsilon < \alpha(\psi)$. Now consider the case $s \in [r,1]$. Again $q_\sigma(\{\phi\} \times \{s\} \times [0,1]) = \{\psi\} \times [0,1]$. Since $\zeta(\{s\} \times [0,1]) \subset [r,2] \times [0,1]$, then $q_T \circ \eta(\{\phi\} \times \{s\} \times [0,1]) = q_T(\{\phi\} \times \zeta(\{s\} \times [0,1])) \subset q_T(\{\phi\} \times [r,2] \times [0,1])$. Therefore, $\theta(\{\psi\} \times [0,1]) \subset q_T(\{\phi\} \times [r,2] \times [0,1])$. At this point, the choice of r made earlier guarantees that $diam_\rho(h \circ \theta(\{\psi\} \times [0,1])) < \alpha(\psi)$. □

THE PROOF THAT **e)**⇒**f)**. Assume **e)**. Suppose M is an ANR, $f : M \to X$ is a map, ρ is a metric on X, and $\alpha : M \to [0,\infty)$ is a map. First suppose $\alpha(M) \subset (0,\infty)$. Since f and α are continuous, there is an open cover $\mathcal{U}$ of M such that for every $U \in \mathcal{U}$, $diam_{\mathbb{R}}(\alpha(U)) < inf(\alpha(U))$ and $diam_\rho(f(U)) < \frac{1}{2} inf(\alpha(U))$. Since M is an ANR, there is an abstract simplicial complex K and maps $\phi : M \to |K|_W$ and $\psi : |K|_W \to M$ such that $\psi \circ \phi$ is $\mathcal{U}$-homotopic to id_M. Since $f \circ \psi : |K|_W \to X$ and $\frac{1}{4}\alpha \circ \psi : |K|_W \to (0,\infty)$ are maps, then by hypothesis $f \circ \psi$ is $\frac{1}{4}\alpha \circ \psi$-near to a map $g : |K|_W \to X - Z$. Thus, $f \circ \psi \circ \phi : M \to X$ is $\frac{1}{4}\alpha \circ \psi \circ \phi$-near to $g \circ \phi : M \to X - Z$.

Next we show that $g \circ \phi$ is α-near to f. Let $y \in M$. There is a $U \in \mathcal{U}$ such that y and $\psi \circ \phi(y) \in U$. Then

$$\begin{aligned}\rho(f \circ \psi \circ \phi(y), g \circ \phi(y)) &< \frac{1}{4}\alpha \circ \psi \circ \phi(y) = \frac{1}{4}\alpha(y) + \frac{1}{4}(\alpha \circ \psi \circ \phi(y) - \alpha(y)) \\ &\leq \frac{1}{4}\alpha(y) + \frac{1}{4} diam_{\mathbb{R}}(\alpha(U)) < \frac{1}{4}\alpha(y) + \frac{1}{4} inf(\alpha(U)) \leq \frac{1}{2}\alpha(y).\end{aligned}$$

Also

$$\rho(f(y), f \circ \psi \circ \phi(y)) \leq diam_\rho(f(U)) < \frac{1}{2} inf(\alpha(U)) \leq \frac{1}{2}\alpha(y).$$

Therefore,

$$\rho(f(y), g \circ \phi(y)) \leq \rho(f(y), f \circ \psi \circ \phi(y)) + \rho(f \circ \psi \circ \phi(y), g \circ \phi(y)) < \alpha(y).$$

This completes the proof in the case that $\alpha(M) \subset (0,\infty)$.

Now suppose $\alpha : M \to [0,\infty)$. Let $M_+ = \alpha^{-1}(0,\infty)$. M_+ is an ANR because it is an open subset of the ANR M. Let $f_+ = f|M_+ : M_+ \to X$ and let $\alpha_+ = \alpha|M_+ : M_+ \to (0,\infty)$. The argument of the preceding two paragraphs proves there is a map $g_+ : M_+ \to X - Z$ which is α_+-near to f_+. We can extend g_+ to $g : M \to X$ by defining $g|\alpha^{-1}(\{0\}) = f|\alpha^{-1}(\{0\})$. Then $g(\alpha^{-1}(0,\infty)) \cap Z = g_+(M_+) \cap Z = \emptyset$. Also, clearly, g is α-near to f. It remains to prove that g is continuous at each point $y \in \alpha^{-1}(\{0\})$. To this end, let $\epsilon > 0$. Then there is an open neighborhood U of y in X such that for all $y' \in U$, $\rho(f(y), f(y')) < \frac{\epsilon}{2}$ and $\alpha(y') = |\alpha(y) - \alpha(y')| < \frac{\epsilon}{2}$. Hence, for each $y' \in U$,

$$\begin{aligned}\rho(g(y), g(y')) &\leq \rho(g(y), f(y)) + \rho(f(y), f(y')) + \rho(f(y'), g(y')) \\ &< 0 + \rho(f(y), f(y')) + \alpha(y') < \epsilon.\end{aligned}$$

This proves the continuity of g at points of $\alpha^{-1}(\{0\})$. □

THE PROOF THAT **f)⇒j)**. Assume **f)**. Let U be an open subset of X and let $k \geq 0$. To prove $\pi_k(U, U-Z) = 0$, suppose $f : (\mathbb{B}^k, \partial\mathbb{B}^k) \to (U, U-Z)$ is a map. It suffices to find a homotopy $h : (\mathbb{B}^k \times [0,1], \partial\mathbb{B}^k \times [0,1]) \to (U, U-Z)$ such that $h_0 = f$ and $h_1(\mathbb{B}^k) \cap Z = \emptyset$. Let $M = \mathbb{B}^k \times [0,1]$. Then M is an ANR. Define $g : M \to X$ by $g(y,t) = f(y)$. Let ρ be a metric on X, let $0 < \epsilon < \rho(f(\mathbb{B}^k), X-U)$, and choose $\alpha : M \to [0,\epsilon]$ to be a map such that $\alpha^{-1}(\{0\}) = (\mathbb{B}^k \times \{0\}) \cup (\partial\mathbb{B}^k \times [0,1])$. (For example, define $\alpha(y,t) = \epsilon(1 - \|y\|)t$ where $\|y\|$ is the Euclidean norm of $y \in \mathbb{R}^k$.) Then, by hypothesis, there is a map $h : M \to X$ that is α-near to g such that $h(\alpha^{-1}(0,\epsilon]) \cap Z = \emptyset$. Then $h(\mathbb{B}^k \times [0,1]) \subset U$ because $f(\mathbb{B}^k) \subset U$ and

$$\rho(h(y,t), f(y)) = \rho(h(y,t), g(y,t)) < \alpha(y,t) \leq \epsilon < \rho(f(\mathbb{B}^k), X-U)$$

for all $(y,t) \in \mathbb{B}^k \times [0,1]$. Also $h(\partial\mathbb{B}^k \times [0,1]) \subset U - Z$ because $\alpha(\partial\mathbb{B}^k \times [0,1]) = \{0\}$ and, hence, $h(\partial\mathbb{B}^k \times [0,1]) = g(\partial\mathbb{B}^k \times [0,1]) = f(\partial\mathbb{B}^k) \subset U - Z$. Since $\alpha(\mathbb{B} \times \{0\}) = \{0\}$, then $h_0 = g_0 = f$. Finally, $h_1(\mathbb{B}^k) \cap Z = \emptyset$ because $h_1(\partial\mathbb{B}^k) \subset h(\partial\mathbb{B}^k \times [0,1]) \subset U - Z$ and

$$h_1(int(\mathbb{B}^k)) \subset h(int(\mathbb{B}^k) \times (0,1]) = h(\alpha^{-1}(0,\epsilon]) \subset U - Z.$$

□

THE PROOF THAT **j)⇒c)**. Assume **j)**. Let $x \in Z$, let U be an open neighborhood of x in X and let $k \geq 0$. Let $f : (\mathbb{B}^k, \partial\mathbb{B}^k) \to (U, U-Z)$ be a map. It suffices to produce a homotopy $h : (\mathbb{B}^k \times [0,1], \partial\mathbb{B}^k \times [0,1]) \to (U, U-Z)$ such that $h_0 = f$ and $h_1(\mathbb{B}^k) \cap Z = \emptyset$. The existence of h follows because, by hypothesis, $\pi_k(U, U-Z) = 0$. □

To complete the proof of **1)**, we must show that **g)** ⇔ **h)**, **g)** ⇒ **i)**, **h)** ⇒ **j)** and **h)** ⇒ **b)** ⇒ **a)**.

The following lemma will be used immediately following its proof to show that **g)** ⇒ **h)** and **g)** ⇒ **i)**.

LEMMA 5.1. *If X is a metrizable space, Y is a topological space, $g : X \times [0,1] \to Y$ is a homotopy and $\mathcal{U}$ is a collections of open subsets of Y, then there is a map $\phi : g_0^{-1}(\bigcup\mathcal{U}) \to (0,1)$ with the property that $\{g(\{x\} \times [0,\phi(x)]) : x \in g_0^{-1}(\bigcup\mathcal{U})\}$ refines $\mathcal{U}$.*

PROOF OF LEMMA 5.1. Let $X_0 = g_0^{-1}(\bigcup\mathcal{U})$. The continuity of g implies that for every $x \in X_0$, there is an open neighborhood V_x of x in X_0, an $r_x \in (0,1)$ and a $U_x \in \mathcal{U}$ such that $g(V_x \times [0,r_x]) \subset U_x$. Let $\{\lambda_x : x \in X_0\}$ be a partition of unity subordinate to $\{V_x : x \in X_0\}$.[49] (In other words, for each $x \in X_0$, $\lambda_x : X_0 \to [0,1]$ is a map such that $\lambda_x^{-1}((0,1]) \subset V_x$, $\{\lambda_x^{-1}((0,1]) : x \in X_0\}$ is a locally finite open cover of X_0 and $\sum_{z \in X_0} \lambda_z(x) = 1$ for every $x \in X_0$.) Define the map $\phi : X_0 \to [0,\infty)$ by $\phi(x) = \sum_{z \in X_0} r_z \lambda_z(x)$. Then $\phi(X_0) \subset (0,1)$. For each $x \in X_0$, there is a $z \in X_0$ such that $\lambda_z(x) > 0$ and $r_z = max\{r_{z'} : z' \in X_0 \text{ and } \lambda_{z'}(x) > 0\}$. Then $\phi(x) \leq r_z$(x). Hence,

$$g(\{x\} \times [0,\phi(x)]) \subset g(\lambda_z^{-1}((0,1]) \times [0,r_z]) \subset g(V_z \times [0,r_z]) \subset U_z \in \mathcal{U}.$$

□

[49] [**9**], Theorem VIII.4.2, page 170.

THE PROOF THAT **g)**$\Rightarrow$**h)**. Assume **g)**. Let $\mathcal{U}$ be an open cover of X. By hypothesis, there is a homotopy $g : X \times [0,1] \to X$ such that $g_0 = id_X$ and $g(X \times (0,1]) \cap Z = \emptyset$. Lemma 5.1 provides a map $\phi : X \to (0,1)$ with the property that $\{g(\{x\} \times [0, \phi(x)]) : x \in X\}$ refines $\mathcal{U}$. Define the homotopy $h : X \times [0,1] \to X$ by $h(x,t) = g(x, \phi(x)t)$ for each $(x,t) \in X \times [0,1]$. Then $h(\{x\} \times [0,1]) = g(\{x\} \times [0, \phi(x)])$ for each $x \in X$. Hence, $h : X \times [0,1] \to X$ is a $\mathcal{U}$-homotopy such that $h_0 = g_0 = id_X$ and $h(X \times (0,1]) \cap Z \subset g(X \times (0,1]) \cap Z = \emptyset$. □

h) obviously implies **g)**. Hence, **g)** $\Leftrightarrow$ **h)**.

THE PROOF THAT **g)**$\Rightarrow$**i)**. Assume **g)**. By hypothesis, there is a homotopy $g : X \times [0,1] \to X$ such that $g_0 = id_X$ and $g(X \times (0,1]) \cap Z = \emptyset$. Let U be an open subset of X and let $i : U - Z \to U$ denote the inclusion map. Lemma 5.1 provides a map $\phi : U \to (0,1)$ such that $g(\{(x,t) \in U \times [0,1] : 0 \le t \le \phi(x)\}) \subset U$. Define the homotopy $h : U \times [0,1] \to U$ by $h(x,t) = g(x, t\phi(x))$ for each $(x,t) \in U \times [0,1]$. Then $h_0 = id_U$ and $h(U \times (0,1]) \cap Z = \emptyset$; in other words, $Z \cap U$ is an instant Z-set in U. Note that h_1 maps U into $U - Z$. Also note that since $h_t \circ i(U - Z) \subset U - Z$ for $t \in [0,1]$, then $h_1 \circ i$ is homotopic to $h_0 \circ i$ in $U - Z$ and $h_0 \circ i = id_{U-Z}$. Finally note that $i \circ h_1 = h_1$ and h_1 is homotopic to id_U in U. Thus, $i : U - Z \to U$ is a homotopy equivalence. □

The equivalence of **g)** and **h)** facilitates a simple proof of a useful fact about instant Z-sets.

COROLLARY 5.2. *If Z is an instant Z-set in a metrizable space X, then X is an ANR if and only if $X - Z$ is and ANR.*

PROOF. Assume $X - Z$ is an ANR. We will prove that X satisfies Hanner's Characterization of ANRs 2.26. Let $\mathcal{U}$ be an open cover of X. Since **g)** implies **h)**, then Z is a controlled instant Z-set in X. Therefore, there is a $\mathcal{U}$-homotopy $h : X \times [0,1] \to X$ such that $h_0 = id_X$ and $h(X \times (0,1]) \cap Z = \emptyset$. Note that $h_1 : X \to X - Z$, and let $j : X - Z \to X$ denote inclusion. Then h is a $\mathcal{U}$-homotopy joining $h_0 = id_X$ to $h_1 = j \circ h_1$. Thus, the ANR $X - Z$ $\mathcal{U}$-dominates X. Hanner's Characterization 2.26 now implies that X is an ANR.

Now assume X is an ANR. We will prove that $X - Z$ satisfies Dowker's Characterization of ANRs 2.27. Let ρ be a metric on X and assign restrict ρ to $X - Z$ to obtain a metric on $X - Z$. Let $\epsilon > 0$. Let $\mathcal{U}$ denote the collection of all (relatively) open subsets of $X - Z$ of ρ-diameter $< \epsilon$. It suffices to prove that $X - Z$ is $\mathcal{U}$-dominated by an ANR. Let $\mathcal{V}$ denote the collection of all open subsets of X of diameter $< \epsilon$, and note that $\mathcal{U} = \{V - Z : V \in \mathcal{V}\}$. Since Z is an instant Z-set in X and, hence, a controlled instant Z-set in X, then there is a $\mathcal{V}$-homotopy $h : X \times [0,1] \to X$ such that $h_0 = id_X$ and $h(X \times (0,1]) \cap Z = \emptyset$. Again let $j : X - Z \to X$ denote inclusion and note that $h_1 : X \to X - Z$. Hence, $h|(X - Z) \times [0,1] \to X - Z$ is a $\mathcal{U}$-homotopy joining $h_0|X - Z = id_{X-Z}$ to $h_1|X - Z = h_1 \circ j$. Thus, the ANR X ϵ-dominates $X - Z$. Dowker's Characterization 2.27 now implies that $X - Z$ is an ANR. □

THE PROOF THAT **h)**$\Rightarrow$**j)**. Assume **h)**. Let U be an open subset of X and let $k \ge 0$. To prove $\pi_k(U, U - Z) = 0$, suppose $f : (\mathbb{B}^k, \partial\mathbb{B}^k) \to (U, U - Z)$ is a map. It suffices to find a homotopy $g : (\mathbb{B}^k \times [0,1], \partial\mathbb{B}^k \times [0,1]) \to (U, U - Z)$ such that $g_0 = f$ and $g_1(\mathbb{B}^k) \cap Z = \emptyset$. Let $\mathcal{U} = \{U, X - f(\mathbb{B}^k)\}$. Then $\mathcal{U}$ is an open cover of X. By hypothesis, there is a $\mathcal{U}$-homotopy $h : X \times [0,1] \to X$ such that $h_0 = id_X$ and $h(X \times (0,1]) \cap Z = \emptyset$. Define the homotopy $g : \mathbb{B}^k \times [0,1] \to X$ by $g(x,t) = h(f(x),t)$ for each $(x,t) \in \mathbb{B}^k \times [0,1]$. Since h is a $\mathcal{U}$-homotopy and $h_0 \circ f = f$, then $h(f(\mathbb{B}^k) \times [0,1]) \subset U$. Also

$$h(f(\partial\mathbb{B}^k) \times [0,1]) \cup h(f(\mathbb{B}^k) \times \{1\}) \subset f(\partial\mathbb{B}^k) \cup h(X \times (0,1]) \subset X - Z.$$

Thus, $g(\mathbb{B}^k \times [0,1], \partial\mathbb{B}^k \times [0,1]) \to (U, U - Z)$, $g_0 = h_0 \circ f = f$ and $g_1(\mathbb{B}^k) \cap Z = h_1(f(\mathbb{B}^k)) \cap Z = \emptyset$. □

THE PROOF THAT **h)⇒b)**. Assume **h)**. Let $\mathcal{U}$ be an open cover of X. By hypothesis, there is a $\mathcal{U}$-homotopy $h : X \times [0,1] \to X$ such that $h_0 = id_X$ and $h(X \times (0,1]) \cap Z = \emptyset$. Then $h_1 : X \to X - Z$ is a map that is $\mathcal{U}$-close to id_X. □

THE PROOF THAT **b)⇒a)**. Assume **b)**. Let $f : \mathbb{Q} \to X$ be a map and let $\mathcal{U}$ be an open cover of X. By hypothesis, there is a map $g : X \to X - Z$ that is $\mathcal{U}$-close to id_X. Then $g \circ f : \mathbb{Q} \to X - Z$ is a map that is $\mathcal{U}$-close to f. □

6. The proof of 2)

Assume that X is an ANR. In the preceding section, we established the equivalence of conditions **c)**, **d)**, **e)**, **f)** and **j)** as well as the relations **g)** ⇔ **h)**, **g)** ⇒ **i)** and **h)** ⇒ **j)**. To complete the proof that conditions **c)** through **j)** are equivalent, we will show that **f)** ⇒ **g)** and **i)** ⇒ **c)**.

THE PROOF THAT **f)⇒g)**. Assume **f)**. Let $M = X \times [0,1]$. Since X is an ANR, then M is an ANR (by Corollary 2.4 and Fact 2.8). Define $\pi : M \to X$ by $\pi(x,t) = x$ and define $\alpha : M \to [0,\infty)$ by $\alpha(x,t) = t$. Let ρ be a metric on X. Then by hypothesis, there is a map $h : M \to X$ that is α-close to π such that $h(\alpha^{-1}(0,\infty)) \cap Z = \emptyset$. Since $\alpha^{-1}(0,\infty) = X \times (0,1]$, then $h : X \times [0,1] \to X$ and $h(X \times (0,1]) \cap Z = \emptyset$. Moreover, for $x \in X$, $\rho(h_0(x), x) = \rho(h(x,0), \pi(x,0)) \leq \alpha(x,0) = 0$. Thus, $h_0 = id_X$. This proves Z is an instant Z-set in X. □

THE PROOF THAT **i)⇒c)**. Assume **i)**. Let $x \in Z$ and let U be an open neighborhood of x in X. Since X is an ANR, Fact 2.10 tells us that X is locally contractible. Hence, there are open neighborhoods V and W of x in X such that V contracts to a point in U and W contracts to a point in V. Let $f : (\mathbb{B}^k, \partial\mathbb{B}^k) \to (W, W - Z)$ be a map. It suffices to produce a homotopy $h : (\mathbb{B}^k \times [0,1], \partial\mathbb{B}^k \times [0,1]) \to (U, U - Z)$ such that $h_0 = f$ and $h_1(\mathbb{B}^k) \cap Z = \emptyset$.

CLAIM. $W - Z$ contracts to a point in $V - Z$.

PROOF. Let $i_W : W - Z \to W$, $i_V : V - Z \to V$, $j_Z : W - Z \to V - Z$ and $j : W \to V$ denote inclusion maps. Then $i_V \circ j_Z = j \circ i_W$. If two maps ϕ and ψ are homotopic, we will denote this by $\phi \sim \psi$. Since W contracts to a point in V, then $j \sim c$ where $c : W \to V$ is a constant map. $i_V : V - Z \to V$ is a homotopy equivalence by hypothesis. Hence, there is a map $r_V : V \to V - Z$ such that $r_V \circ i_V \sim id_{V-Z}$. It follows that

$$j_Z \sim r_V \circ i_V \circ j_Z = r_V \circ j \circ i_W \sim r_V \circ c \circ i_W.$$

Since $r_V \circ c \circ i_W$ is a constant map, we have proved that $W - Z$ contracts to a point in $V - Z$. □

The Claim implies that there is a homotopy $g : \partial\mathbb{B}^k \times [0,1] \to V - Z$ such that $g_0 = f|\partial\mathbb{B}^k$ and $g_1(\partial\mathbb{B}^k) = \{p\}$ for some $p \in V - Z$. Next we define a map $h_\partial : \partial(\mathbb{B}^k \times [0,1]) \to V$ by $h_\partial(x,0) = f(x)$ for $x \in \mathbb{B}^k$, $h_\partial|(\partial\mathbb{B}^k) \times [0,1] = g$ and $h_\partial(\mathbb{B}^k \times \{1\}) = \{p\}$. Since V contracts to a point in U, then h_∂ extends to a map $h : \mathbb{B}^k \times [0,1] \to U$. The proof is finished by observing that $h_0 = f$ and $h(((\partial\mathbb{B}^k) \times [0,1]) \cup (\mathbb{B}^k \times \{1\})) \subset V - Z \subset U - Z$. □

7. The proof of 3)

Assume that X is an ANR and Z is a closed subset of X. The equivalence of conditions **c)** through **j)** and the implications **h)** $\Rightarrow$ **b)** $\Rightarrow$ **a)** were established in the preceding sections. Hence, to prove the equivalence of conditions **a)** through **j)**, it suffices to show that **a)** $\Rightarrow$ **c)**.

THE PROOF THAT **a)**$\Rightarrow$**c)**. Assume **a)**. Let $x \in Z$ and let U be an open neighborhood of x in X. Let $k \geq 0$ and let $f : (\mathbb{B}^k, \partial\mathbb{B}^k) \to (U, U - Z)$ be a map. Let $\mathcal{U} = \{U - f(\partial\mathbb{B}^k), U - Z, X - f(\mathbb{B}^k)\}$. Since Z is a closed subset of X, then $\mathcal{U}$ is an open cover of X. Since X is an ANR, there is an open cover $\mathcal{V}$ of X such that any two $\mathcal{V}$-close maps into X are $\mathcal{U}$-homotopic by Fact 2.16. Embed $\mathbb{B}^k$ in $\mathbb{Q}$. Since $\mathbb{B}^k$ is an AR, then there is a retraction $r : \mathbb{Q} \to \mathbb{B}^k$. Then $f \circ r : \mathbb{Q} \to X$. By hypothesis, $f \circ r$ is $\mathcal{V}$-close to a map $g : \mathbb{Q} \to X - Z$. Hence, $f \circ r|\mathbb{B}^k = f$ is $\mathcal{V}$-close to $g|\mathbb{B}^k$. Thus, there is a $\mathcal{U}$-homotopy $h : \mathbb{B}^k \times [0,1] \to X$ such that $h_0 = f$ and $h_1 = g|\mathbb{B}^k$. Therefore, $h_1(\mathbb{B}^k) = g(\mathbb{B}^k) \subset X - Z$. For $y \in \mathbb{B}^k$, since $h_0(y) = f(y) \in f(\mathbb{B}^k)$ and h is a $\mathcal{U}$-homotopy, then $h(\{y\} \times [0,1])$ is contained in either $U - f(\partial\mathbb{B}^k)$ or $U - Z$. This proves $h(\mathbb{B}^k \times [0,1]) \subset U$. For $y \in \partial\mathbb{B}^k$, since $h_0(y) = f(y) \in f(\partial\mathbb{B}^k) \subset f(\mathbb{B}^k)$ and h is a $\mathcal{U}$-homotopy, then $h(\{y\} \times [0,1]) \subset U - Z$. This proves $h(\partial\mathbb{B}^k \times [0,1]) \subset U - Z$. Thus, $h : (\mathbb{B}^k \times [0,1], \partial\mathbb{B}^k \times [0,1]) \to (U, U - Z)$ is a homotopy such that $h_0 = f$ and $h_1(\mathbb{B}^k) \cap Z = \emptyset$. □

8. Two examples

A) AN EXAMPLE OF A LOCALLY HOMOTOPICALLY NEGLIGIBLE SET Z IN A NON-ANR METRIZABLE SPACE X SUCH THAT Z IS NOT AN INSTANT Z-SET IN X. The example constructed here shows that conclusion **2)** of Torunczyk's Equivalence Theorem relies on the hypothesis that X be an ANR. More specifically, the implication **c)** $\Rightarrow$ **g)** can't be proved if this hypothesis is omitted. It follows that the implication **f)** $\Rightarrow$ **g)** can't be proved in the absence of the ANR hypothesis; and, indeed, the set Z described below is a Z-set for ANRs that is not an instant Z-set in X.

Let $\ell_2 = \ell_2(\mathbb{N})$ which, we recall, is the Hilbert space of all square summable functions from $\mathbb{N} = \{1, 2, 3, \cdots\}$ to $\mathbb{R}$. For each $n \geq 0$, let Σ_n denote an n-sphere of radius $\frac{1}{n}$ centered at $(\frac{1}{n}, 0, 0, 0, \cdots)$ in ℓ_2:

$$\Sigma_n = \{\mathbf{x} = (x_i) \in \ell_2 : \|\mathbf{x} - (\frac{1}{n}, 0, 0, 0, \cdots)\| = \frac{1}{n} \text{ and } x_i = 0 \text{ for } i \geq n + 2\}.$$

Thus, $\mathbf{0} = (0, 0, 0, \cdots) \in \Sigma_n$. Let

$$X = ([0,1] \times \{\mathbf{0}\}) \cup (\bigcup_{n \geq 1} (\{\frac{1}{n}\} \times \Sigma_n)) \subset [0,1] \times \ell_2,$$

and let $Z = \{0\} \times \{\mathbf{0}\} \subset X$. We will show Z is a locally homotopically negligible set in X but not an instant Z-set in X. We also prove that X is not an ANR.

First we prove Z is locally homotopically negligible in X. For each $m \geq 1$, let $V_m = ([0, \frac{1}{m}] \times \{\mathbf{0}\}) \cup (\bigcup_{n \geq m}(\{\frac{1}{n}\} \times \Sigma_n))$. Then $\{V_m : m \geq 1\}$ is a basis of closed neighborhoods of $(0, \mathbf{0})$ in X. Let U be a neighborhood of $(0, \mathbf{0})$ in X. There is an $m \geq 1$ such that $V_m \subset U$. Suppose $k \geq 0$ and $f : (\mathbb{B}^k, \partial\mathbb{B}^k) \to (V_m, V_m - Z)$ is a map. There is a positive integer r such that $r > k$, $r > m$ and $f(\partial\mathbb{B}^k) \cap V_r = \emptyset$. Since $dim(\mathbb{B}^k) < dim(\Sigma_n)$ for all $n \geq r$, then there is a homotopy of f that is stationary on $\partial\mathbb{B}^k$ which for each $n \geq r$: first moves $f(f^{-1}(\{\frac{1}{n}\} \times \Sigma_n))$ to a proper subset T_n of $\{\frac{1}{n}\} \times \Sigma_n$ fixing $\{(\frac{1}{n}, \mathbf{0})\}$ and then contracts T_n to $\{(\frac{1}{n}, \mathbf{0})\}$. Hence, there is a homotopy $(\mathbb{B}^k \times [0,1], \partial\mathbb{B}^k \times [0,1]) \to (V_m, V_m - Z)$ that is stationary on $\partial\mathbb{B}^k$ and moves f to a map whose image intersects V_r in a subset of $[0, \frac{1}{r}] \times \{\mathbf{0}\}$. By concatenating this homotopy with a deformation retraction of $[0, \frac{1}{r}] \times \{\mathbf{0}\}$ onto $\{(\frac{1}{r}, \mathbf{0})\}$, we obtain a homotopy $(\mathbb{B}^k \times [0,1], \partial\mathbb{B}^k \times [0,1]) \to (V_m, V_m - Z)$ that is stationary on $\partial\mathbb{B}^k$ and moves f to a map whose image lies in $V_m - Z$. In other words, we have a homotopy $h : (\mathbb{B}^k \times [0,1], \partial\mathbb{B}^k \times [0,1]) \to (V_m, V_m - Z)$ such that $h_0 = f$ and $h_1(\mathbb{B}^k) \cap Z = \emptyset$. We conclude that Z is locally homotopically negligible in X.

We prove that Z is not an instant Z-set in X by contradiction. Suppose there is a homotopy $h : X \times [0,1] \to X$ such that $h_0 = id_X$ and $h(X \times (0,1]) \cap Z = \emptyset$. Since X is compact, then so is $h_1(X)$. Hence, there is an $n \geq 1$ such that $h_1(X) \cap (\{\frac{1}{n}\} \times \Sigma_n) = \emptyset$. Let $i : (\{\frac{1}{n}\} \times \Sigma_n) \times [0,1] \to X \times [0,1]$ denote the inclusion map and let $r : X \to \{\frac{1}{n}\} \times \Sigma_n$ denote the retraction which maps $X - (\{\frac{1}{n}\} \times \Sigma_n)$ onto $\{(\frac{1}{n}, \mathbf{0})\}$. Since $h_1(X) \cap \Sigma_n = \emptyset$, then $r \circ h \circ i : (\{\frac{1}{n}\} \times \Sigma_n) \times [0,1] \to \{\frac{1}{n}\} \times \Sigma_n$ is a homotopy joining $id_{\{\frac{1}{n}\} \times \Sigma_n}$ to the constant map from $\{\frac{1}{n}\} \times \Sigma_n$ onto $\{(\frac{1}{n}, \mathbf{0})\}$. We have reached a contradiction. Hence, Z is not an instant Z-set in X.

The proof that X is not an ANR is similar to the proof in the preceding paragraph. If X were an ANR, it would be locally contractible by Fact 2.10. Hence, for some $n \geq 1$, there would be a homotopy $h : (\{\frac{1}{n}\} \times \Sigma_n) \times [0,1] \to X$ such that $h_0 = id_{\{\frac{1}{n}\} \times \Sigma_n}$ and $h_1(\{\frac{1}{n}\} \times \Sigma_n)$ is a one-point set. Again let $r : X \to \{\frac{1}{n}\} \times \Sigma_n$ denote the retraction which maps $X - (\{\frac{1}{n}\} \times \Sigma_n)$ onto $\{(\frac{1}{n}, \mathbf{0})\}$. Then $r \circ h : (\{\frac{1}{n}\} \times \Sigma_n) \times [0,1] \to \{\frac{1}{n}\} \times \Sigma_n$ is a homotopy joining $id_{\{\frac{1}{n}\} \times \Sigma_n}$ to the constant map from $\{\frac{1}{n}\} \times \Sigma_n$ to the one-point set $r \circ h_1(\{\frac{1}{n}\} \times \Sigma_n)$, a contradiction. □

B) AN EXAMPLE OF A NON-CLOSED Z-SET (FOR $\mathbb{Q}$) IN $\mathbb{Q}$ THAT IS NOT LOCALLY HOMOTOPICALLY NEGLIGIBLE. The example constructed here shows that conclusion **3)** of Torunczyk's Equivalence Theorem relies on the hypothesis that Z be a closed subset of X. More specifically, the implication **a)** $\Rightarrow$ **c)** can't be proved if this hypothesis is omitted. This example was suggested by Tadek Dobrowolski.

Recall $\mathbb{Q}$ denotes the Hilbert cube $[0,1]^\infty$. Let $C(\mathbb{Q}, \mathbb{Q})$ denote the space of maps from $\mathbb{Q}$ to itself with the compact-open topology. First we sketch a proof that there is a countable dense subset $\{f_i : i \geq 1\}$ of $C(\mathbb{Q}, \mathbb{Q})$ with the property that the collection of images $\{f_i(\mathbb{Q}) : i \geq 1\}$ is pairwise disjoint. Define the metric ρ on $\mathbb{Q}$ by $\rho((x_i), (y_i)) = max\{2^{-i}|x_i - y_i| : i \geq 1\}$, and let σ denote the supremum metric on $C(\mathbb{Q}, \mathbb{Q})$ determined by ρ; i.e., $\sigma(f, g) = sup\{\rho(f(x), f(x)) : x \in \mathbb{Q}\}$ for $f,\ g \in C(\mathbb{Q}, \mathbb{Q})$. Then σ determines the compact-open topology on $C(\mathbb{Q}, \mathbb{Q})$

which makes $C(\mathbb{Q},\mathbb{Q})$ a separable metric space.[50] Hence, $C(\mathbb{Q},\mathbb{Q})$ has a countable dense subset $\{g_j : j \geq 1\}$. For $k \geq 1$ and $a \in [0,1]$, define $\tau_{k,a} : \mathbb{Q} \to \mathbb{Q}$ by $\tau_{k,a}((x_i)) = (x_1, \cdots, x_k, a, a, a, \cdots)$ for $(x_i) \in \mathbb{Q}$. Then $\sigma(id_{\mathbb{Q}}, \tau_{k,a}) \leq 2^{-(k+1)}$. Let $\alpha : \mathbb{N} \times \mathbb{N} \to [0,1]$ be an injective function. For each $(j,k) \in \mathbb{N} \times \mathbb{N}$, define $f_{j,k} \in C(\mathbb{Q},\mathbb{Q})$ by $f_{j,k} = \tau_{k,\alpha(j,k)} \circ g_j$. Then for all $(j,k) \in \mathbb{N} \times \mathbb{N}$, $\sigma(g_j, f_{j,k}) \leq 2^{-(k+1)}$ and $f_{j,k}(\mathbb{Q}) \cap f_{j',k'}(\mathbb{Q}) = \emptyset$ for $(j,k) \neq (j',k') \in \mathbb{N} \times \mathbb{N}$. Thus, $\{f_{j,k} : (j,k) \in \mathbb{N} \times \mathbb{N}\}$ is a countable dense subset of $C(\mathbb{Q},\mathbb{Q})$ with the property that the collection of all images $\{f_{j,k}(\mathbb{Q}) : (j,k) \in \mathbb{N} \times \mathbb{N}\}$ is pairwise disjoint.

Let $\{f_i : i \geq 1\}$ denote a countable dense subset of $C(\mathbb{Q},\mathbb{Q})$ with the property that the collection of images $\{f_i(\mathbb{Q}) : i \geq 1\}$ is pairwise disjoint. Let $Z = \mathbb{Q} - (\bigcup_{i \geq 1} f_i(\mathbb{Q}))$. Then Z is a Z-set (for $\mathbb{Q}$) in $\mathbb{Q}$. Indeed, any map $f : \mathbb{Q} \to \mathbb{Q}$ can be arbitrarily closely approximated by an $f_i : \mathbb{Q} \to \mathbb{Q}$ and $f_i(\mathbb{Q}) \cap Z = \emptyset$.

We will prove Z is not locally homotopically negligible. Let $x \in Z$ and let $U = \mathbb{Q}$. Let V be any open neighborhood of x in $\mathbb{Q}$. We can assume V is path connected. (If not, replace V by the path component of V that contains x.) We will prove there is a map $g : (\mathbb{B}^1, \partial\mathbb{B}^1) \to (V, V - Z)$ for which there is no homotopy $h : (\mathbb{B}^1 \times [0,1], \partial\mathbb{B}^1 \times [0,1]) \to (U, U - Z)$ such that $h_0 = g$ and $h_1(\mathbb{B}^1) \cap Z = \emptyset$. Let W_0 and W_1 be non-empty disjoint open subsets of V. There are integers $j, k \geq 1$ such that $f_j : \mathbb{Q} \to W_0$ and $f_k : \mathbb{Q} \to W_1$. (Let f_j and f_k approximate constant maps of $\mathbb{Q}$ into W_0 and W_1, respectively.) Choose $y_0 \in f_j(\mathbb{Q})$ and $y_1 \in f_k(\mathbb{Q})$. Since V is path connected, there is a map $g : \mathbb{B}^1 \to V$ such that $g(\partial\mathbb{B}^1) = \{y_0, y_1\}$. Thus, $g(\partial\mathbb{B}^1) \subset V - Z$. Assume there is a homotopy $h : (\mathbb{B}^1 \times [0,1], \partial\mathbb{B}^1 \times [0,1]) \to (U, U - Z)$ such that $h_0 = g$ and $h_1(\mathbb{B}^1) \cap Z = \emptyset$. Let $T = (\partial\mathbb{B}^1 \times [0,1]) \cup (\mathbb{B}^1 \times \{1\})$. Then $h(T) \subset U - Z \subset \bigcup_{i \geq 1} f_i(\mathbb{Q})$. Thus, the compact connected metric space $h(T)$ equals the union of the countable pairwise disjoint collection of closed subsets $\{h(T) \cap f_i(\mathbb{Q}) : i \geq 1\}$ that has at least two non-empty elements. This contradicts a theorem of W. Sierpinski.[51] □

9. Strong Z-sets

DEFINITION. Let Z be a subset of a metrizable space X. Z is a *strong Z-set in* X if for every open cover $\mathcal{U}$ of X, id_X is $\mathcal{U}$-close to a map $g : X \to X$ such that $cl(g(X)) \cap Z = \emptyset$.

THEOREM 9.1. *Every Z-set in a locally compact metrizable space is a strong Z-set.*

DEFINITION. A map $f : X \to Y$ between topological spaces is *proper* if $f^{-1}(C)$ is a compact subset of X whenever C is a compact subset of Y.

Our proof of Theorem 9.1 relies on the following two lemmas.

LEMMA 9.2. *If $f : X \to Y$ is a proper map from a topological space to a locally compact Hausdorff space, then f is a closed map.*

PROOF. Let C be a closed subset of X. Suppose $y \in Y - f(C)$. Let U be an open neighborhood of y in Y such that $cl(U)$ is compact. Since $f^{-1}(cl(U))$ is compact and $C \cap f^{-1}(cl(U))$ is a closed subset of $f^{-1}(cl(U))$, then $C \cap f^{-1}(cl(U))$ is compact. Hence, $f(C \cap f^{-1}(cl(U)))$ is compact. Note that $f(C \cap f^{-1}(cl(U))) =$

---

[50] **[27]**, Proposition 1.3.3, pages 31-32.

[51] **[11]**, Theorem 6.1.27, page 358.

$f(C) \cap cl(U)$. (For $A \subset X$ and $B \subset Y$, $f(A \cap f^{-1}(B)) = f(A) \cap B$ is a set-theoretic identity.) Thus, $f(C) \cap cl(U)$ is a compact. Since Y is a Hausdorff space, it follows that $f(C) \cap cl(U)$ is a closed subset of Y. Since $U \cap (f(C) \cap cl(U)) = U \cap f(C)$, then $U - (f(C) \cap cl(U)) = U - f(C)$. Hence, $U - f(C)$ is an open neighborhood of y in Y that is disjoint from $f(C)$. Therefore $y \in Y - cl(f(C))$. This proves $cl(f(C)) = f(C)$. We conclude that $f(C)$ is a closed subset of Y. □

LEMMA 9.3. *Suppose $\mathcal{U}$ is a locally finite open cover of a Hausdorff space Y by sets with compact closures. If $f, g : X \to Y$ are $\mathcal{U}$-close maps of a topological space X into Y and f is proper, then so is g.*

PROOF. Let C be a compact subset of Y. Since $\mathcal{U}$ is locally finite, then the set $\mathcal{F} = \{U \in \mathcal{U} : C \cap U \neq \emptyset\}$ is finite. For each $U \in \mathcal{F}$, since $cl(U)$ is compact and f is proper, then $f^{-1}(cl(U))$ is compact. Hence, $\bigcup\{f^{-1}(cl(U)) : U \in \mathcal{F}\}$ is compact. We claim that $g^{-1}(C) \subset \bigcup\{f^{-1}(U) : U \in \mathcal{F}\}$. To prove this claim, let $x \in g^{-1}(C)$. Then there is a $U \in \mathcal{U}$ such that $\{f(x), g(x)\} \subset U$. Hence, $g(x) \in C \cap U$. Therefore, $U \in \mathcal{F}$. Since $f(x) \in U$, then $x \in f^{-1}(U)$, proving the claim. Since C is compact and Y is Hausdorff, then C is a closed subset of Y. Consequently, $g^{-1}(C)$ is a closed subset of the compact set $\bigcup\{f^{-1}(cl(U)) : U \in \mathcal{F}\}$. Therefore, $g^{-1}(C)$ is compact. This proves g is proper. □

PROOF OF THEOREM 9.1. Let Z be a Z-set in a locally compact metrizable space X. Let $\mathcal{U}$ be an open cover of X. Since X is a locally compact metrizable space, there is a locally finite open cover $\mathcal{V}$ of X by sets with compact closures such that $\mathcal{V}$ refines $\mathcal{U}$. Since Z is a Z-set in X, id_X is $\mathcal{V}$-close to a map $g : X \to X - Z$. Since id_X is a proper map, then Lemma 9.2 implies $g : X \to X$ is proper. Hence, $g : X \to X$ is a closed map by Lemma 9.1. Thus, $g(X)$ is a closed subset of X. So $cl(g(X)) \cap Z = g(X) \cap Z = \emptyset$. □

AN EXAMPLE OF A CLOSED Z-SET IN A NON-LOCALLY COMPACT PLANAR TOPOLOGICALLY COMPLETE AR THAT IS NOT A STRONG Z-SET.[52] For $n \geq 1$, let $S(n) = \{\frac{1}{k} : k \geq n\}$, let

$$X = ([0,1] \times \{0\}) \cup (S(1) \times [0,1]) \subset \mathbb{R}^2,$$

and let $Z = \{(0,0)\}$. X is not locally compact because no neighborhood of $(0,0)$ has compact closure. Since X is locally contractible, 1-dimensional and contractible, then Facts 2.12 and 2.13 imply that X is an AR. X is topologically complete because it is a G_δ-subset of the compact metrizable space $cl_{\mathbb{R}^2}(X)$; indeed,

$$X = \bigcap_{n \geq 1} (cl_{\mathbb{R}^2}(X) - (\{0\} \times [1/n, 1])).$$

Z is a Z-set in X. Indeed, for each $\epsilon > 0$, we now describe a map $f : X \to X - Z$ that is ϵ-close to id_X. Choose $n \geq 1$ so that $\frac{2}{n} < \epsilon$. Choose f to satisfy the following three conditions:

i) f restricts to the identity on

$$([\frac{1}{n}, 1] \times \{0\}) \cup ((S(1) - S(n+1)) \times [0,1]) \cup (S(n+1) \times [\frac{1}{n}, 1]),$$

ii) $f([0, \frac{1}{n}] \times \{0\}) = \{(\frac{1}{n}, 0)\}$ and

[52] [3], Key example on page 56.

iii) for each $k \geq n+1$, f maps $\{\frac{1}{k}\} \times [0, \frac{1}{n}]$ homeomorphically onto $(\{\frac{1}{k}\} \times [0, \frac{1}{n}]) \cup ([\frac{1}{k}, \frac{1}{n}] \times \{0\})$.

We assert that Z is not a strong Z-set in X. To prove this assertion, first note that since X is an ANR, then Fact 2.16 implies there is an open cover $\mathcal{V}$ of X such that any map $g : X \to X$ that is $\mathcal{V}$-close to id_X is $\frac{1}{2}$-homotopic to id_X. Let $g : X \to X$ be any map that is $\mathcal{V}$-close to id_X. We will prove $cl(g(X)) \cap Z \neq \emptyset$, thereby verifying that Z is not a strong Z-set in X. There is a $\frac{1}{2}$-homotopy $h : X \times [0,1] \to X$ such that $h_0 = id_X$ and $h_1 = g$. For each $n \geq 1$, since $h(\{(\frac{1}{n}, 1)\} \times [0,1])$ is a connected subset of X of diameter $\leq \frac{1}{2}$ and $(\frac{1}{n}, 1) = h_0((\frac{1}{n}, 1)) \in h(\{(\frac{1}{n}, 1)\} \times [0,1])$, then $h(\{(\frac{1}{n}, 1)\} \times [0,1]) \subset \{\frac{1}{n}\} \times [\frac{1}{2}, 1]$. Thus, $g((\frac{1}{n}, 1)) = h_1((\frac{1}{n}, 1)) \in \{\frac{1}{n}\} \times [\frac{1}{2}, 1]$. Consequently, $g(X) \cap (\{\frac{1}{n}\} \times [\frac{1}{2}, 1]) \neq \emptyset$ for each $n \geq 1$. X and, hence, $g(X)$ are connected; and for each $n \geq 1$, $\{\frac{1}{n+1}\} \times [\frac{1}{2}, 1]$ and $\{\frac{1}{n}\} \times [\frac{1}{2}, 1]$ lie in distinct components of $X - \{(\frac{1}{n}, 0)\}$. It follows that $(\frac{1}{n}, 0) \in g(X)$ for each $n \geq 1$. Therefore, $(0,0) \in cl(g(X)) \cap Z$. □

10. An exposition of the Guilbault-Moran proof of Ferry's Lemma

Steven Ferry has proved the following result.

LEMMA 10.1 (Ferry's Lemma).[53]

Suppose $f : X \to Y$ and $g : Y \to X$ are maps between metrizable spaces, and $A \subset X$ and $B \subset Y$ such that $f^{-1}(B) = A$ and $g^{-1}(A) = B$.[54] *Furthermore, suppose there is a homotopy $h : X \times [0,1] \to X$ such that $h_0 = id_X$, $h_1 = g \circ f$, h is stationary on A and $h^{-1}(A) = A \times [0,1]$. If B is a closed subset of Y which is an instant Z-set in Y, then A is an instant Z-set in X.*

C. Guilbault and M. Moran found a short simple proof of Ferry's Lemma.[55] We present an exposition of this proof here. However, we first establish:

LEMMA 10.2. *If $f : X \to Y$ is a proper map from a topological space to a first countable Hausdorff space, then f is a closed map.*

PROOF. Let C be a closed subset of X such that $f(C)$ is not a closed subset of Y. Then there is a $y \in Y - f(C)$ such that every open neighborhood of y in Y intersects $f(C)$. Since Y is first countable, there is a sequence $\{x_i : i \geq 1\}$ in C such that $\{f(x_i) : i \geq 1\}$ converges to y. Hence, $K = \{f(x_i) : i \geq 1\} \cup \{y\}$ is compact. Since f is proper, then $f^{-1}(K)$ is also compact. Consequently, $C \cap f^{-1}(K)$ and, therefore, $f(C \cap f^{-1}(K)) = f(C) \cap K = \{f(x_i) : i \geq 1\}$ are compact. Since Y is a Hausdorff space, then it follows that $\{f(x_i) : i \geq 1\}$ is a closed subset of Y. But then $Y - \{f(x_i) : i \geq 1\}$ is an open neighborhood of y that is disjoint from $\{f(x_i) : i \geq 1\}$ – a contradiction. □

THE GUILBAULT-MORAN PROOF OF FERRY'S LEMMA. Assume B is a closed subset of Y that is an instant Z-set in Y. Then there is a homotopy $\zeta : Y \times [0,1] \to Y$ such that $\zeta_0 = id_Y$ and $\zeta(Y \times (0,1]) \subset Y - B$. Define a homotopy $\eta : X \times [0,1] \to X$ by

$$\eta_t = h_{2t} \text{ for } t \in [0, \frac{1}{2}] \text{ and } \eta_t = g \circ \zeta_{2t-1} \circ f \text{ for } t \in [\frac{1}{2}, 1].$$

[53] [**12**], Proposition 1.6, pages 289-290.

[54] To compare this version of Ferry's Lemma with the versions in [**12**] and [**13**], observe that the statement $f^{-1}(B) = A$ is equivalent to the statement $f(A) \subset B$ *and* $f(X - A) \subset Y - B$.

[55] [**13**], Lemma 3.1, page 299.

Then $\eta_0 = id_X$, $\eta_t|A = id_A$ for $t \in [0, \frac{1}{2}]$ and $\eta(((X - A) \times [0,1]) \cup (X \times (\frac{1}{2}, 1])) \subset X - A$.

The goal of the remainder of the proof is to modify η by "squeezing" subsets of its domain of the form $\{x\} \times [0, \frac{1}{2}]$ where $x \in A$ to single points. The map $q : X \times [0,1] \to X \times [0,1]$, which we are about to define, accomplishes this "squeezing" process. Since B is a closed subset of Y and $A = f^{-1}(B)$, then A is a closed subset of X. Hence, there is a map $\lambda : X \to [0, \frac{1}{2}]$ such that $\lambda^{-1}(0) = A$. (For example, define $\lambda(x) = min\{\rho(x, A), \frac{1}{2}\}$ where ρ is a metric on X.) Define $q : X \times [0,1] \to X \times [0,1]$ by

$$q(x,t) = (x, 2\lambda(x)t) \text{ for } (x,t) \in X \times [0, \frac{1}{2}]$$

and

$$q(x,t) = (x, 2\lambda(x)(1-t) + 2t - 1) \text{ for } (x,t) \in X \times [\frac{1}{2}, 1].$$

We assert that $q : X \times [0,1] \to X \times [0,1]$ is a proper map. Indeed, if K is a compact subset of $X \times [0,1]$, then $q^{-1}(K)$ is a closed subset of the compact set $\pi(K) \times [0,1]$, where $\pi : X \times [0,1] \to X$ denotes projection. Hence, $q^{-1}(K)$ is compact. Now Lemma 10.2 implies that $q : X \times [0,1] \to X \times [0,1]$ is a closed map.

Define $\theta : X \times [0,1] \to X$ by $\theta = \eta \circ q^{-1}$; in other words, $\theta(x,t) = \eta(q^{-1}(x,t))$. θ is the desired modification of η which shows that A is an instant Z-set in X. We must first verify that θ is a well-defined function. Specifically, we must show that whenever q^{-1} "blows up" (x,t) to a set with more than one element, then η "squeezes" this set back to a one-element set. An examination of the definition of q shows that q^{-1} "blows up" (x,t) if and only if $(x,t) \in \lambda^{-1}(\{0\}) \times \{0\} = A \times \{0\}$; and, in this case, $q^{-1}(x,t) = \{x\} \times [0, \frac{1}{2}]$. Since $\eta(\{x\} \times [0, \frac{1}{2}]) = \{x\}$ for every $x \in A$, then we conclude that if q^{-1} "blows up" (x,t), then $\eta(q^{-1}(x,t)) = x$. Thus, $\theta = \eta \circ q^{-1}$ is a well-defined function. To verify that θ is continuous, observe that if C is a closed subset of X, then $\theta^{-1}(C) = q(\eta^{-1}(C))$ is a closed set because η is continuous and q is a closed map. It is easily verified that $\eta_0 = id_X$. (We have already checked this equation at points of A.) Since $q^{-1}(X \times (0,1]) \subset ((X - A) \times (0,1]) \cup (A \times (\frac{1}{2}, 1])$ and η maps the latter set into $X - A$, then $\theta(X \times (0,1]) \subset X - A$. This proves A is an instant Z-set in X. □

Acknowledgements

The author gratefully acknowledges the referee's careful reading of the manuscript and insightful comments which led to significant improvement in the logical organization of the paper.

References

[1] F. D. Ancel, *The role of countable dimensionality in the theory of cell-like relations*, Trans. Amer. Math. Soc. **287** (1985), no. 1, 1–40, DOI 10.2307/2000395. MR766204

[2] R. D. Anderson, *On topological infinite deficiency*, Michigan Math. J. **14** (1967), 365–383. MR214041

[3] M. Bestvina, P. Bowers, J. Mogilski, and J. Walsh, *Characterization of Hilbert space manifolds revisited*, Topology Appl. **24** (1986), no. 1-3, 53–69, DOI 10.1016/0166-8641(86)90049-0. MR872478

[4] M. Bestvina and G. Mess, *The boundary of negatively curved groups*, J. Amer. Math. Soc. **4** (1991), no. 3, 469–481, DOI 10.2307/2939264. MR1096169

[5] K. Borsuk, *Theory of retracts*, Monografie Matematyczne [Mathematical Monographs], Tom 44, Państwowe Wydawnictwo Naukowe, Warsaw, 1967. MR216473

[6] T. A. Chapman, *Lectures on Hilbert cube manifolds*, Regional Conference Series in Mathematics, No. 28, American Mathematical Society, Providence, RI, 1976. MR423357

[7] C. H. Dowker, *Topology of metric complexes*, Amer. J. Math. **74** (1952), 555–577, DOI 10.2307/2372262. MR48020

[8] C. H. Dowker, *Homotopy extension theorems*, Proc. London Math. Soc. (3) **6** (1956), 100–116, DOI 10.1112/plms/s3-6.1.100. MR73982

[9] J. Dugundji, *Topology*, Allyn and Bacon, Inc., Boston, MA, 1966. MR193606

[10] R. D. Edwards, *Characterizing infinite-dimensional manifolds topologically (after Henryk Toruńczyk)*, Séminaire Bourbaki (1978/79), Lecture Notes in Math., vol. 770, Springer, Berlin, 1980, pp. Exp. No. 540, pp. 278–302. MR572429

[11] R. Engelking, *General topology*, 2nd ed., Sigma Series in Pure Mathematics, vol. 6, Heldermann Verlag, Berlin, 1989. MR1039321

[12] S. C. Ferry, *Stable compactifications of polyhedra*, Michigan Math. J. **47** (2000), no. 2, 287–294, DOI 10.1307/mmj/1030132534. MR1793625

[13] C. R. Guilbault and M. A. Moran, *Proper homotopy types and $\mathcal{Z}$-boundaries of spaces admitting geometric group actions*, Expo. Math. **37** (2019), no. 3, 292–313, DOI 10.1016/j.exmath.2018.03.004. MR4007577

[14] A. Hatcher, *Algebraic topology*, Cambridge University Press, Cambridge, 2002. MR1867354

[15] D. W. Henderson, *Infinite-dimensional manifolds are open subsets of Hilbert space*, Topology **9** (1970), 25–33, DOI 10.1016/0040-9383(70)90046-7. MR250342

[16] D. W. Henderson, *Z-sets in ANR's*, Trans. Amer. Math. Soc. **213** (1975), 205–216, DOI 10.2307/1998044. MR391008

[17] D. W. Henderson and R. Schori, *Topological classification of infinite dimensional manifolds by homotopy type*, Bull. Amer. Math. Soc. **76** (1970), 121–124, DOI 10.1090/S0002-9904-1970-12392-8. MR251749

[18] S.-t. Hu, *Theory of retracts*, Wayne State University Press, Detroit, MI, 1965. MR181977

[19] M. Mather, *Counting homotopy types of manifolds*, Topology **3** (1965), 93–94, DOI 10.1016/0040-9383(65)90050-9. MR176470

[20] J. Milnor, *On spaces having the homotopy type of a* CW*-complex*, Trans. Amer. Math. Soc. **90** (1959), 272–280, DOI 10.2307/1993204. MR100267

[21] J.-i. Nagata, *Modern general topology*, Second revised edition, Bibliotheca Mathematica [Mathematics Library], Vol. VII, North-Holland Publishing Co., Amsterdam-London; Wolters-Noordhoff Publishing, Groningen American Elsevier Publishing Co., Inc., New York, 1974. MR474164

[22] K. Sakai, *The completions of metric ANR's and homotopy dense subsets*, J. Math. Soc. Japan **52** (2000), no. 4, 835–846, DOI 10.2969/jmsj/05240835. MR1774631

[23] K. Sakai, *Small subdivisions of simplicial complexes with the metric topology*, J. Math. Soc. Japan **63** (2011), no. 3, 789–800. MR2836743

[24] K. Sakai, *Geometric aspects of general topology*, Springer Monographs in Mathematics, Springer, Tokyo, 2013, DOI 10.1007/978-4-431-54397-8. MR3099433

[25] K. Sakai, *Topology of infinite-dimensional manifolds*, Springer Monographs in Mathematics, Springer, Singapore, 2020, DOI 10.1007/978-981-15-7575-4. MR4179591

[26] J. van Mill, *Infinite-dimensional topology*, North-Holland Mathematical Library, vol. 43, North-Holland Publishing Co., Amsterdam, 1989. MR977744

[27] J. van Mill, *The infinite-dimensional topology of function spaces*, North-Holland Mathematical Library, Vol. 64, Elsevier, Amsterdam, 2001.

[28] H. Toruńczyk, *Absolute retracts as factors of normed linear spaces*, Fund. Math. **86** (1974), 53–67, DOI 10.4064/fm-86-1-53-67. MR365471

[29] H. Toruńczyk, *Concerning locally homotopy negligible sets and characterization of l_2-manifolds*, Fund. Math. **101** (1978), no. 2, 93–110, DOI 10.4064/fm-101-2-93-110. MR518344

[30] H. Toruńczyk, *Characterization of infinite-dimensional manifolds*, Proceedings of the International Conference on Geometric Topology (Warsaw, 1978), PWN, Warsaw, 1980, pp. 431–437. MR656785

[31] H. Toruńczyk, *Characterizing Hilbert space topology*, Fund. Math. **111** (1981), no. 3, 247–262, DOI 10.4064/fm-111-3-247-262. MR611763

[32] J. E. West, *Compact ANR's have finite type*, Bull. Amer. Math. Soc. **81** (1975), 163–165, DOI 10.1090/S0002-9904-1975-13689-5. MR358791

[33] J. H. C. Whitehead, *Simplicial Spaces, Nuclei and m-Groups*, Proc. London Math. Soc. (2) **45** (1939), no. 4, 243–327, DOI 10.1112/plms/s2-45.1.243. MR1576810

Department of Mathematical Sciences, University of Wisconsin-Milwaukee
Email address: ancel@uwm.edu

Contemporary Mathematics
Volume **812**, 2025
https://doi.org/10.1090/conm/812/16261

The End Sum of Surfaces

Liam Axon and Jack Calcut

Abstract. End sum is a natural operation for combining two noncompact manifolds and has been used to construct various manifolds with interesting properties. The uniqueness of end sum has been well-studied in dimensions three and higher. We study end sum—and the more general notion of adding a 1-handle at infinity—for surfaces and prove uniqueness results. The result of adding a 1-handle at infinity to distinct ends of a surface with compact boundary is uniquely determined by the chosen ends and the orientability of the 1-handle. As a corollary, the end sum of two surfaces with compact boundary is uniquely determined by the chosen ends. Unlike uniqueness results in higher dimensions, which rely on isotopy uniqueness of rays, our results rely fundamentally on a classification of noncompact surfaces.

1. Introduction

End sum is the analogue for open manifolds of the boundary sum of manifolds with boundary. It was introduced by Gompf [**Gom83, Gom85**] in the 1980's to construct smooth manifolds homeomorphic to $\mathbb{R}^4$. Gompf [**Gom83**, p. 322] colloquially described end sum as gluing together two noncompact manifolds using a piece of tape. More formally, the piece of tape is a 1-handle at infinity. Since that time, several authors have used end sum to construct manifolds with interesting properties. Recently, Bennett [**Ben16**] used end sum to produce new smooth structures on open 4-manifolds and Sparks [**Spa18**] used it to construct 4-dimensional splitters. For further examples, see the second author and Gompf [**CG19**] and the second author, Haggerty, and Guilbault [**CGH20**].

A 1-handle at infinity is attached to a manifold M along a chosen pair of disjoint, properly embedded rays pointing to ends of M. End sum is the special case where M has two components and the 1-handle connects them. The dependence on ray choice of adding a 1-handle at infinity has been well-studied. Gompf [**Gom85**] first showed that end sums of manifolds homeomorphic to $\mathbb{R}^4$ are independent of ray choice. Myers [**Mye99**] showed that end summing two copies of $\mathbb{R}^3$ using knotted rays yields uncountably many homeomorphism types of contractible, open 3-manifolds. The second author and Haggerty [**CH14**] constructed examples of pairs of connected, open, oriented, one-ended n-manifolds for each $n \geq 3$ that may be end summed using various rays to produce manifolds that are not proper

2020 *Mathematics Subject Classification.* Primary 57K20; Secondary 57Q99.
Key words and phrases. End, end sum, 1-handle at infinity, noncompact surface, classification of surfaces, proper ray.

homotopy equivalent. The latter examples arise from complicated fundamental group behavior at even just one of the ends. In dimension $n = 1$, adding a 1-handle along specified ends is always unique.

For manifolds of any dimension, the *Mittag-Leffler* condition (also called *semistability*) on an end is a necessary and sufficient condition for any two proper rays pointing to that end to be properly homotopic. In dimensions four and higher, such a proper homotopy may be upgraded to an ambient isotopy. That yields uniqueness results for end sums and adding 1-handles at infinity as proved by the second author and Gompf [**CG19**].

THEOREM 1.1 (Calcut and Gompf). *Let X be a (possibly disconnected) n-manifold, $n \geq 4$. Then the result of attaching a (possibly infinite) collection of 1-handles at infinity to some oriented Mittag-Leffler ends of X depends only on the pairs of ends to which each 1-handle is attached, and whether the corresponding orientations agree.*

An immediate corollary is the uniqueness of oriented end sums of two n-manifolds with $n \geq 4$ along oriented Mittag-Leffler ends.

In the present paper, we study end sum for surfaces. Note from the outset that for each end of a *surface with compact boundary*, the Mittag-Leffler condition is actually equivalent to the end being collared. An end of a manifold is *collared* provided it has a neighborhood of the form $Z \times [0, \infty)$ for some connected, compact manifold Z. The authors plan to address the classification of Mittag-Leffler ends of general surfaces in a future paper. Observe now that a Mittag-Leffler end of a surface with noncompact boundary components need not be collared as shown by the end η in Figure 1.1. Properly embedded rays in $\mathbb{R}^2$, $S^1 \times [0, \infty)$, and closed

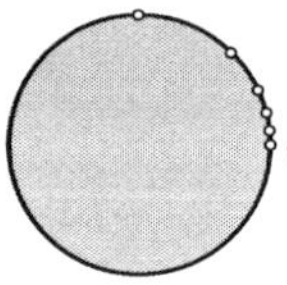

FIGURE 1.1. Surface M obtained from the closed disk by removing a sequence of boundary points and the single limit point of that sequence.

half space $\mathbb{R}^2_+$ may be straightened by ambient isotopy [**CKS12**, pp. 1845–1852]. So, up to isotopy, there are unique proper rays in Mittag-Leffler (=collared) ends of surfaces with compact boundary. In general, ends of noncompact surfaces may have infinite genus, complicated fundamental group behavior, and fail to be Mittag-Leffler. Thus, one might suspect that end sums of surfaces depend on the choices of rays within specified ends. In fact, the contrary is true. The following is our main result.

THEOREM 1.2. *Let M be a (possibly disconnected) surface with compact boundary. Then the result of attaching a 1-handle at infinity to distinct ends of M depends only on the ends to which the 1-handle is attached and orientations. If the chosen ends lie in different components of M or they lie in the same non-orientable component of M, then orientation is irrelevant.*

A more precise statement is given in Theorem 5.1 below. An immediate corollary is uniqueness of end sums of two connected surfaces with compact boundary along chosen ends (irrespective of orientations).

Contrasting Theorems 1.1 and 1.2, we see that for surfaces the relevant ends need not be Mittag-Leffler and there is greater flexibility with orientations. On the other hand, our arguments for surfaces assume: (i) M has compact boundary, (ii) a single 1-handle is attached to M, and (iii) the relevant ends of M are distinct. Throughout, we opt to work with a single 1-handle for simplicity. Applying our results iteratively yields results for attaching finitely many 1-handles, and our results likely carry over to appropriate settings involving infinitely many 1-handles. Before we discuss assumptions (i) and (iii), we introduce some useful terminology.

Consider the general case of attaching a single 1-handle at infinity to an n-manifold M. Let r and r' denote the rays in M along which the 1-handle is attached, let ε and ε' denote the ends of M to which r and r' point (respectively), and let $N = M \cup (\text{1-handle})$ denote the resulting n-manifold. In this unrestricted setting, we allow M to be disconnected, ∂M to be compact (possibly empty) or noncompact (possibly with noncompact boundary components), ε and ε' to be equal or distinct, and the 1-handle attachment to respect or ignore any possible given orientations. An end η of N is *ordinary* provided it has a neighborhood disjoint from the 1-handle. Otherwise, η is *extraordinary*. Intuitively, the extraordinary end(s) of N are those involved in the attachment of the 1-handle. For a space X, let $\mathcal{E}(X)$ denote the *space of ends* of X, and let $|\mathcal{E}(X)|$ denote the *number of ends* of X. The possible number of extraordinary ends of N varies with the dimension n. If $n = 1$, then there are no extraordinary ends—attaching a 1-handle at infinity to a 1-manifold simply eliminates two ends of M and $|\mathcal{E}(N)| = |\mathcal{E}(M)| - 2$. If $n \geq 3$, then there is one extraordinary end, and $|\mathcal{E}(N)| = |\mathcal{E}(M)| - 1$ (when $\varepsilon \neq \varepsilon'$) or $|\mathcal{E}(N)| = |\mathcal{E}(M)|$ (when $\varepsilon = \varepsilon'$). For surfaces, the situation is more complicated—ultimately due to the fact that a surface may be separated by a 1-dimensional submanifold. There may be one or two extraordinary ends, and $|\mathcal{E}(N)|$ may equal $|\mathcal{E}(M)| - 1$, $|\mathcal{E}(M)|$, or $|\mathcal{E}(M)| + 1$. Predicting which occur is subtle, especially in the presence of noncompact boundary components. The following examples exhibit that subtlety and others. Basic examples are included for context and comparison. In each example, $n = 2$ and $N = M \cup (\text{1-handle})$.

(1) Let $M = \mathbb{R}^2 \sqcup \mathbb{R}^2$ be the disjoint union of two copies of $\mathbb{R}^2$ as in Figure 1.2. Regardless of the orientability of the 1-handle, N is homeomorphic to $\mathbb{R}^2$

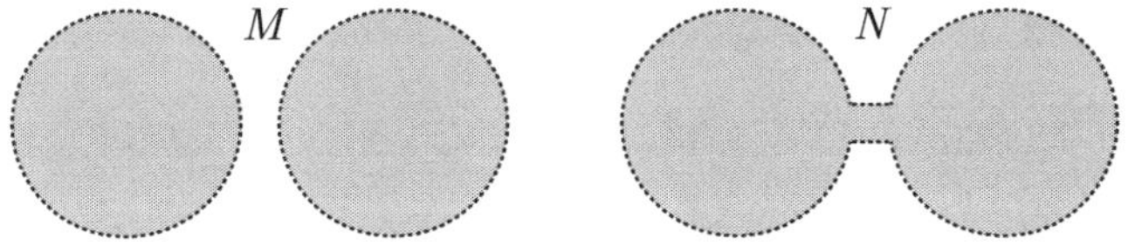

FIGURE 1.2. Surface $M = \mathbb{R}^2 \sqcup \mathbb{R}^2$ (left) and end sum N of M (right).

and has one end (extraordinary). Here, $|\mathcal{E}(N)| = |\mathcal{E}(M)| - 1$.

(2) Let $M = \mathbb{R}^2$ as in Figure 1.3. Let N be the result using an oriented 1-handle, and let N' be the result using a non-oriented 1-handle. Then, N is an open cylinder with two ends (both extraordinary), and N' is an open

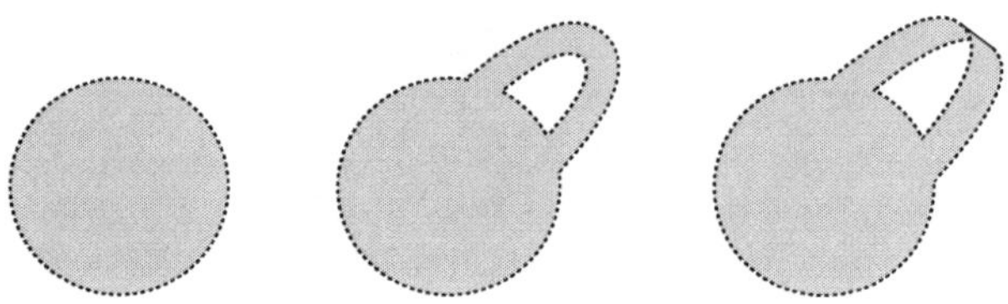

FIGURE 1.3. Surface $M = \mathbb{R}^2$ (left), result N of attaching an oriented 1-handle (middle), and result N' of attaching a non-oriented 1-handle (right).

Möbius band with one end (extraordinary). Here, $|\mathcal{E}(N)| = |\mathcal{E}(M)| + 1$ and $|\mathcal{E}(N')| = |\mathcal{E}(M)|$.

(3) Let M be an open annulus as in Figure 1.4. Attach a 1-handle to the

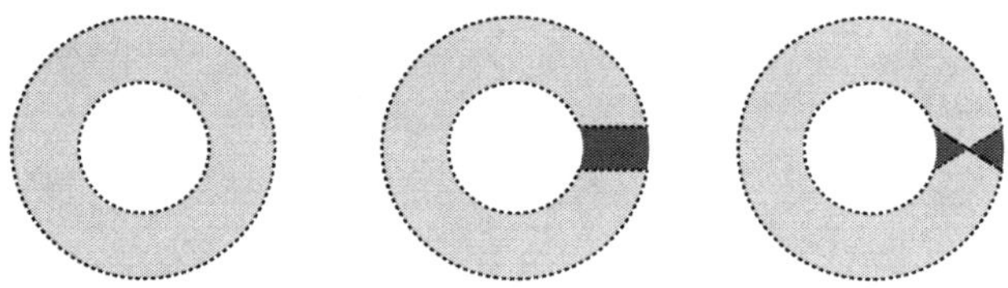

FIGURE 1.4. Open annulus M (left), result N of attaching an oriented 1-handle (middle), and result N' of attaching a non-oriented 1-handle (right).

distinct ends of M. Let N be the result using an oriented 1-handle, and let N' be the result using a non-oriented 1-handle. Then, N is a punctured torus with one end (extraordinary), and N' is a punctured Klein bottle with one end (extraordinary). Here, $|\mathcal{E}(N)| = |\mathcal{E}(M)| - 1 = |\mathcal{E}(N')|$.

(4) Let $M = \mathbb{R}^2_+ \sqcup \mathbb{R}^2_+$ be the disjoint union of two copies of closed half-space as in Figure 1.5. Regardless of the orientability of the 1-handle, N is

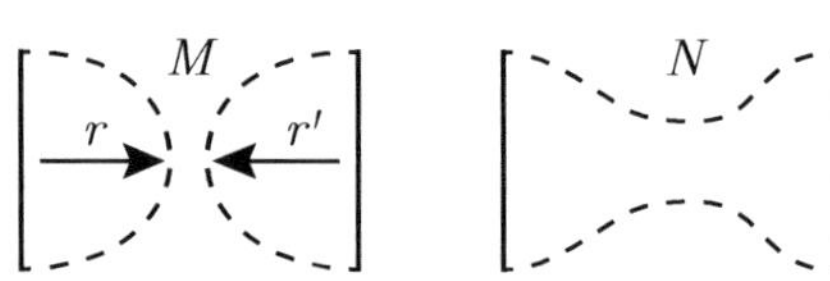

FIGURE 1.5. Surface $M = \mathbb{R}^2_+ \sqcup \mathbb{R}^2_+$ and end sum N of M.

homeomorphic to $[0,1] \times \mathbb{R}$ and has two ends (both extraordinary). Here, $|\mathcal{E}(N)| = |\mathcal{E}(M)|$ in both cases. In a sense, the noncompact boundary components of M clog up the ends of N.

(5) Let M be the open, oriented, one-ended surface with infinite genus as in Figure 1.6. Phillips and Sullivan [**PS81**] referred to M as the *Infinite Loch Ness monster*; see also Aramayona and Vlamis [**AV20**, p. 463]. Attach an oriented 1-handle at infinity to M. Let N be the result using the non-parallel rays r and s, and let N' be the result using the parallel rays t and u. Then N has one end (infinite genus and extraordinary), and N' has two ends (one of genus zero, the other of infinite genus, and both

FIGURE 1.6. One-ended, infinite genus surface M containing two non-parallel rays r and s (left) and two parallel rays t and u (right).

extraordinary). Here, $|\mathcal{E}(N)| = |\mathcal{E}(M)|$ and $|\mathcal{E}(N')| = |\mathcal{E}(M)|+1$. Thus, the hypothesis in Theorem 1.2 that the 1-handle is attached to distinct ends may not be omitted. In fact, the distinct end hypothesis is required even when M has finite genus. For a planar example, let M be $\mathbb{R}^2$ with the integer points on the x-axis removed. So, M has infinitely many ends, exactly one of which is not isolated. Let r and s be non-parallel rays in the positive and negative y-axes respectively. Let t and u be parallel rays in the upper half-plane. Attach an oriented 1-handle at infinity to M. Let N be the result using the non-parallel rays r and s, and let N' be the result using the parallel rays t and u. Then N and N' are not homeomorphic since N' has one nonisolated end whereas N has two.

(6) Let X be the 2-disk D^2 with three points removed from its boundary. So, X has three noncompact boundary components and three ends. Each end of X has genus zero and an open neighborhood homeomorphic to $[0,1]\times\mathbb{R}$. Let Y be obtained from X by connect summing a sequence of tori converging to one end. The surface Y is depicted on the left in Figure 1.7 where the zeros indicate genus zero ends, ∞ indicates the infinite genus end, and the segment indicates a ray. The surface $M = Y \sqcup Y$

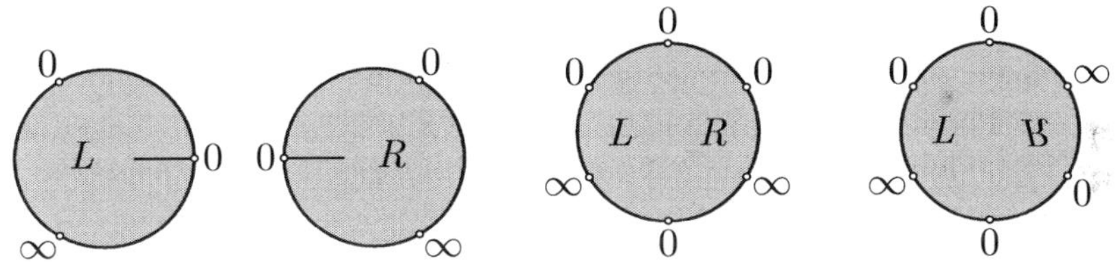

(a) Surface $M = Y \sqcup Y$ to be end summed along indicated rays.

(b) End sum N using an oriented 1-handle.

(c) End sum N' using a non-oriented 1-handle.

FIGURE 1.7. Orientation of the 1-handle at infinity is relevant for end sums of surfaces with noncompact boundary components.

is the disjoint union of two copies of Y. End sum M along the indicated rays. Let N be the result using an oriented 1-handle, and let N' be the result using a non-oriented 1-handle. The surfaces N and N' are shown in Figure 1.7, where the letters L and R are included to display orientations. Both N and N' have six ends (four of genus zero, two of infinite genus, and two extraordinary both of genus zero). Here, $|\mathcal{E}(N)| = |\mathcal{E}(M)| = |\mathcal{E}(N')|$. Notice that N and N' are not homeomorphic—N has two noncompact boundary components that point only to genus zero ends of M and point to a common end of M, whereas N' does not have such boundary components. Up to homeomorphism, there are exactly two end sums, namely N and N', of M along the specific ends just used since ray choice is irrelevant at those ends. To obtain similar examples with M non-orientable, remove an open disk from the interior of Y and glue in a crosscap. Define N and N' as before. Though both components of M are non-orientable, and often orientability of the 1-handle is not relevant when M is non-orientable, the end sums N and N' remain non-homeomorphic (the same argument still applies). Thus, for end sums of surfaces with noncompact boundary components, orientability of the 1-handle is relevant.

Example 5 showed that Theorem 1.2 is false without the distinct ends hypothesis. To remove that hypothesis, appropriate replacement hypotheses would be necessary. Our proof of Theorem 1.2 proceeds by studying end invariants of the extraordinary end of N and then applying the classification of noncompact surfaces with compact boundary as proved by Richards [**Ric60**]. One may attempt to remove from Theorem 1.2 the hypothesis that ∂M is compact by instead using the classification of noncompact surfaces with possibly noncompact boundary due to Brown and Messer [**BM79**]. Even with that approach to the case where ∂M is noncompact, one must understand how ray choice affects the number of extraordinary ends and the number of ends of N, and how the orientation of the 1-handle affects N. The examples above indicate that those may be subtle questions.

To circumvent those nuances, the present paper focuses on the case where a single 1-handle at infinity is attached to a surface M with compact boundary along rays pointing to distinct ends of M. In that case, there is a unique extraordinary end of $N = M \cup (\text{1-handle})$ and $|\mathcal{E}(N)| = |\mathcal{E}(M)| - 1$. We allow M to be disconnected and ∂M to be empty. Without loss of generality, it suffices to consider two cases: (i) M is connected, and (ii) M has two connected components and the 1-handle connects them. The latter operation is the end sum of the two components of M.

This paper is organized as follows. Section 2 defines an end and recalls the theory of ends sufficient for our purposes. We have included that material to help make this paper more self-contained and accessible. Section 3 defines end sum and 1-handle addition at infinity, sets up notation for those operations, and proves that those operations indeed yield manifolds (we are unaware of a published proof of this fact). Section 4 recalls the classification of noncompact surfaces with compact boundary, including generalized genus, parity, end invariants, and orientability. Section 5 proves our main result—Theorem 5.1—for PL surfaces by studying how each of the following end invariants are affected by the addition of a 1-handle at infinity: the space of ends, boundary, orientability, genus, and parity. Several results in Section 5 are proved more generally for n-manifolds. In particular, Lemma 5.9 shows that the space of ends of $N = M \cup (\text{1-handle})$ is the quotient space of the space of ends of M by identifying the ends of M along which the 1-handle is attached (see Lemma 5.9 for a more precise statement). Section 6 provides an alternative proof of Theorem 5.1 using Brown and Messer's [**BM79**] classification of noncompact surfaces. That approach also yields a ray uniqueness result for surfaces—see Theorem 6.10. Lastly, Section 7 explains how to extend Theorem 5.1 to TOP and DIFF surfaces.

We use the following conventions where X is a topological space, A is any subspace of X (denoted $A \subset X$), and M is a manifold.

- $\overline{A}$ denotes the topological **closure** of A in X.
- A° denotes the topological **interior** of A in X.
- $\mathbb{R}_+$ denotes $[0, \infty)$ and $\mathbb{R}^n_+$ denotes $[0, \infty) \times \mathbb{R}^{n-1}$.
- D^n denotes the closed n-disk.
- CAT denotes one of the manifold categories: TOP, PL, or DIFF.
- CAT manifolds are Hausdorff and paracompact, possibly with boundary.
- A connected manifold without boundary is **closed** provided it is compact and is **open** provided it is noncompact.
- $M \cong M'$ denotes **isomorphism** of CAT manifolds.
- ∂M denotes the **manifold boundary** of M.
- $\mathcal{E}(X)$ denotes the **space of ends** of X.
- If α is an end of X and K is a compact subspace of X, then $\mathcal{N}(\alpha; K)$ denotes the set of all ends β of X for which $\beta(K) = \alpha(K)$.
- $b(M) = |\pi_0(\partial M)|$ denotes the **number of boundary components** of M.
- $g(M)$ denotes the **genus** of a surface M and $\mathcal{P}(M)$ denotes the parity of M.

2. Ends of Spaces

Loosely speaking, an end of a space may be thought of as an infinity of the space. For example, the closed interval has no ends, a ray has one end, and the real line has two ends. Figure 2.1 shows three manifolds and their ends. Each (suitably

(a) The real line has two ends.

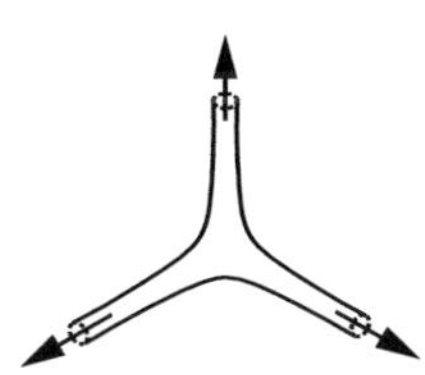

(b) The thrice punctured sphere has three ends.

(c) The closed disk has no ends.

FIGURE 2.1. Manifolds with their ends indicated by arrows.

nice) space is compact if and only if it has no ends. The set of all ends of a space may be equipped with a natural topology. Ends and the space of ends play fundamental roles in the study of noncompact spaces. A prime example is Whitehead's open, contractible 3-manifold, the first example of an open, contractible manifold not isomorphic to Euclidean space. It may be distinguished from $\mathbb{R}^3$ by properties of its end (see Guilbault [**Gui16**, pp. 6–7]). The space of ends is also an essential ingredient in Richards' [**Ric60**] classification of noncompact surfaces which we review in Section 4.

The theory of ends dates back to Freudenthal and Hopf in the 1930's. For interesting further reading, see Freudenthal's original paper [**Fre31**], Siebenmann's thesis [**Sie65**, pp. 8–12], and Guilbault's chapter [**Gui16**]. As we are focused on manifolds, we restrict our study to spaces that satisfy the following condition.

DEFINITION 2.1. A topological space X is **nice for ends** provided X is Hausdorff, locally compact, σ-compact, connected, and locally connected.

We adopt the convention that a space X is **connected** provided it has exactly two subspaces that are both open and closed in X, namely X and $\varnothing$. In particular, the empty space is neither connected nor disconnected.

Connected manifolds are nice for ends, as are connected, locally finite simplicial complexes. Each space that is nice for ends is necessarily paracompact. However, it is not necessarily separable, metrizable, or even first-countable. For example, the product of uncountably many copies of $[0, 1]$ is compact and nice for ends, but does not satisfy those three properties.

Throughout this section—unless explicitly stated otherwise—X is a space nice for ends.

DEFINITION 2.2. An **end** of X is any function ϵ defined on the collection of compact subspaces of X such that: for each compact $K \subset X$, the output $\epsilon(K)$ is a connected component (hence nonempty) of $X - K$, and if $K_1 \subset K_2$, then $\epsilon(K_2) \subset \epsilon(K_1)$. Let $\mathcal{E}(X)$ denote the set of all ends of X.

For example, $\mathbb{R}$ has two ends ϵ_- and ϵ_+. Given a compact $K \subset \mathbb{R}$, $\epsilon_-(K)$ is the connected component of $\mathbb{R} - K$ that contains elements less than K, and $\epsilon_+(K)$ is the connected component of $\mathbb{R} - K$ that contains elements greater than K. One may verify that ϵ_- and ϵ_+ are ends of $\mathbb{R}$ and, in fact, are the only ends of $\mathbb{R}$.

We recall some fundamental properties of ends. Several proofs will be left to the reader.

LEMMA 2.3. *If ϵ is an end of X, and K and K' are compact subspaces of X, then $\epsilon(K) \cap \epsilon(K') \neq \varnothing$.*

LEMMA 2.4. *If X is compact, then X has no ends.*

The converse of Lemma 2.4 also holds—see Corollary 2.17 below.

DEFINITION 2.5. A subspace of X is **bounded** provided its closure in X is compact, and **unbounded** otherwise.

This terminology aligns with the definition of a bounded subset of Euclidean space, although in general metric spaces the notions can differ.

LEMMA 2.6. *Let ϵ be an end of X. If $K \subset X$ is compact, then $\epsilon(K)$ is an unbounded subspace of X.*

LEMMA 2.7. *Let $K \subset X$ be compact. Then, $X - K$ has finitely many unbounded connected components. Furthermore, if A is the union of all bounded connected components of $X - K$, then $K \cup A$ is compact and $X - (K \cup A)$ has only unbounded connected components.*

For the proof of Lemma 2.7, we offer this hint: use local compactness to construct a bounded open neighborhood V of K. Then, look at the open cover of $\overline{V}$ consisting of V and all components of $X - K$.

EXAMPLE 2.8. Let $X = D^2$. For each integer $n \geq 1$, let $p_n = (0, 2^{-n})$ and let U_n be an open disk of radius 4^{-n} centered at p_n. The space $K = X - \cup_n U_n$ (a nonsurface) is compact, but each U_n is a bounded connected component of $X - K$. Thus, although there are always finitely many unbounded connected components of $X - K$, there may be infinitely many bounded components.

The notion of a *neighborhood of an end* is used to define a topology on the set of ends of X.

DEFINITION 2.9. Let ϵ be an end of X. A **neighborhood** of ϵ is a subspace $A \subset X$ such that there exists some compact subspace $K \subset X$ for which $\epsilon(K) \subset A$.

For example, consider $X = \mathbb{R}$ with ends ϵ_- and ϵ_+. A subspace of $\mathbb{R}$ is a neighborhood of ϵ_- if and only if it contains $(-\infty, a)$ for some $a \in \mathbb{R}$. Similarly, a subspace of $\mathbb{R}$ is a neighborhood of ϵ_+ if and only if it contains (b, ∞) for some $b \in \mathbb{R}$.

LEMMA 2.10. *If $K \subset X$ is compact, then $X - K$ is a neighborhood of every end of X.*

LEMMA 2.11. *If ϵ_1 and ϵ_2 are distinct ends of X, then there exist disjoint neighborhoods A_1 of ϵ_1 and A_2 of ϵ_2 (so, $A_1 \cap A_2 = \varnothing$).*

An alternative definition of an end may given using a *compact exhaustion.* This important and equivalent definition is useful for constructing and visualizing ends.

DEFINITION 2.12. A **compact exhaustion** of a topological space X is a sequence (K_i) of compact subspaces $K_i \subset X$ such that $K_i \subset K_{i+1}$ for all i and

$$\bigcup_{i=1}^{\infty} K_i^\circ = X$$

For example, $K_i = [-i, i]$ is a compact exhaustion of $\mathbb{R}$. See Figure 2.2 for two more examples. To make use of compact exhaustions, we need their existence.

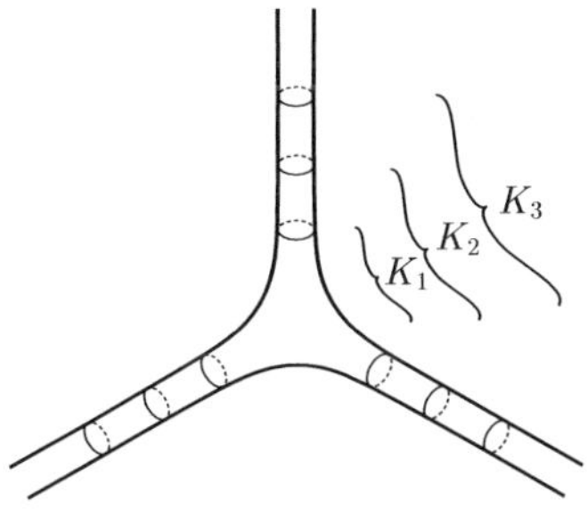

(a) Thrice punctured sphere.

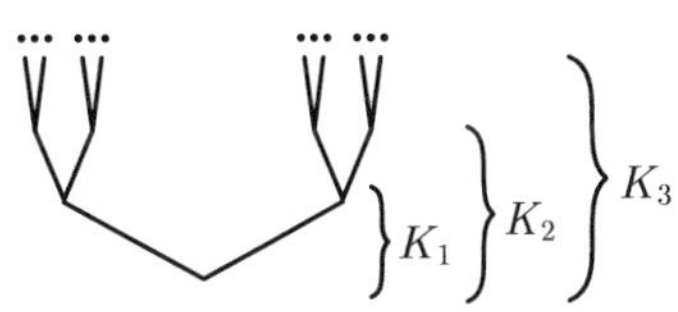

(b) Infinite tree.

FIGURE 2.2. Compact exhaustions.

LEMMA 2.13. *The (nice for ends) space X has a compact exhaustion.*

PROOF. As X is σ-compact, there exist compact subspaces L_1, L_2, ... of X such that $\cup_j L_j = X$. As X is locally compact, each L_j has a compact neighborhood N_i in X (meaning $L_j \subset N_i^\circ$). Define $K_i = \cup_{j \le i} N_j$, which is compact. For each $x \in X$, there exists j such that $x \in L_j$. So, $x \in K_i^\circ$ and (K_i) is a compact exhaustion of X. □

A compact exhaustion (K_i) of X describes X as a limit of the compact sets K_i. It may also be used to define an end of X as a limit of unbounded components of the complements $X - K_i$.

Fix a compact exhaustion (K_i) of X. Define $\mathcal{E}'(X)$ to be the set of all sequences (V_i) such that V_i is a connected (hence nonempty) component of $X - K_i$, and $V_{i+1} \subset V_i$.

LEMMA 2.14. *There is a one-to-one correspondence between $\mathcal{E}(X)$ and $\mathcal{E}'(X)$ that associates to each end ϵ the sequence $V_i = \epsilon(K_i)$.*

PROOF. If $\epsilon \in \mathcal{E}(X)$, then the sequence $V_i = \epsilon(K_i)$ is in fact in $\mathcal{E}'(X)$. On the other hand, given any sequence $(V_i) \in \mathcal{E}'(X)$, one may associate an end ϵ as follows. For each compact $L \subset X$, choose i such that $L \subset K_i$. Then, V_i is a connected component of $X - K_i$ and is contained in a unique connected component W of $X - L$. Define $\epsilon(L) = W$. One may verify that ϵ is well-defined, and that these associations are inverses. □

When a compact exhaustion (K_i) is given for X, we will often conflate an end of X with the sequence $(V_i) \in \mathcal{E}'(X)$. Some authors—for example Guilbault [**Gui16**, § 3.3]—take this sequence as the definition of an end of X. For alternative definitions, see Geoghegan [**Geo08**, § 13.4] and Aramayona and Vlamis [**AV20**]. As an example, consider $X = \mathbb{R}$ with the compact exhaustion $K_i = [-i, i]$. The only sequences in $\mathcal{E}'(X)$ are $V_i = (-\infty, i)$ and $V_i = (i, \infty)$. These represent the two ends of $\mathbb{R}$. Using compact exhaustions, we give a straightforward proof that each space with no ends is compact.

LEMMA 2.15. *Fix a compact exhaustion (K_i) of X. If $V_1 \supset V_2 \supset \cdots \supset V_n$ is a finite sequence where each V_i is an unbounded connected component of $X - K_i$, then this sequence extends to an infinite sequence (V_i) which represents an end of X.*

PROOF. It suffices to prove that if V_i is an unbounded connected component of $X - K_i$, then there exists an unbounded connected component of $X - K_{i+1}$ contained in V_i. Using this, any finite initial sequence may be extended inductively to an infinite sequence.

Let L be the union of K_{i+1} and all of the bounded components of $X - K_{i+1}$. By Lemma 2.7, L is compact. As V_i is unbounded, $V_i - L \neq \varnothing$. Let $y \in V_i - L$. Then y is contained in some component V_{i+1} of $X - K_{i+1}$. As $y \notin L$, this component must be unbounded. □

COROLLARY 2.16. *If $K \subset X$ is compact and V is an unbounded connected component of $X - K$, then there exists an end $\epsilon \in \mathcal{E}(X)$ such that $\epsilon(K) = V$.*

PROOF. Let (K_i) be any compact exhaustion of X such that $K_1 = \varnothing$, and let $K_i' = K_i \cup K$ for each i. Note that (K_i') is a compact exhaustion of X and $K_1' = K$. Apply Lemma 2.15 to the single-term initial sequence (V). □

COROLLARY 2.17. *If X is noncompact, then X has at least one end.*

PROOF. Apply Corollary 2.16 to $K = \varnothing$ and $V = X$. □

Compact exhaustions also provide an upper bound on the number of ends of a space.

LEMMA 2.18. *The number of ends of X is at most the continuum.*

PROOF. Fix a compact exhaustion (K_i) of X. There exists one end for each choice of sequence (V_i) where V_i is a connected component of $X - K_i$ and $V_i \supset V_{i+1}$. By Lemma 2.7, there exist only finitely many unbounded connected components of $X - K_i$. There are at most continuum many ways to pick one element from each of a countable number of finite sets. □

The set of ends of X admits a natural topology. The resulting *space of ends* is essential for the classification of noncompact surfaces.

DEFINITION 2.19. Let ϵ be an end of X, and let $K \subset X$ be compact. Let $\mathcal{N}(\epsilon; K)$ denote the set of all ends β of X such that $\epsilon(K) = \beta(K)$. We refer to $\mathcal{N}(\epsilon; K)$ as a **basic end-space neighborhood** of ϵ.

Equip $\mathcal{E}(X)$ with the topology generated by

$$\{\mathcal{N}(\epsilon; K) : \epsilon \in \mathcal{E}(X), K \subset X \text{ is compact}\}$$

Note that if $\mathcal{N}(\epsilon; K)$ and $\mathcal{N}(\eta; L)$ both contain an end α, then they both contain $\mathcal{N}(\alpha; K \cup L)$. Thus, this collection indeed forms a basis for a topology on $\mathcal{E}(X)$.

LEMMA 2.20. *The space of ends $\mathcal{E}(X)$ is Hausdorff, compact, separable, and totally disconnected.*

PROOF. Fix a compact exhaustion (K_i) of X. If ϵ, η are two distinct ends of X, then there exists an n such that $\epsilon(K_n) \neq \eta(K_n)$. $\mathcal{N}(\epsilon; K_n)$ and $\mathcal{N}(\eta; K_n)$ are two disjoint open sets containing ϵ and η, respectively. Thus, $\mathcal{E}(X)$ is Hausdorff. Furthermore, the complement of $\mathcal{N}(\epsilon; K_n)$ is the union of $\mathcal{N}(\rho; K_n)$ for all ends ρ for which $\rho(K_n) \neq \epsilon(K_n)$. Thus, the complement of $\mathcal{N}(\epsilon; K_n)$ is open, so $\mathcal{N}(\epsilon; K_n)$ is clopen. As any two ends are separated by clopen subsets, $\mathcal{E}(X)$ is totally disconnected.

For each fixed n, there are finitely many unbounded components of $X - K_n$, and so finitely many distinct sets $\mathcal{N}(\epsilon; K_n)$ for $\epsilon \in \mathcal{E}(X)$. Because sets of the form $\mathcal{N}(\epsilon; K_n)$ form a basis for the topology of $\mathcal{E}(X)$, and there are a finite number of these sets for each n, $\mathcal{E}(X)$ is second-countable, hence separable.

Lastly, we prove that $\mathcal{E}(X)$ is compact. Suppose for contradiction that $\mathcal{E}(X)$ has an open cover $\mathcal{C}$ with no finite subcover. Say that a subset of $\mathcal{E}(X)$ is not finitely coverable if no finite subset of $\mathcal{C}$ covers it. Suppose that $\{\epsilon : \epsilon(K_n) = V\}$ is not finitely coverable for some $V \subset X$. Then, let $W_1, W_2, \ldots, W_k$ be the unbounded connected components of $X - K_{n+1}$ contained in V. We can see that $\{\epsilon : \epsilon(K_n) = V\}$ is the union of $\{\epsilon : \epsilon(K_{n+1}) = W_i\}$ over all W_i. As $\{\epsilon : \epsilon(K_n) = V\}$ is not finitely coverable, there exists some j such that $\{\epsilon : \epsilon(K_{n+1}) = W_j\}$ is not finitely coverable. As $\mathcal{E}(X)$ is not finitely coverable, we obtain a sequence $W_1, W_2, \ldots$ where W_i is a connected component of $X - K_i$, $\{\epsilon : \epsilon(K_i) = W_i\}$ is not finitely coverable, and $W_i \supset W_{i+1}$. This determines an end, α, of X. The end α is in some open subset $U \in \mathcal{C}$, which contains a subset of the form $\mathcal{N}(\alpha; K_n)$ for some n. But, then, $\mathcal{N}(\alpha; K_n)$ would be finitely coverable, which we know is not the case. This is a contradiction, so $\mathcal{E}(X)$ is compact. □

As ends are defined using compact subspaces, the theory of ends naturally utilizes *proper maps*. Recall that a continuous function $f : X \to Y$ is **proper** provided the inverse image of each compactum is compact. Topological spaces and proper maps form a category that is well-suited to the study of ends. Given a proper map $f : X \to Y$, we may extend f to a (proper) map $\mathcal{E}(f) : \mathcal{E}(X) \to \mathcal{E}(Y)$. In other words, $\mathcal{E}(-)$ is a functor.

DEFINITION 2.21. Let $\mathcal{E}(f)(\epsilon)$ denote the unique end $\eta \in \mathcal{E}(Y)$ such that $f^{-1}(\eta(K)) \supset \epsilon(f^{-1}(K))$ for all compact $K \subset Y$.

It may be shown that $\mathcal{E}(f)$ is well-defined and continuous. For example, let $\mathbb{R}_+ = [0, \infty) \subset \mathbb{R}$. A **ray** in X is a proper embedding $r : \mathbb{R}_+ \to X$. There is a simple description of $\mathcal{E}(r)$. Let ϵ be the unique end of $\mathbb{R}_+$, and let $\eta = \mathcal{E}(r)(\epsilon) \in \mathcal{E}(X)$. Then, $\eta(K)$ is the unique component of $X - K$ which contains $r((a, \infty))$ for some $a \in \mathbb{R}_+$. In this case, we say that r **points to** η.

Although arguments are cleaner for connected spaces, the theory of ends applies to *disconnected* topological spaces that are otherwise nice for ends. This allows us to define the ends of any manifold, which we will use frequently when adding 1-handles at infinity.

DEFINITION 2.22. Let X be Hausdorff, locally compact, locally connected, and σ-compact, but not necessarily connected. An **end** of X is an end of a connected component of X.

When X is disconnected, we may still define a natural topology on $\mathcal{E}(X)$ as above. As a topological space, $\mathcal{E}(X)$ is the disjoint union of $\mathcal{E}(C)$ over all connected components C of X. If X has only finitely many connected components, then $\mathcal{E}(X)$ is still compact. If X has infinitely many components, then $\mathcal{E}(X)$ may be noncompact. Our focus is the effect of adding a single 1-handle at infinity to a manifold M. This operation involves either one or two components of M, and any other components may be safely ignored.

3. End Sum and 1-handles at Infinity

The basic idea of the end sum operation is to combine two noncompact manifolds of the same dimension along a proper ray in each. For example, the interior of a boundary sum of two manifolds with boundary is an end sum of their interiors. In general, end sum is more complicated than boundary sum as it applies to noncompact manifolds that are not necessarily the interior of any compact manifold, and it requires a choice of ray in each manifold. End sum also requires a choice of a tubular neighborhood map for each ray, though different choices yield isomorphic manifolds. End sum is a special case of the more general operation of attaching a 1-handle at infinity to a possibly disconnected manifold (see also [**CG19**, pp. 1303–1305]). We define these operations simultaneously for CAT=TOP, PL, and DIFF. For the remainder of this section, embeddings and homeomorphisms are assumed to be CAT.

Let M be a (possibly disconnected) n-manifold. A **ray** in M is a proper, locally flat embedding $r : \mathbb{R}_+ \to M$ with image in the manifold interior of M. We often conflate a ray with its image in M. Applying the end functor $\mathcal{E}$ to r (see Definition 2.21), we have that $\mathcal{E}(r)$ picks out a single end $\eta \in \mathcal{E}(M)$. In this case, we say that r **points to** η. Define a **tubular neighborhood map** of r to be a proper, locally flat embedding $\nu : [0, \infty) \times D^{n-1} \to M$ with image in the manifold interior of M and such that $\nu(x, 0) = r(x)$.

Let r and r' be disjoint rays in M, and let ν and ν' be tubular neighborhood maps of r and r' (respectively) with disjoint images. Let $U \subset M$ be the image

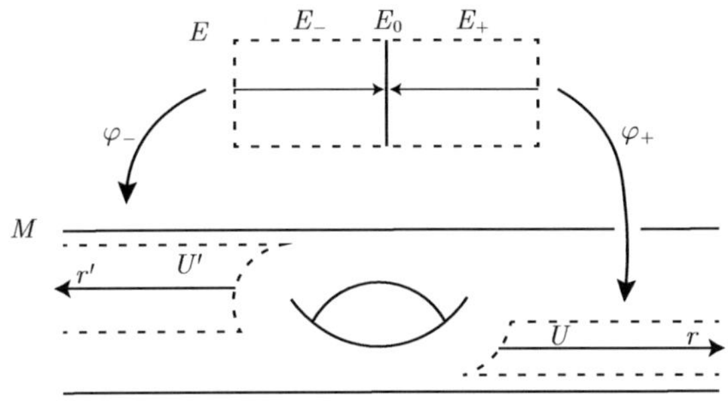

FIGURE 3.1. Data for the addition of a 1-handle at infinity to M.

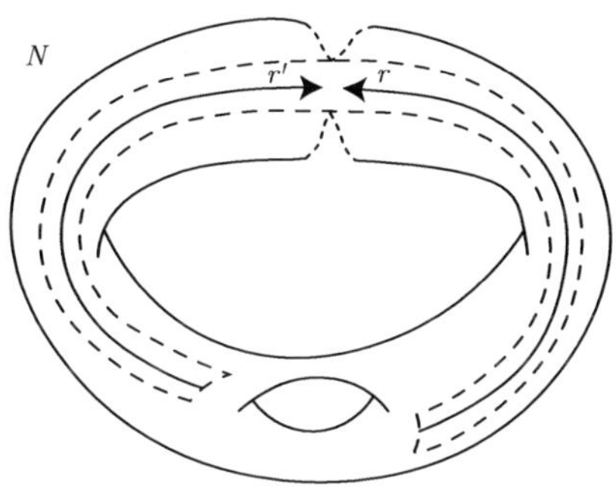

FIGURE 3.2. The result N of adding a 1-handle at infinity to M.

of ν restricted to the manifold interior of its domain. Similarly, define U' for ν'. As U and U' are locally flat embeddings of n-manifolds without boundary into the manifold interior of M, U and U' are open subsets of the manifold interior of M (hence, also of M).

Let $E = \mathbb{R} \times \mathbb{R}^{n-1}$, which is the 1-handle at infinity that will be attached to M. As in Figure 3.1, we partition E into $E_- = (-\infty, 0) \times \mathbb{R}^{n-1}$, $E_0 = \{0\} \times \mathbb{R}^{n-1}$, and $E_+ = (0, \infty) \times \mathbb{R}^{n-1}$. The restriction of ν to U gives a homeomorphism $(0, \infty) \times (D^{n-1})^\circ \cong U$ where $(D^{n-1})^\circ$ denotes the manifold interior of the disk. By reversing the first factor and radially expanding the second, we get a homeomorphism $\varphi_+ : E_+ \to U$. Similarly, the restriction of ν' to U' gives a homeomorphism $(0, \infty) \times (D^{n-1})^\circ \cong U'$. By reversing *and negating* the first factor and radially expanding the second, we get a homeomorphism $\varphi_- : E_- \to U'$.

We glue E to M using the maps φ_+ and φ_-. Explicitly, let $\sim$ be the equivalence relation on $M \sqcup E$ generated by $x \sim \varphi_+(x)$ for $x \in E_+$ and $x \sim \varphi_-(x)$ for $x \in E_-$. Let N be the quotient space of $M \sqcup E$ by this equivalence relation as in Figure 3.2. Let $q : M \sqcup E \to N$ be the associated quotient map. We call N a **result of attaching a 1-handle at infinity** to M. If $M = M_1 \sqcup M_2$ has exactly two components M_1 and M_2, each containing one of the rays r or r', then we call N an **end sum** of M_1 and M_2 (or simply an end sum of M). When M is an oriented manifold and φ_+ and φ_- are both orientation-preserving, we call this construction an **oriented 1-handle**.

Below, we prove that N is naturally an n-manifold (we are not aware of a proof of this fact in the literature). Define the composite maps $i_M : M \hookrightarrow M \sqcup E \xrightarrow{q} N$

and $i_E : E \hookrightarrow M \sqcup E \xrightarrow{q} N$. As the tubular neighborhood maps ν and ν' have disjoint images, the maps i_M and i_E are injective.

LEMMA 3.1. *The quotient map $q : M \sqcup E \to N$ is open.*

PROOF. The quotient map is open if and only if the saturation of any open set in $M \sqcup E$ with respect to $\sim$ is still open in $M \sqcup E$. Let V be open in $M \sqcup E$ and let $W = q^{-1}(q(V))$ be the saturation of V with respect to $\sim$. We have:

$$W = V \cup \varphi_+(V \cap E_+) \cup \varphi_+^{-1}(V \cap U) \cup \varphi_-(V \cap E_-) \cup \varphi_-^{-1}(V \cap U')$$

As φ_+ and φ_- are homeomorphisms between open subspaces of $M \sqcup E$, W is open. □

COROLLARY 3.2. *The maps $i_M : M \to N$ and $i_E : E \to N$ are open.*

PROOF. The inclusions $M \hookrightarrow M \sqcup E$ and $E \hookrightarrow M \sqcup E$ are open by the definition of the disjoint union topology. Now, apply Lemma 3.1. □

The following is the key step where properness of the rays and tubular neighborhood maps are required.

LEMMA 3.3. *The quotient space N is Hausdorff.*

PROOF. Let a and b be distinct points in N. If $a, b \in \operatorname{Im}\, i_M$, then $i_M^{-1}(a)$ and $i_M^{-1}(b)$ are separated by disjoint open sets A and B in M. The sets $i_M(A)$ and $i_M(B)$ are disjoint since i_M is injective, open in N by Corollary 3.2, and thus separate a and b. If $a, b \in \operatorname{Im}\, i_E$, then similarly $i_E^{-1}(a)$ and $i_E^{-1}(b)$ are separated by disjoint open sets in E whose images under i_E separate a and b. Lastly, consider the case where $a \in \operatorname{Im}\, i_M$ and $b \in \operatorname{Im}\, i_E$. We may assume $b \in i_E(E_0)$, as otherwise $b \in \operatorname{Im}\, i_M$. Let A be a bounded open subset of M containing a. As the tubular neighborhood maps ν and ν' are proper, there exists some $s > 0$ such that $\nu((s, \infty) \times D^{n-1})$ and $\nu'((s, \infty) \times D^{n-1})$ are disjoint from A. So, there exists some $t > 0$ such that $i_E((-t, t) \times \mathbb{R}^{n-1})$ is disjoint from $i_M(A)$. The sets $i_E((-t, t) \times \mathbb{R}^{n-1})$ and $i_M(A)$ are open by Corollary 3.2 and separate a and b. □

LEMMA 3.4. *The quotient space N is naturally an n-dimensional CAT manifold. If M is oriented and the 1-handle is oriented, then N is naturally an oriented manifold. The maps i_M and i_E are open embeddings, and if the 1-handle is oriented, then i_M and i_E are orientation-preserving.*

PROOF. The quotient space N is Hausdorff by Lemma 3.3. As $M \sqcup E$ is second-countable and the open quotient of a second-countable space is second-countable, N is also separable. Every point of N has a neighborhood homeomorphic to a subset of M or E, so in particular, N is locally Euclidean. Thus, N is a TOP manifold. If CAT=PL or DIFF, then N inherits a natural PL or DIFF structure using an atlas generated by the charts of M and E.

If in addition M is an oriented manifold and the 1-handle is oriented, then the atlas generated by the oriented charts of M and E defines an orientation of N.

Lastly, i_M and i_E are injective, open by Corollary 3.2, and thus are open embeddings. If in addition M is an oriented manifold and the 1-handle is oriented, then i_M and i_E respect the orientation of N. □

There is an equivalent operation to adding a 1-handle at infinity that is sometimes useful. In this alternative formulation, disjoint closed regular neighborhoods A and A' of r and r' are chosen in the interior of M. Then, N is defined by removing the manifold interiors of A and A' from M and then gluing together the resulting boundary components—both are copies of $\mathbb{R}^{n-1}$—by an orientation reversing homeomorphism. For more on this approach, see [**CKS12**, pp. 1813–1818].

4. Classification of Surfaces with Compact Boundary

The classification of noncompact surfaces with compact boundary is essential for our end sum uniqueness results. This contrasts with end sum uniqueness results in higher dimensions which rely on isotopy uniqueness of rays [**CG19**]. Manifolds in this section are assumed to be PL although the results hold for DIFF and TOP. Following Richards [**Ric60**], we define the genus and parity of any surface with compact boundary and extend those concepts to end invariants. Using those invariants, we present a classification theorem of surfaces with compact boundary.

4.1. Compact Surfaces. We begin with the well-known classification of compact surfaces. For a proof, see [**Mun00**, pp. 446–476]. Given a compact surface M, let $\chi(M)$ denote its Euler characteristic, and let $b(M)$ denote the number of boundary components of M.

THEOREM 4.1 (Classification of compact surfaces). *Let M be a compact, connected surface. If M is orientable, then for some non-negative integer g, M is isomorphic to a sphere with g handles and $b(M)$ holes, and $\chi(M) = 2 - 2g - b(M)$. If M is non-orientable, then for some positive integer k, M is isomorphic to the sphere with k cross-caps and $b(M)$ holes, and $\chi(M) = 2 - k - b(M)$.*

It will be more convenient for us to classify surfaces by their genera. We use the following convention (sometimes called the "generalized genus") to extend the concept of genus to any compact surface. Given a compact surface M, let $|\pi_0(M)|$ denote the number of connected components of M.

DEFINITION 4.2. Let M be a compact surface. The **genus** of M, denoted $g(M)$, is an integer or half-integer defined by the following.

$$g(M) := |\pi_0(M)| - \frac{b(M) + \chi(M)}{2}$$

For example, consider the surfaces described in Theorem 4.1; in the orientable case $g(M) = g$, and in the non-orientable case $g(M) = k/2$. Observe that the genus is unchanged if a surface is punctured (meaning an open disk is removed). Further, genus is additive over disjoint union since it is a linear combination of functions that are each additive over disjoint union. These properties make the genus a user-friendly alternative to the Euler characteristic. In a sense, the genus measures the complexity of a surface. This idea is made concrete by the following.

THEOREM 4.3. *If N is a compact surface, then $g(N) \geq 0$. If M is a subsurface of N, then $g(M) \leq g(N)$.*

PROOF. The classification of compact surfaces (Theorem 4.1) implies that every compact, connected surface has non-negative genus. As genus is additive over disjoint union, every compact surface has non-negative genus.

Now consider a compact surface N with a compact subsurface M. It will be convenient for M to avoid the boundary of N. So, we let N' be the result of gluing an external collar $\partial N \times [0,1]$ to N along ∂N. Note that N and N' are isomorphic and that M lies in the manifold interior of N'.

As M is a subsurface of N' and is disjoint from ∂N, we have $E = \overline{N' - M}$ is a subsurface of N'. Note that N' is obtained by gluing E and M together along their boundaries. More precisely, each boundary component of M is glued to a unique boundary component of E. We have

$$b(N') = b(E) - b(M)$$
$$\chi(N') = \chi(M) + \chi(E) - \chi(M \cap E) = \chi(M) + \chi(E)$$

Beginning with $|\pi_0(M)| + |\pi_0(E)|$ components of M and E, observe that each time a boundary component of M is glued to one of E, the total number of components is reduced by 0 or 1. Thus, we have

$$|\pi_0(N')| \geq |\pi_0(M)| + |\pi_0(E)| - b(M)$$

Therefore

$$\begin{aligned} g(N') &= |\pi_0(N')| - \frac{b(N') + \chi(N')}{2} \\ &\geq g(M) + g(E) \end{aligned}$$

As $g(N') = g(N)$ and $g(E) \geq 0$, we get $g(N) \geq g(M)$. $\square$

Define the **parity** of a compact surface M, denoted $\mathcal{P}(M)$, to be $2g(M) \pmod 2$. We call a compact surface **even** if its parity is zero and **odd** otherwise. Parity is strongly related to orientability by the following lemma.

LEMMA 4.4. *If N is a compact, orientable surface, then $\mathcal{P}(N) = 0$. If N is a compact surface and $M \subset N$ is a subsurface such that $\partial N \subset M$ and $N - M$ is orientable, then $\mathcal{P}(N) = \mathcal{P}(M)$*

PROOF. By the classification of compact surfaces (Theorem 4.1), it can be checked that every compact, connected, orientable surface has even parity. As the parity of a disconnected surface is the sum of the parities of its connected components, every compact, orientable surface has parity zero.

Now, consider a compact surface N with compact subsurface M such that $\partial N \subset M$ and $N - M$ is orientable. Let $E = \overline{N - M}$. Because $\partial N \subset M$, E is actually a (compact) subsurface of N. As $N - M$ is orientable, E is orientable. Thus, E has even parity.

We have

$$b(N) = b(M) - b(E)$$
$$\chi(N) = \chi(M) + \chi(E) - \chi(M \cap E) = \chi(M) + \chi(E)$$

Thus

$$\begin{aligned} g(N) - g(M) &= \left(|\pi_0(N)| - \frac{b(N) + \chi(N)}{2}\right) - \left(|\pi_0(M)| - \frac{b(M) + \chi(M)}{2}\right) \\ &= |\pi_0(N)| - |\pi_0(M)| + \frac{b(E)}{2} - \frac{\chi(E)}{2} \\ &= |\pi_0(N)| - |\pi_0(M)| - |\pi_0(E)| + b(E) + g(E) \end{aligned}$$

As $g(E)$ is an integer, $g(N) - g(M)$ is an integer. Hence, $\mathcal{P}(N) = \mathcal{P}(M)$. $\square$

4.2. Noncompact Surfaces with Compact Boundary. To extend our definitions from compact surfaces to noncompact surfaces, we will require the existence of arbitrarily large compact subsurfaces. The key theorem we need for these arguments is the following.

THEOREM 4.5. *Let M be a surface and let $K \subset M$ be compact. Then there exists a compact, locally flatly embedded subsurface $S \subset M$ such that $K \subset S$.*

PROOF. Fix a triangulation Σ of M. Let L be the union of all triangles in Σ that intersect K. Let Σ' be the barycentric subdivision of Σ and let S be the union of all triangles in Σ' that intersect L. It can be checked that S is a PL subsurface of M. □

As a consequence of Theorem 4.3, we see that for a *compact* surface N, $g(N)$ is the supremum of $g(M)$ over all compact subsurfaces M. This suggests a way to define the genus for arbitrary (noncompact) surfaces in terms of their compact subsurfaces. This generalized genus will play a vital role in our classification of noncompact surfaces with compact boundary.

DEFINITION 4.6. Let N be a surface with compact boundary. The **genus** of N, denoted $g(N)$, is the supremum of $g(M)$ over all compact subsurfaces $M \subset N$ (and may be infinite).

Definition 4.6 is not circular because the genus of an arbitrary surface is defined in terms of the genus of compact surfaces, which we have already defined. By Theorem 4.5, we know that there exist arbitrarily large compact subsurfaces of N, so $g(N)$ is actually the *limit* of $g(M)$ for compact subsurfaces M as M gets arbitrarily large. Similar to the compact case, we may think of the genus of a noncompact surface as a measure of complexity by the following extension of Theorem 4.3.

THEOREM 4.7. *If N is a surface, then $g(N) \geq 0$. If M is a subsurface of N, then $g(M) \leq g(N)$.*

Parity applies to noncompact surfaces with compact boundary provided the surface is orientable outside of a compact subset. When N is connected, this is equivalent to N having no non-orientable ends (see Definition 4.9). If N is orientable outside of a compact subset, then by Theorem 4.5, there exists a compact subsurface $M \subset N$ such that $\partial N \subset M$ and $N - M$ is orientable. Following Lemma 4.4, we would like to define $\mathcal{P}(N)$ to equal $\mathcal{P}(M)$.

To prove this is well-defined, consider two compact subsurfaces $M_1, M_2 \subset N$ such that $N - M_1$ and $N - M_2$ are both orientable. Let $S \subset N$ be a subsurface such that $M_1 \cup M_2 \subset S$. By Lemma 4.4, $\mathcal{P}(M_1) = \mathcal{P}(E)$ and $\mathcal{P}(M_2) = \mathcal{P}(E)$. Thus, $\mathcal{P}(M_1) = \mathcal{P}(M_2)$. As a consequence, the following is well-defined.

DEFINITION 4.8. Let N be a surface with compact boundary that is orientable outside of a compact subset. The **parity** of N, denoted $\mathcal{P}(N)$, is the parity of any compact subsurface $M \subset N$ such that $\partial M \subset N$ and $N - M$ is orientable.

When $g(N)$ is finite, we can see that $\mathcal{P}(N) \equiv 2g(N) \pmod 2$. However, the parity of a surface may exist even when the genus is infinite.

4.3. End-Invariants. To continue our exploration of the properties of noncompact surfaces, we will need to classify the ends of a surface. We will do this through the idea of an "end-invariant", a concept that can actually be applied to any topological space which is nice for ends.

Let X and X' be nice for ends. Define two ends $\eta \in \mathcal{E}(X)$ and $\eta' \in \mathcal{E}(X')$ to be **isomorphic** if there exist closed neighborhoods A of η and A' of η' and a homeomorphism $f : A \to A'$ such that $\mathcal{E}(f)(\eta) = \eta'$. It can be checked that isomorphism of ends is an equivalence relation. An **end-invariant** is any property or quantity associated to an end that is invariant under end isomorphism. One particularly important end-invariant is its orientability.

DEFINITION 4.9. An end η of a manifold M is **orientable** if η has a closed neighborhood A such that A° is orientable. Otherwise, we say that η is **non-orientable**.

Examples of orientable and non-orientable ends are shown in Figure 4.1. If η

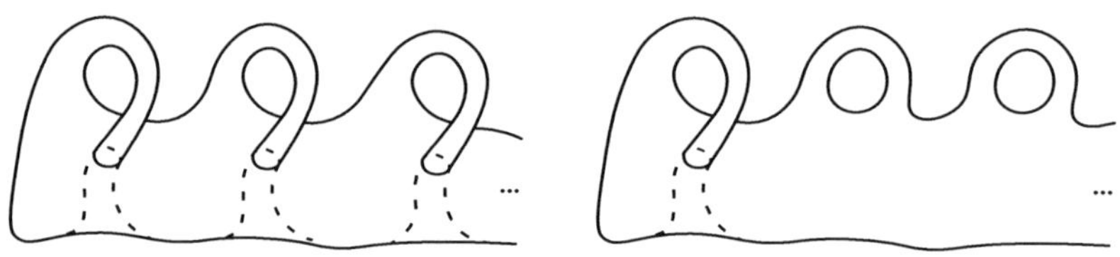

(a) Non-orientable end. (b) Orientable end.

FIGURE 4.1. One-ended, non-orientable surfaces.

has an orientable neighborhood $\eta(K)$ for some compact $K \subset M$, then every end in $\mathcal{N}(\eta; K)$ is also orientable. Thus, the orientable ends of M form an open subset of $\mathcal{E}(M)$. Let M be connected. If M is orientable outside of a compact set, then all of its ends are orientable. Conversely, suppose that every end of M is orientable. As every end of M has an orientable neighborhood and $\mathcal{E}(M)$ is compact, M is orientable outside of a compact set.

Another important end-invariant is the genus of an end.

DEFINITION 4.10. The **genus** of an end η of a surface M with compact boundary is defined as the infimum of $g(A^\circ)$ over all closed neighborhoods A of η.

For example, any end collared by S^1 has genus zero. If A and B are two neighborhoods of η with $A \subset B$, then $g(A^\circ) \leq g(B^\circ)$. Thus, the genus of an end can also be considered the limit of $g(U)$ for arbitrarily small neighborhoods U of η. In fact, this limit can only have two possible values.

LEMMA 4.11. *An end η of a surface with compact boundary M either has zero genus or infinite genus.*

PROOF. Let η be an end of M with finite genus $h > 0$. Let $K \subset M$ be compact such that $g(\eta(K)) = h$. By definition, there exists a compact subsurface $L \subset \eta(K)$ such that $g(L) = h$. Now, $g(\eta(K \cup L))$ must still be h, so there exists a compact subsurface $L' \subset \eta(K \cup L)$ such that $g(L') = h$. However, L and L' are disjoint, so $g(L \cup L') = g(L) + g(L') = 2h$. In addition, $L \cup L'$ is a subset of $\eta(K)$. Thus, $g(\eta(K)) \geq g(L \cup L') = 2h$. This is a contradiction. $\square$

Let M be a surface with compact boundary. If η is genus zero, then $g(\eta(K)) = 0$ for some compact $K \subset M$. Thus, every end in $\mathcal{N}(\eta; K)$ has genus zero. Thus, the genus zero ends of M form an open subset of $\mathcal{E}(M)$.

Let M be a connected surface with compact boundary. If M has finite genus, then every end of M has finite genus (hence zero genus). Conversely suppose that every end of M has finite genus. As every end of M has a genus zero neighborhood and $\mathcal{E}(M)$ is compact, M is genus zero outside of a compact set.

The genus and orientability of an end are related. If an end η of a surface has genus zero, then it has a genus zero neighborhood. As every genus zero surface is orientable, this neighborhood is also orientable. Thus, every genus zero end of a surface with compact boundary is also an orientable end.

Beyond these more geometric end-invariants, there are a whole host of algebraic end-invariants. The most important of these are the homotopy, homology, and cohomology groups at infinity [**Geo08**, pp. 229–281, 369–401]. These are incredibly important, and provide very powerful tools for distinguishing various ends, and by extension, distinguishing various manifolds. Like the respective invariants of algebraic topology (that is, the homotopy, homology, and cohomology groups), the algebraic end-invariants listed so far are all functorial. But whereas the homotopy, homology, and cohomology groups are invariant under homotopy, their end-invariant counterparts are invariant only under proper homotopy.

The cohomology groups at infinity have proven particularly useful in investigating the uniqueness of end sums. In forthcoming work, Calcut and Guilbault prove that the cohomology groups at infinity of an end sum are independent of which choices are made in the end sum. However, the ring structure of the cohomology at infinity is crucial in work thus far to distinguish between non-isomorphic end sums of two manifolds. In some classifications of noncompact surfaces, the cohomology ring at infinity plays a crucial role, carrying effectively the same information as the space of ends and the geometric invariants of the ends [**Gol71**].

4.4. The Classification Theorem. We may now state the classification of surfaces without boundary. For a proof, see Richards' thesis [**Ric60**].

THEOREM 4.12. *Let M and N be two connected surfaces without boundary with the same genus, the same orientability class, and homeomorphic end-space considered as a topological triplet (A, B, C) where A is the space of non-orientable ends, B is the space of infinite genus ends, and C is the space of ends. Then, M and N are isomorphic.*

We also have a natural extension to surfaces with compact boundary.

COROLLARY 4.13. *If M and N are two connected surfaces with compact boundary with the same invariants used in Theorem 4.12, and in addition $b(M) = b(N)$, then M and N are isomorphic.*

5. Main Theorem

Our main theorem is the following.

THEOREM 5.1. *Let M be a (possibly disconnected) surface with compact boundary, and let η and η' be distinct ends of M. Let N be a result of adding a 1-handle at infinity to M along η and η'. If η and η' are ends of distinct connected components* *of M, then N is unique up to isomorphism. If η and η' are ends of the same connected, non-orientable component M_1 of M, then N is unique up to isomorphism. Lastly, suppose that η and η' are ends of the same connected, orientable component M_1 of M. If the 1-handle is oriented, then N is unique up to isomorphism. If the 1-handle is not oriented, then N is unique up to isomorphism.*

This has the following immediate corollary for the end sum of surfaces.

COROLLARY 5.2. *Let M and M' be distinct connected surfaces with compact boundary. Let η be an end of M and let η' be an end of M'. Then, the end sum of M and M' along η and η' is uniquely determined up to isomorphism.*

The main tool we will use to prove Theorem 5.1 is the classification of surfaces with compact boundary (Corollary 4.13). That classification is based on the following characteristics of a connected surface A: boundary, orientability, genus, parity, space of ends, subspace of infinite genus ends, and subspace of non-orientable ends. In each of the next few subsections, we examine one or more of these characteristics and their behavior under the addition of a 1-handle at infinity.

CONVENTIONS 5.3. For the remainder of this paper, unless otherwise stated, we will use the following conventions. Some of these conventions are reused from Section 3.

- All manifolds are PL.
- Let M be an n-manifold, and let η and η' be ends of M (not necessarily distinct). Let r and r' be disjoint rays in M pointing to η and η' respectively. Let ν and $\nu' : \mathbb{R}_+ \times D^{n-1} \to M$ be tubular neighborhood maps of r and r' respectively with disjoint images.
- Let $E = \mathbb{R}^n$, and define the subspaces $E_- = (-\infty, 0) \times \mathbb{R}^{n-1}$, $E_0 = \{0\} \times \mathbb{R}^{n-1}$, and $E_+ = (0, \infty) \times \mathbb{R}^{n-1}$.
- Let N be the result of adding a 1-handle at infinity to M along the tubular neighborhood maps ν and ν' by attaching the strip E as described in Section 3. Let $\varphi_- : E_- \to M$ and $\varphi_+ : E_+ \to M$ be the gluing maps used in the construction of N. Let i_M be the natural embedding of M into N, and let i_E be the natural embedding of E into N.
- Let $R_m = [-m, m]^n$. We will use (R_m) as a compact exhaustion of E.
- (K_m) is a compact exhaustion of M as specified in Theorem 5.5.
- Let $L_m = i_M(K_m) \cup i_E(R_m)$. We will use (L_m) as a compact exhaustion of N.

In our study of the ends of M and N, it will be helpful to have exhaustions that are situated nicely with respect to both the rays r and r' and the tubular neighborhood maps ν and ν'.

LEMMA 5.4. *For every compact subset $K \subset M$, there exists a compact n-manifold $A \subset M$ such that:*

(i) $K \subset A^\circ$,
(ii) $\nu^{-1}(A) = [0, a] \times D^{n-1}$, *and*
(iii) $\nu'^{-1}(A) = [0, a'] \times D^{n-1}$ *for some $a, a' > 1$. This situation is presented in Figure* 5.1.

PROOF. Pick c such that $\nu^{-1}(K) \subset [0, c] \times D^{n-1}$. Similarly for c' and ν'. Set

$$L = K \cup \nu([0, c] \times D^{n-1}) \cup \nu'([0, c'] \times D^{n-1})$$

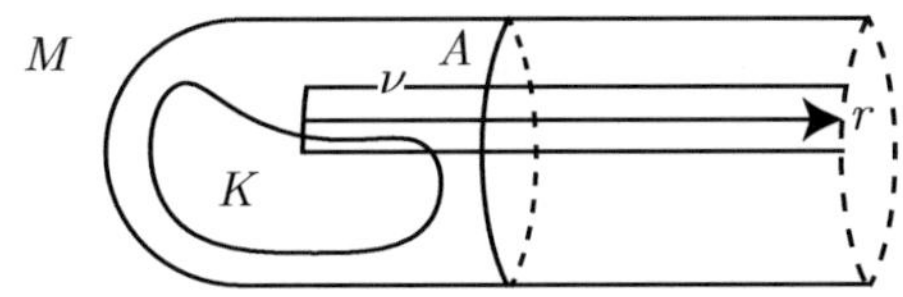

FIGURE 5.1. Surface M containing a compact set K, ray r, and tubular neighborhood ν of r. The compact subsurface A of M contains K in its interior and meets ν nicely.

Let T be a regular neighborhood of L in M (see Rourke and Sanderson [**RS72**, Ch. 3] and Scott [**Sco67**] for the PL theory of regular neighborhoods). Importantly for us, $K \subset T^\circ$ and T is a compact n-dimensional submanifold of M [**RS72**, p. 34].

Construct an ambient isomorphism ψ with support disjoint from L and Im ν' such that $\psi(\partial T)$ is transverse to ∂Im ν [**AZ67**, p. 185]. Replace T by $\psi(T)$. Note that T is still a regular neighborhood of L, but now intersects Im ν transversely. Similarly, modify T to be transverse to Im ν' as well.

Pick $d > 1$ such that $\nu^{-1}(T)$ is contained in $[0, d] \times D^{n-1}$. Similarly, find d'. Set

$$A = T \cup \nu([0, d] \times D^{n-1}) \cup \nu'([0, d'] \times D^{n-1})$$

It is clear that A is locally Euclidean at all points except $\partial T \cap \partial \text{Im } \nu$ and $\partial T \cap \partial \text{Im } \nu'$. By transversality, for every $p \in \partial T \cap \partial \text{Im } \nu$, there exists a neighborhood U of p and a coordinate map $\varphi : U \to \mathbb{R}^n$ where $\varphi(p) = 0$, $\varphi(T) = [0, \infty) \times \mathbb{R} \times \mathbb{R}^{n-2}$, and $\varphi(\text{Im } \nu) = \mathbb{R} \times [0, \infty) \times \mathbb{R}^{n-2}$. In this local coordinate map, it is clear that A is locally Euclidean at p. Thus, A is a n-dimensional submanifold of M. Since $K \subset T^\circ$, $K \subset A^\circ$. By construction, $\nu^{-1}(A) = [0, d] \times D^{n-1}$ and $\nu'^{-1}(A) = [0, d'] \times D^{n-1}$. □

By constructing larger and larger submanifolds using Lemma 5.4, we can create a nice compact exhaustion for M.

THEOREM 5.5. *Let $C \subset M$ be compact. There exists a compact exhaustion (K_m) of M such that*

(i) *$C \subset K_m$ for all m,*
(ii) *K_m is an n-manifold for all m,*
(iii) *$\nu^{-1}(K_m) = [0, a_m] \times D^{n-1}$ and $\nu'^{-1}(K_m) = [0, a'_m] \times D^{n-1}$ for some sequences $a_m, a'_m > 1$.*

PROOF. Start with any compact exhaustion (A_m) of M. Inductively build (K_m) as follows: set $K_0 = C$. For all $m > 0$, let K_m be a compact submanifold of M containing $A_m \cup K_{m-1}$ that has nice intersection with ν and ν', as given by Lemma 5.4. □

Using a compact exhaustion of M provided by Theorem 5.5, we may construct a nice compact exhaustion of N as well. Define $L_m = R_m \cup K_m$. Because K_m and R_m have transverse intersection, (L_m) is a compact exhaustion of N by n-manifolds. In this section, we use the compact exhaustions: (K_m) of M, (R_m) of E, and (L_m) of N. These exhaustions are depicted in Figure 5.2.

5.1. Space of Ends. In this section, we tackle the end-space of N. In the process, we will encounter ordinary and extraordinary ends of N, and the map $h : \mathcal{E}(N) \to \mathcal{E}(M) / (\eta \equiv \eta')$.

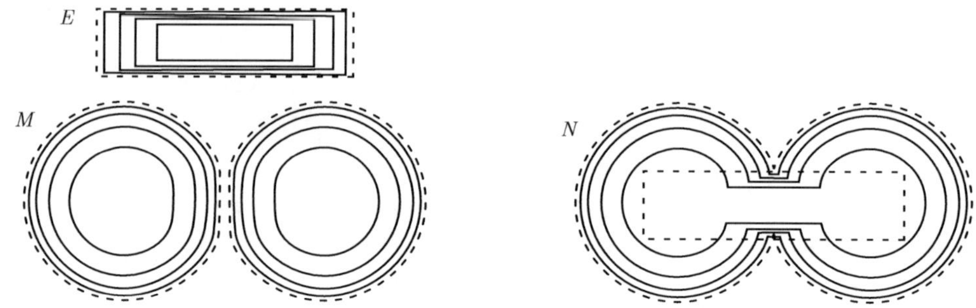

(a) Manifolds M and E with compact exhaustions (K_m) and (R_m) respectively

(b) End sum N with compact exhaustion (L_m). The image of E, $i_E(E)$, is dashed for clarity

FIGURE 5.2. Constructing the compact exhaustion (L_m) of the end sum N.

Call an end α of N **ordinary** provided that it has a neighborhood contained in $i_M(M)$, or equivalently, that it has a neighborhood disjoint from $i_E(E_0)$. Otherwise, call α **extraordinary**. An ordinary end is essentially not involved in the 1-handle construction, and is the simplest to describe.

LEMMA 5.6. *If α is an ordinary end of N, then $\alpha(i_M(K))$ is disjoint from $i_E(E_0)$ for sufficiently large compact $K \subset M$.*

PROOF. Because α is ordinary, $\alpha(L_m)$ is disjoint from $i_E(E_0)$ for some m. Consider $U_a = i_E((-a,a) \times \mathbb{R}^{n-1})$. For small enough a, $L_m \cap U_a = i_E((-a,a) \times [-m,m]^{n-1})$ by construction of L_m. If so, then all components of $U_a - L_m$ intersect $i_E(E_0)$. So, $\alpha(L_m)$ is disjoint from U_a, and therefore $\alpha(L_m)$ is a connected component of $N - (L_m - U_a)$. Thus, $\alpha(L_m - U_a) = \alpha(L_m)$ is disjoint from $i_E(E_0)$. □

LEMMA 5.7. *If α is an extraordinary end, then $\alpha(i_M(K))$ is the disjoint union $i_M(\eta(K)) \cup i_M(\eta'(K)) \cup i_E(E_0)$ for all compact $K \subset M$.*

PROOF. Since α is extraordinary, $\alpha(i_M(K))$ must not be disjoint from $i_E(E_0)$. Thus, $\alpha(i_M(K))$ contains $i_E(E_0)$. Every neighborhood of $i_E(E_0)$ intersects $i_M(\eta(K))$ and $i_M(\eta'(K))$. Furthermore, $i_M(\eta(K)) \cup i_M(\eta'(K)) \cup i_E(E_0)$ is an open set. Its complement in $N - i_M(K)$ is a union of open sets of the form $i_M(U)$, where U is a component of $M - K$. It follows that $i_M(\eta(K)) \cup i_M(\eta'(K)) \cup i_E(E_0)$ is a component of $N - i_M(K)$, hence equals $\alpha(i_M(K))$. □

LEMMA 5.8. *To every ordinary end α of N, we can assign a unique end β of M such that $i_M(\beta(K)) \subset \alpha(i_M(K))$ for all compact $K \subset M$. Furthermore, α and β are isomorphic as ends, and $\beta \neq \eta, \eta'$.*

PROOF. By Lemma 5.6, $\alpha(i_M(K_m))$ is disjoint from $i_E(E_0)$ for some m. For any compact $A \subset M$, consider $L_A = i_M(A \cup K_m)$. This is a compact subset of N, and $\alpha(L_A)$ is disjoint from $i_E(E_0)$. Define $\beta(A)$ to be the unique connected component of $M - A$ containing $i_M^{-1}(\alpha(L_A))$. Note that β is a well-defined end of M. So, $i_M(\beta(A))$ is the connected component of $i_M(M) - i_M(A)$ containing $\alpha(L_A)$, and $\alpha(i_M(A))$ is the connected component of $N - i_M(A)$ containing $\alpha(L_A)$. Thus, $i_M(\beta(A)) \subset \alpha(i_M(A))$.

Suppose β' is another end satisfying that condition. For large enough compact $K \subset M$, $\alpha(i_M(K))$ is disjoint from $i_E(E_0)$, so $i_M(\beta(K)) = \alpha(i_M(K)) = i_M(\beta'(K))$. Thus, $\beta' = \beta$.

For large enough compact $K \subset M$, $i_M(\beta(K)) = \alpha(i_M(K))$. Thus, α and β are isomorphic as ends. □

When α is an ordinary end of N, let $h(\alpha)$ denote the associated end of M, as given by Lemma 5.8. Note that with this definition, h is a function from the ordinary ends of N to $\mathcal{E}(M) - \{\eta, \eta'\}$. Extend h to a map from $\mathcal{E}(N)$ to $\mathcal{E}(M)/(\eta \equiv \eta')$ by setting $h(\alpha) = \eta \equiv \eta'$ whenever α is an extraordinary end of N. By $\mathcal{E}(M)/(\eta \equiv \eta')$, we mean the topological quotient of $\mathcal{E}(M)$ by the equivalence relation $\equiv$ generated by $\eta \equiv \eta'$. This quotient map is closed (since it is a continuous function from a compact space to a Hausdorff space), but is not necessarily open (consider two copies of the surface M above in Figure 1.1 and end sum along their nonisolated ends). It will be shown in Lemma 5.9 that in many cases, this h is also a homeomorphism, completely describing the end space of N.

LEMMA 5.9. *The canonical map $h : \mathcal{E}(N) \to \mathcal{E}(M)/(\eta \equiv \eta')$ is continuous. If $n \geq 2$, then h is surjective. If (i) $n = 2$, ∂M is compact, and $\eta \neq \eta'$, or (ii) $n \geq 3$, then h is a homeomorphism.*

PROOF. First, we prove that h is continuous. Let α be an end of N, and let $\beta = h(\alpha)$. Let U be an open neighborhood of β in $\mathcal{E}(M)/(\eta \equiv \eta')$. We will show that $h(V) \subset U$ for some open neighborhood V of α. If α is ordinary, then we may take $U = \mathcal{N}(\beta; K)$ for some compact $K \subset M$, and by Lemma 5.6, we can assume that $\alpha(i_M(K))$ is disjoint from $i_E(E_0)$. For every end $\tau \in \mathcal{N}(\alpha; i_M(K))$, $\tau(K)$ is disjoint from $i_E(E_0)$, so τ is ordinary. In fact, $h(\tau)(K) = \beta(K)$, so $h(\tau) \in U$. If α is extraordinary, then we may take $U = \mathcal{N}(\eta; K) \cup \mathcal{N}(\eta'; K)$ for some compact $K \subset M$. For every extraordinary end $\tau \in \mathcal{N}(\alpha; i_M(K))$, $h(\tau) = \beta \in U$. For every ordinary end $\tau \in \mathcal{N}(\alpha; i_M(K))$,

$$i_M(h(\tau)(K)) \subset \tau(i_M(K)) = \alpha(i_M(K)) = i_M(\eta(K)) \cup i_M(\eta'(K)) \cup i_E(E_0)$$

The last equality is true by Lemma 5.7. Thus, $h(\tau) = \eta(K)$ or $\eta'(K)$. So, $h(\tau) \in U$. Thus, h is continuous.

In the other direction, let $\beta \neq \eta, \eta'$ be an end of M. Pick compact $K \subset M$ such that $\beta(K) \neq \eta(K), \eta'(K)$. Note that $i_M(\beta(K))$ is also a connected component of $N - i_M(K)$. Given compact $L \subset N$, set $L' = K \cup (\beta(K) \cap i_M^{-1}(L))$. Note that L' is a compact subset of M. Define $\alpha(L)$ as the unique component of $N - L$ containing $i_M(\beta(L'))$. Then, α is the unique end of M such that $h(\alpha) = \beta$.

The map h is surjective provided there is at least one extraordinary end of N. Furthermore, as a continuous map between compact spaces, h is a homeomorphism as long as it is bijective, that is, as long as there is one extraordinary end of N. An end α of N is extraordinary provided that $\alpha(L)$ intersects $i_E(E_0)$ for all compact $L \subset N$. If $n \geq 2$, then $i_E(E_0)$ is noncompact, so it has at least one end ρ. Let $\alpha(L)$ be the unique component of $N - L$ containing $\rho(i_E(E_0) \cap L)$. Then α is an extraordinary end of N.

If $n \geq 3$, then $i_E(E_0) \cong \mathbb{R}^{n-1}$ has a single end ρ, and for any extraordinary end ζ, $\zeta(L) \supset \rho(i_E(E_0) \cap L)$. Thus, N has exactly one extraordinary end.

If $n = 2$, then $i_E(E_0) \cong \mathbb{R}$ has two ends, ρ_- and ρ_+. For any extraordinary end ζ of N, $\zeta(L)$ contains at least one of $\rho_-(i_E(E_0) \cap L)$ or $\rho_+(i_E(E_0) \cap L)$. Thus, N

has at most two extraordinary ends, and N has a single extraordinary end precisely when $\rho_-(i_E(E_0) \cap L)$ and $\rho_+(i_E(E_0) \cap L)$ are always contained in the same connected component of $N - L$.

Suppose that $n = 2$, $\eta \neq \eta'$, and ∂M is compact. It suffices to prove that $\rho_+(i_E(E_0) \cap L_m)$ and $\rho_-(i_E(E_0) \cap L_m)$ are connected in $N - L_m$ for large enough m. Starting at $\rho_+(i_E(E_0) \cap L_m)$, trace along the boundary of $i_E(R_m)$ until it intersects $i_M(K_m)$. Let T be the boundary component of K_m that we intersect. Since ∂M is compact, $\partial M \subset K_m$ for large enough m. Thus, staying within $N - L_m$, we can traverse $i_M(T)$ until we again intersect with $i_E(R_m)$. Traverse the boundary of $i_E(R_m)$ until we return to $\rho_+(i_E(E_0) \cap L_m)$ or arrive at $\rho_-(i_E(E_0) \cap L_m)$. If we return to $\rho_+(i_E(E_0) \cap L_m)$, then T must be a boundary component of both $\eta(K_m)$ and $\eta'(K_m)$. Since $\eta \neq \eta'$, $\eta(K_m) \neq \eta'(K_m)$ for large enough m, in which case this situation is impossible. We have traced a path from $\rho_+(i_E(E_0) \cap L_m)$ to $\rho_-(i_E(E_0) \cap L_m)$ within $N - L_m$. Thus, they are connected in $N - L_m$, and N has exactly one extraordinary end. □

In the future, whenever there is a unique extraordinary end of N, we will call it ζ. While the ordinary ends of N are isomorphic to their corresponding ends in M and easy to classify, understanding ζ takes more effort.

5.2. Boundary. In this section, we briefly consider the boundary of N.

THEOREM 5.10. *There is a canonical isomorphism $\partial N \cong \partial M$.*

PROOF. As i_M and i_E are open embeddings, $\partial N = i_M(\partial M) \cup i_E(\partial E)$. As E has no boundary components, $\partial N = i_M(\partial M)$. □

5.3. Orientability. In this section, we tackle the orientability of N. The orientability of N depends on the orientability of M, which ends are used for the 1-handle, and the orientation of the 1-handle. For examples, see Figures 1.3 and 1.4 in Section 1 above. To prove that N is orientable, it suffices to exhibit an orientation on N. To prove that N is non-orientable, we will use the notion of an orientation-reversing loop.

DEFINITION 5.11. A loop γ in a manifold M is **orientation-preserving** if γ lifts to a loop in the oriented double-cover of M. Define two paths Γ_1 and Γ_2 in M with the same endpoints to have the **same effect on orientation** if the concatenation of Γ_1 by the reverse of Γ_2 is an orientation-preserving loop.

For a description of the oriented double-cover of a manifold, see [**Hat02**, pp. 233–235]. As a homotopy of a loop in the base space lifts to a homotopy in the covering space, two homotopic loops are either both orientation-preserving or both orientation-reversing. Importantly for us, every loop in M is orientation-preserving if and only if M is orientable.

THEOREM 5.12. *If M is connected, then N is orientable if and only if M is orientable and the 1-handle is oriented. If $M = M_1 \cup M_2$ where each M_i is a connected component of M, η is an end of M_1 and η' is an end of M_2, then N is orientable if and only if both M_1 and M_2 are orientable.*

PROOF. First consider the case where M is connected. Suppose that M is non-orientable. As $i_M(M)$ is an open subset of N, N is also non-orientable. Next, suppose that M is orientable. Fix an orientation of M. If the 1-handle is oriented, then by Lemma 3.4, N is orientable. If the 1-handle is not oriented, then we may assume without loss of generality that φ_+ is orientation-preserving, and φ_- is orientation-reversing. Fix $a \in i_E(E_-)$ and $b \in i_E(E_+)$. Choose a path within $i_E(E)$ from a to b and a path within $i_M(M)$ from b to a. Concatenating these paths gives an orientation-reversing loop. Thus, N is non-orientable.

Second, consider the case where $M = M_1 \cup M_2$ where each M_i is a connected component of M, η is an end of M_1, and η' is an end of M_2. Suppose that M_1 is non-orientable. As $i_M(M_1)$ is an open subset of N, N is also non-orientable. Similarly, if M_2 is non-orientable, then so is N. Next, suppose that both M_1 and M_2 are orientable. By reversing the orientations on M_1 and M_2 if needed, we may assume that the 1-handle is oriented. By Lemma 3.4, N is orientable. □

5.4. Orientability of the Ends. Thus far in Section 5, we have given general results on 1-handles and end-sum for n-manifolds. We now narrow our focus to the conditions of Theorem 5.1. From here on, we assume that M is a surface with compact boundary and $\eta \neq \eta'$. In particular, Lemma 5.9 implies that N has a unique extraordinary end ζ.

In this subsection, we study the orientability of the ends of N. Recall that we define an end of M to be orientable provided that it has an orientable neighborhood, and the set of orientable ends is open in the space of all ends.

THEOREM 5.13. *For every ordinary end α of N, $h(\alpha)$ has the same orientability as α. The extraordinary end ζ of N is orientable if and only if both η and η' are orientable.*

PROOF. From Theorem 5.8, we know that for every ordinary end α of N, $h(\alpha)$ is isomorphic to α as ends. Hence, $h(\alpha)$ and α have the same orientability.

Suppose first that η is non-orientable. Let m be large enough so that $\eta(K_m) \neq \eta'(K_m)$. Let Γ_m be an orientation-reversing loop in $\eta(K_m)$. From the construction of R_m and K_m, we know that K_m splits R_m in two. We may homotope any segment of Γ_m that intersects (but does not cross) R_m to one that is disjoint from R_m. We may replace any segment of Γ_m that crosses R_m by a path that follows ∂R_m until it meets ∂K_m, then follows ∂K_m until it meets ∂R_m on the other side, then follows back up ∂R_m. Note that this path has the same effect on orientation as the original segment of Γ_m. The modified loop created in this way will still be orientation-reversing, but will also be disjoint from $\varphi_+(R_m)$. Thus, $\eta(K_m) - \varphi_+(R_m)$ is non-orientable. Thus, $\zeta(L_m)$ is non-orientable. Since this is true for sufficiently large m, ζ is non-orientable.

Now suppose that both η and η' are orientable. We can pick m large enough that $\eta(K_m)$, $\eta'(K_m)$ are distinct and orientable. Choose orientations on $\eta(K_m)$ and $\eta'(K_m)$ that are compatible with the 1-handle. These orientations define an orientation on $\zeta(i_M(K_m))$. Thus, ζ is orientable. □

5.5. Genus-Related Properties. Lastly, we study the genus-related properties of N. Specifically, we determine the genus of N, the parity of N (if all ends are orientable), and which ends of N have infinite genus.

5.5.1. *Genus and Parity.* Recall that we use $g(M)$ to denote the genus of M and $\mathcal{P}(M)$ to denote the parity of M (as an element of $\mathbb{Z}/2\mathbb{Z}$).

THEOREM 5.14. *If η and η' lie in distinct components of M, then $g(N) = g(M)$. If η and η' lie in the same component of M, then $g(N) = g(M) + 1$. If M is* *orientable outside of a compact set, then N is orientable outside of a compact set and $\mathcal{P}(N) = \mathcal{P}(M)$.*

PROOF. We may assume that N is connected. If M is orientable outside of a compact set, then M has no non-orientable ends and Theorem 5.12 implies that N also has no non-orientable ends. Thus, the parity $\mathcal{P}(N)$ is well-defined and equals the limit of $\mathcal{P}(L_m)$ as $m \to \infty$. On the other hand, the genus $g(M)$ is always well-defined (possibly infinite) and equals the limit of $g(L_m)$ as $m \to \infty$. We will compute the genus of L_m and use that to compute the genus and parity of N.

Note that R_m meets the boundary of K_m in two intervals. Since $\eta \neq \eta'$ and ∂M is compact, these intervals lie in distinct components of ∂K_m for all sufficiently large m. If so, then $b(L_m) = b(K_m) - 1$. For the Euler characteristic, we have $\chi(L_m) = \chi(K_m) + \chi(R_m) - \chi(K_m \cap R_m) = \chi(K_m) - 1$.

If M is connected, then for all sufficiently large m, both components of $K_m \cap R_m$ lie in the same connected component of K_m and so $|\pi_0(L_m)| = |\pi_0(K_m)|$. Thus

$$\begin{aligned} g(L_m) &= |\pi_0(L_m)| - \frac{b(L_m) + \chi(L_m)}{2} \\ &= |\pi_0(K_m)| - \frac{b(K_m) + \chi(K_m)}{2} + 1 \\ &= g(K_m) + 1 \end{aligned}$$

In this case, $g(N) = g(M) + 1$ and $\mathcal{P}(N) = \mathcal{P}(M)$ (if applicable).

If M is disconnected, then one component of $K_m \cap R_m$ lies in each connected component of M and so $|\pi_0(L_m)| = |\pi_0(K_m)| - 1$. Thus

$$\begin{aligned} g(L_m) &= |\pi_0(L_m)| - \frac{b(L_m) + \chi(L_m)}{2} \\ &= |\pi_0(K_m)| - 1 - \frac{b(K_m) + \chi(K_m)}{2} + 1 \\ &= g(K_m) \end{aligned}$$

In this case, $g(N) = g(M)$ and $\mathcal{P}(N) = \mathcal{P}(M)$ (if applicable). □

5.5.2. *Genus of the Ends.* Recall that the genus of an end $\tau \in \mathcal{E}(M)$ is the limit of $g(K_n)$ as $n \to \infty$ for any compact exhaustion K_n of M. The genus of an end is either zero or infinity.

THEOREM 5.15. *If α is an ordinary end of N, then α has the same genus as $h(\alpha)$. Furthermore, ζ has infinite genus if and only if either η or η' has infinite genus.*

PROOF. To simplify the argument, assume without loss of generality that $\eta(K_m) \neq \eta'(K_m)$ for all m. If α is an ordinary end of N, then α and $h(\alpha)$ are isomorphic as ends. Thus, α has the same genus as $h(\alpha)$.

Next, consider the extraordinary end ζ. To compute $g(\zeta)$, we introduce some subsurfaces of N as depicted in Figure 5.3.

Here, $A_m = \overline{\zeta(K_m)}$, $B_m = \overline{\zeta(L_m)}$, and A_m is the union of B_m and a 2-disk $C_m \subset R_m$. As A_m and B_m are connected, we have $|\pi_0(A_m)| = |\pi_0(B_m)| = 1$. Gluing C_m onto B_m separates one boundary component of B_m into two, so $b(A_m) = b(B_m) + 1$.

$$\chi(A_m) = \chi(B_m) + \chi(C_m) - \chi(B_m \cap C_m) = \chi(B_m) - 1$$

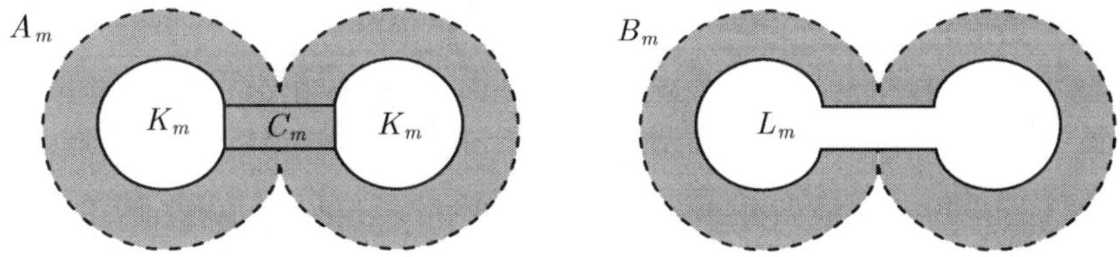

FIGURE 5.3. Subsurfaces of N: A_m (shaded at left) and B_m (shaded at right). Here, $C_m \subset R_m$ is a 2-disk, $A_m = B_m \cup C_m$, and the unshaded dumbbell (at right) is L_m.

It follows that $g(A_m) = g(B_m)$.

Note that A_m is an end sum of $\overline{\eta(K_m)}$ and $\overline{\eta'(K_m)}$. By Theorem 5.14, $g(A_m) = g\left(\overline{\eta(K_m)}\right) + g\left(\overline{\eta'(K_m)}\right)$. In the limit as $m \to \infty$, $g(\eta(K_m))$ converges to $g(\eta)$, $g(\eta'(K_m))$ converges to $g(\eta')$, and $g(B_m)$ converges to $g(\zeta)$. Thus, $g(\zeta) = g(\eta) + g(\eta')$. In other words, ζ has infinite genus if and only if one of η or η' does. $\square$

5.6. Summary. Let M be a PL surface with compact boundary. Let N be a result of adding a 1-handle at infinity to M along distinct ends η and η'. The surface M has connected components M_i. If a component M_i does not contain η or η', then M_i is unchanged by the addition of the 1-handle. So, it suffices to consider only the component of N which contains the 1-handle. From now on, assume N is connected.

By Lemma 5.9, there is a canonical homeomorphism $h : \mathcal{E}(N) \cong \mathcal{E}(M)/(\eta \equiv \eta')$. For all ordinary ends α of N, α and $h(\alpha)$ are isomorphic as ends, so they have the same invariants. The single extraordinary end ζ of N is orientable if and only if both η and η' are orientable; ζ has zero genus if and only if both η and η' have zero genus.

If η and η' lie in distinct components of M, then N is orientable if and only if that M is orientable. Otherwise, N is orientable if and only if M is orientable and the 1-handle is oriented.

If η and η' lie in distinct components of M, then $g(N) = g(M)$. Otherwise, $g(N) = g(M) + 1$. In either case, $\mathcal{P}(N) = \mathcal{P}(M)$.

We conclude: if M is non-orientable, then N is unique up to PL isomorphism; if M is orientable and the 1-handle is oriented, then N is unique up to PL isomorphism; and if M is non-orientable and the 1-handle is not oriented, then N is unique up to PL isomorphism. That completes the proof of Theorem 5.1 in PL. The TOP and DIFF versions of Theorem 5.1 are proved below in Section 7.

6. Main Theorem by the Uniqueness of Rays

In this section, we prove a ray uniqueness result for surfaces. Namely, that all rays within a surface with compact boundary that point to a given end are related by a global isomorphism of the surface. Using that result, we give an alternate and relatively straightforward proof of our main theorem. The authors thank Ric Ancel for suggesting this strategy.

6.1. Classification of Surfaces with Noncompact Boundary. Our first objective is to study and classify all rays r in a given surface with compact boundary up to global isomorphism. We will do that indirectly by removing the interior of

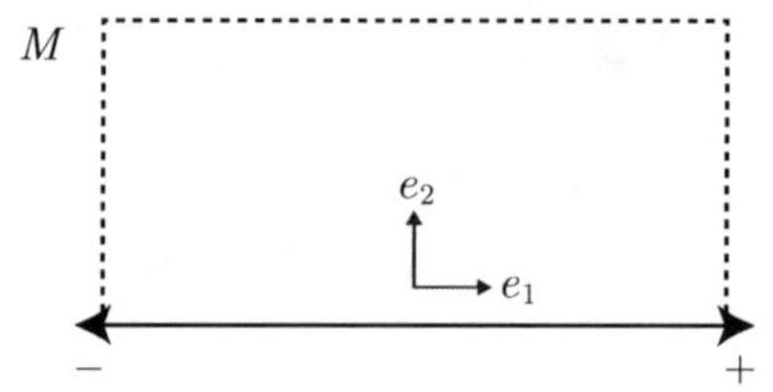

FIGURE 6.1. Surface $M = \mathbb{R}^2_+$ with one noncompact boundary component having a positive end + and a negative end −.

a closed regular neighborhood of r to produce a surface with a single noncompact boundary component, and then classify the resulting surface. We use Brown and Messer's classification of surfaces with possibly noncompact boundary [**BM79**]. We recount the invariants in their classification using our notation.

Let M be a connected surface. As in the case of surfaces with compact boundary, the following "global invariants" are essential: orientability, genus, parity, and compact boundary components. These invariants are defined exactly as in the compact boundary case, although now the number of compact boundary components may be infinite.

The rest of the invariants required to classify M deal with the ends of M and how those ends interact with its boundary components. As in the case of compact boundary, and end may be zero genus or infinite genus, and it may be orientable or non-orientable.

In surfaces with infinitely many compact boundary components, boundary components may accumulate in certain ends. Define an end of M to be **without compact boundary** provided that a neighborhood of that end contains no compact boundary components of M. It can be shown that M has finitely many compact boundary components provided that every end of M is without compact boundary. The surfaces we are interested in have finitely many boundary components, and thus have no ends with compact boundary.

Let $\mathrm{d}M$ denote the union of all noncompact boundary components of M. The inclusion map $\mathrm{d}M \to M$ induces a map $v : \mathcal{E}(\mathrm{d}M) \to \mathcal{E}(M)$. Each connected component B of $\mathrm{d}M$ is a copy of $\mathbb{R}$ and has two ends, say α and β. The map $e : \mathcal{E}(\mathrm{d}M) \to \pi_0(\mathrm{d}M)$ is defined by sending α and β to B.

We adopt the usual outward normal first convention for orienting the boundary of an oriented manifold. For example, Figure 6.1 shows $M = \mathbb{R}^2_+$ equipped with the standard orientation corresponding to the ordered basis (e_1, e_2). The **positive end** of the noncompact boundary component of M is the end labeled $+$. The **negative end** of the noncompact boundary component of M is the end labeled $-$. The same conventions and terminology apply to ends of noncompact boundary components of general surfaces in which some disjoint neighborhoods of ends are oriented.

Following Brown and Messer [**BM79**, pp. 379–381], we define an **orientation of** $\mathcal{E}(\mathrm{d}M)$ to be a subset $\mathcal{P} \subset \mathcal{E}(\mathrm{d}M)$ as follows. Given a compact subsurface $C \subset M$, a **complementary domain of** C is the closure of one of the components of $M - C$.

- If M is orientable, then fix an orientation on M. Then, $\mathcal{P}$ consists of the resulting positive ends of $\mathrm{d}M$.

- If M is non-orientable, then there exists a (nonunique) sequence $Q_1, Q_2, \ldots$ of disjoint orientable complementary domains of compact subsurfaces of M such that every orientable end of M has some (unique) Q_i as a neighborhood. Choose an orientation on each Q_i. Then, $\mathcal{P}$ consists of the resulting positive ends of $\mathrm{d}M$.

It is important to note that there may be many valid orientations of $\mathcal{E}(\mathrm{d}M)$ for a given surface M. All of that end-related data can be combined to form a **diagram** as in (6.1). Here, $\mathcal{K}$ denotes the set of non-orientable ends, $\mathcal{H}$ denotes the set of non-planar ends, and $\mathcal{S}$ denotes the set of ends with compact boundary.

$$\begin{array}{ccccccccc} \pi_0(\mathrm{d}M) & \xleftarrow{e} & \mathcal{E}(\mathrm{d}M) & \xrightarrow{v} & \mathcal{E}(M) & \longleftarrow & \mathcal{H} & \longleftarrow & \mathcal{K} \\ & & \uparrow & & \uparrow & & & & \\ & & \mathcal{P} & & \mathcal{S} & & & & \end{array} \tag{6.1}$$

Two diagrams Δ of M and Δ' of M' are **isomorphic** if there are bijective maps from each object in Δ to the corresponding object in Δ' which commute with the arrows of each diagram. The map from $\mathcal{E}(M)$ to $\mathcal{E}(M')$ must also be a topological homeomorphism. It is important to note that, as $\mathcal{P}$ is not uniquely defined, it is possible for one surface to have multiple non-isomorphic diagrams. Written in full, the classification theorem [**BM79**, p. 388] is as follows.

THEOREM 6.1 (Classification of Surfaces). *Let M and M' be two connected surfaces. If M and M' have the same genus, parity, orientability, and number of compact boundary components, and there exist diagrams Δ of M and Δ' of M' such that $\Delta \cong \Delta'$, then M and M' are isomorphic.*

We will need the following strengthening of Theorem 6.1.

THEOREM 6.2 (Addendum to the Classification of Surfaces). *Suppose furthermore that we are given an isomorphism $h : \Delta \to \Delta'$. Then, there is an isomorphism $\psi : M \to M'$ which induces the given isomorphism h of diagrams. If M and M' are oriented surfaces, and their orientations agree with the orientations of Δ and Δ' respectively, then ψ may be chosen to be orientation-preserving.*

The Addendum follows by Brown and Messer's proof of Theorem 6.1. Their proof [**BM79**, p. 388–389] is iterative and begins with the empty function f_0. Given a homeomorphism $f_k : C_k \to C'_k$ (with certain properties) of compact subsurfaces of M and M' respectively, they extend f_k to a homeomorphism $f_{k+1} : C_{k+1} \to C'_{k+1}$ of larger compact subsurfaces. If M and M' are oriented, then the first nonempty function f_1 may be chosen to respect orientation. In general, their proof imposes compatibility conditions between the constructed homeomorphisms and isomorphisms of diagrams (see [**BM79**, pp. 383–388], especially Lemma 2.1, the paragraph on p. 384 before the proof of Lemma 2.1, and the proof of Lemma 2.1). The present authors found it instructive to run their proof on various examples to gain familiarity with their orientation and compatibility conventions.

In the remainder of this subsection, we present a useful lemma and some applications of the Addendum to demonstrate its utility. In the next subsection, we prove a ray unknotting theorem (Theorem 6.10) for rays in certain surfaces. Recall from Section 2 that a precise definition was given for a proper ray r to **point to** an end η of M. We now expand this definition. Say an end τ of a (noncompact) boundary component B of M **points to** and end η of M if $\mathcal{E}(i)(\tau) = \eta$, where

where $i : B \to M$ is the inclusion map. Equivalently, τ points to η if every ray in B that points to τ also points to η.

LEMMA 6.3. *Let M be a connected surface and let η be an end of M. If M has finitely many noncompact boundary components, then the number of ends of noncompact boundary components of M that point to η is even. In other words, if $|\pi_0(\mathrm{d}M)|$ is finite, then $\left|v^{-1}(\eta)\right|$ is even. In particular, if M has exactly one noncompact boundary component B, then both ends of B point to the same end of M.*

PROOF. Let $\eta, \tau_1, \ldots, \tau_k$ be all the ends of M pointed to by ends of $\mathrm{d}M$. Choose a large compact subsurface $K \subset M$ such that $\overline{M - K}$ is also a subsurface of M, and $\eta(K) \neq \tau_i(K)$ for any i. Consider $\overline{\eta(K)}$. This is a subsurface of M, and the ends of noncompact boundary components of $\overline{\eta(K)}$ pointing to η are in one-to-one correspondence with ends of noncompact boundary components of M pointing to η. Since every end of a noncompact boundary component of $\overline{\eta(K)}$ points to η and there are an even number of ends of noncompact boundary components of $\overline{\eta(K)}$, then an even number of ends of noncompact boundary components of M point to η. That proves the first conclusion. The second conclusion follows immediately from the first. $\square$

REMARK 6.4. The first conclusion of Lemma 6.3 also holds provided η is an isolated end of M and at most finitely many noncompact boundary components of M point to η. The proof is similar. The conclusions of Lemma 6.3 are false without some restrictions on M or η. Consider the surface M depicted in Figure 6.2 that is obtained from the closed disk by removing a sequence of boundary points and the single limit point of that sequence. The sequence converges to the limit point

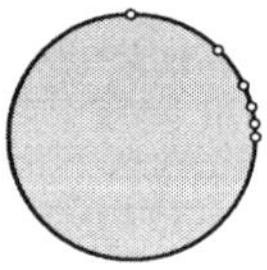

FIGURE 6.2. Surface M obtained from the closed disk by removing a sequence of boundary points and the single limit point of that sequence.

from one side. The single nonisolated end of M is denoted η. That end η of M is pointed to by exactly one end of a noncompact boundary component of M.

COROLLARY 6.5. *Let M be a connected surface with exactly one noncompact boundary component B. Then, there exists an automorphism $\psi : M \to M$ that induces the identity on $\mathcal{E}(M)$ and interchanges the two ends of B.*

PROOF. By the previous Lemma 6.3, both ends of B point to the same end η of M. First, consider the case where M is orientable. Fix an orientation on M and consider the corresponding diagram Δ of M. Let Δ' be the diagram for M with the opposite orientation. We have an isomorphism of diagrams $h : \Delta \to \Delta'$ that swaps the ends of B and is otherwise the identity. The desired conclusion now follows by the Addendum (Theorem 6.2). Second, consider the case where M is not orientable, but η is orientable. Then the proof from the first case also applies. Third, consider the case where η is not orientable. Note that Δ has empty orientation. We have an isomorphism $h : \Delta \to \Delta$ that swaps the ends of B, and the conclusion follows by the Addendum. □

REMARK 6.6. If M has more than one noncompact boundary component, then it may not be possible to interchange the ends of a given noncompact boundary component B of M by an automorphism of M. Consider the strip $[0, 1] \times \mathbb{R}$ and connect sum a sequence of projective planes or tori off to one end. A planar example is in Figure 6.2 above (see also Dickmann [**Dic23**, Ex. 3.5]).

A curious question arises: which oriented surfaces admit an orientation reversing automorphism? The classification of compact, connected, oriented surfaces (see Theorem 4.1 in Section 4) implies that each such surface admits an orientation reversing automorphism. (The question has also been studied for closed manifolds in higher dimensions by Müllner [**Mül10**].) For noncompact surfaces, the situation is more complicated. We give a positive result as well as two surfaces not admitting an orientation reversing automorphism.

COROLLARY 6.7. *Let M be a connected, oriented, noncompact surface with zero or one noncompact boundary component(s). Then M admits an orientation reversing automorphism.*

PROOF. First, consider the case where M has zero noncompact boundary components. Let M' be M with the opposite orientation. By the Addendum (Theorem 6.2), there is an orientation-preserving isomorphism from M to M'. That is an orientation-reversing automorphism of M. Second, consider the case where M has one noncompact boundary component B. By the previous corollary, there is an automorphism $\psi : M \to M$ that interchanges the ends of B. Evidently, ψ is locally orientation reversing near B, and so ψ is globally orientation reversing. □

REMARK 6.8. Figure 6.3 depicts two oriented surfaces that do not admit an orientation reversing automorphism. The surface M is obtained from the closed

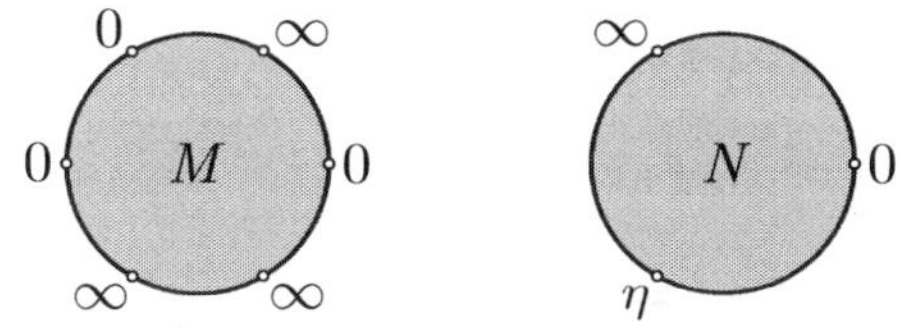

FIGURE 6.3. Surfaces M and N that do not admit orientation reversing automorphisms.

disk (with its standard orientation) by removing six boundary points and connect summing a sequence of tori at each end marked ∞. So, M has six ends: three of genus zero and three of infinite genus. The cyclic order of the genera of the ends of M prevent the existence of an orientation reversing automorphism of M. The surface N is obtained from the disjoint union of the closed disk and closed upper half space as follows. Remove three boundary points from the closed disk, connect sum a sequence of tori at the end marked ∞, then glue in a sequence of tubes between the end marked η and the end of the copy of closed upper half space. The resulting surface has three ends: one of genus zero, one of infinite genus, and η. The first two ends each have two ends of noncompact boundary components pointing to them, while η is pointed to by four ends of noncompact boundary components. Those properties prevent the existence of an orientation reversing automorphism of N. It seems interesting to ask whether there is a reasonable classification of connected, oriented surfaces that admit an orientation reversing automorphism.

6.2. Uniqueness of Rays in a Surface. Although we will use a more technical result to prove the main theorem, it is worth reporting Theorem 6.10, which gives a clean result on the uniqueness of rays in a surface. Many of the ideas in this proof will be used in Section 6.3.

LEMMA 6.9. *Let M be a surface with compact boundary, and A be a closed subset of M isomorphic to $\mathbb{R}^2_+$ that is disjoint from ∂M. Let $N = \overline{M - A}$. Then, the inclusion map $i : N \to M$ induces a homeomorphism on the space of ends.*

PROOF. Identify A with $\mathbb{R}^2_+$. Let F_j be the closed half-disk of radius j centered at $(0,0)$. Note that (F_j) is a compact exhaustion of A. Let (G_j) be a compact exhaustion of N by subsurfaces such that $F_j \cap \partial A = G_j \cap \partial A$, and let $H_j = F_j \cup G_j$.

To prove that the end map is bijective, it suffices to prove that for j large enough, each unbounded connected component of $M - H_j$ contains exactly one unbounded connected component of $N - G_j$. Since end spaces are compact and Hausdorff, any bijective map between end spaces is automatically a homeomorphism.

Let U be any unbounded connected component of $M - H_j$. If U is disjoint from A, then U is already an unbounded connected component of $N - G_j$. Assume then that U intersects A. Then, U contains $A - F_j$. Since $A - F_j$ borders $N - G_j$, U contains at least one connected component of $N - G_j$. To prove that there is exactly one, it suffices to prove that the two components of $\partial A - F_j$ are connected via a path in $N - G_j$. Consider the boundary of the closed subset $\overline{N - G_j}$. This boundary has only two ends, corresponding to the two ends of $\partial A - F_j$. Thus, those ends lie in the same connected component (which is a copy of $\mathbb{R}$). Thus, $\partial A - F_j$ is connected via paths in $\overline{N - G_j}$. Thus, it is connected via paths in $N - G_j$. $\square$

THEOREM 6.10 (Ray Uniqueness for Surfaces). *Let M be a surface with compact boundary, and let η be an end of M. If r and r' are rays in M pointing to η, then there is an automorphism $\psi : M \to M$ which sends r to r'.*

PROOF. It suffices to consider the case where M is connected. Begin by removing regular neighborhoods A of r and A' of r'. Let $N = \overline{M - A}$ and let $N' = \overline{M - A'}$. We first exhibit an isomorphism from N to N' using the classification of surfaces. Then, we will find a compatible isomorphism from A to A'. Lastly, we will combine these to define $\psi : M \to M'$. It is clear that M, N, and N' have the same number of compact boundary components. If M is orientable, then this orientation defines an orientation on N as well. Suppose that N is oriented. There there is an induced orientation on $\partial A \subset \partial N$. Choose an orientation on A that induces the same orientation on ∂A. Gluing these orientations together yields an orientation on M. Similarly, N' is orientable if and only if M is.

The genus and parity are both defined in terms of a compact exhaustion by subsurfaces. As in Lemma 6.9, identify A with $\mathbb{R}^2_+$. Let F_j be the closed half-disk of radius j centered at $(0,0)$, let (G_j) be a compact exhaustion of N by subsurfaces such that $F_j \cap \partial A = G_j \cap \partial A$, and let $H_j = F_j \cup G_j$. So, (H_j) is a compact exhaustion of M by subsurfaces.

$$g(H_j) = |\pi_0(H_j)| - \frac{b(H_j) + \chi(H_j)}{2} = |\pi_0(G_j)| - \frac{b(G_j) + \chi(G_j)}{2} = g(G_j)$$

As a result, the genus and parity of N equal those of M. Similarly, the genus and parity of N' equal those of M.

By Lemma 6.9, the inclusion maps $i : N \to M$ and $i' : N' \to M$ induce homeomorphisms on the end spaces. By looking within a neighborhood disjoint from A and A', it can be shown that for every end of M other than η, the corresponding ends in N and N' have the same end characteristics.

Given any end $\tau \neq \eta$, the corresponding ends in N and N' are isomorphic. This can be shown by looking in a neighborhood of τ disjoint from A and A'.

It was shown in Section 5.4 that removing a strip does not change the orientation of η. Those arguments apply to this situation as well. It was shown in Section 5.5.2 that removing a strip does not change the genus of η. Those arguments apply to this situation as well.

Recall from Section 6.1 above that an orientation $\mathcal{P}$ of $\mathcal{E}(\mathrm{d}N)$ is a certain subset of $\mathcal{E}(\mathrm{d}N)$. The surface N has a single noncompact boundary component B, and both ends of B point to the same end β of N by Lemma 6.3. If β is nonorientable, then $\mathcal{P} = \emptyset$. If β is orientable, then choose an orientation on an orientable complementary domain that contains β; this yields $\mathcal{P} = \{\tau\}$ where τ is one of the two ends of B. Similarly, an orientation $\mathcal{P}'$ of $\mathcal{E}(\mathrm{d}N')$ is defined. Note that η is orientable if and only if β is orientable, and η is orientable if and only if β'. Thus, $\mathcal{P}$ and $\mathcal{P}'$ are either both empty or are both singletons. In either case, there is a unique bijection $\mathcal{P} \to \mathcal{P}'$. That bijection extends to an isomorphism $\mathcal{E}(\mathrm{d}N) \to \mathcal{E}(\mathrm{d}N')$. Altogether, this data can be used to construct an isomorphism of diagrams for N and N'.

By the Classification of Surfaces (Theorem 6.1), there exists an isomorphism $\varphi : N \to N'$. This isomorphism must send ∂A to $\partial A'$. To extend φ to an automorphism of M, it suffices to find a compatible isomorphism from A to A'. But any isomorphism $\partial\mathbb{R}^2_+ \to \partial\mathbb{R}^2_+$ extends to an isomorphism $\mathbb{R}^2_+ \to \mathbb{R}^2_+$. The resulting automorphism of M sends r to r'. □

REMARK 6.11. The conclusion of Theorem 6.10 does not hold in general if M has noncompact boundary components. Consider the surface M which is a closed disk with one boundary point removed and a sequence of 1-handles attached as in Figure 6.4. (Dickmann **[Dic23]** called M a *sliced Loch Ness monster* and studied

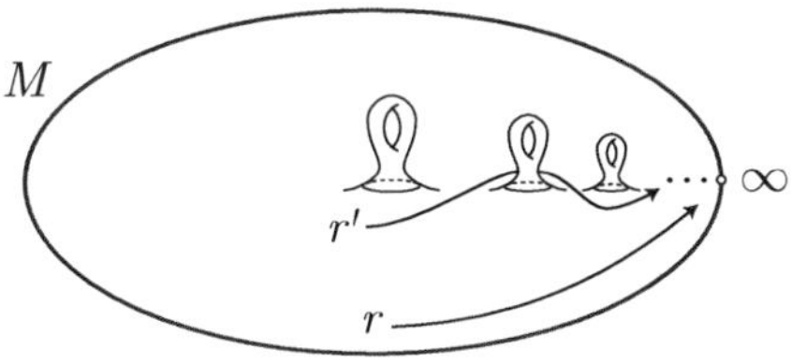

FIGURE 6.4. Surface M containing rays r and r'.

the mapping class groups of M and similar surfaces.) So, M has one end η and that end has infinite genus. Let r be a ray parallel to ∂M, and let r' be a ray that winds around the glued in 1-handles as in Figure 6.4. Suppose there is an automorphism

ψ of M that sends r to r'. Let A and A' be closed regular neighborhoods of r and r', respectively. By the uniqueness of regular neighborhoods, ψ can be modified so that it also sends A to A'. Thus, $N = \overline{M - A}$ is isomorphic to $N' = \overline{M - A'}$. However, N has one zero-genus end and one infinite-genus end, while N' has two infinite-genus ends as in Figure 6.5. This is a contradiction, so r and r' are not

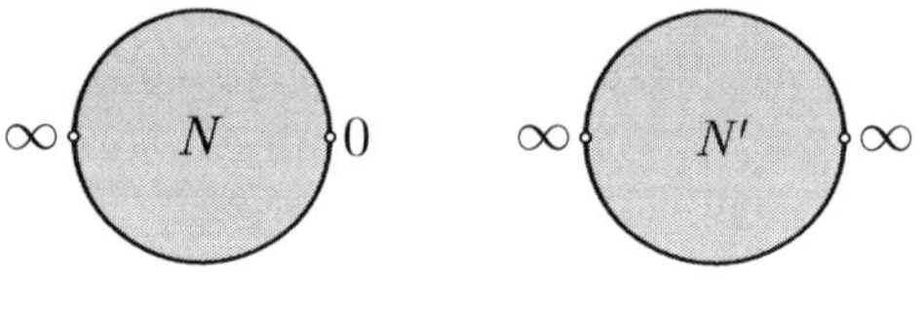

FIGURE 6.5. Nonisomorphic surfaces N and N' each obtained from M by removing the interior of a regular neighborhood of a ray.

related by an automorphism of M.

6.3. Main Theorem, Re-Proved. Using the classification of all surfaces, with our addendum, we are ready to dive into the main theorem once again.

LEMMA 6.12. *Let M be a surface with compact boundary. Let η and η' be distinct ends of M. For $i = 1, 2$, let A_i, A'_i be disjoint closed subsurfaces of M isomorphic to $\mathbb{R}^2_+$ and disjoint from ∂M. Suppose as well that that A_i points to η and A'_i points to η'. Let $\mu : A_1 \to A_2$ and $\mu' : A'_1 \to A'_2$ be isomorphisms. If any of the following conditions hold, then there is an automorphism ψ of M such that the restriction of ψ to A_1 is μ and the restriction of ψ to A'_1 is μ':* (i) *M is connected and orientable, and μ, μ' are both orientation-preserving,* (ii) *M is connected and non-orientable, or* (iii) *η and η' lie in distinct connected components of M.*

PROOF. Let $N_i = \overline{M - (A_i \cup A'_i)}$. We will show that N_1 and N_2 are isomorphic, and that this isomorphism is compatible with μ and μ'.

Using the same strategies as Theorem 6.10, it can be shown that N_1 and N_2 have the same genus, parity, orientability, and number of compact boundary components as M.

Using the same strategies as Theorem 6.10, it can be shown that the inclusion maps $i_i : N_i \to M$ induce homeomorphisms on the space of ends, and that corresponding ends have the same genus, orientability, and number of compact boundary components.

By restricting μ and μ' to ∂A_1 and $\partial A'_1$, we obtain a map from $\partial A_1 \cup \partial A'_1 \to \partial A_2 \cup \partial A'_2$. Define $h : \mathcal{E}(\mathrm{d}N_1) \to \mathcal{E}(\mathrm{d}N_2)$ to be the corresponding map of ends.

Suppose first that M is connected and orientable. Fix an orientation on M, which induces orientations on N_1 and N_2. Let $\mathcal{P}_i$ be the orientation of $\mathcal{E}(\mathrm{d}N)$ induced by the orientation on N_i. Since μ and μ' are orientation-preserving, the map h sends $\mathcal{P}_1$ to $\mathcal{P}_2$. Using h, we can construct an isomorphism of diagrams. By the Addendum (Theorem 6.2), there is an orientation-preserving isomorphism $\varphi : N_1 \to N_2$ that induces this isomorphism of diagrams.

Since μ and μ' are both orientation-preserving, φ has the same effect on the orientations of ∂A_1 and $\partial A'_1$ as μ and μ' do. Because of this, φ can be isotoped in a neighborhood of ∂A_1 and $\partial A'_1$ to become equal to μ and μ' on ∂A_1 and $\partial A'_1$.

Now suppose that M is connected and non-orientable. If η is orientable, then add an end of ∂A_1 to $\mathcal{P}_1$, and add the corresponding end of ∂A_2 (under μ) to $\mathcal{P}_1$.

Do the same for η'. Now, h sends $\mathcal{P}_1$ to $\mathcal{P}_2$, so we can again obtain an isomorphism $\varphi : N_1 \to N_2$ that is compatible with h. And, φ can be isotoped in a neighborhood of ∂A_1 and $\partial A_1'$ to become equal to μ and μ' on ∂A_1 and $\partial A_1'$.

Now suppose that η and η' lie in distinct connected components of M. As before, construct $\mathcal{P}_1$ and $\mathcal{P}_2$, and obtain an isomorphism $\varphi : N_1 \to N_2$ that is compatible with h. Isotope φ in a neighborhood of A_1 and A_1' to become equal to μ and μ' on ∂A_1 and $\partial A_1'$.

Once φ agrees with with μ and μ' on A_1 and A_1', all three maps can be glued together to form an isomorphism $\psi : M \to M$. □

As a corollary, we recover the main theorem.

THEOREM 6.13. *Let M be a PL surface with compact boundary, and let η, η' be distinct ends of M. Let N be a result of adding a 1-handle at infinity to M along η and η'. If η and η' are ends of distinct connected components of M, then N is unique up to isomorphism. If η and η' are ends of the same connected component M_1 of M, and M_1 is non-orientable, then N is unique up to isomorphism. Lastly, suppose that η and η' are ends of the same connected component M_1 of M, and M_1 is orientable. If the 1-handle is oriented, then N is unique up to isomorphism. If the 1-handle is not oriented, then N is unique up to isomorphism.*

PROOF. Consider two end sums, N_1 and N_2, along η and η', satisfying the above criteria. Let ν_i, ν_i' be the tubular neighborhood maps used to construct N_i, and let A_i, A_i' be images of ν_i, ν_i' respectively. Let $\psi : A_1 \to A_2$ be the canonical isomorphism between regular neighborhoods, defined by $\psi = \nu_2 \circ \nu_1^{-1}$. Similarly, define $\psi' : A_1' \to A_2'$.

Using Lemma 6.12, we extend ψ and ψ' to a global automorphism $\chi : M \to M$. Since χ preserves all data used in the 1-handle construction, χ extends to an isomorphism from N_1 to N_2. □

7. Extension of the Main Theorem to TOP and DIFF

Our main theorem—Theorem 5.1—also holds true in the TOP and DIFF categories. Hence, Corollary 5.2 does as well. This is an aspect of a general theme that often statements about surfaces are true independent of category. An important rationale for this theme, and a key reason why we can generalize in this case, is that every surface has a unique TOP, PL, and DIFF structure up to isomorphism. See, for example, Moise [**Moi77**, Preface & Ch. 8], Thurston [**Thu97**, §3.10], and Hatcher [**Hat13**]. In this section, we show that Theorem 5.1 for PL implies the corresponding TOP and DIFF analogues.

First, suppose we are given TOP data for the addition of a 1-handle at infinity as in Theorem 5.1. We reuse the notation from Conventions 5.3 in Section 5. In particular, N is the TOP surface that results from adding a TOP 1-handle at infinity to the surface M along ends η and η'. The 1-handle is added using disjoint rays r and r', pointing to the ends η and η' respectively, and disjoint tubular neighborhood maps ν and ν'. By the triangulation of surfaces, there exists a PL surface M_{PL} and a homeomorphism $j : M \to M_{\text{PL}}$. So, $j \circ r$ and $j \circ r'$ are disjoint TOP rays in M_{PL}, and $j \circ \nu$ and $j \circ \nu'$ are disjoint TOP tubular neighborhood maps. We may adjust M_{PL} by a small ambient homeomorphism with with support in any prescribed open neighborhood of $\operatorname{Im} j \circ \nu \cup \operatorname{Im} j \circ \nu'$ such that the images of the tubular neighborhood maps $j \circ \nu$ and $j \circ \nu'$ are PL. That is, by Moise [**Moi77**, Thm. 10.13], there is a homeomorphism $k : M_{\text{PL}} \to M_{\text{PL}}$ with support in any prescribed open neighborhood of $\text{Im } j \circ \nu \cup \text{Im } j \circ \nu'$ such that the homeomorphism $h = k \circ j$ satisfies: (i) the disjoint rays $h \circ r$ and $h \circ r'$ are PL embeddings and (ii) the disjoint tubular neighborhood maps $h \circ \nu$ and $h \circ \nu'$ are PL embeddings. As j and k are homeomorphisms, each induces a homeomorphism on the spaces of ends; note that k induces the identity on the space of ends of M_{PL}. Let η_{PL} and η'_{PL} denote the ends of M_{PL} corresponding under j to η and η' respectively.

We now have PL data for Theorem 5.1 consisting of the surface M_{PL}, disjoint rays $h \circ r$ and $h \circ r'$, pointing to the ends η_{PL} and η'_{PL} respectively, and disjoint tubular neighborhood maps $h \circ \nu$ and $h \circ \nu'$. Let N_{PL} be the PL surface that results from adding a PL 1-handle at infinity to M_{PL} according to this data. Note that we have an induced homeomorphism $\alpha : N \to N_{\text{PL}}$.

Now, suppose we have possibly different TOP data pointing to the same ends of M. That is, we have disjoint rays s and s', pointing to the *same* ends as before η and η' respectively, and disjoint tubular neighborhood maps μ and μ'. Let P be the TOP surface that results from adding a TOP 1-handle at infinity to M according to this data. We must show that N and P are homeomorphic. We have the homeomorphism $j : M \to M_{\text{PL}}$. As above, there is a homeomorphism $l : M_{\text{PL}} \to M_{\text{PL}}$ with support in any prescribed open neighborhood of $\text{Im } j \circ \mu \cup \text{Im } j \circ \mu'$ such that the homeomorphism $g = l \circ j$ satisfies: (i) the disjoint rays $g \circ s$ and $g \circ s'$ are PL embeddings and (ii) the disjoint tubular neighborhood maps $g \circ \mu$ and $g \circ \mu'$ are PL embeddings. Let P_{PL} be the PL surface that results from adding a PL 1-handle at infinity to M_{PL} according to this data. Note that we have an induced homeomorphism $\beta : P \to P_{\text{PL}}$.

Observe that in M_{PL}, the rays $h \circ r$ and $g \circ s$ both point to η_{PL}, and the rays $h \circ r'$ and $g \circ s'$ both point to η'_{PL}. Therefore, the PL version of Theorem 5.1 yields a PL homeomorphism $\gamma : N_{\text{PL}} \to P_{\text{PL}}$. Hence, $\beta^{-1} \circ \gamma \circ \alpha : N \to P$ is a homeomorphism as desired. This completes our proof of the TOP version of Theorem 5.1.

Second, suppose we are given two collections of DIFF data for the addition of a 1-handle at infinity as in Theorem 5.1. One collection is the surface M, disjoint rays r and r', pointing to the ends η and η' respectively, and disjoint tubular neighborhood maps ν and ν'. Let N be the DIFF surface that results by adding a DIFF 1-handle at infinity to M according to this data. The other collection is the surface M, disjoint rays s and s', pointing to the *same* ends as before η and η' respectively, and disjoint tubular neighborhood maps μ and μ'. Let P be the DIFF surface that results by adding a DIFF 1-handle at infinity to M according to this data. We must show that N and P are diffeomorphic. Ignoring DIFF structures for the moment, the TOP version of Theorem 5.1—proved above—implies that N and P are homeomorphic. As N and P are homeomorphic DIFF surfaces, they are diffeomorphic as desired. That completes our proof of the DIFF version of Theorem 5.1.

Acknowledgments

The authors are grateful to Craig Guilbault for suggesting the main problem studied herein and to Ric Ancel for suggesting a fruitful alternative approach to prove the main theorem based on uniqueness of rays in a surface. The authors also thank an anonymous referee for several useful comments.

References

[AV20] J. Aramayona and N. G. Vlamis, *Big mapping class groups: an overview*, In the tradition of Thurston—geometry and topology, Springer, Cham, 2020, pp. 459–496, DOI 10.1007/978-3-030-55928-1_12. MR4264585

[AZ67] M. A. Armstrong and E. C. Zeeman, *Piecewise linear transversality*, Bull. Amer. Math. Soc. **73** (1967), 184–188, DOI 10.1090/S0002-9904-1967-11708-7. MR206964

[Ben16] J. Bennett, *Exotic smoothings via large* $\mathbb{R}^4$*'s in Stein surfaces*, Algebr. Geom. Topol. **16** (2016), no. 3, 1637–1681, DOI 10.2140/agt.2016.16.1637. MR3523051

[BM79] E. M. Brown and R. Messer, *The classification of two-dimensional manifolds*, Trans. Amer. Math. Soc. **255** (1979), 377–402, DOI 10.2307/1998182. MR542887

[CG19] J. S. Calcut and R. E. Gompf, *On uniqueness of end sums and 1-handles at infinity*, Algebr. Geom. Topol. **19** (2019), no. 3, 1299–1339, DOI 10.2140/agt.2019.19.1299. MR3954283

[CGH20] J. S. Calcut, C. R. Guilbault, and P. V. Haggerty, *Extreme nonuniqueness of end-sum*, J. Topol. Anal. **14** (2022), no. 2, 461–503, DOI 10.1142/S179352532150014X. MR4446916

[CH14] J. S. Calcut and P. V. Haggerty, *Connected sum at infinity and 4-manifolds*, Algebr. Geom. Topol. **14** (2014), no. 6, 3281–3303, DOI 10.2140/agt.2014.14.3281. MR3302962

[CKS12] J. S. Calcut, H. C. King, and L. C. Siebenmann, *Connected sum at infinity and Cantrell-Stallings hyperplane unknotting*, Rocky Mountain J. Math. **42** (2012), no. 6, 1803–1862, DOI 10.1216/RMJ-2012-42-6-1803. MR3028764

[Dic23] R. Dickmann, *Mapping class groups of surfaces with noncompact boundary components*, Algebr. Geom. Topol. **23** (2023), no. 6, 2777–2821, DOI 10.2140/agt.2023.23.2777. MR4640140

[Fre31] H. Freudenthal, *Über die Enden topologischer Räume und Gruppen* (German), Math. Z. **33** (1931), no. 1, 692–713, DOI 10.1007/BF01174375. MR1545233

[Geo08] R. Geoghegan, *Topological methods in group theory*, Graduate Texts in Mathematics, vol. 243, Springer, New York, 2008, DOI 10.1007/978-0-387-74614-2. MR2365352

[Gol71] M. E. Goldman, *An algebraic classification of noncompact 2-manifolds*, Trans. Amer. Math. Soc. **156** (1971), 241–258, DOI 10.2307/1995610. MR275436

[Gom83] R. E. Gompf, *Three exotic* $\mathbf{R}^4$*'s and other anomalies*, J. Differential Geom. **18** (1983), no. 2, 317–328. MR710057

[Gom85] R. E. Gompf, *An infinite set of exotic* $\mathbf{R}^4$*'s*, J. Differential Geom. **21** (1985), no. 2, 283–300. MR816673

[Gui16] C. R. Guilbault, *Ends, shapes, and boundaries in manifold topology and geometric group theory*, Topology and geometric group theory, Springer Proc. Math. Stat., vol. 184, Springer, [Cham], 2016, pp. 45–125, DOI 10.1007/978-3-319-43674-6_3. MR3598162

[Hat13] A. Hatcher, *The Kirby torus trick for surfaces*, `arXiv:1312.3518` [math.GT], 2022, 1–10.

[Hat02] A. Hatcher, *Algebraic topology*, Cambridge University Press, Cambridge, 2002. MR1867354

[Moi77] E. E. Moise, *Geometric topology in dimensions* 2 *and* 3, Graduate Texts in Mathematics, Vol. 47, Springer-Verlag, New York-Heidelberg, 1977. MR488059

[Mül10] D. Müllner, *Orientation reversal of manifolds*, Algebr. Geom. Topol. **9** (2009), no. 4, 2361–2390, DOI 10.2140/agt.2009.9.2361. MR2558314

[Mun00] J. R. Munkres, *Topology*, Prentice Hall, Inc., Upper Saddle River, NJ, 2000. Second edition of [MR0464128]. MR3728284

[Mye99] R. Myers, *End sums of irreducible open* 3*-manifolds*, Quart. J. Math. Oxford Ser. (2) **50** (1999), no. 197, 49–70, DOI 10.1093/qjmath/50.197.49. MR1673256

[PS81] A. Phillips and D. Sullivan, *Geometry of leaves*, Topology **20** (1981), no. 2, 209–218, DOI 10.1016/0040-9383(81)90039-2. MR605659

[Ric60] J. I. Richards, *A Classification of Noncompact Surfaces*, ProQuest LLC, Ann Arbor, MI, 1960. Thesis (Ph.D.)–Harvard University. MR2939141

[RS72] C. P. Rourke and B. J. Sanderson, *Introduction to piecewise-linear topology*, Ergebnisse der Mathematik und ihrer Grenzgebiete [Results in Mathematics and Related Areas], Band 69, Springer-Verlag, New York-Heidelberg, 1972. MR350744

[Sco67] A. Scott, *Infinite regular neighborhoods*, J. London Math. Soc. **42** (1967), 245–253, DOI 10.1112/jlms/s1-42.1.245. MR212807

[Sie65] L. C. Siebenmann, *The Obstruction to Finding a Boundary for An Open Manifold of Dimension Greater Than Five*, ProQuest LLC, Ann Arbor, MI, 1965. Thesis (Ph.D.)–Princeton University. MR2615648

[Spa18] P. Sparks, *The double n-space property for contractible n-manifolds*, Algebr. Geom. Topol. **18** (2018), no. 4, 2131–2149, DOI 10.2140/agt.2018.18.2131. MR3797063

[Thu97] W. P. Thurston, *Three-dimensional geometry and topology. Vol. 1*, Princeton Mathematical Series, vol. 35, Princeton University Press, Princeton, NJ, 1997. MR1435975

Newton, Massachusetts 02459
Email address: laxon26@gmail.com

Department of Mathematics, Oberlin College, Oberlin, Ohio 44074
Email address: jcalcut@oberlin.edu

Contemporary Mathematics
Volume **812**, 2025
https://doi.org/10.1090/conm/812/16262

A survey on classifying hyperbolic actions of groups

Sahana H. Balasubramanya

ABSTRACT. This paper is a survey of results proved in recent years that pertain to classifying cobounded hyperbolic actions of any group G. In other words, we discuss results that allow us to describe the partially ordered set $\mathcal{H}(G)$, first introduced in by Abbott-Balasubramanya-Osin. In certain cases, a complete classification of the poset is possible. In cases where this seems out of reach, we provide descriptions of large subsets of the poset. This covers a wide range of groups, including acylindrically hyperbolic groups, nilpotent groups, many solvable groups and "Thompson-like" groups. We also cover the range of strategies utilized to obtain these classification results. As a result, we produce some new examples of groups with interesting $\mathcal{H}(G)$ structure.

1. Introduction

Isometric actions on hyperbolic metric spaces, termed hyperbolic actions, have proven to be important tools in geometric group theory. They have been used to study the coarse geometry of a wide variety of groups, bounded cohomology ([**BF02**]), quotients of groups ([**DGO17**]), and decision problems for groups ([**Bri02**]). It is thus natural to consider the problem of classifying *all* of the hyperbolic actions of a group.

Although natural, requesting a complete classification of hyperbolic actions for any group G is a highly complex problem to tackle. For instance, any group can be made to act trivially on any hyperbolic space. One can also use a combinatorial horoball construction from [**GM08**] to construct many parabolic actions of a countable group G on a hyperbolic space. However, these actions retain almost no information about the algebra of the group, and so it is reasonable to impose conditions that exclude these actions from consideration.

It turns out that studying *cobounded* hyperbolic actions eliminates the problem posed by parabolic and elliptic actions. Additionally, it makes sense to consider actions up to *quasi-isometric* equivalence. i.e. spaces which share the same coarse geometry, and where the actions are effectively the same up to some bounded error. This is especially helpful as hyperbolicity is preserved under quasi-isometries. Together, these conditions rule out all trivial actions except on a point and all parabolic actions. They also allow us to introduce a partial order on the actions, and arrange them into the poset of *hyperbolic structures on* G, denoted $\mathcal{H}(G)$. Roughly speaking, we consider one action to be smaller than another when the smaller action

2020 *Mathematics Subject Classification.* Primary 20F65, 20F16.

may be obtained by collapsing an equivariant family of subspaces in the larger action (see section 2 for details). Intuitively, the larger action retains more information about the group; the action on a point (called the trivial action/structure) is thus always the smallest in this poset.

Despite these conditions, many groups classically studied in geometric group theory admit uncountably many distinct hyperbolic actions (see section 3), making the classification problem intractable even for well-understood groups. Although we can still describe large sections of the poset $\mathcal{H}(G)$ (which has been useful in these cases nonetheless), it urges us to to attempt to solve this problem for groups with naturally fewer hyperbolic actions. The author made the first step towards this goal in [**Bal20**], which was followed by the papers [**AR23, ABR23, ABAP22**]. These papers completely classify $\mathcal{H}(G)$ of certain classically studied solvable groups, which prompted the authors to together look into the classification problem for more general solvable groups. This work is covered in Section 4, with the goal to develop tools that describes $\mathcal{H}(G)$ for arbitrary metabelian, and eventually, solvable groups.

In another related direction, one can ask which groups, besides finite groups, admit only the trivial hyperbolic structure. Such groups are natural examples of groups that are inaccessible for study from the perspective of hyperbolic actions. Nonetheless, it is interesting to study the common elements in the behavior and structure of such groups, especially those that obstruct the group from acting on a hyperbolic space in a meaningful way. The paper [**BFGS22**] formalized the study of such groups; which cover many Thompson-like groups (see section 5). Other (amenable) groups are also studied and the paper provides ways to check when these can admit hyperbolic actions without explicitly constructing one.

More recently, the paper [**BCFS22**] explores this classification problem in the setting of lattices in semi-simple groups. When the simple factors have higher rank, the lattice was known to have no non-trivial hyperbolic actions by [**Hae20**]. However, the question remained unexplored in the case when the simple factors have rank one. It is natural to expect that the action of the factors will have connections to the actions of the lattices, as we can always create actions by factoring through the rank one factors. [**BCFS22**] shows that, under mild conditions, these completely classify the actions of the lattice; providing a new source of groups for which $\mathcal{H}(G)$ is completely understood; see Section 6.

The present paper was born from a desire to collect all these results into a common body of work. We hope such an overview will make it easier to reference and connect the different pieces of the larger puzzle for those interested in an overview or understanding this area of research, especially since it incorporates tools from commutative algebra, bounded cohomology, valuations and Lie groups.

2. Definitions and Preliminaries

In this section, we collect the necessary background information and formalize definitions and theorems that will be used throughout the paper. We start with the precise description of $\mathcal{H}(G)$ and its poset relation.

2.1. Comparing generating sets and group actions. Throughout this paper, all group actions on hyperbolic spaces are assumed to be isometric, unless otherwise stated. Given a metric space X, let d_X denote the distance function on X. If G is a group and S is a generating set of G, then $\|\cdot\|_S$ denotes the word length on G associated to S and d_S denotes the corresponding word metric $d_S(g,h) = \|gh^{-1}\|_S$. $\Gamma(G,S)$ denotes the Cayley graph of G with respect to S.

DEFINITION 2.1 ([**ABO19**, Definition 1.1]). Let S, T be two (possibly infinite) generating sets of a group G. We say that S is *dominated* by T, written $S \preceq T$, if

$$\sup_{t\in T} \|t\|_S < \infty.$$

The relation $\preceq$ is a preorder on the set of generating sets of G, and therefore it induces an equivalence relation in the standard way: $S \sim T \quad \Leftrightarrow \quad S \preceq T$ and $T \preceq S$. This is equivalent to the condition that the spaces $\Gamma(G,S)$ and $\Gamma(G,T)$ are G–equivariantly quasi-isometric. We denote by $[S]$ the equivalence class of a generating set S. The preorder $\preceq$ extends to a partial order on the equivalence classes in the standard way: $[S] \preccurlyeq [T] \quad \Leftrightarrow \quad S \preceq T$.

For example, all finite generating sets of a finitely generated group are equivalent and the equivalence class of a finite generating set is the largest element. For every group G, the smallest element is $[G]$. Note also that this order is "inclusion reversing": if S and T are generating sets of G such that $S \subseteq T$, then $T \preceq S$.

To define a hyperbolic structure on a group, we first recall the definition of a hyperbolic space, for which we use the Rips condition.

DEFINITION 2.2. A metric space X is called *δ–hyperbolic* if it is geodesic and for any geodesic triangle Δ in X, each side of Δ is contained in the union of the closed δ–neighborhoods of the other two sides.

DEFINITION 2.3 ([**ABO19**, Definition 1.2]). A *hyperbolic structure* on G is an equivalence class $[S]$ such that the Cayley graph $\Gamma(G,S)$ is hyperbolic. Since hyperbolicity of geodesic metric spaces is quasi-isometry invariant, this definition is independent of the choice of the representative in $[S]$. We denote the set of hyperbolic structures by $\mathcal{H}(G)$. It is a poset with the same partial order induced from the one on the equivalence classes.

An equivalent description of $\mathcal{H}(G)$ can be made in terms of cobounded hyperbolic actions of G up to coarsely equivariant quasi-isometry, which we now summarize.

DEFINITION 2.4. The action $G \curvearrowright X$ is *cobounded* if for some (equivalently any) $x \in X$ there exists $R > 0$ such that every point of X is distance at most R from some point of the orbit Gx.

Given a cobounded hyperbolic action $G \curvearrowright X$, there is an associated hyperbolic structure given by the Schwarz-Milnor Lemma.

LEMMA 2.5 ([**ABO19**, Lemma 3.11]). *Let $G \curvearrowright X$ be a cobounded hyperbolic action of G. Let $B \subseteq X$ be a bounded subset such that $\bigcup_{g\in G} gB = X$. Let $D =$ $\operatorname{diam}(B)$ and let $x \in B$. Then G is generated by the set*

$$S = \{g \in G : d_X(x, gx) \leq 2D + 1\},$$

and X is G–equivariantly quasi-isometric to $\Gamma(G,S)$.

Thus, cobounded hyperbolic actions of G correspond to actions of G on its hyperbolic Cayley graphs. There is an equivalence relation on actions which respects the poset structure: two actions $G \curvearrowright R$ and $G \curvearrowright S$ are equivalent if there exists a coarsely G-equivariant quasi-isometry $R \to S$. So, classifying the cobounded hyperbolic actions of G is tantamount to classifying the elements of $\mathcal{H}(G)$.

We will sometimes refer to representatives of hyperbolic structures: if $[S] \in \mathcal{H}(G)$, we say that $[S]$ is represented by a cobounded hyperbolic action $G \curvearrowright X$ if $G \curvearrowright X$ is equivalent to the action $G \curvearrowright \Gamma(G, S)$.

2.2. Actions on hyperbolic spaces. Let G be a group acting on a hyperbolic space X, and let ∂X denote the Gromov boundary of X. X is not assumed to be proper, and its boundary is defined as the set of equivalence classes of sequences convergent at infinity. Let $\Lambda(G)$ denote the set of limit points of G on ∂X. That is, $\Lambda(G) := \partial X \cap \overline{Gx}$, where $\overline{Gx}$ denotes the closure of a G–orbit in $X \cup \partial X$, for any choice of basepoint $x \in X$. This definition is independent of the choice of $x \in X$.

A loxodromic element g in a hyperbolic action on X is an infinite order element such that the map $n \to g^n x$ is a quasi-isometric embedding for some (equivalently any) $x \in X$. Every loxodromic element $g \in G$ has exactly 2 limit points $g^{\pm\infty}$ on ∂X. Loxodromic elements $g, h \in G$ are *independent* if $\{g^{\pm\infty}\} \cap \{h^{\pm\infty}\} = \emptyset$.

The following theorem summarizes the classification of group actions on hyperbolic spaces due to Gromov [**Gro87**, Section 8.2] and the some related results from [**CCMT15**, Propositions 3.1 and 3.2].

THEOREM 2.6. *Let G be a group acting on a hyperbolic space X. Then exactly one of the following conditions holds.*

1) *The action of G is* elliptic. *i.e.* $|\Lambda(G)| = 0$. *Equivalently, G has bounded orbits.*
2) *The action of G is parabolic. i.e.* $|\Lambda(G)| = 1$. *A parabolic action cannot be cobounded.*
3) *The action of G is lineal. i.e.* $|\Lambda(G)| = 2$. *Equivalently, G contains a loxodromic element and any two loxodromic elements have the same limit points on ∂X. A lineal action $G \curvearrowright X$ is said to be* orientable *if G fixes its limit points on ∂X* pointwise, *and* non-orientable *otherwise.*
4) *The action of G is non-elementary. i.e.* $|\Lambda(G)| = \infty$. *Then G always contains loxodromic elements. In turn, this case breaks into two subcases.*
   (a) *The action is* quasi-parabolic. *i.e. G fixes a point of ∂X. Equivalently, any two loxodromic elements of G have a common limit point on the boundary.*
   (b) *The action of G is of* general type. *i.e. G does not fix any point of ∂X.*

It follows immediately from the above that for any group G,

$$\mathcal{H}(G) = \mathcal{H}_e(G) \sqcup \mathcal{H}_\ell(G) \sqcup \mathcal{H}_{qp}(G) \sqcup \mathcal{H}_{gt}(G)$$

where the sets of elliptic, lineal, quasi-parabolic, and general type hyperbolic structures on G are denoted by $\mathcal{H}_e(G)$, $\mathcal{H}_\ell(G)$, $\mathcal{H}_{qp}(G)$, and $\mathcal{H}_{gt}(G)$ respectively. Namely, $[S] \in \mathcal{H}(G)$ lies in $\mathcal{H}_\ell(G)$ if the action $G \curvearrowright \Gamma(G, S)$ is lineal, and similar definitions hold for the posets $\mathcal{H}_e$, $\mathcal{H}_{qp}$, and $\mathcal{H}_{gt}$. Note that $\mathcal{H}_e(G) = \{[G]\}$ always. We direct the reader to [**ABO19**, Section 4] and [**ABO19**, Theorem 4.6] for details.

2.3. The Busemann pseudocharacter. A function $q\colon G \to \mathbb{R}$ is a *quasi-character* (or *quasi-morphism*) if there exists a constant D such that

$$|q(gh) - q(g) - q(h)| \leq D$$

for all $g, h \in G$. We say that q has *defect at most D*. If, in addition, the restriction of q to every cyclic subgroup of G is a homomorphism, then q is called a *pseudocharacter* (or *homogeneous quasi-morphism*).

Given any action of a group G on a hyperbolic space X fixing a point on ∂X, one can associate a natural pseudocharacter β called the *Busemann pseudocharacter*. As we do not require the exact definition of β, we refer the reader to [**Gro87**, Sec. 7.5.D] and [**Man08**, Sec. 4.1] for details. An element $g \in G$ is loxodromic with respect to the action $G \curvearrowright X$ if and only if $\beta(g) \neq 0$. In particular, β is not identically zero whenever $G \curvearrowright X$ is quasi-parabolic or orientable lineal. Conversely, given a pseudocharacter on a group G, one can always construct an orientable lineal action, as evidenced by the following lemma, which can also be deduced from [**Man06**, Proposition 3.1].

LEMMA 2.7 ([**ABO19**, Lemma 4.15]). *Let $p\colon G \to \mathbb{R}$ be a non-zero pseudocharacter. Let C be any constant such that the defect of p is at most $C/2$ and there exists a value of p in the interval $(0, C/2)$. Let*

$$X = X_{p,C} = \{g \in G : |p(g)| < C\}.$$

Then X generates G and the map $p\colon (G, d_X) \to \mathbb{R}$ is a quasi-isometry. In particular, $[X]$ is orientable lineal.

2.4. Acylindrical actions. There are many groups arising from geometric and topological settings that admit natural actions on hyperbolic spaces; yet are not hyperbolic or relatively hyperbolic groups. Nonetheless, they do share many interesting properties with hyperbolic groups and this initiated the study of non-geometric actions, including WPD actions. In a continued vein, acylindrical actions were first defined by Sela for groups acting on trees [**Sel97**] and later generalized by Bowditch for general metric spaces [**Bow08**].

DEFINITION 2.8. An action of a group G on a metric space X is called acylindrical if for every $\varepsilon > 0$, there exist constants $R, N \geq 0$ (depending only on ε) such that for any $x, y \in X$ satisfying $d(x, y) \geq R$, we have that

$$\#\{g \in G \mid d(x, gx) \leq \varepsilon \text{ and } d(y, gy) \leq \varepsilon\} \leq N.$$

Acylindricity can be thought of as a kind of properness of the action on $X \times X$ minus a "thick diagonal". However, acylindricity is not a true generalization of the notion of a proper action. For instance, one can use the combinatorial horoball construction from [**GM08**] to construct proper actions that are not acylindrical.

In contrast to hyperbolic actions, acylindrical actions are much more restrictive, as evidenced by the following theorem of Osin [**Osi16**, Theorem 1.1].

THEOREM 2.9. *Let G be a group acting acylindrically on a hyperbolic space. Then G satisfies exactly one of the following three conditions.*

(a) *G has bounded orbits.*
(b) *G is virtually cyclic and contains a loxodromic element.*
(c) *G contains infinitely many independent loxodromic elements.*

In other words, acylindrical actions can be either elliptic, lineal or of general type. The following may then be taken as a definition – a group G is acylindrically hyperbolic if it is not virtually cyclic and admits an acylindrical action on a hyperbolic space with unbounded orbits. For such groups, one can ask if it is possible to achiev a complete classification of acylindrical actions on hyperbolic spaces. This allows us to define the corresponding subposet $\mathcal{AH}(G)$ of $\mathcal{H}(G)$ as

$$\mathcal{AH}(G) = \{[X] \in \mathcal{H}(G) \mid G \curvearrowright \Gamma(G, X) \text{ is acylindrical}\}.$$

In the next section, we begin the survey of results that study the structure of $\mathcal{AH}(G)$, and how this influenced the direction of subsequent research on solvable groups.

3. Acylindrically hyperbolic groups

It follows easily from Theorem 2.9 that if the group is not acylindrically hyperbolic, then either $\mathcal{AH}(G) = \{[G]\}$ or G is virtually cyclic and $\mathcal{AH}(G) = \{[G], [X]\}$, where X is a finite generating set for G. Similarly, it follows that for an acylindrically hyperbolic group G,

$$\mathcal{AH}(G) = \mathcal{AH}_e(G) \sqcup \mathcal{AH}_{gt}(G) = \{[G]\} \sqcup \mathcal{AH}_{gt}(G),$$

and so the classification problem in this case reduces to studying acylindrical general type actions of the group. In practice, however, this might be very hard to do due to the large cardinality of this poset.

THEOREM 3.1 ([**ABO19**, Theorem 2.6]). *For an acylindrically hyperbolic group, $|\mathcal{AH}(G)| \geq 2^{\aleph_0}$. Further, $\mathcal{AH}(G)$ contains a copy of $P(\omega)$; the poset of all subsets of $\mathbb{N}$, ordered by inclusion.*

The proof of this statement makes use of hyperbolic structures on groups induced from hyperbolic structures on subgroups. Specifically, one can show that if H is a hyperbolically embedded subgroup of G, then $\mathcal{AH}(H)$ embeds into $\mathcal{AH}(G)$. Hyperbolically embedded subgroups are a generalization of peripheral subgroups of relatively hyperbolic groups; we omit going to into details here and refer the curious reader to [**ABO19**, Section 5]. However, the result is applied to prove that $\mathcal{AH}(G)$ is sufficiently complicated as follows: every acylindrically hyperbolic group G contains a hyperbolically embedded subgroup isomorphic to $F_2 \times K$, where F_2 is free of rank 2 and K is a finite group [**DGO17**, Theorem 2.24]. It then suffices to show that $\mathcal{AH}(F_2 \times K)$ is sufficiently complicated, and the latter poset is much easier to understand. To this end, we use small cancellation techniques to construct many general type actions for any non-elementary hyperbolic group [**ABO19**, Theorem 2.11], thereby also generalizing a former result of Ivanov.

The statement of [**ABO19**, Theorem 2.11] is infact much stronger. Indeed, it additionally shows that hyperbolic structures are not determined by the set of their loxodromic elements, even though equivalent actions have the same sets of loxodromic elements. The theorem shows that for every non-elementary $[X] \in \mathcal{AH}(G)$, there are uncountably many inequivalent acylindrically hyperbolic structures with the same set of loxodromic elements.

Besides the classification problem, another motivation for studying $\mathcal{AH}(G)$ comes from the desire to understand *universal* actions. An element $g \in G$ is called a *generalized loxodromic* if g acts as a loxodromic element in *some* acylindrical action of G on a hyperbolic space. An acylindrical action of G on a hyperbolic space X is said to be *universal* if every generalized loxodromic is a loxodromic for $G \curvearrowright X$. Not every acylindrically hyperbolic group admits a universal action (eg: Dunwoody's group [**Abb16**]). In the context of $\mathcal{AH}(G)$, one can consider the related question of when the poset is *accessible*.

DEFINITION 3.2. We say that a group G is *$\mathcal{AH}$-accessible* if $\mathcal{AH}(G)$ contains the largest element.

Note that if a group is not acylindrically hyperbolic, then it must be $\mathcal{AH}-$accessible by Theorem 2.9, with the largest acylindrical action on a quasi-line (if the group is virtually cyclic) or on a point. Thus the question of $\mathcal{AH}-$accessibility only makes sense in the context of acylindrically hyperbolic groups.

If a largest acylindrical action exists, it is necessarily universal. The converse, however, is not true. [**ABO19**, Theorem 7.3] shows that there exists a finitely presented group N that is not $\mathcal{AH}-$accessible, but does have a universal action; the group N is a finitely presented non-hyperbolic normal subgroup of a hyperbolic group G such that $G/N \cong \mathbb{Z}$. Further, using small cancellation techniques, [**AH21**] produces uncountably many quasi-isometry classes of finitely generated groups that are not $\mathcal{AH}-$accessible, but have uncountably many distinct universal actions. However, being $\mathcal{AH}-$accessible does not imply the existence of a unique universal action either – it follows from [**ABO19**, Theorem 2.11] that there are uncountable many elements of $\mathcal{AH}(G)$ with the same set of loxodromic elements. Thus any $\mathcal{AH}-$accessible group has uncountably many inequivalent universal actions.

On the other hand, interestingly, many groups traditionally studied in the geometric group theory are $\mathcal{AH}$-accessible.

THEOREM 3.3 ([**ABO19**, Theorem 2.18]). *The following groups are $\mathcal{AH}$-accessible.*

(a) *Finitely generated relatively hyperbolic groups whose parabolic subgroups are not acylindrically hyperbolic.*
(b) *Mapping class groups of punctured closed surfaces. The largest action is the action on the associated Curve complex.*
(c) *Right-angled Artin groups. The largest action is the action on the extension graph.*
(d) *Fundamental groups of compact orientable 3-manifolds with empty or toroidal boundary. The largest action is either on a point, the Bass-Serre tree associated to the JSJ decomposition of the space if M only has Seifert-fibered pieces, or a "blow up" of the Bass-Serre tree where vertices corresponding to hyperbolic components of M are replaced by the universal covers of the pieces.*

In fact, the work in [**ABO19**, Section 7] actually proves something stronger. It shows that the groups listed in the above theorem are *strongly* $\mathcal{AH}$-accessible. i.e. they contain a largest acylindrical action that dominates all other (not necessarily cobounded) acylindrical actions, with the actions listed above being the largest in this weaker setting.

Following an early draft of [**ABO19**], parts (b), (c), and the special case of part (d) when the 3-manifold has no Nil or Sol in its prime decomposition of the above theorem are also proven in [**ABB$^+$21**]. [**ABB$^+$21**] also adds to the list of examples of $\mathcal{AH}-$accessible groups using different methods, as summarized in the following theorem.

THEOREM 3.4 ([**ABB$^+$21**, Theorem A]). *Every hierarchically hyperbolic group admits a largest acylindrical action.*

In particular, this includes groups that act properly and cocompactly on a special CAT(0) cube complex, and more generally any cubical group which admits a factor system. This includes right angled Artin groups, right-angled Coxeter groups, and many examples from [**HS20**].

The result about right angled Artin groups has also been generalized in another direction to graph products of groups.

THEOREM 3.5 ([**Val21**, Corollary D]). *Let Γ be a finite simplicial graph and let $\mathcal{G} = \{G_v \mid v \in V(\Gamma)\}$ be a collection of infinite groups. Suppose that for each isolated vertex $v \in V(\Gamma)$, the group G_v is strongly $\mathcal{AH}$-accessible. Then the associated graph product of group $\Gamma\mathcal{G}$ is strongly $\mathcal{AH}$-accessible. Furthermore, suppose that*

(1) *Γ has no isolated vertices,*
(2) *X is the Cayley graph of $\Gamma\mathcal{G}$ with respect to the generating set $S = \{G_v \mid v \in V(\Gamma)\}$,*
(3) *$\mathcal{C}X$ is the contact graph associated to X.*

Then the action $\Gamma\mathcal{G} \curvearrowright \mathcal{C}X$,is the largest acylindrical action of $\Gamma\mathcal{G}$ on a hyperbolic metric space.

The above theorem relies on work from [**Gen19b**], which shows that the space X in the statement above is a quasi-median space, allowing for a well defined contact graph $\mathcal{C}X$ to be constructed in [**Val21**].

Recently, yet another class of groups was proven to have a largest acylindrical action. In the following theorem, the term *admissible graph of groups* refers to a class of groups introduced by Croke and Kleiner to abstract the structure of $\pi_1 M$, when M is a non-geometric graph manifold [**CK02**]. Roughly speaking, an admissible graph of groups is a nontrivial finite graph of groups $\mathcal{G}$ where each edge group is $\mathbb{Z}^2$ and each vertex group G_μ has infinite cyclic center Z_μ with quotient a non-elementary hyperbolic group. Additionally, the various edge groups need to be pairwise non-commensurable inside each vertex group.

THEOREM 3.6 ([**HRSS23**, Corollary 5]). *Let $\mathcal{G}$ be an admissible graph of groups, and let $G = \pi_1\mathcal{G}$. Then G is $\mathcal{AH}-$ accessible and the largest action is the action on the associated Bass-Serre tree.*

We note that [**ABO19**] also considers the notion of $\mathcal{H}-$ *accessibility* : A group G is $\mathcal{H}-$accessible if the poset $\mathcal{H}(G)$ has a largest action. In contrast to the acylindrical case, most groups are $\mathcal{H}-$inaccessible. Indeed, the only known examples of $\mathcal{H}-$accessible groups are hyperbolic groups. In this way, the poset $\mathcal{H}(G)$ is even wilder than its subposet $\mathcal{AH}(G)$. We refer the interested reader to [**ABO19, ARar, ANR23**] for further details and results pertaining to $\mathcal{H}-$accessibility.

4. Solvable groups

As seen from the results stated in the previous section, the class of acylindrically hyperbolic groups has a "wild" behaviour, which makes obtaining a complete description of $\mathcal{H}(G)$ (even $\mathcal{AH}(G)$) too intractable. Further, the proof of Theorem 3.1 suggests that free subgroups of rank ≥ 2 could be a source of similar issues in classifying hyperbolic actions of any group that admits a non-elementary hyperbolic action.

It is therefore, natural, to restrict our focus to the smaller class of groups where such issues do not arise – this makes the class of solvable groups a natural candidate to consider. Indeed, solvable actions contain no non-elementary free subgroups, and thus admit no general type actions, this follows from the well known Ping-Pong Lemma.

Another motivation to focus on solvable groups comes from the following fact: For every $n \in \mathbb{N}$, the authors of [**ABO19**] succeeded in producing examples of groups H_n such that $|\mathcal{H}_\ell(H_n)| = n$; as well as examples of groups G_n such that $|\mathcal{H}_{gt}(G_n)| = n$ (see [**ABO19**, Theorems 4.22 and 4.28]). However, producing examples of groups with finitely many quasi-parabolic structures remained elusive. As solvable groups can only admit quasi-parabolic non-elementary actions, studying the structure of $\mathcal{H}(G)$ for a solvable group G could provide an indication to a source of examples in this vein.

4.1. Abelian-by-cyclic and metabelian groups. The first step in this direction was taken by the author in [**Bal20**], which describes a part of the poset for groups of the form $G = H \operatorname{wr} \mathbb{Z}$; see [**Bal20**, Theorem 1.5]. In particular, the work from [**Bal20**] describes the poset $\mathcal{H}(G)$ completely for the Lamplighter groups $\mathcal{L}_n = (\mathbb{Z}/n\mathbb{Z}) \operatorname{wr} \mathbb{Z}$ for $n \geq 2$, and shows that $|\mathcal{H}_{qp}(\mathcal{L}_n)|$ is twice the number of proper divisors of n. To the best of our knowledge, these were the first examples of groups known to have finitely many quasi-parabolic structures produced in the literature.

The main tool used in this work was the machinery of confining subsets from [**CCMT15**], which allows us to classify quasi-parabolic actions of G in terms of *confining* subsets of H; see [**CCMT15**, Section 4 and Proposition 4.6]. This machinery applies best in the case when the group has abelianization of rank 1, and was also used in [**AR23**, **ARar**] to describe $\mathcal{H}(G)$ in the cases for solvable Baumslag Solitar groups $BS(1,n) = \mathbb{Z}\left[\frac{1}{n}\right] \rtimes \mathbb{Z}$ and fundamental groups of the Anosov mapping torus $\mathbb{Z}^2 \rtimes_\phi \mathbb{Z}$, for $\phi \in SL(2,\mathbb{Z})$; see [**AR23**, Theorem 1.1] and [**ARar**, Theorem 1.3]. If $G = H \rtimes_\varphi \mathbb{Z}$, then $Q \subset H$ is *confining* if the following three conditions hold.

(1) $\varphi(Q) \subset Q$;
(2) For each $h \in H$, there is an $n \geq 0$ such that $\varphi^n(h) \in Q$;
(3) There is an $n_0 \geq 0$ such that $\varphi^{n_0}(Q.Q) \subset Q$.

In this case, if t is the generator of $\mathbb{Z}$, then the Cayley graph $\Gamma(G, Q \cup \{t^{\pm 1}\})$ is a hyperbolic metric space.

EXAMPLE 4.1. Consider a lamplighter group $(\mathbb{Z}/n\mathbb{Z}) \wr \mathbb{Z}$, where $n \geq 2$. Then $\mathcal{H}((\mathbb{Z}/n\mathbb{Z}) \wr \mathbb{Z})$ is as pictured in Figure 1: there is a unique lineal structure and two copies of the poset $\operatorname{Sub}(\mathbb{Z}/n\mathbb{Z})$ of subgroups of $\mathbb{Z}/n\mathbb{Z}$ intersecting in the unique lineal structure; all other structures from $\operatorname{Sub}(\mathbb{Z}/n\mathbb{Z})$ represent quasi-parabolic action. Every structure can be represented by action on a tree or on a point. This description was first obtained in [**Bal20**].

EXAMPLE 4.2. Consider the solvable Baumslag-Solitar groups $BS(1,n) = \langle t, a : tat^{-1} = a^n \rangle$ for $n \in \mathbb{Z} \backslash \{-1, 0, 1\}$, and let $n = \pm p_1^{n_1} \cdots p_r^{n_r}$ be the prime factorization of n. Then $\mathcal{H}(BS(1,n))$ is as pictured in Figure 2: there is a unique lineal structure, a copy of the poset $2^{\{1,\ldots,r\}}$ of subsets of $\{1, \ldots, r\}$ containing the single lineal structure, while all other structures are quasi-parabolic and represented by actions on trees. There is an additional quasi-parabolic structure represented by the action on the hyperbolic plane $\mathbb{H}^2$. This description was first obtained in [**AR23**].

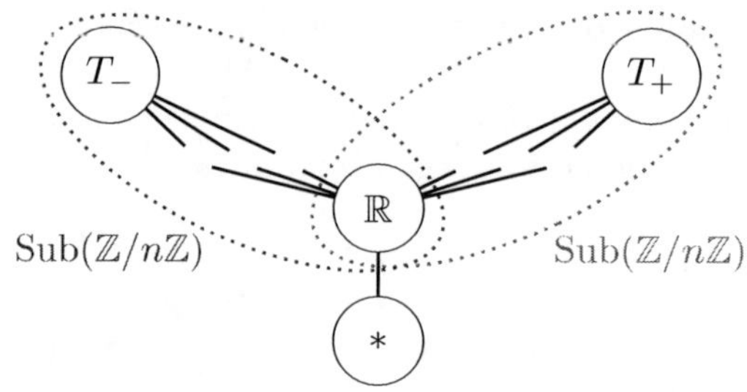

FIGURE 1. $\mathcal{H}((\mathbb{Z}/n\mathbb{Z}) \wr \mathbb{Z})$

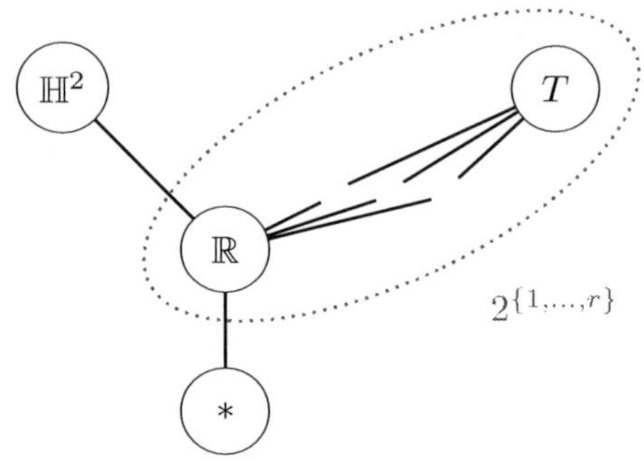

FIGURE 2. $\mathcal{H}(BS(1,n))$

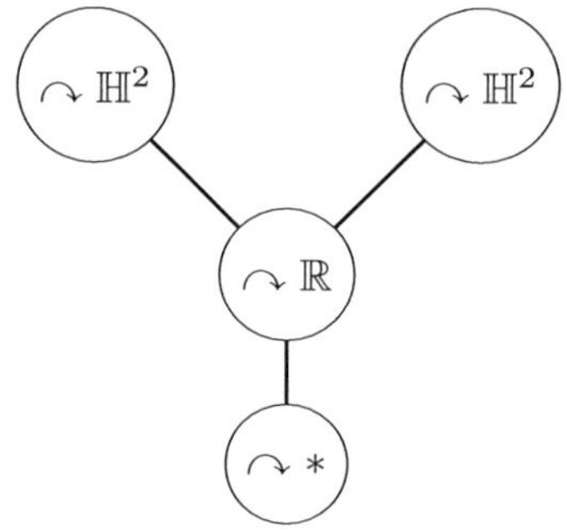

FIGURE 3. $\mathcal{H}(G)$

EXAMPLE 4.3. Consider an Anosov mapping torus $G = \mathbb{Z}^2 \rtimes_\phi \mathbb{Z}$, where $\phi \in SL(2,\mathbb{Z})$. Then $\mathcal{H}(G)$ is as pictured in Figure 3 : there is a unique lineal structure, dominated by two inequivalent quasi-parabolic actions. The confining subsets associated to the parabolic actions may be described by neighborhoods of the attracting and repelling eigenlines of ϕ in $\mathbb{R}^2$. Each quasi-parabolic action may be represented by an action on the hyperbolic plane $\mathbb{H}^2$. This description was obtained in [**ARar**].

The paper [**ABAP22**] followed from the realization that the classification results contained in [**Bal20, AR23, ARar**] fit within a common algebraic framework. By identifying this common framework and developing tools for working with groups

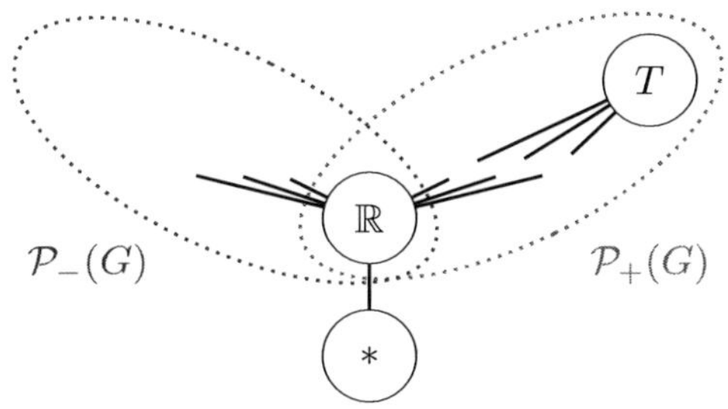

FIGURE 4. $\mathcal{H}(G)$

that fit within it, the authors were able to classify the hyperbolic actions of numerous abelian-by-cyclic groups, and recover the key results from [**Bal20, AR23**]. The main idea introduced in [**ABAP22**] was a method to convert the classification problem into a commutative algebra problem. The work utilizes confining subsets, radix representations, completions, and valuations on commutative rings, thereby surprisingly building strong relations between hyperbolic actions and commutative geometry. Before stating the main results, we describe the common framework in which they apply.

Consider a ring R that is generated as a $\mathbb{Z}$–algebra by an element γ. Assuming that γ is neither a unit nor a zero divisor, we may define an ascending HNN extension $G\colon = G(R,\gamma)$ of R by the endomorphism of (the abelian group) R defined by multiplication by γ. Assuming some natural restrictions on the algebra of R, one may write elements of R as *radix representations* of the form $a_0 + a_1\gamma + a_2\gamma^2 + \cdots$ and ensure that these are somewhat well-behaved. Whenever this setup holds, we can derive a description of the poset $\mathcal{H}(G)$ in terms of the (γ)*-adic completion* $\widehat{R}$ *of* R.

THEOREM 4.4 ([**ABAP22**, Theorem 1.1]). *Under the assumptions on G as above, the poset $\mathcal{H}(G)$ consists of the following structures (see Figure 4): a single elliptic structure, which is dominated by a single lineal structure, and two subposets $\mathcal{P}_-(G)$ and $\mathcal{P}_+(G)$ that intersect in the single lineal structure. All other structures from $\mathcal{P}_+(G)$ and $\mathcal{P}_-(G)$ are quasi-parabolic. The subposet $\mathcal{P}_+(G)$ is isomorphic to the poset of ideals of the (γ)-adic completion $\widehat{R}$ considered up to multiplication by γ. The poset $\mathcal{P}_-(G)$ is a lattice. Moreover, each element of $\mathcal{P}_+(G)$ can be represented by a quasi-parabolic action on a simplicial tree.*

The proof of Theorem 4.4 first associates to each hyperbolic action a confining subset (unique up to multiplication by γ), and then describes the confining subsets in terms of ideals of the completion; this strategy is used for other results contained in the paper as well. The theorem broadly expands our understanding of $\mathcal{H}(G)$ for a large class of groups G. In particular, roughly half the work of describing $\mathcal{H}(G)$ is reduced to the algebraic problem of classifying the ideals of the completion $\widehat{R}$ up to equivalence of ideals under multiplication by γ. In general though, the lattice $\mathcal{P}_-(G)$ described in Theorem 4.4 can be hard to get a handle on. However, the work in [**ABAP22**] also completely describes $\mathcal{P}_-(G)$ for a wide class of metabelian groups G, under some additional assumptions.

An *abelian-by-cyclic* group is one of the form $G = A \rtimes \mathbb{Z}$ where A is abelian. We focus on torsion-free, finitely presented abelian-by-cyclic groups. An abelian-by-cyclic group G can be described as an ascending HNN extension of a free abelian group $\mathbb{Z}^n$ [**BS78**]. That is, there is an endomorphism of $\mathbb{Z}^n$, represented by a matrix $\gamma \in M_n(\mathbb{Z})$ with $\det \gamma \neq 0$, such that:

$$G = \langle \mathbb{Z}^n, t : tzt^{-1} = \gamma z, \forall z \in \mathbb{Z}^n \rangle .$$

In this case, results from [**ABAP22**] completely describe $\mathcal{H}(G)$ under two hypotheses on the matrix γ; namely that γ is expanding (i.e. all eigenvalues of γ lie outside the unit disk in $\mathbb{C}$) and that $\mathbb{Z}^n$ is a cyclic $\mathbb{Z}[x]$-module, where the action of x is induced by that of γ. We call a matrix γ with these properties *admissible.*

We consider the prime factorization of the minimal polynomial (which, in this case, is also the characteristic polynomial [**HK71**, Section 7.1]) $p = up_1^{n_1} \cdots p_r^{n_r}$ in the formal power series ring $\mathbb{Z}[[x]]$, where $u \in \mathbb{Z}[[x]]$ is a unit and each $p_i \in \mathbb{Z}[[x]]$ is an irreducible power series.

THEOREM 4.5 ([**ABAP22**, Theorem 1.2]). *Let G be an ascending HNN extension of $\mathbb{Z}^n$ by an admissible matrix γ. Then $\mathcal{H}(G)$ is as described in Theorem 4.4. Further,*

(1) *$\mathcal{P}_+(G)$ is isomorphic to to the poset of divisors of $p = up_1^{n_1} \cdots p_r^{n_r}$ considered up to multiplication by a unit;*
(2) *$\mathcal{P}_-(G)$ is isomorphic to the poset of subspaces of $\mathbb{R}^n$ that are invariant under γ; and*
(3) *each element of $\mathcal{P}_-(G)$ is represented by an action on a quasi-convex subspace of a Heintze group (negatively-curved groups of the form $N \rtimes \mathbb{R}$, where N is a nilpotent Lie group).*

Theorem 4.5 reduces the classification of hyperbolic structures of abelian-by-cyclic groups based on admissible matrices γ to two algebraic computations: computing the invariant subspaces of γ and computing the prime factorization of p in the formal power series ring $\mathbb{Z}[[x]]$. Computing the invariant subspaces is straightforward using Jordan normal forms (see, e.g., [**GLR06**]), while computing factorizations in formal power series rings can be solved algorithmically (see, e.g., [**Ell14**]).

Theorems 4.4 and 4.5 completely recover the main theorems of [**AR23**] and [**Bal20**]. Moreover, they can be applied to describe the poset $\mathcal{H}(G)$ for numerous new abelian-by-cyclic groups; we give an example below. To further substantiate the extent to which the result can be applied, we note that admissible matrices are very common. In particular, the result applies to an abundant family of examples, including when p is monic, irreducible and expanding or has a square-free constant term. Exact descriptions of the poset in these cases may be found in [**ABAP22**, Corollaries 1.6 and 1.7].

EXAMPLE 4.6 (New example). Consider a group $G(\gamma) = \langle \mathbb{Z}^n, t : tzt^{-1} = \gamma z$ for all $z \in \mathbb{Z}^n \rangle$, where $\gamma \in M_n(\mathbb{Z})$ is the admissible matrix

$$\gamma = \begin{pmatrix} 0 & 0 & -210 \\ 1 & 0 & -1 \\ 0 & 1 & 0 \end{pmatrix}.$$

Theorem 4.5 can be used to show that $\mathcal{H}(G(\gamma))$ is as pictured in Figure 5.

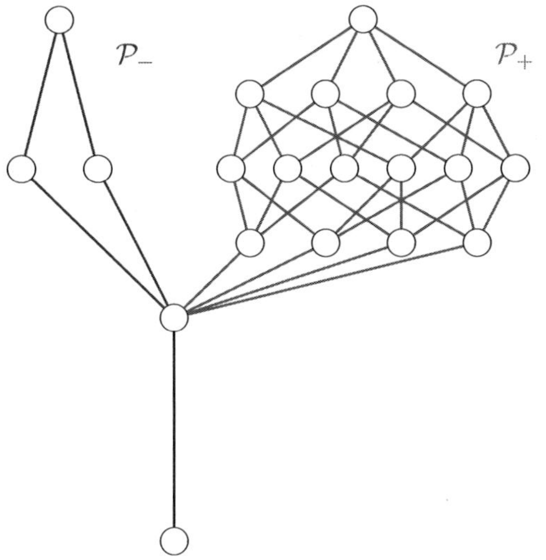

FIGURE 5. Poset $\mathcal{H}(G(\gamma))$ for the chosen admissible matrix γ. Every equivalence class to the left of the figure is represented by an action on a Heintze group, while every class to the right is represented by an action on a tree.

Part of the description of Theorem 4.5 holds in greater generality. In particular, we can drop the requirement that $\mathbb{Z}^n$ is a cyclic module, while recovering much of the information about the structure of $\mathcal{H}(G)$: the general structure of $\mathcal{H}(G)$ and parts (2) and (3) of Theorem 4.5 continue to hold. While we lose the exact description of $\mathcal{P}_+(G)$, it continues to remain a lattice (see [**ABAP22**, Theorem 1.8]).

4.2. Groups with higher rank abelianizations. While the machinery of [**ABAP22**] applies to many groups, it is not specific enough to classify the hyperbolic actions of solvable groups whose abelianizations have higher rank. Thus, the techniques described in the section above do not immediately extend to such groups. The paper [**ABR23**] began the work necessary to extend the theory in [**CCMT15**] to general finitely generated solvable groups. In particular, [**ABR23**] develops a more general definition of confining subsets for semidirect products of the form $H \rtimes \mathbb{Z}^n, n \geq 2$, which are (up to equivalence) in on-to-one correspondence with hyperbolic actions of such groups when they are solvable (also more generally, when H is amenable).

In particular, this allows us to completely describe $\mathcal{H}(G)$ of certain solvable groups that were previously out of reach. In contrast to the examples from the previous sections, whose posets are finite, these groups always admit uncountably many inequivalent actions on lines, because $\mathbb{Z}^n$ does when $n \geq 2$ (see [**ABO19**, Example 4.23]). However, the remaining hyperbolic structures can be understood for certain solvable groups.

The two main theorems of [**ABR23**], namely [**ABR23**, Theorems 1.1 and 1.2], give a correspondence between *higher rank* confining subsets and quasi-parabolic actions for groups $H \rtimes \mathbb{Z}^n$. This should be compared to [**CCMT15**, Theorem 4.1] in the case $n = 1$.

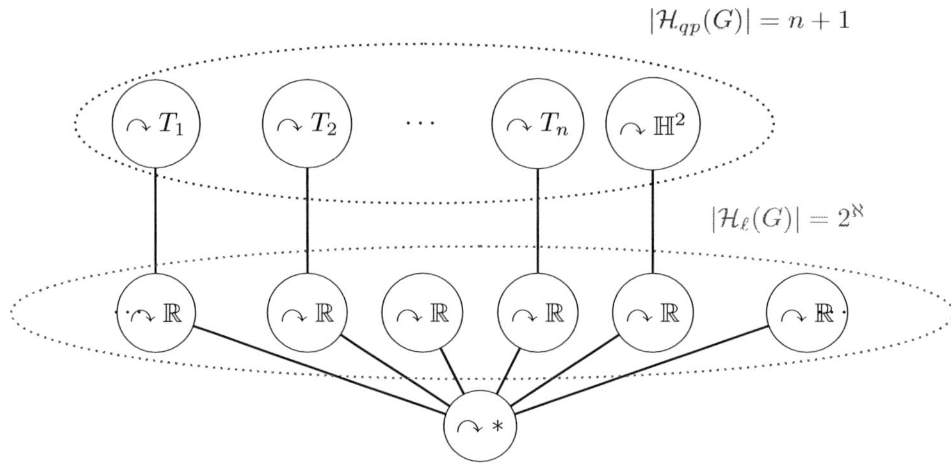

FIGURE 6. $\mathcal{H}((\mathbb{Z}/n\mathbb{Z}) \wr \mathbb{Z})$

To illustrate the use of this upgraded machinery, [**ABR23**] gives a complete description of $\mathcal{H}(G)$ for a class of groups related to solvable Baumslag-Solitar groups. If $k = p_1^{m_1} \cdots p_n^{m_n}$ is the prime factorization of k with $n \geq 2$, then the *generalized solvable Baumslag-Solitar group* G_k is defined as $G_k := \mathbb{Z}\left[\frac{1}{k}\right] \rtimes_\gamma \mathbb{Z}^n$, where the image under γ of the ith generator of $\mathbb{Z}^n$ acts on $\mathbb{Z}[\frac{1}{k}]$ by multiplication by $p_i^{m_i}$. This group has a presentation

$$G_k = \left\langle a, t_1, \ldots, t_n \mid [t_i, t_j] = 1, t_i a t_i^{-1} = a^{p_i^{m_i}} \text{ for all } i, j \right\rangle,$$

where a corresponds to a normal generator of the subgroup $\mathbb{Z}[\frac{1}{k}]$. When k is a power of a single prime number, the structure of $\mathcal{H}(G)$ can be recovered from the results of [**AR23**] discussed in the previous section.

THEOREM 4.7. *For any $k \geq 2$ which is not a power of a prime, the poset $\mathcal{H}(G_k)$ has the following structure: $\mathcal{H}_{qp}(G_k)$, the subposet of quasi-parabolic actions, consists of $n+1$ incomparable elements. Each quasi-parabolic action dominates a single lineal action; there are uncountably many lineal actions; and all lineal actions dominate a single elliptic action (see Figure 6). Moreover, every element of $\mathcal{H}(G_k)$ is represented either by an action on a tree, the hyperbolic plane, or a point.*

Theorem 4.7 reveals that the natural actions of $\mathrm{PSL}_2\left(\mathbb{Z}\left[\frac{1}{k}\right]\right)$ on the hyperbolic plane and Bruhat-Tits trees give rise to all of the non-elementary hyperbolic actions of G_k. Using a characterization of the BNS invariant of a finitely generated group in terms of its actions on trees due to Brown [**Bro87**], Theorem 4.7 can be used to compute the BNS invariants of the groups G_k, recovering a result of Sgobbi and Wong [**SW21**]. Thus, in theory, the sturcture of $\mathcal{H}(G)$ could be used to compute the BNS invariant of new groups.

4.3. Amenable groups. In this subsection, we focus on results pertaining to different types of *amenable groups* (including nilpotent and solvable groups) and when they admit hyperbolic actions. An amenable group has no free subgroups, and so their non-elementary actions are necessarily quasi-parabolic. The work contained in this section comes from [**BFGS22**], which introduced and initiates the study of Property (NL).

DEFINITION 4.8. A group G satisfies Property (NL) – standing for *No Loxodromics* – if no action by isometries of G on a hyperbolic space admits a loxodromic element. A group has Property *Hereditary* (NL) if all of its finite-index subgroups have Property (NL).

In other words, a group is (NL) if its only possible isometric actions on hyperbolic spaces are elliptic or horocyclic. These are the only types of actions we cannot rule out since every group admits elliptic actions (e.g. the trivial action on a single point) and every countably infinite group admits horocyclic actions (e.g. combinatorial horoballs on Cayley graphs [**GM08**] or trees in case the group is not finitely generated [**Ser77**]).

The motivation to study property (NL) comes from the natural question to study groups that are "inaccessible" from the viewpoint of hyperbolic actions. i.e. groups that do not admit any interesting hyperbolic actions. The approach can be compared with the study of Kazhdan's Property (T), which imposes rigidity conditions on the group by forbidding isometric actions on Hilbert spaces without a global fixed point.

Besides the obvious finite groups, several families of groups were known to have few or no interesting actions on hyperbolic spaces, such as Burnside groups, Tarski monsters, Grigorchuk's group (as they are infinite torsion groups) and Thompsons group V [**Gen19a**]. The paper [**BFGS22**] initiates a formal systematic study of this property, which interestingly, has connections to other fixed point properties (see [**BFGS22**, Section 6.1]). One important connection is to the structure of $\mathcal{H}(G)$, which follows from the appendix of [**BFGS22**].

PROPOSITION 4.9 ([**BFGS22**, Corollary 6.7]). *A group G has Property* (NL) *if and only if $\mathcal{H}(G) = \{*\}$.*

In this setting, the structure of amenable groups allows us to prove many strong results, starting with the following.

PROPOSITION 4.10 ([**BFGS22**, Proposition 3.1]). *An amenable group G has Property* (NL) *(resp. hereditary* (NL)*) if and only if every homomorphism from G (resp. a finite-index subgroup of G) to $\mathbb{R} \rtimes \mathbb{Z}/2\mathbb{Z}$ has image of order at most 2. If G is finitely generated, this amounts to saying that it does not surject (resp. virtually surject) onto $\mathbb{Z}$ nor $\mathbb{D}_\infty$.*

Abelian and nilpotent groups are widely studied classes of groups among amenable groups. The rigid structure of both these classes of groups allows us to prove even stronger results about them and Property (NL).

PROPOSITION 4.11 ([**BFGS22**, Proposition 3.2]). *Let G be a group with no non-abelian free sub-semigroup. Then G cannot admit a non-elementary action on a hyperbolic space. If G is amenable and admits a lineal action, then the action factors through a homomorphism $G \to \mathbb{R} \rtimes \mathbb{Z}/2\mathbb{Z}$.*

The above proposition is interesting as it involves the introduction and construction of a *Buseman quasi-cocylce.* The Busemann quasimorphism is classically defined only for actions fixing a point in the boundary; but here we must introduce a generalization to non-orientable lineal actions, where instead two points on the boundary are fixed setwise. Consequently, an analogue of Lemma 2.7 is also developedfor non-orientable lineal actions , see [**BFGS22**, Corollary 2.11 and Lemma 2.14] for details.

As a direct consequence of the above result, abelian groups, nilpotent groups, groups of sub-exponential growth, virtually nilpotent groups and supramenable groups satisfy $\mathcal{H}(G) = \mathcal{H}_e(G) \sqcup \mathcal{H}_\ell(G)$. Moreover, every lineal action of an abelian or nilpotent group is necessarily oriented, and factors through a homomorphism $G \to \mathbb{R}$. This leads to the following results.

COROLLARY 4.12 ([**BFGS22**, Corollary 3.5 and 3.6]). *An abelian group G has property* (NL) *(or hereditary* (NL)*) if and only if it is a torsion group. A nilpotent group G has property* (NL) *if and only if every homomorphism to $\mathbb{R}$ is trivial.*

Lastly, we discuss property (NL) among solvable groups. Unlike abelian and nilpotent groups which have no non-elementary actions, these can admit (many) quasi-parabolic actions, as seen in Section 4. The following result clarifies which solvable groups admit quasi-parabolic actions.

PROPOSITION 4.13 ([**BFGS22**, Proposition 3.7]). *A finitely generated solvable group is either virtually nilpotent or contains a finite-index subgroup admitting a quasi-parabolic action.*

COROLLARY 4.14 ([**BFGS22**, Corollary 3.8]). *A finitely generated solvable group does not admit non-elementary actions if and only if it is virtually nilpotent.*

A (virtually) nilpotent group G is (virtually) cyclic if and only if its abelianization is cyclic, in which case $\mathcal{H}(G) = \{[G], [X]\}$, where X is a finite generating set for G. If the abelianization of a nilpotent group G has higher rank, then the group admits uncountably many pairwise inequivalent actions on lines (which can be classified by inequivalent homomorphisms $G \to \mathbb{R}$) in addition to the trivial structure. This completely classifies the poset structure for virtually nilpotent groups.

5. Thompson like groups

While the work contained in [**BFGS22**] additionally covers the stability of property (NL) under group operations and its connections to other fixed point properties, one of its main results produces many new interesting examples of groups with property (NL). The main tool used to produce new examples is a dynamical criterion for groups with "rich" actions by *homeomorphisms* on compact Hausdorff spaces. Interestingly, the actions by homeomorphims record important information about the group and its subgroups, which in turn restricts the isometric hyperbolic actions such a group can admit. In particular, the criterion applies to many groups acting on the circle or a Cantor set. The result draws inspiration from the strategy used in [**Gen19a**] for Thompsons group V, but applies to a much larger class of "Thompson-like" groups.

THEOREM 5.1 ([**BFGS22**, Theorem 5.1]). *Let G be a group acting faithfully on a compact Hausdorff space X. Suppose that there is a basis $\mathcal{I}$ of non-dense and non-empty open sets in X, such that the following holds:*

1. (**C**: Complements) *For every $I \in \mathcal{I}$ there exists $J \in \mathcal{I}$ such that $I^c \subset J$.*
2. (2**T**: Double transitivity) *G preserves $\mathcal{I}$ and acts* doubly transitively *on it: for every two pairs $(I, J), (I', J') \in \mathcal{I}^{(2)}$, there exists $g \in G$ such that $(gI, gJ) = (I', J')$.*
3. (3**T**: Weak triple transitivity) *For every $g, h \in G$ there exist $(M, N, P) \in \mathcal{I}^{(3)}$ such that (M, N, P) and (gM, hN, P) are in the same G-orbit in $\mathcal{I}^{(3)}$.*
4. (**L**: Local action) *Let $(I, J, K) \in \mathcal{I}^{(3)}$, and let $g, h \in G$ be such that $gI = I$ and $hJ = J$. Then there exists $b \in G$ such that $b|_I = g|_I$ and $b|_J = h|_J$ and $b|_K = id|_K$.*

Then G has no general type hyperbolic actions.

Combined with some well known properties about the existence (or lack thereof) of unbounded quasimorphisms, [**BFGS22**] produces the following list of examples of groups with property (NL).

THEOREM 5.2 ([**BFGS22**, Theorem 1.6]). *The following groups are hereditary* (NL)*:*

(1) *Thompsons group T*
(2) *Higman–Thompson groups $V_n(r)$. As a special case, this recovers Thompsons group $V = V_2(1)$*
(3) *Some Röver–Nekrashevych groups, including Neretin's group*
(4) *Twisted Brin–Thompson groups SV_Γ whenever S is a countable faithful Γ-set*
(5) *Higman–Stein–Thompson groups $T_{n_1,\ldots,n_k}$, with $k \geq 1$ and $n_1 = 2, n_2 = 3$*
(6) *The golden ratio Thompson group T_τ*
(7) *The finitely presented group S of piecewise projective homeomorphisms of the circle, from* [**Lod19**]
(8) *Symmetrizations of Higman–Thompson groups $QV_n(r)$ and QT*

In particular, the result on twisted Brin–Thompson groups shows that groups with property (NL) are ubiquitous, in the following sense.

COROLLARY 5.3 ([**BFGS22**, Corollary 1.7]). *Every finitely generated group quasi-isometrically embeds into a finitely generated simple group, which is hereditary* (NL).

We warn the reader that our methods ensure Property (NL) under the existence of flexible enough actions on compact Hausdroff spaces, and in no way claims that all "Thompson-like" groups have Property (NL). Indeed, Braided Thompson groups have rich actions on hyperbolic spaces [**FYZ22**], and serve as a counterexample to this notion.

6. Higher rank lattices

Recently, work has emerged pertaining to the hyperbolic actions of irreducible lattices in a semi-simple group G of rank ≥ 2, which we cover in this section. Note that it follows from [**Hae20**] that if all simple factors of the ambient product have rank ≥ 2, then the lattice has property (NL). The case when G has simple factors of rank 1 is covered in [**BCFS22**]. This is an interesting case to consider as rank 1 simple groups have a natural geometric action on a hyperbolic space, so the lattices in G admit non-trivial actions on hyperbolic spaces via their projections on the rank 1 simple factors.

The setup considered in [**BCFS22**] is slightly different from that of cobounded hyperbolic actions. Given an action on a hyperbolic space, one can make a larger hyperbolic space containing the first one as a quasiconvex subspace while maintaining important properties of the original action. For instance, this can be achieved by attaching geodesic rays equivariantly. To take this possibility into account, the work in [**BCFS22**] focuses on *coarsely minimal* actions.

DEFINITION 6.1. We say that an action $G \curvearrowright X$ on a hyperbolic space is coarsely minimal if X is unbounded, the limit set of G on ∂X is not a single point, and every quasi-convex G-invariant subset of X is coarsely dense.

Although coarsely minimal actions provide a broader setup than cobounded actions, it closely links to the structure of $\mathcal{H}(G)$. Indeed, for any hyperbolic structure, the action is either coarsely minimal or elliptic. Thus, the number of hyperbolic structures is one more than the number of equivalence classes of coarsely minimal actions; where the equivalence on coarsely minimal actions is provided roughly by $G-$equivariant quasi-isometries; we refer the reader to [**BCFS22**, Definition 3.5] for a precise description. Consequently, if one classifies all coarsely minimal actions up to equivalence, the elements of $\mathcal{H}(G)$ can be easily deduced.

In what follows, the notion of a standard rank one group is taken from [**CCMT15**, Theorem D] – a locally compact group G is a standard rank one group if it has no nontrivial compact normal subgroups and either

(1) G is the group of isometries or orientation-preserving isometries of a rank one symmetric space X of non-compact type, or
(2) G has a continuous, proper, faithful action by automorphisms on a locally finite non-elementary tree T , without inversions and with exactly two orbits of vertices, such that the action of G on the set of ends ∂T is 2-transitive.

The symmetric space X in case (1) and the tree T in case (2) are called the model spaces for the standard rank one group. While standard rank one groups of type (1) correspond to real Lie groups of rank one, type (2) includes simple algebraic groups over non-archimedean local fields of rank one. With this terminology, the main result of [**BCFS22**] is as follows.

THEOREM 6.2 ([**BCFS22**, Theorem 1.1]). *Let $N \geq n \geq 0$ be integers. Let $G = \prod_{i=1}^{N} G_i$ be a product of N locally compact groups, where G_i is a standard rank one group for all $i \in \{1, \cdots, n\}$, and G_j is a simple algebraic group defined over a local field k_j with $\mathrm{rk}_{k_j}(G_j) \geq 2$ for all $j \in \{n+1, \cdots, N\}$. Let $\Gamma < G$ be a lattice. Assume that $n \geq 2$ or that $N > n$. If $N > 1$, assume in addition that Γ has a dense projection to each proper sub-product. Then any coarsely minimal action of Γ on a geodesic hyperbolic space is equivalent to one of the actions*

$$\Gamma \to G \xrightarrow{pr_i} G_i \to Isom(X_i, d_i) \quad (\text{for } 1 \leq i \leq n)$$

where each X_i is a rank-one symmetric space or a tree, corresponding to the standard rank one factor G_i being of type (1) or (2).

In other words, the only possible actions on hyperbolic spaces of such lattices arise from their projections to the rank 1 simple factors. Therefore, in the language of [**ABO19**], the result implies that the lattices under consideration have exactly $n+1$ inequivalent hyperbolic structures: n from the projections to the factors, which are pairwise inequivalent and all of which dominate the trivial structure. The result allows for the classification of the poset $\mathcal{H}(G)$ for new groups, including the famous Burger-Mozes groups, which are irreducible lattices Γ in a product $G_1 \times G_2$ of two standard rank one groups $G_i < Aut(T_i), i = 1, 2$. [**BM97**].

COROLLARY 6.3 ([**BCFS22**, Example 1.5]). *Every Burger-Mozes group has two non-trivial hyperbolic structures. Specifically, the only non-trivial hyperbolic structures of* Γ *come from the actions on the trees* T_1 *and* T_2.

Other new examples from [**BCFS22**] include producing groups with exactly n hyperbolic structures, for each $n \geq 1$; different from the ones previously produced in [**ABO19**]. These are given by an irreducible lattice in the semi-simple real Lie group $\mathrm{SO}(4,1) \times \mathrm{SO}(3,2)$ (for $n = 1$) and the groups $G = \mathrm{SL}_2\left(\mathbb{Z}\left[\frac{1}{p_2}, \cdots, \frac{1}{p_n}\right]\right)$ for $n-1$ distinct primes $p_2, \cdots, p_n$ for $n \geq 2$; see [**BCFS22**, Examples 1.3 and 1.4] for details.

7. New examples and open questions

In this last section, we record some examples of interest and pose related open problems that could inspire future directions of research.

7.1. Groups with finitely many quasi-parabolic actions. By Example 4.1, the lamplighter group $\mathcal{L}_2 = (\mathbb{Z}/2\mathbb{Z}) \wr \mathbb{Z}$ has exactly 2 quasi-parabolic structures. It follows from [**ABO19**, Lemma 4.20], that the group $H_i = \bigoplus_{j=1}^{i} \mathcal{L}_2$ has exactly $2i$ quasi-parabolic structures for all $i \geq 1$.

Further, the group G_6 from Theorem 4.7 has exactly 3 quasi-parabolic structures. Consequently, the groups $K_i = H_i \oplus G_6$ have exactly $2i + 3$ quasi-parabolic structures for all $i \geq 1$. This produces groups with exactly n quasi-parabolic structures for every $n \geq 2$, almost completely answering an open problem from [**ABO19**]. It remains to consider the following.

QUESTION 7.1. Does there exist a group G such that $|\mathcal{H}_{qp}(G)| = 1$?

Such an example, if it exists, could lead to an interesting study of inherent properties that a group must have in order for it to admit only one-quasi-parabolic action. If such a group cannot exist, then the proof of this fact would also be interesting.

7.2. Embeddings into groups with property (NL). Recall that Corollary 5.3 allows us to quasi-isometrically embed any countable group into a finitely generated, simple group with property (NL). The result uses a quasi-isometric embedding into twisted Brin-Thompson groups; however this quasi-isometric embedding is actually an injective group homomorphism. So, we actually can *embed* every countable group into a finitely generated, simple group with (NL).

We give a new construction that also allows us to embed any countable group into a simple group with property (NL). But this new result uses groups where any two elements of the same order are conjugate, and twisted Brin-Thompson groups do not have this property. For this, we need the following result by Osin.

THEOREM 7.2 ([**Osi10**, Theorem 1.1]). *Any countable group G can be embedded into a 2-generated group C such that any two elements of the same order are conjugate in C.*

The above immediately allows us to prove the following.

PROPOSITION 7.3. *Any countable group G can be embedded into a finitely generated group with property* (NL).

PROOF. Given G, we use Theorem 7.2 to embed G into the group C. We claim that C has property (NL). Indeed, any element c of infinite order is conjugate to c^2 in C (since they are both infinite order). It follows that the group $\langle c \rangle$ cannot be quasi-isometrically embedded in any action on a hyperbolic space; which implies that c cannot be a loxodromic element. $\square$

REMARK 7.4. Observe that the above proof can be adapted to see that any group with finitely many conjugacy classes also has property (NL). In particular, the torsion free finitely generated group with exactly two conjugacy classes constructed in [**Osi10**] has property (NL).

We can further utilize the result by Osin to prove the claimed result.

PROPOSITION 7.5. *Every countable group G with an infinite order element can be embedded into a finitely generated, simple group with property* (NL).

PROOF. Given a finitely generated group G, consider the countable group $H = \bigoplus_n (\mathbb{Z}/n\mathbb{Z}) \oplus G$. We first embed H into a countable, simple group S. Next, we embed S into the group C from Theorem 7.2. Then G embeds into C as well. Clearly, C is finitely generated and has property (NL); this follows from the proof of the previous proposition. So, it remains to show that C is simple in this case.

To this end, consider any element $a \in C$. Let a^C denote the conjugates of a by an element of C. Then $a^C \cap S \neq \{1\}$, as $S \leq C$ contains elements of all orders. This means that all elements of C are conjugate to an element of S. Suppose that C contains a non-trivial normal subgroup N, we consider the quotient C/N. If a is the kernel of this quotient, then so is an element of S. As S is simple, this means that S is in the kernel of the quotient, but then so is the entire group C. Thus $N = C$, and so C is simple. $\square$

These different embedding results prompt the following questions.

QUESTION 7.6. Given a finitely generated group G, in how many distinct ways can G (quasi-isometrically) embed into a group with property (NL)?

QUESTION 7.7. For a countable group G with infinitely many conjugacy classes, what "algebraic" conditions are sufficient to imply property (NL)?

7.3. Anosov mapping torii. Recall that [**ARar**] obtained the classification of $\mathcal{H}(G)$ for the fundamental group of the Anosov mapping torus $\mathbb{Z}^2 \rtimes_\phi \mathbb{Z}$, where ϕ is a matrix in $SL(2, \mathbb{Z})$; see Figure 3.

The work in [**ABAP22**] does not currently subsume this result, as it deals with the case when ϕ is expanding, which a matrix from $SL(2, \mathbb{Z})$ fails to satisfy. However, it is reasonable, especially in light of part (2) of Theorem 4.5, if the condition that requires an expanding matrix can be dropped. To be precise, we believe that the following question can be answered affirmatively.

QUESTION 7.8. Consider the group $G = \mathbb{Z}^n \rtimes_\phi \mathbb{Z}$, where $\phi \in SL(n, \mathbb{Z})$. Are the quasi-parabolic structures of G are in one-to-one correspondence with the proper invariant subspaces of ϕ in $\mathbb{R}^n$?

7.4. Further classification results. We expect that the theory developed in [**ABAP22**, **ABR23**] will lay the groundwork for future, more general classifications of hyperbolic actions of metabelian groups. In particular, we may consider the following question.

QUESTION 7.9. Can Theorems 4.4 and the techniques developed in [**Bal20**, **AR23**, **ARar**] be applied to classify the hyperbolic actions of wreath products $A \wr B$ and extensions $A \rtimes B$ when A and B are finitely generated abelian groups?

As the work contained in [**ABR23**, **ABAP22**] heavily relies on the semi-direct product structure of the group, we may also ask the following.

QUESTION 7.10. Can the theory of confining subsets be extended to solvable groups that do not necessarily admit a semi-direct product decomposition?

References

[ABAP22] C.R. Abbott, S.H. Balasubramanya, A.J.Rasmussen, and S. Payne, *Valuations, completions, and hyperbolic actions of metabelian groups*, arXiv:2207.12945 (2022).

[Abb16] C. R. Abbott, *Not all finitely generated groups have universal acylindrical actions*, Proc. Amer. Math. Soc. **144** (2016), no. 10, 4151–4155, DOI 10.1090/proc/13101. MR3531168

[ABB+21] C. Abbott, J. Behrstock, and M. G. Durham, *Largest acylindrical actions and stability in hierarchically hyperbolic groups*, Trans. Amer. Math. Soc. Ser. B **8** (2021), 66–104, DOI 10.1090/btran/50. With an appendix by Daniel Berlyne and Jacob Russell. MR4215647

[ABO19] C. Abbott, S. H. Balasubramanya, and D. Osin, *Hyperbolic structures on groups*, Algebr. Geom. Topol. **19** (2019), no. 4, 1747–1835, DOI 10.2140/agt.2019.19.1747. MR3995018

[ABR23] C. R. Abbott, S. H. Balasubramanya, and A. J. Rasmussen, *Higher rank confining subsets and hyperbolic actions of solvable groups*, Adv. Math. **424** (2023), Paper No. 109045, 47, DOI 10.1016/j.aim.2023.109045. MR4585997

[AH21] C. R. Abbott and D. Hume, *Actions of small cancellation groups on hyperbolic spaces*, Geom. Dedicata **212** (2021), 325–363, DOI 10.1007/s10711-020-00561-3. MR4251676

[ANR23] C. R. Abbott, H. T. Nguyen, and A. J. Rasmussen, *Largest hyperbolic actions of 3-manifold groups*, 2023.

[AR23] C. R. Abbott and A. J. Rasmussen, *Actions of solvable Baumslag-Solitar groups on hyperbolic metric spaces*, Algebr. Geom. Topol. **23** (2023), no. 4, 1641–1692, DOI 10.2140/agt.2023.23.1641. MR4602410

[ARar] ———, *Largest hyperbolic actions and quasi-parabolic actions in groups*, J. Topol. Anal. (To appear).

[Bal20] S. H. Balasubramanya, *Hyperbolic structures on wreath products*, J. Group Theory **23** (2020), no. 2, 357–383, DOI 10.1515/jgth-2018-0176. MR4069979

[BCFS22] U. Bader, P. E. Caprace, A. Furman, and A. Sisto, *Hyperbolic actions of higher-rank lattices come from rank-one factors*.

[BF02] M. Bestvina and K. Fujiwara, *Bounded cohomology of subgroups of mapping class groups*, Geom. Topol. **6** (2002), 69–89, DOI 10.2140/gt.2002.6.69. MR1914565

[BFGS22] S. H. Balasubramanya, F. Fournier, A. Genevois, and A. Sisto, *Property* NL *for group actions on hyperbolic spaces*, Accepted in Groups, Geometry and Dynamics (2022).

[BM97] M. Burger and S. Mozes, *Finitely presented simple groups and products of trees* (English, with English and French summaries), C. R. Acad. Sci. Paris Sér. I Math. **324** (1997), no. 7, 747–752, DOI 10.1016/S0764-4442(97)86938-8. MR1446574

[Bow08] B. H. Bowditch, *Tight geodesics in the curve complex*, Invent. Math. **171** (2008), no. 2, 281–300, DOI 10.1007/s00222-007-0081-y. MR2367021

[Bri02] M. R. Bridson, *The geometry of the word problem*, Invitations to geometry and topology, Oxf. Grad. Texts Math., vol. 7, Oxford Univ. Press, Oxford, 2002, pp. 29–91. MR1967746

[Bro87] K. S. Brown, *Trees, valuations, and the Bieri-Neumann-Strebel invariant*, Invent. Math. **90** (1987), no. 3, 479–504, DOI 10.1007/BF01389176. MR914847

[BS78] R. Bieri and R. Strebel, *Almost finitely presented soluble groups*, Comment. Math. Helv. **53** (1978), no. 2, 258–278, DOI 10.1007/BF02566077. MR498863

[CCMT15] P.-E. Caprace, Y. Cornulier, N. Monod, and R. Tessera, *Amenable hyperbolic groups*, J. Eur. Math. Soc. (JEMS) **17** (2015), no. 11, 2903–2947, DOI 10.4171/JEMS/575. MR3420526

[CK02] C. B. Croke and B. Kleiner, *The geodesic flow of a nonpositively curved graph manifold*, Geom. Funct. Anal. **12** (2002), no. 3, 479–545, DOI 10.1007/s00039-002-8255-7. MR1924370

[DGO17] F. Dahmani, V. Guirardel, and D. Osin, *Hyperbolically embedded subgroups and rotating families in groups acting on hyperbolic spaces*, Mem. Amer. Math. Soc. **245** (2017), no. 1156, v+152, DOI 10.1090/memo/1156. MR3589159

[Ell14] J. Elliott, *Factoring formal power series over principal ideal domains*, Trans. Amer. Math. Soc. **366** (2014), no. 8, 3997–4019, DOI 10.1090/S0002-9947-2014-05903-5. MR3206450

[FYZ22] F. Fournier, Y. Lodha, and M. C. B. Zaremsky, *Braided Thompson groups with and without quasimorphisms*, Algebr. Geom. Topol. To appear. `arXiv:2204.05272`, 2022.

[Gen19a] A. Genevois, *Hyperbolic and cubical rigidities of Thompson's group V*, J. Group Theory **22** (2019), no. 2, 313–345, DOI 10.1515/jgth-2018-0103. MR3918481

[Gen19b] A. Genevois, *Hyperbolicities in cat(0) cube complexes*, 2019.

[GLR06] I. Gohberg, P. Lancaster, and L. Rodman, *Invariant subspaces of matrices with applications*, Classics in Applied Mathematics, vol. 51, Society for Industrial and Applied Mathematics (SIAM), Philadelphia, PA, 2006. Reprint of the 1986 original, DOI 10.1137/1.9780898719093. MR2228089

[GM08] D. Groves and J. F. Manning, *Dehn filling in relatively hyperbolic groups*, Israel J. Math. **168** (2008), 317–429, DOI 10.1007/s11856-008-1070-6. MR2448064

[Gro87] M. Gromov, *Hyperbolic groups*, Essays in group theory, Math. Sci. Res. Inst. Publ., vol. 8, Springer, New York, 1987, pp. 75–263, DOI 10.1007/978-1-4613-9586-7_3. MR919829

[Hae20] T. Haettel, *Hyperbolic rigidity of higher rank lattices* (English, with English and French summaries), Ann. Sci. Éc. Norm. Supér. (4) **53** (2020), no. 2, 439–468, DOI 10.24033/asens.2425. With an appendix by Vincent Guirardel and Camille Horbez. MR4094562

[HK71] K. Hoffman and R. Kunze, *Linear algebra*, 2nd ed., Prentice-Hall, Inc., Englewood Cliffs, NJ, 1971. MR276251

[HRSS23] M. Hagen, J. Russell, A. Sisto, and D. Spriano, *Equivariant hierarchically hyperbolic structures for 3-manifold groups via quasimorphisms*, 2023.

[HS20] M. F. Hagen and T. Susse, *On hierarchical hyperbolicity of cubical groups*, Israel J. Math. **236** (2020), no. 1, 45–89, DOI 10.1007/s11856-020-1967-2. MR4093881

[Lod19] Y. Lodha, *A finitely presented infinite simple group of homeomorphisms of the circle*, J. Lond. Math. Soc. (2) **100** (2019), no. 3, 1034–1064, DOI 10.1112/jlms.12254. MR4048731

[Man06] J. F. Manning, *Quasi-actions on trees and property (QFA)*, J. London Math. Soc. (2) **73** (2006), no. 1, 84–108, DOI 10.1112/S0024610705022738. With an appendix by N. Monod and B. Rémy. MR2197372

[Man08] J. F. Manning, *Actions of certain arithmetic groups on Gromov hyperbolic spaces*, Algebr. Geom. Topol. **8** (2008), no. 3, 1371–1402, DOI 10.2140/agt.2008.8.1371. MR2443247

[Osi10] D. Osin, *Small cancellations over relatively hyperbolic groups and embedding theorems*, Ann. of Math. (2) **172** (2010), no. 1, 1–39, DOI 10.4007/annals.2010.172.1. MR2680416

[Osi16] D. Osin, *Acylindrically hyperbolic groups*, Trans. Amer. Math. Soc. **368** (2016), no. 2, 851–888, DOI 10.1090/tran/6343. MR3430352

[Sel97] Z. Sela, *Acylindrical accessibility for groups*, Invent. Math. **129** (1997), no. 3, 527–565, DOI 10.1007/s002220050172. MR1465334

[Ser77] J.-P. Serre, *Arbres, amalgames,* SL_2 (French), Astérisque, No. 46, Société Mathématique de France, Paris, 1977. Avec un sommaire anglais; Rédigé avec la collaboration de Hyman Bass. MR476875

[SW21] W. Sgobbi and P. Wong, *The BNS invariants of the generalized solvable Baumslag-Solitar groups and of their finite index subgroups*, Comm. Algebra **51** (2023), no. 8, 3354–3370, DOI 10.1080/00927872.2023.2183028. MR4585886

[Val21] M. Valiunas, *Acylindrical hyperbolicity of groups acting on quasi-median graphs and equations in graph products*, Groups Geom. Dyn. **15** (2021), no. 1, 143–195, DOI 10.4171/ggd/595. MR4235751

Current address: Department of Mathematics, Lafayette College, Easton, Pennsylvania 18042

Email address: `hassanba@lafayette.edu`

Contemporary Mathematics
Volume **812**, 2025
https://doi.org/10.1090/conm/812/16263

Stallings's Fibring Theorem and PD^3-pairs

Martin R. Bridson, Dawid Kielak, and Monika Kudlinska

ABSTRACT. We give a relatively self-contained proof that if a group G fibres algebraically and is part of a PD^3-pair, then G is the fundamental group of a fibred compact aspherical 3-manifold. This yields a homological proof of a classical theorem of Stallings: if $G = \pi_1(M^3)$ is the fundamental group of a compact irreducible 3-manifold M^3 and $\phi\colon G \to \mathbb{Z}$ is a surjective homomorphism with finitely generated kernel, then ϕ is induced by a topological fibration of M^3 over the circle.

In this note we present a homological proof of the following result, which was proved by Stallings [**Sta62**] in the case when M^3 is a compact irreducible manifold and the kernel K is not isomorphic to $\mathbb{Z}/2\mathbb{Z}$.

THEOREM 1. *Let M^3 be a compact 3-manifold with aspherical (possibly empty) boundary. Suppose that there exists a short exact sequence*

$$1 \to K \to \pi_1(M^3) \xrightarrow{\phi} \mathbb{Z} \to 1$$

with K finitely generated. Then M^3 is the total space of a fibre bundle

$$\Sigma \to M^3 \xrightarrow{\Phi} \mathbb{S}^1$$

with Φ inducing ϕ.

Let G be a group and $\mathcal{H}$ a family of (not necessarily distinct) subgroups of G. Following Bieri–Eckmann [**BE78**], we say that $(G, \mathcal{H})$ is an n-dimensional *Poincaré duality pair*, abbreviated to PD^n-pair, if there exists a G-module C which is isomorphic to $\mathbb{Z}$ as an abelian group, called the *dualising module*, and natural isomorphisms

$$\begin{aligned} H^k(G; M) &\simeq H_{n-k}(G, \mathcal{H}; C \otimes M) \\ H^k(G, \mathcal{H}; M) &\simeq H_{n-k}(G; C \otimes M) \end{aligned} \tag{1}$$

for all G-modules M and all $k \in \mathbb{Z}$, where we tensor over $\mathbb{Z}$, and the action of G on $C \otimes M$ is the diagonal action. We refer the reader to [**BE78**] for the definition of relative (co)homology of groups. When $\mathcal{H} = \emptyset$, the above reduces to the definition of a PD^n group.

2020 *Mathematics Subject Classification.* Primary 20J05, 57K30.

This work has received funding from the European Research Council (ERC) under the European Union's Horizon 2020 research and innovation programme (Grant agreement No. 850930).

The combined work of Eckmann, Müller and Linnell in [**EM80, EL83**] shows that every PD^2 group is the fundamental group of a closed aspherical 2-manifold. (There are at least two alternate, more modern proofs of this fact, by Bowditch [**Bow04**] and by P. Kropholler and the second author [**KK21**].) From this, one can deduce that any PD^2-pair $(G, \mathcal{H})$ is *geometric* (see the discussion at the start of Section 11 in [**BE78**]): that is, G is the fundamental group of a compact 2-manifold Σ and $\mathcal{H}$ is a (possibly empty) collection of infinite cyclic subgroups of G which correspond to the boundary components of Σ.

Famously, for $n \geqslant 3$ it is conjectured that every finitely presented PD^n group is the fundamental group of an aspherical n-manifold. Theorem 2 provides some evidence for this conjecture, and is part of a more general attempt to find homological alternatives to topological arguments. Another interesting example of this approach is a recent article of Reeves, Scott and Swarup [**RSS20**]. We refer the reader to Hillman's book [**Hil20**] for an encyclopaedic treatment of this topic.

THEOREM 2. *Let $(G, \mathcal{H})$ be a PD^3-pair. If G admits a homomorphism onto $\mathbb{Z}$ with finitely generated kernel then G is the fundamental group of an aspherical 3-manifold M and $\mathcal{H}$ is the family of the fundamental groups of the boundary components of M. Moreover, the manifold M fibres over the circle and each of the boundary components is either a torus or a Klein bottle.*

When $\mathcal{H} = \emptyset$ and the kernel is of type FP_2, this result was proved by Thomas [**Tho84**, Theorem 5], and Hillman later showed [**Hil02**, Theorem 1.19] that it suffices to know that the kernel is finitely generated.

The case with boundary, that is, $\mathcal{H} \neq \emptyset$, is more tricky. Hillman [**Hil87**, Theorem 2] recovers most of Theorem 2 when the kernel is again of type FP_2. He relies on the following result of Bieri [**Bie81**, Corollary 8.6]: if G is a group of cohomological dimension 2, then any normal infinite-index subgroup $K \leqslant G$ of type FP_2 must be free. The analogue of Bieri's result is false if one assumes only that K is finitely generated, even in the case $G/K \simeq \mathbb{Z}$, with the Rips construction [**Rip82**] providing an ample source of counterexamples.

To circumvent this finiteness issue, one can use the theory of Novikov homology. This was done by Hillman [**Hil20**, Theorem 9.9] via the work of Kochloukova [**Koc06**]. We also use Novikov rings in our proof, but in a more direct fashion.

The *Novikov ring* $\widehat{\mathbb{Z}G}^{\phi}$ associated to a non-trivial homomorphism $\phi\colon G \to \mathbb{Z}$ is the ring obtained by taking all functions $G \to \mathbb{Z}$ whose support intersected with $\phi^{-1}((-\infty, \kappa])$ is finite for every $\kappa \in \mathbb{Z}$. Since $\widehat{\mathbb{Z}G}^{\phi}$ contains $\mathbb{Z}G$, multiplication turns it into a $\mathbb{Z}G$-module. The only fact about Novikov rings that we are going to use is the generalised Sikorav theorem:

$$\begin{gathered}\text{The group } K = \ker\phi \text{ is of type } \mathrm{FP}_m \text{ if and only if} \\ H_i(G; \widehat{\mathbb{Z}G}^{\phi}) = 0 = H_i(G; \widehat{\mathbb{Z}G}^{-\phi}) \text{ for every } i \leqslant m.\end{gathered} \tag{2}$$

The case $m = 1$ was proved by Sikorav in his thesis [**Sik87**]; the more general version above is a combination of theorems of Schweitzer [**Bie07**, Theorm A.1] and Bieri–Renz [**BR88**, Theorem B]. For a yet more general statement, see the work of S. Fisher [**Fis21**].

PROOF OF THEOREM 2. Let $(G, \mathcal{H})$ be as in the statement, with $\mathcal{H} = \{H_i\}_{i \in I}$. Let K be the kernel of the epimorphism $\phi\colon G \to \mathbb{Z}$ and suppose that K is finitely generated. We start by showing that K is of type FP_2. When $\mathcal{H} = \emptyset$, this is a special case of Hillman's result [**Hil02**, Theorem 1.19]; another argument can be extracted from the proof of [**Kie20**, Theorem 5.32]. When $\mathcal{H} \neq \emptyset$, the possibility of some element of $\mathcal{H}$ lying entirely in K causes a certain amount of annoyance, and stands in the way of using Hillman's proof. The second argument does however generalise, in a rather non-direct way, as follows.

Since we are proving that K is of type FP_2, it is enough to prove the statement for a finite-index subgroup of K. Hence we will pass to a subgroup of G of index at most 2 that, together with a suitable collection of peripheral subgroups consisting of subgroups of index 2 of groups in $\mathcal{H}$ or two copies of such groups, forms an *orientable* PD^3-pair. We do this in order to make the dualising module trivial, and hence to avoid worries about tensoring with it. This subgroup of index at most 2 inherits an epimorphism to $\mathbb{Z}$ with finitely generated kernel, and hence we will lose no generality by assuming that $(G,\mathcal{H})$ is an orientable PD^3-pair for this part of the argument.

We now consider a part of the long exact sequence for the group pair $(G,\mathcal{H})$:

$$H^0(\mathcal{H};\widehat{\mathbb{Z}G}^{\phi}) \to H^1(G,\mathcal{H};\widehat{\mathbb{Z}G}^{\phi}) \to H^1(G;\widehat{\mathbb{Z}G}^{\phi}) \to H^1(\mathcal{H};\widehat{\mathbb{Z}G}^{\phi}) \to H^2(G,\mathcal{H};\widehat{\mathbb{Z}G}^{\phi}).$$

Since K is finitely generated, that is, of type FP_1, we know that $H_i(G;\widehat{\mathbb{Z}G}^{-\phi}) = 0$ for $i \in \{0,1\}$. We claim that the cohomology groups $H^i(G;\widehat{\mathbb{Z}G}^{\phi})$ also vanish when $i \in \{0,1\}$. Indeed, after picking a resolution of $\mathbb{Z}$ by free finitely generated $\mathbb{Z}G$-modules, this allows us to construct a partial chain contraction over $\widehat{\mathbb{Z}G}^{-\phi}$ witnessing the vanishing of homology up to degree one. Dualising, this gives us a partial chain contraction over $\widehat{\mathbb{Z}G}^{\phi}$ that witnesses the vanishing of $H^i(G;\widehat{\mathbb{Z}G}^{\phi})$ for $i \in \{0,1\}$. We plug this information back into the long exact sequence and obtain

$$H^0(\mathcal{H};\widehat{\mathbb{Z}G}^{\phi}) \to H^1(G,\mathcal{H};\widehat{\mathbb{Z}G}^{\phi}) \to 0 \to H^1(\mathcal{H};\widehat{\mathbb{Z}G}^{\phi}) \to H^2(G,\mathcal{H};\widehat{\mathbb{Z}G}^{\phi}).$$

The last term is isomorphic to $H_1(G;\widehat{\mathbb{Z}G}^{\phi}) = 0$ by Poincaré duality, forcing

$$0 = H^1(\mathcal{H};\widehat{\mathbb{Z}G}^{\phi}) = \bigoplus_{H\in\mathcal{H}} H^1(H;\widehat{\mathbb{Z}G}^{\phi}).$$

Now, the groups $H \in \mathcal{H}$ are themselves orientable PD^2 groups by [**BE78**, Theorem 4.2(iii)], and hence are fundamental groups of closed orientable aspherical surfaces. Poincaré duality gives us

$$0 = H^1(H;\widehat{\mathbb{Z}G}^{\phi}) \simeq H_1(H;\widehat{\mathbb{Z}G}^{\phi}).$$

If ϕ does not vanish on H, then $H_0(H;\widehat{\mathbb{Z}G}^{\phi}) = 0$ as well, and therefore the generalised Sikorav theorem tells us that $K \cap H$ is finitely generated. This implies that $H \simeq \mathbb{Z}^2$, since no other closed orientable aspherical surface has fundamental group that algebraically fibres.

Now let us focus on the initial part of the long exact sequence above. Poincaré duality tells us that

$$H^0(\mathcal{H};\widehat{\mathbb{Z}G}^{\phi}) = \bigoplus_{H\in\mathcal{H}} H^0(H;\widehat{\mathbb{Z}G}^{\phi}) \simeq \bigoplus_{H\in\mathcal{H}} H_2(H;\widehat{\mathbb{Z}G}^{\phi}).$$

If ϕ does not vanish on H, then we have just shown that $\ker \phi|_H \simeq \mathbb{Z}$, which is of type FP_2, and hence $H_2(H; \widehat{\mathbb{Z}G}^{\phi}) = 0$. If $H \leqslant K$, then $\widehat{\mathbb{Z}G}^{\phi}$ is a flat $\mathbb{Z}H$-module, and so $H_2(H; \widehat{\mathbb{Z}G}^{\phi}) = H_2(H; \mathbb{Z}H) \otimes_{\mathbb{Z}H} \widehat{\mathbb{Z}G}^{\phi} = 0 \otimes_{\mathbb{Z}H} \widehat{\mathbb{Z}G}^{\phi} = 0$. We conclude that

$$H^0(\mathcal{H}; \widehat{\mathbb{Z}G}^{\phi}) = 0,$$

and plugging this back into the long exact sequence yields $H^1(G, \mathcal{H}; \widehat{\mathbb{Z}G}^{\phi}) = 0$. But by Poincaré duality, this abelian group is isomorphic to $H_2(G; \widehat{\mathbb{Z}G}^{\phi})$. An analogous argument yields that $H_2(G; \widehat{\mathbb{Z}G}^{-\phi}) = 0$, and the generalised Sikorav theorem (2) tells us that $K = \ker \phi$ is of type FP_2, as claimed.

We now return to the general case, that is, we no longer assume $(G, \mathcal{H})$ to be orientable; we retain however the newly acquired information that K is of type FP_2.

For each $i \in I$, let X_i be a system of double coset representatives of $K \backslash G / H_i$. Fix $t \in G$ to be a lift of a generator of $\mathbb{Z}$. Note that for every $H_i \in \mathcal{H}$, the cardinality of X_i is given by the index n_i of $\phi(H_i)$ in $\mathbb{Z}$, and we can take $X_i = \{1, t, \ldots, t^{n_i - 1}\}$ if $n_i < \infty$, and $X_i = \{t^i \mid i \in \mathbb{Z}\}$ otherwise. Define the system of subgroups

$$\mathcal{H}_K = \bigcup_{i \in I} \{K \cap x^{-1} H_i x \mid x \in X_i\}.$$

We claim that $(K, \mathcal{H}_K)$ is a PD^2-pair. When $\mathcal{H} = \emptyset$, this follows from [**Hil02**, Theorem 1.19]; when $\mathcal{H} \neq \emptyset$, it follows from [**Hil20**, Theorem 9.9]. We will give a direct proof of the claim following the outline of [**Bie81**, Theorem 9.11].

If $\mathcal{H} \neq \emptyset$, then $\mathrm{cd}(G) = 2$ and hence $\mathrm{cd}(K) \leqslant 2$. If $\mathcal{H} = \emptyset$, then $\mathrm{cd}(K) \leqslant 2$ by [**Str77**]. In either case, by the argument above, K is of type FP_2 and thus it is of type FP.

When $\mathcal{H} = \emptyset$, set $\Delta_{K/\mathcal{H}_K}$ to be the augmentation ideal in $\mathbb{Z}K$. When $\mathcal{H} \neq \emptyset$, let $\Delta_{K/\mathcal{H}_K}$ denote the kernel of the augmentation map

$$\bigoplus_{i \in I,\, x \in X_i} \mathbb{Z}[K / K \cap x^{-1} H_i x] \to \mathbb{Z}$$

defined by linearly extending the map which sends each coset $g(K \cap x^{-1} H_i x)$ to a fixed generator of $\mathbb{Z}$. In either case, write $\widetilde{\Delta}_{K/\mathcal{H}_K} = C \otimes \Delta_{K/\mathcal{H}_K}$, where C is the dualising module of the PD^3-pair $(G, \mathcal{H})$ and the action of $K \leqslant G$ on C is the restricted action; the tensor product is taken over $\mathbb{Z}$, and the action of K on it is diagonal.

Consider the Lyndon–Hochschild–Serre spectral sequence

$$E_2^{pq} : H^p(G/K; H^q(K, \mathbb{Z}G)) \Rightarrow H^{p+q}(G; \mathbb{Z}G).$$

Since $G/K \simeq \mathbb{Z}$, the only values of p yielding groups that are potentially non-zero are 0 and 1, and hence the spectral sequence collapses on page 2. The only terms of the filtration that require further investigation are of the form $H^0(\mathbb{Z}; H^i(K; \mathbb{Z}G))$ and $H^1(\mathbb{Z}; H^{i-1}(K; \mathbb{Z}G))$. Since K is of type FP, we have $H^i(K; \mathbb{Z}G) \simeq H^i(K; \mathbb{Z}K) \otimes_{\mathbb{Z}K} \mathbb{Z}G$ as $\mathbb{Z}G$-modules. In order to compute the cohomology groups above, we need to understand the action of $G/K \simeq \mathbb{Z}$ on $H^i(K; \mathbb{Z}K) \otimes_{\mathbb{Z}K} \mathbb{Z}G$. To do this, it helps to view $H^i(K; \mathbb{Z}K) \otimes_{\mathbb{Z}K} \mathbb{Z}G$ as $H^i(K; \mathbb{Z}K) \otimes_{\mathbb{Z}} \mathbb{Z}(K \backslash G)$. The action of G/K is then diagonal, with the action on $\mathbb{Z}(K \backslash G)$ being multiplication. We can now easily compute the G/K invariants and coinvariants of $H^i(K; \mathbb{Z}G)$. Indeed, the invariants form the zero module and the coinvariants are a copy of $H^i(K;\mathbb{Z}K)$, with a $\mathbb{Z}G$-module structure that restricts to the usual $\mathbb{Z}K$-module structure. Using the fact that $\mathbb{Z}$ is an orientable PD^1-group, we conclude from the spectral sequence that

$$H^i(G;\mathbb{Z}G) \simeq H^{i-1}(K;\mathbb{Z}K)$$

as $\mathbb{Z}G$-modules.

When $\mathcal{H} = \emptyset$, the above immediately tells us that K is a PD^2-group, as claimed. When $\mathcal{H} \neq \emptyset$, the pair $(G, \mathcal{H})$ is a PD^3-pair, and so [**BE78**, Proposition 6.1] tells us that G is a 2-dimensional duality group with dualising module $H^2(G;\mathbb{Z}G)$ isomorphic to the kernel of the augmentation map

$$\bigoplus_{i\in I} \mathbb{Z}[G/H_i] \to \mathbb{Z}$$

tensored over $\mathbb{Z}$ with C. This tells us that K is a 1-dimensional duality group, and a trivial check verifies that the dualising module $H^2(G;\mathbb{Z}G)$ restricted to being a $\mathbb{Z}K$-module is precisely $\widetilde{\Delta}_{K/\mathcal{H}_K}$. This allows us to conclude that $(K, \mathcal{H}_K)$ is a PD^2-pair using [**BE78**, Proposition 6.1] again.

Since PD^2-pairs are geometric, K is the fundamental group of a compact aspherical surface Σ with boundary components in one-to-one correspondence with the elements in $\mathcal{H}_K$, and there exists $\alpha \in \mathrm{Out}(\pi_1(\Sigma))$ such that

$$G \simeq \pi_1(\Sigma) \rtimes_\alpha \mathbb{Z}.$$

It remains to show that the outer automorphism α of $\pi_1(\Sigma)$ is induced by a homeomorphism of Σ. When Σ is a closed surface, this is a classical result, usually attributed to Dehn, that was proved by Nielsen in [**Nie27**] in the orientable case and by Mangler [**Man39**] in the non-orientable case.

Suppose now that Σ has non-empty boundary. It is clear that the conjugation action of t permutes the K-conjugacy classes of the subgroups in $\mathcal{H}_K$, and thus the outer automorphism α permutes the conjugacy classes of the elements of $\pi_1(\Sigma)$ corresponding to the boundary components of Σ. Now by Theorems 5.7.1 and 5.7.2 in [**ZVC80**], α is induced by a homeomorphism $f\colon \Sigma \to \Sigma$. Hence G is isomorphic to the fundamental group of the mapping torus of f. The boundary components of the mapping torus are tori or Klein bottles in one-to-one correspondence with the set I, and this correspondence identifies their fundamental groups with the groups in $\mathcal{H}$. This proves the theorem. □

We say a 3-manifold M^3 is *sporadic* if it is an $\mathbb{S}^2$- or $\mathbb{RP}^2$-bundle over the circle.

PROOF OF THEOREM 1. Let M^3 be a compact 3-manifold with aspherical (possibly empty) boundary, and suppose that there exists a surjective homomorphism $\phi\colon \pi_1(M^3) \to \mathbb{Z}$ with finitely generated kernel K. In particular, this implies that if $\pi_1(M^3) = A * B$ or $\pi_1(M^3) = (A * B) \rtimes \mathbb{Z}/2\mathbb{Z}$, then at least one of A or B is the trivial group.

If M^3 is sporadic then there are exactly two epimorphisms $\pi_1(M^3) \to \mathbb{Z}$ and the result obviously holds. Hence, from now on we assume that M^3 is not sporadic. We claim that it suffices to consider 3-manifolds M^3 with incompressible boundary that are *P^2-irreducible*, meaning irreducible and not containing an embedded two-sided projective plane.

Indeed, if M^3 is non-sporadic and reducible then M^3 splits as a connected sum $M^3 \simeq M_1 \# M_2$, where neither M_1 nor M_2 is the 3-sphere (see [**Hem76**,

Lemma 3.13]). Then, $\pi_1(M) \simeq \pi_1(M_1) * \pi_1(M_2)$ and by Perelman's solution of the Poincaré conjecture, it follows that $\pi_1(M_1)$ and $\pi_1(M_2)$ are non-trivial.

If M^3 contains a compressible boundary component, then a standard argument using Papakyriakopoulos' Loop Theorem (see e.g. [**AFW15**, Lemma 1.5]) shows that again $\pi_1(M^3)$ admits a non-trivial free product decomposition.

Finally, if a non-sporadic M^3 is irreducible and contains a two-sided embedded projective plane, then by [**Hei73**] its fundamental group is of the form $(A * B) \rtimes \mathbb{Z}/2\mathbb{Z}$ with A and B non-trivial groups. Hence, we may assume that M^3 is P^2-irreducible and thus by the Sphere Theorem [**Pap57**] and the Projective Plane Theorem [**Eps61**], combined with the fact that $\pi_1(M^3)$ is infinite, we conclude that M^3 is aspherical.

Let $\mathcal{H} = \{H_i\}_{i \in I}$ be the family of subgroups of $\pi_1(M^3)$ corresponding to the boundary components of M^3. Then $(\pi_1(M^3), \mathcal{H})$ is a PD^3-pair. Hence by the proof of Theorem 2, $K \simeq \pi_1(\Sigma)$ for some compact surface Σ, and there exists a homeomorphism $f\colon \Sigma \to \Sigma$ such that $\pi_1(M^3)$ is isomorphic to the fundamental group of the mapping torus M_f of f,

$$\pi_1(M^3) \simeq \pi_1(M_f) = \pi_1(\Sigma) \rtimes_{f_*} \mathbb{Z}. \tag{3}$$

Now M_f and M^3 are P^2-irreducible, M_f contains an incompressible surface, and the isomorphism (3) preserves the peripheral structure of the fundamental groups of the 3-manifolds M_f and M^3. Thus by [**Hei69**], there exists a homeomorphism $M^3 \to M_f$ which induces the isomorphism in (3). We compose this with the fibration $M_f \to \mathbb{S}^1$ induced by the mapping torus structure, to obtain a topological fibration

$$\Sigma' \to M^3 \to \mathbb{S}^1$$

with Σ' homeomorphic to the surface Σ. This proves Theorem 1. □

Acknowledgments

The authors thank Jonathan Hillman, Peter Scott, and Gadde Swarup for helpful correspondence.

References

[AFW15] M. Aschenbrenner, S. Friedl, and H. Wilton, *3-manifold groups*, EMS Series of Lectures in Mathematics, European Mathematical Society (EMS), Zürich, 2015, DOI 10.4171/154. MR3444187

[BE78] R. Bieri and B. Eckmann, *Relative homology and Poincaré duality for group pairs*, J. Pure Appl. Algebra **13** (1978), no. 3, 277–319, DOI 10.1016/0022-4049(78)90012-9. MR509165

[Bie81] R. Bieri, *Homological dimension of discrete groups*, 2nd ed., Queen Mary College Mathematics Notes, Queen Mary College, Department of Pure Mathematics, London, 1981. MR715779

[Bie07] R. Bieri, *Deficiency and the geometric invariants of a group*, J. Pure Appl. Algebra **208** (2007), no. 3, 951–959, DOI 10.1016/j.jpaa.2006.02.003. MR2283437

[Bow04] B. H. Bowditch, *Planar groups and the Seifert conjecture*, J. Reine Angew. Math. **576** (2004), 11–62, DOI 10.1515/crll.2004.084. MR2099199

[BR88] R. Bieri and B. Renz, *Valuations on free resolutions and higher geometric invariants of groups*, Comment. Math. Helv. **63** (1988), no. 3, 464–497, DOI 10.1007/BF02566775. MR960770

[EL83] B. Eckmann and P. Linnell, *Poincaré duality groups of dimension two. II*, Comment. Math. Helv. **58** (1983), no. 1, 111–114, DOI 10.1007/BF02564628. MR699010

[EM80] B. Eckmann and H. Müller, *Poincaré duality groups of dimension two*, Comment. Math. Helv. **55** (1980), no. 4, 510–520, DOI 10.1007/BF02566702. MR604709

[Eps61] D. B. A. Epstein, *Projective planes in 3-manifolds*, Proc. London Math. Soc. (3) **11** (1961), 469–484, DOI 10.1112/plms/s3-11.1.469. MR152997

[Fis21] Sam P. Fisher, *Improved algebraic fibrings*, arXiv:2112.00397 [math.GR], 2021.

[Hei69] W. Heil, *On P^2-irreducible 3-manifolds*, Bull. Amer. Math. Soc. **75** (1969), 772–775, DOI 10.1090/S0002-9904-1969-12283-4. MR251731

[Hei73] W. Heil, *Testing 3-manifolds for projective planes*, Pacific J. Math. **44** (1973), 139–145. MR324704

[Hem76] J. Hempel, *3-Manifolds*, Annals of Mathematics Studies, No. 86, Princeton University Press, Princeton, NJ; University of Tokyo Press, Tokyo, 1976. MR415619

[Hil87] J. A. Hillman, *Three-dimensional Poincaré duality groups which are extensions*, Math. Z. **195** (1987), no. 1, 89–92, DOI 10.1007/BF01161601. MR888129

[Hil02] J. A. Hillman, *Four-manifolds, geometries and knots*, Geometry & Topology Monographs, vol. 5, Geometry & Topology Publications, Coventry, 2002. MR1943724

[Hil20] J. A. Hillman, *Poincaré duality in dimension 3*, The Open Book Series, vol. 3, Mathematical Sciences Publishers, Berkeley, CA, 2020. MR4235127

[Kie20] D. Kielak, *The Bieri-Neumann-Strebel invariants via Newton polytopes*, Invent. Math. **219** (2020), no. 3, 1009–1068, DOI 10.1007/s00222-019-00919-9. MR4055183

[KK21] D. Kielak and P. Kropholler, *Isoperimetric inequalities for Poincaré duality groups*, Proc. Amer. Math. Soc. **149** (2021), no. 11, 4685–4698, DOI 10.1090/proc/15596. MR4310095

[Koc06] D. H. Kochloukova, *Some Novikov rings that are von Neumann finite and knot-like groups*, Comment. Math. Helv. **81** (2006), no. 4, 931–943, DOI 10.4171/CMH/81. MR2271229

[Man39] W. Mangler, *Die Klassen von topologischen Abbildungen einer geschlossenen Fläche auf sich* (German), Math. Z. **44** (1939), no. 1, 541–554, DOI 10.1007/BF01210672. MR1545786

[Nie27] J. Nielsen, *Untersuchungen zur Topologie der geschlossenen zweiseitigen Flächen* (German), Acta Math. **50** (1927), no. 1, 189–358, DOI 10.1007/BF02421324. MR1555256

[Pap57] C. D. Papakyriakopoulos, *On Dehn's lemma and the asphericity of knots*, Ann. of Math. (2) **66** (1957), 1–26, DOI 10.2307/1970113. MR90053

[Rip82] E. Rips, *Subgroups of small cancellation groups*, Bull. London Math. Soc. **14** (1982), no. 1, 45–47, DOI 10.1112/blms/14.1.45. MR642423

[RSS20] Lawrence Reeves, Peter Scott, and Gadde A. Swarup, *A deformation theorem for Poincaré duality pairs in dimension 3*, arXiv:2006.15684 June 2020.

[Sik87] J.-C. Sikorav, *Homologie de Novikov associée à une classe de cohomologie réelle de degré un*, Thèse Orsay, 1987.

[Sta62] J. Stallings, *On fibering certain 3-manifolds*, Topology of 3-manifolds and related topics (Proc. The Univ. of Georgia Institute, 1961), Prentice-Hall, Inc., Englewood Cliffs, NJ, 1961, pp. 95–100. MR158375

[Str77] R. Strebel, *A remark on subgroups of infinite index in Poincaré duality groups*, Comment. Math. Helv. **52** (1977), no. 3, 317–324, DOI 10.1007/BF02567371. MR457588

[Tho84] C. B. Thomas, *Splitting theorems for certain PD^3-groups*, Math. Z. **186** (1984), no. 2, 201–209, DOI 10.1007/BF01161804. MR741302

[ZVC80] H. Zieschang, E. Vogt, and H.-D. Coldewey, *Surfaces and planar discontinuous groups*, Lecture Notes in Mathematics, vol. 835, Springer, Berlin, 1980. Translated from the German by John Stillwell. MR606743

Mathematical Institute, Andrew Wiles Building, Observatory Quarter, University of Oxford, Oxford, OX2 6GG, United Kingdom
Email address: bridson@maths.ox.ac.uk

Mathematical Institute, Andrew Wiles Building, Observatory Quarter, University of Oxford, Oxford, OX2 6GG, United Kingdom
Email address: kielak@maths.ox.ac.uk

Mathematical Institute, Andrew Wiles Building, Observatory Quarter, University of Oxford, Oxford, OX2 6GG, United Kingdom
Email address: kudlinska@maths.ox.ac.uk

Contemporary Mathematics
Volume **812**, 2025
https://doi.org/10.1090/conm/812/16264

Totally Symmetric Sets

Noah Caplinger and Dan Margalit

Abstract. We survey the theory of totally symmetric sets, with applications to homomorphisms of symmetric groups, braid groups, linear groups, and mapping class groups.

1. Introduction

The theory of totally symmetric sets is a tool that has been proven to be useful in classifying homomorphisms between certain types of groups. The basic definitions were introduced by Kordek and the second author in their study of homomorphisms between braid groups [**19**]. Since that work, the theory has been used in the study of homomorphisms between symmetric groups, braid groups, linear groups, and mapping class groups.

Here is the definition. A totally symmetric set in a group G is a subset

$$X = \{x_1, \ldots, x_k\} \subset G$$

with the following property: for every $\sigma \in \Sigma_k$, there is a $g_\sigma \in G$ such that

$$g_\sigma x_i g_\sigma^{-1} = x_{\sigma(i)}$$

for all i. It follows from the definition that the elements of a totally symmetric set lie in a single conjugacy class.

Evidently, if $X \subseteq G$ is a totally symmetric set and $f : G \to H$ is a homomorphism, then $f(X)$ is a totally symmetric set in H. As we show in Section 2.1, a much stronger condition is true: $f(X)$ is either a totally symmetric set of size $|X|$ or it is a singleton. In the phrasing of Salter and the first author: collision implies collapse. This is the fundamental property of totally symmetric sets.

For groups G and H, the collision-implies-collapse property yields an (unreasonably effective) blueprint for classifying homomorphisms $f : G \to H$, as follows:

Step 1. Find a large totally symmetric set $X \subset G$.
Step 2. Classify the large totally symmetric sets in H.
Step 3. Deduce properties of $f(X)$ and draw conclusions about f.

2020 *Mathematics Subject Classification.* Primary 20E34, 20F65, 20F36, 20B05.

This material is based upon work supported by the National Science Foundation under Grant No. DMS-2203431.

For instance, if G has a totally symmetric set X with $|X| = k$, and H has no totally symmetric set of cardinality k, then any homomorphism $f : G \to H$ must collapse X. Moreover, for any $x_i, x_j \in X$ the normal closure of $x_i x_j^{-1}$ lies in the kernel of f. In particular, if the $x_i x_j^{-1}$ are normal generators for G, then f is the trivial map.

Step 2 of the blueprint is generally the most challenging (and interesting). For a given group H, the approach is to choose a space Y on which H acts. As explained in Section 2.2, totally symmetric sets in H correspond to totally symmetric geometric configurations in Y. These configurations can be, for example, configurations of eigenspaces for linear maps or canonical reduction systems for mapping classes. The main strategy is to classify the totally symmetric configurations, and then use this to classify the totally symmetric sets.

Overview. In Section 2 we introduce the basic notions and examples in the theory of totally symmetric sets. We also prove the collision-implies-collapse property. We then use the blueprint to give an intuitive explanation for why the outer automorphism group of the symmetric group is (usually) trivial.

In Sections 3 and 4 we explain how the blueprint is applied in the cases of the general linear group and the braid group. The former case was addressed in a paper by the first author and Salter [**9**] and the latter in a paper by the second author and Kordek [**19**]. As per the blueprint, the strategy in both cases is to classify large totally symmetric configurations and then to promote this to a classification of large totally symmetric sets. For the general linear group, the configurations are configurations of subspaces. We then use the theory of Jordan normal form to do the promotion. For the braid group the configurations are configurations of multicurves. In this case we use Nielsen–Thurston theory to do the analogous promotion.

In Section 5 we prove a theorem of Kolay, which says that the standard map $B_n \to \Sigma_n$ gives the smallest non-cyclic quotient of the braid group. To streamline Kolay's argument, we first introduce a variation on totally symmetric sets, namely, collapsing sets. These are exactly the sets that satisfy the collision-implies-collapse property. We then present Kolay's proof of the theorem.

Finally, in Section 6 we make an explicit analogy between the collision-implies-collapse property and Schur's lemma from representation theory. We discuss several questions that arise from this analogy.

2. Totally symmetric sets and the blueprint

The three subsections in this section correspond to the three steps of the blueprint for totally symmetric sets. In Section 2.1 we give some basic examples of totally sets, and state and prove the collision-implies-collapse lemma. In Section 2.2, we define totally symmetric configurations, and use them to give upper bounds on the sizes of totally symmetric sets in certain groups. Finally, in Section 2.3 we use the results of Sections 2.1 and 2.2 to give a conceptual explanation for the classification of automorphisms of the symmetric group Σ_n.

2.1. Totally symmetric sets and collision implies collapse.

Examples of totally symmetric sets. Among the most basic examples of totally symmetric sets are:

$$\{(1\ 2),(3\ 4),\ldots\} \subseteq \Sigma_n,\quad \{(1\ i),\ldots,(1\ n)\} \subseteq \Sigma_n,\quad \{E_{1,2},\ldots,E_{1,n}\} \subseteq \mathrm{GL}_n(\mathbb{Z}),$$
$$\{E_1,\ldots,E_n\} \subseteq \mathrm{GL}_n(\mathbb{Z}),\quad \text{and } \{\sigma_1,\sigma_3,\ldots\} \subseteq B_n.$$

where the $(i\ j)$ are transpositions in the symmetric group Σ_n, the σ_i are the standard half-twist generators for the braid group B_n, the $E_{i,j}$ are elementary matrices in the general linear group $\mathrm{GL}_n(\mathbb{Z})$, and the E_i are the elements of $\mathrm{GL}_n(\mathbb{Z})$ obtained from the identity by negating the ith diagonal entry. We leave it as an exercise to verify that these are all totally symmetric sets.

Collision implies collapse. The following lemma, mentioned in the introduction, encapsulates the fundamental property of totally symmetric sets. The lemma originally appears in the work of Kordek and the second author of this paper [**19**, Lemma 2.1].

LEMMA 2.1. *Let $f : G \to H$ be a homomorphism of groups. If $X \subseteq G$ is a totally symmetric set then $f(X)$ is either a totally symmetric set of cardinality $|X|$ or a singleton.*

PROOF. Let $x, y, z \in X$ and suppose $f(x) = f(y)$. Total symmetry guarantees some $g \in G$ such that $g(x,y,z)g^{-1} = (x,z,y)$. We have:

$$f(xz^{-1}) = f(gxy^{-1}g^{-1}) = f(g)f(xy^{-1})f(g)^{-1} = 1.$$

Thus $f(z) = f(x)$ and the lemma follows. □

In the original paper by Kordek and the second author, totally symmetric sets were assumed to have the additional property that the elements commute pairwise. So in that paper, the set $\{(1\,2),\ (3\,4),\ldots\}$ would be considered as a totally symmetric set in Σ_n, whereas $\{(1\ i),\ldots,(1\ n)\}$ would not. The commutativity condition was included because it simplifies the classification of totally symmetric configurations for braid groups. Since the more general totally symmetric sets considered here still satisfy Lemma 2.1, we will henceforth use the term "commuting totally symmetric set" to refer to a totally symmetric set with the additional property that the elements commute pairwise.

In defining totally symmetric sets, Kordek and the second author were directly inspired by the work of Aramayona–Souto, who used a symmetric group action on a collection of Dehn twists to similar effect [**1**, Section 5].

2.2. Totally symmetric configurations and upper bounds on totally symmetric sets. Recall that Step 2 of the blueprint concerns the classification of large totally symmetric sets in a given group G (in the blueprint the group is called H). After the fashion in geometric group theory, we approach this problem by considering a suitable action of G on a space Y. Given such an action, we often have at least one natural choice of function:

$$G \to \text{subsets of } Y$$
$$g \mapsto Y_g$$

For a given $g \in G$, the subset Y_g might be the fixed set, or an eigenspace, or an invariant axis, etc. As long as the association $g \mapsto Y_g$ is natural, it will satisfy the equation

$$Y_{hgh^{-1}} = h \cdot Y_g$$

for all $h \in G$ (we can take this equivariance condition as the definition of naturality). In particular, if $\{x_1, \ldots, x_k\}$ is a totally symmetric subset of G, then $\{Y_{x_1}, \ldots, Y_{x_k}\}$ is a totally symmetric configuration in Y in the sense that any permutation of the Y_{x_i} can be realized by the action of G.

Totally symmetric configurations. Motivated by this discussion, we can give the definition of a totally symmetric configuration. Suppose that a group G acts on a space Y. Let

$$\{Y_1, \ldots, Y_k\}$$

be a collection of subspaces of Y. We say that $\{Y_i\}$ is a totally symmetric configuration if for each $\sigma \in \Sigma_k$, there is a $g_\sigma \in G$ so that

$$g_\sigma \cdot (Y_i) = Y_{\sigma(i)}$$

for all i. Again, the point of the definition is that, as long as the association of a subspace to a group element satisfies the naturality property

$$Y_{hgh^{-1}} = h \cdot Y_g,$$

the configuration associated to a totally symmetric set is a totally symmetric configuration.

Unifying the definitions. Our definitions of totally symmetric sets and totally symmetric configurations are almost identical. As observed by the first author and Salter [**9**, Definition 2.1], they can be combined into one definition as follows:

> *Let G act on a set Z. A subset $X = \{x_1, \ldots, x_k\} \subset Z$ is totally symmetric if for all $\sigma \in \Sigma_k$, there is some $g_\sigma \in G$ such that $g_\sigma \cdot x_i = x_{\sigma(i)}$*

The definition of a totally symmetric set is recovered by considering the action of G on itself by conjugation, and the definition of a totally symmetric configuration is recovered by considering the action of G on a set of subsets of a space Y that carries an action of G. Even in this general setting, totally symmetric sets obey the collision-implies-collapse principle where the homomorphism in Lemma 2.1 is replaced by a G-equivariant map.

Example: Dihedral groups. We will use the notion of totally symmetric configurations to prove the following fact:

> *A totally symmetric set $X \subseteq D_n$ has $|X| \le 3$.*

The first step is to prove that a totally symmetric set of rotations has cardinality at most two (exercise). Suppose then that X is a totally symmetric set consisting of reflections. To each reflection in X we can associate the corresponding line of reflection in the plane. As above, this gives a totally symmetric configuration of lines in the plane. The largest such configuration has three lines (another exercise). Since reflections are determined by the corresponding lines, the desired statement follows.

This argument shows more:

> *If X is a totally symmetric set of D_n with $|X| = 3$ then $3 \mid n$ and X consists of reflections about lines that pairwise form an angle of $\pi/3$.*

As a sample consequence, we have the following fact:

> *If $n \geq 8$ and $m \geq 3$, then every homomorphism $B_n \to D_m$ has cyclic image.*

While this fact is not difficult to prove directly, the theory of totally symmetric sets gives a natural, conceptual explanation.

Example: Free groups. We now use totally symmetric configurations to prove the following:

> *A totally symmetric set in $X \subseteq F_2$ has $|X| \leq 1$.*

Consider the action of F_2 on its Cayley graph, the regular four-valent tree T_4. Each element $x \in F_2$ acts on T_4 by translating along an axis, which is a bi-infinite geodesic in T_4. Again, this gives a totally symmetric configuration of geodesics.

Within a given conjugacy class in F_2, an element is determined by its axis. Therefore, it suffices to show that there is no totally symmetric configuration consisting of two bi-infinite geodesics in T_4.

Let $Y \subset T_4$ be a totally symmetric configuration of bi-infinite geodesics. If Y_1 and Y_2 are distinct elements of Y, then by total symmetry there is an element of F_2 interchanging Y_1 and Y_2. This is impossible, since (using the usual embedding of T_4 in the plane) the elements of F_2 act on T_4 by orientation-preserving planar automorphisms.

An upper bound for all groups. The first author proved the following result [**7**, Theorem 1], which gives an upper bound for the cardinality of a totally symmetric set in an arbitrary group.

THEOREM 2.2 (Caplinger). *Let X be a totally symmetric set in a group G. If $|X| \geq 4$, then*

$$|G| \geq (|X| + 1)!$$

Equality is attained only when $G = \Sigma_n$.

As a sample application, any totally symmetric set in the monster group M has cardinality less than 44, since $44! > |M|$.

Analogous (but not sharp) upper bounds on the cardinalities of commuting totally symmetric sets were proved by Chudnovsky–Kordek–Li–Partin [**12**, Proposition 2.2] and by Scherich–Verberne [**22**, Theorem A].

Other upper bounds on commuting totally symmetric sets. Kordek–Li–Partin [**18**] provide a suite of upper bounds for the cardinality of a commutative totally symmetric set. For instance they show that the largest cardinality of a commutative totally symmetric set in the dihedral group is 2 [**18**, Theorem 3.3] and that the largest cardinality of a commutative totally symmetric set in the Baumslag–Solitar group $BS(1,n)$ with $n \neq -1$ is 1. They also prove that the largest cardinality of a commutative totally symmetric set in a product $G \times H$ or $G * H$ is the supremum of the cardinalities for totally symmetric sets in a single factor. They also prove that a solvable group cannot have a commutative totally symmetric set with 5 elements. We refer the reader to their paper for a full accounting of their results.

2.3. Application to the symmetric group. Using the basic theory of totally symmetric sets already established, we can give a conceptual explanation of the following classical theorem. This argument originally appeared in the work of the first author [**7**].

THEOREM 2.3. *For $n \geq 7$, the outer automorphism group of Σ_n is trivial.*

Let Z_n denote the totally symmetric set

$$Z_n = \{(1\ i) \mid i \geq 2\} \subseteq \Sigma_n.$$

To prove Theorem 2.3 we use the following auxiliary result [**7**, Theorem 2], which is a classification of large totally symmetric sets in Σ_n:

> *Let $n \geq 7$. If $X \subset \Sigma_n$ is a totally symmetric set with $|X| \geq n-1$. Then X is conjugate to Z_n.*

From this fact, the proof of Theorem 2.3 proceeds as follows. Let $f : \Sigma_n \to \Sigma_n$ be an automorphism. Then $f(Z_n)$ is equivalent to Z_n in the following sense: there exists $\tau \in \Sigma_n$ with $\tilde{\tau} f(Z_n) = Z_n$, where $\tilde{\tau}$ is the inner automorphism corresponding to τ. Then $\tilde{\tau} \circ f$ permutes Z_n, so total symmetry gives some $\sigma \in \Sigma_n$ so that

$$\tilde{\sigma} \circ \tilde{\tau} \circ f = \widetilde{\sigma\tau} \circ f$$

is the identity on Z_n. Since Z_n generates Σ_n, we conclude that $\widetilde{\sigma\tau} \circ f = \mathrm{id}$, that is, $f = (\widetilde{\sigma\tau})^{-1}$. In particular, f is an inner automorphism, completing the proof of the theorem.

A slight modification of the above argument yields the following (well-known) generalization of Theorem 2.3.

THEOREM 2.4. *Let $7 \leq n \leq m$, and let $f : \Sigma_m \to \Sigma_n$ be a homomorphism whose image is not cyclic. Then $m = n$ and f is an inner automorphism.*

To obtain this stronger theorem, the only additional observation required is that—by the classification of large totally symmetric sets in Σ_n—the restriction $f|Z_n$ cannot be injective when $n < m$. Thus it must be trivial, and so the image is cyclic (of order at most 2).

We would be remiss not to describe the situation for Σ_6, which does have a nontrivial outer automorphism. From the perspective of totally symmetric sets, the reason why this outer automorphism exists is that Σ_6 has two conjugacy classes of totally symmetric sets with five elements: the standard one and its image under the nontrivial outer automorphism.

The proof of Theorem 2.3 given here is not simpler than the classical proof. However, it gives a conceptually simple, structural explanation. Also, the classification of large totally symmetric sets creates a broad tool for studying any homomorphism to or from Σ_n. We will return to this theme several times in what follows.

3. Totally symmetric sets in the general linear group

In this section we turn our attention to the general linear group. The first author and Salter give a classification of large totally symmetric sets in $\mathrm{GL}_n(\mathbb{C})$, Theorem 3.1 below. They used this classification to give a new, conceptual proof of the following classical fact:

> *Any non-abelian representation of Σ_n has dimension at least $n-1$.*

We will start by describing the largest totally symmetric sets in $\mathrm{GL}_n(\mathbb{C})$, then state the classification theorem, and then explain the applications to representation theory. Following the blueprint, we then classify large totally symmetric configurations in $\mathbb{C}^n$, before using this classification to prove the classification of totally symmetric sets in $\mathrm{GL}_n(\mathbb{C})$.

Standard totally symmetric sets. Consider a regular n-simplex $\Delta \subset \mathbb{R}^n$ centered at the origin. The set of vertices $v_1, \ldots, v_{n+1}$ of Δ is a totally symmetric configuration in the sense that for all $\sigma \in \Sigma_{n+1}$, there is an A_σ in $\mathrm{GL}_n(\mathbb{C})$ (in fact in $O(n)$) such that $A_\sigma v_i = v_{\sigma(i)}$. The hyperplanes $v_i^\perp$ also form a totally symmetric configuration with the same choice of A_σ. This allows us to form a totally symmetric set in $\mathrm{GL}_n(\mathbb{C})$ as follows. Pick $\lambda, \mu \in \mathbb{C}$ distinct and non-zero, and define $A_i \in \mathrm{GL}_n(\mathbb{C})$ by declaring $\mathbb{C}v_i$ and $v_i^\perp$ to be λ- and μ-eigenspaces respectively. The set $\mathcal{A}_n = \{A_1, \ldots, A_{n+1}\}$ is then totally symmetric. We refer to (any conjugate of) any such $\mathcal{A}_n$ as a standard totally symmetric set in $\mathrm{GL}_n(\mathbb{C})$.

The classification of large totally symmetric sets. The following theorem says that the above construction is the only construction of totally symmetric sets in $\mathrm{GL}_n(\mathbb{C})$ of cardinality $n+1$.

THEOREM 3.1 (Caplinger–Salter). *Let $n \neq 5$ and let $X \subset \mathrm{GL}_n(\mathbb{C})$ be a totally symmetric set. Then $|X| \leq n+1$, and equality is achieved exactly when X is standard.*

This theorem immediately applies to bound the dimension of a faithful representation of any group: if G contains a totally symmetric subset of size n, then G has no faithful representations in dimension less than $n-1$. In the case of Σ_n, we can say more.

Application to representations of the symmetric group. Building on the last idea, let $\rho\colon \Sigma_n \to \mathrm{GL}_m(\mathbb{C})$ be a non-abelian representation of Σ_n. We would like to show $m \geq n-1$.

For the standard totally symmetric set Z_n in Σ_n we have that $\rho(Z_n) \subset \mathrm{GL}_m(\mathbb{C})$ is a totally symmetric set. We show that if $m < n-1$ then $\rho(Z_n)$ is a singleton, implying that ρ has cyclic image. We treat the cases $m < n-2$ and $m = n-2$ in turn.

If $m < n-2$ then Theorem 3.1 gives that there is no totally symmetric subset of $\mathrm{GL}_m(\mathbb{C})$ of size $n-1$. Thus by Lemma 2.1, $\rho(Z_n)$ is a singleton, as desired.

If $m = n-2$, Theorem 3.1 gives that $\rho(Z_n)$ is a singleton or is standard. But no two distinct elements of the standard totally symmetric set $\mathcal{A}_n$ satisfy the braid relation, so $\rho(Z_n)$ must be a singleton, as desired.

Application to algebraic geometry. Let $\mathcal{U}_{n,d}$ denote the space of smooth degree d hypersurfaces in $\mathbb{CP}^n$. We can continue the above line of reasoning in order to constrain the dimension of a representation of $\pi_1(\mathcal{U}_{n,d})$:

> *If $\rho : \pi_1(\mathcal{U}_{n,d}) \to \mathrm{GL}_m(\mathbb{C})$ is a non-cyclic representation with $d \geq 5$, then*
> $$m \geq \left\lceil \frac{d-1}{2} \right\rceil^n - 1.$$

We now outline the proof of this fact. Lönne gave a presentation of $\pi_1(\mathcal{U}_{n,d})$ generalizing the standard presentation of the braid group [**21**]. From this presentation, we can see that $\pi_1(\mathcal{U}_{n,d})$ has a totally symmetric set that is analogous to the standard totally symmetric set in the braid group (see Section 4), and has cardinality $\lceil \frac{d-1}{2} \rceil^n$. Just like in the braid group, when $d \geq 5$ this set collapses if and only if ρ has cyclic image, so Theorem 3.1 gives the desired bound. (In fact, a slightly better bound can be obtained from a related result in the paper of the first author and Salter [**9**, Theorem A] that classifies commutative totally symmetric sets in $\mathrm{GL}_n(\mathbb{C})$.)

Totally symmetric configurations. As per Step 2 of the blueprint, we now explain the classification of large totally symmetric configurations used in the classification of large totally symmetric sets in $\mathrm{GL}_n(\mathbb{C})$.

The first author and Salter give the following classification [**9**, Theorem A]. In the statement, a standard totally symmetric configuration is the collection of 1-dimensional eigenspaces of the elements of the standard totally symmetric set, or the image of this configuration under any element of $\mathrm{GL}_n(\mathbb{C})$. The dual of such a configuration is the set of orthogonal complements (in the paper by the first author and Salter these are referred to as simplex configurations and their duals).

PROPOSITION 3.2 (Caplinger–Salter). *Let $\mathcal{W}$ be a totally symmetric configuration of subspaces in $\mathbb{C}^n$. Then $|\mathcal{W}| \leq n+1$, and when $n \neq 5$, equality is realized only by a standard configuration or the dual of such.*

In the work of the first author and Salter, Proposition 3.2 is proved inductively, in tandem with Theorem 3.1. Here, we assume the proposition without proof, and show how the inductive step for Theorem 3.1 proceeds.

Proof of Theorem 3.1 *assuming Proposition* 3.2. We discuss the two statements in turn, namely, the upper bound on the size of a totally symmetric set and the classification of large totally symmetric sets.

We proceed by induction on n, with base case $n = 1$. In this base case we are considering $\mathrm{GL}_1(\mathbb{C}) \cong \mathbb{C}^*$. Since the latter is abelian, the largest totally symmetric set is a singleton. As $1 \leq 2$ the first statement of the proposition is verified. The second statement is vacuous since the upper bound $n+1$ is not realized in this case.

Let $X = \{A_1, \ldots, A_k\} \subset \mathrm{GL}_n(\mathbb{C})$ be a totally symmetric set. Consider the generalized eigenspaces
$$E^i_{\lambda,j} = \ker\,(A_i - \lambda I)^j.$$
If for any λ and j the arrangement of subspaces
$$\{E^1_{\lambda,j}, \ldots E^k_{\lambda,j}\}$$

is non-degenerate (i.e. not a singleton), then Lemma 2.1 (really, the version for configurations) and Proposition 3.2 give the bound $k \le n+1$. Thus, we may henceforth assume that all arrangements $\{E^i_{\lambda,j}\}_{i=1}^k$ are degenerate. In other words, the A_i share a common Jordan filtration

$$E^i_{\lambda,1} \subset E^i_{\lambda,2} \subset \cdots$$

for every eigenvalue λ. We may therefore drop the superscript and write $E_{\lambda,j}$ for $E^i_{\lambda,j}$.

Restricting the A_i to any $E_{\lambda,j}$ gives a totally symmetric set in some $\mathrm{GL}_d(\mathbb{Z})$ with $d < n$. If this restricted totally symmetric set is non-degenerate, then by induction we have $k \le d+1 < n+1$, as desired. Similarly, the maps induced by A_i on the quotients $\mathbb{C}^n/E_{\lambda,j}$ are totally symmetric, and if they are nondegenerate, induction applies.

We are thus left with the case where all restrictions and quotients associated to each $E_{\lambda,j}$ are identical. This is a strong condition. From here, the first author and Salter use a variety of techniques to coax out totally symmetric sets of smaller dimension. They then apply induction to obtain the bound $k \le n$.

We now turn to the second statement of the theorem, the classification of totally symmetric sets of size $n+1$. The above argument shows that if $k = n+1$, there must be an eigenvalue λ and an index j so that the eigenspace arrangement $\{E^i_{\lambda,j}\}$ corresponding to $X = \{A_i\}$ is nondegenerate. Moreover, Proposition 3.2 implies that this arrangement must be the standard totally symmetric configuration or the dual to such.

Let $\phi : \Sigma_{n+1} \to \mathrm{GL}_n(\mathbb{C})$ be a realization map for $X = \{A_i\}$, by which we mean that

$$\phi(\sigma) A_i \phi(\sigma)^{-1} = A_{\sigma(i)}$$

for all $\sigma \in \Sigma_{n+1}$ and all $i \in \{1, \dots, k\}$. The elements of the image of ϕ must permute the set $\{E^i_{\lambda,j}\}_{i=1}^k$ accordingly.

The first author and Salter show that, up to scaling, the only realization map for the standard totally symmetric subspace configuration (or its dual) is the standard representation $\Sigma_{n+1} \to \mathrm{GL}_n(\mathbb{C})$. Since the standard representation is irreducible, no (proper) arrangement of eigenspaces can be degenerate, as such an arrangement would give an invariant subspace.

The next step is to show that each A_i has two eigenvalues. There are two cases, namely, where $\{E^i_{\lambda,j}\}_{i=1}^k$ is the standard totally symmetric configuration and its dual. To illustrate the idea, we treat the former case. Assume for the purposes of contradiction that λ is the only eigenvalue for the A_i. In this case, $\mathbb{C}^n$ is a single Jordan block, and $\{E_{\lambda,j}\}$ does not stabilize until $j = n$. But then if $n > 2$, the generalized eigenspace $\{E_{\lambda,2}\}$ would be a non-degenerate totally symmetric collection of 2-dimensional subspaces with $n+1$ elements. This is impossible by Proposition 3.2.

By the previous paragraph, we may assume that each A_i has two distinct eigenvalues, λ and μ. The standard representation $\phi : \Sigma_{n+1} \to \mathrm{GL}_n(\mathbb{C})$ must be a realization map for both $\{E_{\lambda,1}\}$ and $\{E_{\mu,1}\}$. Then both the λ- and μ-eigenspaces of A_i must be stabilized by $\phi(\mathrm{Stab}(i))$, whose fixed subspaces are D_i and $\mathbb{C}v_i$. Since $\mathrm{Stab}(i) \to D_i$ has no subrepresentations, the eigenspaces must be exactly D_i and $\mathbb{C}v_i$. The theorem follows.

4. Totally symmetric sets in braid groups

In this section we use the theory of totally symmetric sets to outline a proof of the following result originally due to Dyer–Grossman [**14**]. Our argument is a modification of the one used by Kordek and the second author [**19**], simplified for this special case.

For the statement, let $\sigma_1, \ldots, \sigma_{n-1}$ denote the standard half-twist generators for the braid group B_n, and let ϵ be the automorphism of B_n given by $\sigma_i \mapsto \sigma_i^{-1}$ for all i.

THEOREM 4.1 (Dyer–Grossman). *For $n \geq 3$ the automorphism ϵ represents the unique nontrivial outer automorphism of B_n. In particular,*

$$\mathrm{Aut}(B_n) \cong B_n/Z(B_n) \rtimes \mathbb{Z}/2.$$

The original proof of this theorem by Dyer–Grossman has an algebraic flavor. The proof we outline here, while only valid as stated for $n \geq 5$, uses combinatorial topology and the theory of mapping class groups.

In what follows we consider B_n as the mapping class group of the disk D_n with n marked points in the interior (not to be confused with the dihedral group!). We will use a number of aspects of the theory of mapping class groups, including the Nielsen–Thurston classification theorem, the canonical reduction systems of Birman–Lubotzky–McCarthy, and the change of coordinates principle. We refer the reader to the book by Farb and the second author of this article for background on these topics [**15**]. From the point of view of mapping class groups, the Dyer–Grossman theorem can be stated as: every automorphism of B_n is induced by a homeomorphism of D_n.

The outline for this section mirrors the one for Section 3. This stands to reason, as we will be following the same blueprint.

A large commutative totally symmetric set. As per Step 1 of the blueprint, we will require the services of a large (commutative) totally symmetric set in B_n. The desired set is:

$$\{\sigma_1, \sigma_3, \sigma_5, \ldots, \sigma_m\}$$

where m is $n-1$ or $n-2$, according to whether n is even or odd, respectively. As such, the cardinality of this set is $\lfloor n/2 \rfloor$. That this set is a totally symmetric set is an application of the change of coordinates principle from the theory of mapping class groups. We refer to any B_n-conjugate of this totally symmetric set as a standard totally symmetric set in B_n.

Crash course in Nielsen–Thurston theory. Briefly, the Nielsen–Thurston classification gives that every braid is periodic, pseudo-Anosov, or reducible. Periodic braids have powers that are central in B_n; they correspond to rotations of D_n. Each reducible braid preserves a multicurve, that is, the isotopy class of a collection of pairwise disjoint and pairwise non-homotopic simple closed curves in D_n. Any such multicurve is called a reduction system for the braid. Pseudo-Anosov braids do not preserve any multicurve.

For a reducible braid b, we may restrict b to the complementary components, and inductively apply the classification. Thus, there exists a reduction system with the property that the associated restrictions are all periodic or pseudo-Anosov.

There is in fact a unique minimal such reduction system, called the canonical reduction system $\mathrm{CRS}(b)$. We will use the following properties of canonical reduction systems:

(1) for $a, b \in B_n$ we have $\mathrm{CRS}(aba^{-1}) = a\,\mathrm{CRS}(b)$, and
(2) if a and b commute then $\mathrm{CRS}(a)$ and $\mathrm{CRS}(b)$ have trivial geometric intersection.

For the proof, the main points to keep in mind are that there is a map

$$\mathrm{CRS} : B_n \to \{\text{multicurves in } D_n\},$$

and that this map satisfies the above two properties.

Totally symmetric configurations of multicurves. The configurations we will use to address Step 2 of the blueprint are collections of multicurves in the disk D_n. To each element g of B_n we associate its canonical reduction system $\mathrm{CRS}(g)$. Then, to a totally symmetric set $X = \{g_1, \dots, g_k\}$ we can associate the collection of multicurves

$$\{\mathrm{CRS}(g_1), \dots, \mathrm{CRS}(g_k)\}.$$

Because X is totally symmetric, this multicurve configuration is totally symmetric in the sense that for any $\sigma \in \Sigma_k$ there is a braid g_σ so that

$$g_\sigma \cdot (\mathrm{CRS}(g_1), \dots, \mathrm{CRS}(g_k)) = (\mathrm{CRS}(g_{\sigma(1)}), \dots, \mathrm{CRS}(g_{\sigma(k)}))$$

(using the first property of CRS above). When X is a commutative totally symmetric set, the multicurves $\mathrm{CRS}(g_i)$ have trivial intersection pairwise (using the second property); in what follows, we refer to such a collection as a noncrossing multicurve configuration.

In the original work of Kordek and the second author, they associate a single labeled multicurve to X instead of a configuration of multicurves. This is equivalent, but we take this point of view to make the analogy with $\mathrm{GL}_n(\mathbb{C})$ more clear.

Large noncrossing totally symmetric multicurve configurations. We now turn towards Step 2 of the blueprint. Let us realize D_n at the closed unit disk in the complex plane, with all marked points on the real axis. For $1 \le i \le n-1$, let c_i be the isotopy class of curves corresponding to a round circle surrounding the ith and $(i+1)$st marked points. The curves are chosen precisely so that $\mathrm{CRS}(\sigma_i)$ is equal to c_i. We define the following noncrossing totally symmetric multicurve in D_n:

$$M_n = \{c_1, c_3, \dots\}.$$

This is the noncrossing totally symmetric multicurve configuration associated to the totally symmetric set $\{\sigma_1, \sigma_3, \dots\}$ in B_n.

There are two variations on M_n that we will need to consider. First, we have the dual totally symmetric configuration

$$M_n^* = \{\{c_1\}^{\mathsf{C}}, \{c_3\}^{\mathsf{C}}, \dots\}$$

where $\{c_i\}^{\mathsf{C}}$ is the complementary configuration to c_i in M_n. In other words, $\{c_i\}^{\mathsf{C}}$ is the multicurve whose components are all the curves appearing in M_n except for c_i.

Second, when n is odd, we have the totally symmetric labeled multicurve

$$\widehat{M_n} = \{\{c_1, d\}, \{c_3, d\}, \dots\},$$

where d is represented by the round curve surrounding the first $n-1$ marked points. Finally, we can combine these two variations in order to obtain

$$\widehat{M}_n^* = \{\{c_1\}^{\mathsf{C}} \cup \{d\}, \{c_3\}^{\mathsf{C}} \cup \{d\}, \dots\}.$$

That all of these multicurve configurations are totally symmetric follows again from the change of coordinates principle.

Kordek and the second author prove [**19**, Lemma 2.3] that in fact these are the only examples of large noncrossing totally symmetric multicurve configurations in D_n. In the statement, a configuration of multicurves $\{m_1, \dots, m_k\}$ is degenerate if two m_i are equal.

PROPOSITION 4.2 (Kordek–Margalit). *Let $M = \{m_1, \dots, m_k\}$ be a nondegenerate, noncrossing, totally symmetric multicurve configuration in D_n with $k = \lfloor n/2 \rfloor$. Then M is B_n-equivalent to one of M_n, M_n^*, $\widehat{M}_n$, or $\widehat{M}_n^*$.*

The main idea of the proof is as follows. Suppose that some m_i contains a curve c_i^p surrounding p marked points. By total symmetry each of the m_i contains such a curve. Then if $p > 2$ it must be that these c_i^p are not distinct (otherwise the noncrossing condition would be violated). We are then led to consider the case that some curve c surrounds exactly p marked points and lies in exactly d of the m_i. Again applying total symmetry, there must be $\binom{k}{d}$ such curves, all with pairwise trivial geometric intersection, and all surrounding p marked points. But for $d < k$, the quantity $\binom{k}{d}$ is quadratic in $k = \lfloor n/2 \rfloor$, hence quadratic in n. Again, this violates the noncrossing condition (there are in fact at most $n - 2$ pairwise non-isotopic curves in D_n with pairwise trivial geometric intersection). It follows that the only possibilities are that each curve appearing in an m_i surrounds exactly two marked points, or it lies in all the m_i. From here the proof is straightforward.

Large commutative totally symmetric sets. Continuing with Step 2 of the blueprint, we now explain how Proposition 4.2 is used to classify large commutative totally symmetric sets in B_n.

To each noncrossing totally symmetric multicurve configuration M_n, M_n^*, $\widehat{M}_n$ and $\widehat{M}_n^*$, there is an associated commutative totally symmetric set in B_n, namely:

$$Z_n = \{\sigma_1, \sigma_3, \dots\}, \quad Z_n^* = \{\sigma_1^*, \sigma_3^*, \dots\},$$
$$\widehat{Z}_n = \{\sigma_1 T_d, \sigma_3 T_d, \dots\}, \text{ and } \widehat{Z}_n^* = \{\sigma_1^* T_d, \sigma_3^* T_d, \dots\}.$$

Here σ_i^* is the product of the elements of Z_n not equal to σ_i and T_d is the Dehn twist about the curve used in the definitions of $\widehat{M}_n$ and $\widehat{M}_n^*$. We refer to any B_n-conjugate of any of these as a standard commutative totally symmetric set in B_n.

We can modify any of the standard totally symmetric sets by raising all elements to the same nonzero power. We can also modify them by multiplying all elements of the set by the same power of z, a generator for the (cyclic) center of B_n. We refer to any totally symmetric set obtained in this way as a modification of a standard commutative totally symmetric set in B_n. Kordek and the second author prove the following [**19**, Lemma 2.6].

PROPOSITION 4.3. *Let $n \geq 3$ and let $k = \lfloor n/2 \rfloor$. Any commutative totally symmetric set X in B_n with $|X| = k$ is a modification of a standard one.*

To prove the proposition, we of course first use the fact that, through canonical reduction systems, X gives rise to a noncrossing totally symmetric multicurve configuration. We then use Proposition 4.2 to reduce to the four cases of multicurve configurations given there.

We then treat the four cases in turn. For the case $\mathrm{CRS}(X) = M_n$, the idea is as follows. Say that X is $\{g_1, \dots, g_k\}$. Up to conjugation in B_n we may assume that $\mathrm{CRS}(g_i)$ is equal to c_{2i-1}. Note that this is the canonical reduction system of σ_{2i-1}, the ith element of Z_n.

We would like to show that X is equal to Z_n. On the exterior of c_1 the element g_1 is either the identity, periodic, or pseudo-Anosov. But since g_1 commutes with the other g_i, it must fix the curves $c_3, c_5, \dots$ There is no (nontrivial) periodic or pseudo-Anosov map that can fix these curves. It follows that this exterior component of g_1 is trivial, and hence that g_1 is equal to $\sigma_1^\ell z^s$ for some nonzero ℓ and some s. The other three cases are similar.

Proof of Theorem 4.1 *assuming Proposition* 4.3. Let $\rho : B_n \to B_n$ be an automorphism. By Lemma 2.1, the image of $Z_n = \{\sigma_1, \sigma_3, \dots\}$ is either a singleton or a commutative totally symmetric set of the same size. In the first case, it follows that ρ has cyclic image. Indeed, for $n \geq 5$ the normal closure of $\sigma_1\sigma_3^{-1}$ is B_n' and $B_n/B_n' \cong \mathbb{Z}$. Thus, we may henceforth assume that $\rho(Z_n)$ is a commutative totally symmetric set of cardinality $|Z_n|$.

By Proposition 4.3, $\rho(Z_n)$ is conjugate to a modification of Z_n, Z_n^*, $\widehat{Z}_n$, or $\widehat{Z}_n^*$. Let us consider the first case. Up to conjugating ρ, we may assume that $\rho(Z_n)$ is exactly a modification of Z_n, that is,

$$\rho(\sigma_i) = \sigma_i^\ell z^s$$

for all odd i.

For i even, we then have that $\rho(\sigma_i)$ is conjugate to $\sigma_i^\ell z^s$. So each such $\rho(\sigma_i)$ is equal to $H_{a_i}^\ell z^s$, where H_{a_i} is the half-twist about a curve a_i.

It is a fact that if H_a and H_b are the half-twists about curves a and b in D_n, and they satisfy the braid relation

$$H_a^\ell H_b^\ell H_a^\ell = H_b^\ell H_a^\ell H_b^\ell,$$

then $i(a, b) = 2$ and $\ell = \pm 1$; see [**5**, Lemma 4.9]. Up to the exceptional automorphism ϵ, we may assume that $\ell = 1$. It then further follows that $s = 0$, since an automorphism of B_n must preserve word length, that is, it respects the abelianization $B_n \to \mathbb{Z}$.

It also follows that the sequence of curves

$$c_1, a_2, c_3, a_4, \dots$$

is a chain, meaning that consecutive curves intersect twice and all other pairs of curves have trivial geometric intersection. Up to automorphisms of B_n, we then have (by change of coordinates)

$$\rho(\sigma_i) = \sigma_i$$

for all i. In other words, up to modifying ρ by automorphisms, it is the identity. This completes the proof in the first case. Using similar reasoning, we rule out the other three possibilities for $\rho(Z_n)$, completing the proof.

From braid groups to mapping class groups. Chen–Mukherjea [**11**] use a similar approach to classify homomorphisms from the braid group B_n to the mapping class group $\mathrm{Mod}(S_g)$ when $g < n - 2$. As a corollary, they partially recover the result of Aramayona–Souto [**1**] classifying homomorphisms $\mathrm{Mod}(S_g) \to \mathrm{Mod}(S_h)$ for $h < 2g$.

5. Finite quotients of braid groups and mapping class groups

In 1947, Emil Artin [**3**] proved that for $n \geq 5$ every non-cyclic homomorphism $B_n \to \Sigma_n$ is standard. This means that up to conjugacy, the map sends σ_i to the transposition $(i\ i{+}1)$ for all i. His proof uses Bertrand's postulate, a deep fact from number theory which states that every interval $[n, 2n]$ contains a prime number. Artin wrote: "it would be preferable if a proof could be found that does not make use of this fact."

Kolay [**17**] found in 2021 a short, elementary proof of Artin's theorem, and in fact proved more. In the statement, we say that a quotient map is minimal if there is no quotient map whose codomain has smaller cardinality.

THEOREM 5.1 (Kolay). *Let $n \geq 3$. Up to conjugacy, there is a unique minimal non-cyclic quotient of B_n, namely, the standard map $B_n \to \Sigma_n$ for $n \neq 4$ or the standard map $B_4 \to \Sigma_3$.*

In 2019 the second author of this paper had asked: *What is the smallest non-cyclic quotient of B_n? Is it Σ_n?* Kolay's theorem answers this in the affirmative.

We give Kolay's stunningly simple proof below. The main ingredients are (1) a large collapsing set in B_n and (2) the orbit-stabilizer theorem. While Artin never defined collapsing sets, he certainly had all of the tools to prove Kolay's theorem. It is remarkable that 74 years passed in between the two works.

Before Kolay's work, partial answers to the second author's question were given by Chudnovsky–Kordek–Li–Partin [**12**], Caplinger–Kordek [**8**], and Scherich–Verberne [**22**].

5.1. Braid groups.

In this section we explain Kolay's proof of Theorem 5.1, and in the next we explain how Kolay applied the same ideas to the case of the mapping class group.

Collapsing sets. Let G be a group. We say that a subset $X \subseteq G$ is a collapsing set if for every group homomorphism $f : G \to H$ the restriction $f|X$ is either injective or constant. This notion is a generalization of totally symmetric sets. Indeed, Lemma 2.1 implies that every totally symmetric set is a collapsing set.

We also remark that under any homomorphism, a collapsing set maps to a collapsing set. Therefore, there is an analogous blueprint for collapsing sets, an idea that seems to be unexplored.

Strong collapsing sets. Let G be a group and let $X = \{x_1, \ldots, x_k\} \subseteq G$ be a subset. We say that X is a strong collapsing set if

$$G/\langle\!\langle x_i x_j^{-1} \rangle\!\rangle$$

is abelian for all pairs $\{i, j\}$. This is the same as saying that the normal closure of each $x_i x_j^{-1}$ contains the commutator subgroup $[G, G]$. If the x_i are all conjugate, the $x_i x_j^{-1}$ lie in $[G, G]$ and so each $G/\langle\!\langle x_i x_j^{-1} \rangle\!\rangle$ must exactly be the abelianization of G. It follows from this that a strong collapsing set of conjugate elements is a collapsing set, since conjugate elements in an abelian group are equal.

A large collapsing set. Similar to Step 1 of the totally symmetric set blueprint, we will make use of a large strong collapsing set in B_n for $n \geq 5$.

For each unordered pair $I = \{i, j\} \subseteq \{1, \ldots, n\}$ let σ_I denote the half-twist in B_n given by the counter-clockwise exchange of the ith and jth marked points in the upper half of D_n. If I and J are distinct ordered pairs then $\sigma_I \sigma_J^{-1}$ is conjugate in B_n to exactly one of the following: $\sigma_1\sigma_2^{-1}$, $\sigma_1\sigma_3^{-1}$, or $b = \sigma_{\{1,3\}}\sigma_{\{2,4\}}^{-1}$.

We claim that for $n \geq 5$ the normal closure of any of these three elements in B_n is the commutator subgroup B_n'. For the first two elements, this is a standard fact. Similarly, the commutator $[b, \sigma_4]$ both lies in the normal closure of b and is conjugate to $\sigma_1\sigma_2^{-1}$. It follows that the normal closure of b is again B_n'. (An alternate, but equivalent, proof of the claim is given by the well-suited arc criterion of Lanier and the second author [**10**, Lemma 6.2].)

It follows from the claim that the set of all σ_I is a strong collapsing set for B_n. We refer to this as the standard strong collapsing set for B_n and denote it X_n.

Two basic group theory facts. In the proof of Theorem 5.1 we will use the following fact:

> *(1) If $f : \mathbb{Z} \times H \to G$ is a group homomorphism, and t denotes a generator of $\mathbb{Z}$, then $f(t) \notin f(H)$ if and only if*
> $$|f(\mathbb{Z} \times H)| \geq 2|f(H)|.$$

This fact is true because both conditions are equivalent to the statement that $f(t)f(H)$ is a nontrivial coset of $f(H)$ in $f(\mathbb{Z} \times H)$. We will also use the following:

> *(2) If G is a group, then $Z(G)$ is nontrivial if and only if*
> $$|G| \geq 2|G/Z(G)|.$$

This is true because if $Z(G)$ is nontrivial then each coset has at least two elements. While both statements make sense for infinite groups, we will only apply them when G is finite.

Base cases. We can prove the $n = 3$ and $n = 4$ cases of Theorem 5.1 by direct inspection. Because the abelianization of B_n is cyclic, a non-cyclic quotient of B_n is non-abelian. The only non-abelian group of order 6 or less is Σ_3. Thus, all other finite non-abelian quotients of B_3 and B_4 have order strictly greater than 6.

To see that the standard maps $B_3 \to \Sigma_3$ and $B_4 \to \Sigma_3$ are unique up to automorphisms of Σ_3, we simply check that (up to automorphisms of Σ_3) the only ordered pair of elements of Σ_3 satisfying the braid relation is $((1\ 2), (2\ 3))$.

Extension of the $n = 4$ case. We also will require the following statement:

> If $f : B_4 \to G$ is a quotient map that is injective on X_4 then $|G| \geq 4!$. Further, if $|G| = 4!$ then $G = \Sigma_4$ and f is standard.

Since this can be easily proved with a computer, we omit the proof (although it is a fun exercise to do it by hand!).

KOLAY'S PROOF. We prove the theorem by induction on n, with the base cases $n = 3$ and $n = 4$ (and the extension of the latter) handled as above.

Let $f : B_n \to G$ be a non-cyclic quotient map. Let X_n be the standard strong collapsing set in B_n. The group $G = f(B_n)$ acts by conjugation on the conjugacy class of $f(\sigma_1)$ in G. To prove that $|G| \geq n!$, we will apply the orbit-stabilizer theorem to this action. A further analysis will give the statement that f is conjugate to the standard map to Σ_n.

For the orbit-stabilizer argument there are, naturally, two steps. Specifically, if $\mathcal{O} \subseteq G$ and $\mathcal{S} \subseteq G$ are the orbit and stabilizer of $f(\sigma_1)$ then we will show that

$$|\mathcal{O}| \geq \binom{n}{2} \quad \text{and} \quad |\mathcal{S}| \geq 2 \cdot (n-2)!.$$

Orbit. Since X_n is a strong collapsing set, and since G is not cyclic, it follows that $|f(X_n)| = |X_n| = \binom{n}{2}$. In particular $|\mathcal{O}| \geq \binom{n}{2}$.

Stabilizer. We consider the action of B_n on itself by conjugation. The stabilizer of σ_1 in B_n contains a subgroup

$$\langle \sigma_1, \sigma_3, \sigma_4, \ldots, \sigma_{n-1} \rangle \cong \mathbb{Z} \times B_{n-2}.$$

The image $f(\mathbb{Z} \times B_{n-2})$ is a subgroup of $\mathcal{S}$. We would like to bound the cardinality of this image from below. We treat two cases, according to whether $f(\sigma_1)$ lies in $f(B_{n-2})$.

By induction we may assume that $|f(B_{n-2})| \geq (n-2)!$. Indeed, if $f(B_{n-2})$ were cyclic, then f would be cyclic, contrary to assumption. We will use this assumption in both cases.

CASE 1 ($f(\sigma_1) \notin f(B_{n-2})$). By the first basic group theory fact above, we have

$$|\mathcal{S}| \geq |f(\mathbb{Z} \times B_{n-2})| \geq 2|f(B_{n-2})| \geq 2(n-2)!$$

as desired.

CASE 2 ($f(\sigma_1) \in f(B_{n-2})$). In this case $f(\mathbb{Z} \times B_{n-2}) = f(B_{n-2})$. Since f is nontrivial, $f(\sigma_1)$ is nontrivial. Since σ_1 lies in the centralizer of B_{n-2} it must be that $f(\sigma_1)$ lies in the center of $f(B_{n-2})$. In particular, $Z(f(B_{n-2}))$ is nontrivial.

We claim that $f(B_{n-2})/Z(f(B_{n-2}))$ is not cyclic. Indeed, if it were cyclic then $f(B_{n-2})$ would be abelian (for any group G, if $G/Z(G)$ is abelian then G is). The abelianization of B_{n-2} is cyclic, and so any abelian quotient of it is cyclic. In particular, $f(B_{n-2})$ is cyclic. It then follows that $f(B_n)$ is cyclic, contrary to assumption.

By the claim, the group $f(B_{n-2})/Z(f(B_{n-2}))$ is a non-cyclic quotient of B_{n-2}. By induction, its order is bounded below by $(n-2)!$. By the second basic group theory fact, we have

$$|\mathcal{S}| \geq |f(\mathbb{Z} \times B_{n-2})| = |f(B_{n-2})| \geq 2|f(B_{n-2})/Z(f(B_{n-2}))| \geq 2(n-2)!$$

We may now complete the proof of the first statement. By the orbit-stabilizer theorem, we have

$$|G| \geq f(B_n) \geq |\mathcal{O}||\mathcal{S}| \geq \binom{n}{2} 2 \cdot (n-2)! = \frac{n(n-1)}{2} \cdot 2 \cdot (n-2)! = n!$$

The first statement, $n = 6$ case. For $n = 6$ the argument is the same, except we must use the extension of the base case $n = 4$. Since we may assume that f is injective on X_6, it is injective on the copy of X_4 associated to $B_4 \leqslant B_6$. Hence the size of the stabilizer $\mathcal{S}$ is bounded below by $2 \cdot 4!$.

The second statement. To prove the stronger statement that any quotient of B_n with order $n!$ is the standard one, it suffices to show that the $f(\sigma_i)$ have order 2 (because the quotient $B_n \to B_n/\langle\!\langle \sigma_1^2 \rangle\!\rangle \cong \Sigma_n$ is the standard quotient). But this is true because in order to realize the lower bound $|\mathcal{S}| \geq 2 \cdot (n-2)!$ it must be true by induction that $f(B_{n-2})$ is the standard quotient. □

5.2. Mapping class groups. We now turn our attention to the analogue of Theorem 5.1 for mapping class groups. The natural action of $\mathrm{Mod}(S_g)$ on $H_1(S_g; \mathbb{F}_2)$ gives rise to a representation

$$\mathrm{Mod}(S_g) \to \mathrm{Sp}_{2g}(\mathbb{F}_2).$$

The order of the latter group is

$$|\,\mathrm{Sp}_{2g}(\mathbb{F}_2)| = 2^{g^2} \prod_{i=1}^{g} (2^{2i} - 1).$$

Remarkably, we again have that the most natural small quotient is the smallest.

THEOREM 5.2 (Kielak–Pierro). *Let $g \geq 1$. Up to conjugacy, there is a unique minimal non-cyclic quotient of* $\mathrm{Mod}(S_g)$*, namely, the standard map* $\mathrm{Mod}(S_g) \to \mathrm{Sp}_{2g}(\mathbb{F}_2)$.

This theorem was conjectured by Zimmermann **[23]** in 2012. Kielak–Pierro **[16]** proved it in 2019, using the approach established by Baumeister–Kielak–Pierro **[4]** in their work on the analogous problem about outer automorphisms of free groups.

The Kielak–Pierro proof of Theorem 5.2 relies on the classification of finite simple groups and the representation theory of the mapping class group, as well as the deep work of Berrick–Gebhardt–Paris **[6]**, which itself uses the Matsumoto presentation of the mapping class group. It is astonishing that Kolay's argument for the braid group applies with little modification to prove the same theorem.

Again, the keys to Kolay's proof of Theorem 5.2 are the construction of a large strong collapsing set in $\mathrm{Mod}(S_g)$, and an orbit-stabilizer argument.

The general outline is closely analogous to the argument for the braid group. Even the description of the large collapsing set constructed in Step 1 is similar. Several new tools are required. We introduce these in turn as we go.

Reduction to the open case. Let S_g^1 denote the surface with one boundary component obtained from S_g by removing the interior of an embedded disk. Since $H_1(S_g^1; \mathbb{F}_2)$ is naturally isomorphic to $H_1(S_g; \mathbb{F}_2)$ we also have a natural map

$$\mathrm{Mod}(S_g^1) \to \mathrm{Sp}_{2g}(\mathbb{F}_2).$$

By filling the disk back in, we also obtain a quotient map

$$\mathrm{Mod}(S_g^1) \to \mathrm{Mod}(S_g).$$

In general if (within some class of groups) G is the smallest quotient of a group M_1, and M is a quotient of M_1 that also has G as a quotient, then G is the smallest quotient of M (in that class of groups). Thus to prove Theorem 5.2 it suffices to prove the analogous statement for $\mathrm{Mod}(S_g^1)$.

The base case. The base case for the $\mathrm{Mod}(S_g^1)$-version of Theorem 5.2 is the case $g = 1$. In this case Dehn [**13**, p. 172] proved that $\mathrm{Mod}(S_g^1) \cong B_3$. By Theorem 5.1, the smallest non-cyclic quotient of this group is $\Sigma_3 \cong \mathrm{Sp}_2(\mathbb{F}_2)$, as desired. (One way to prove the last isomorphism is to use the formula for the cardinality of $\mathrm{Sp}_2(\mathbb{F}_2)$ and apply Theorem 5.1!).

A well-suited curve criterion. We now turn our attention to Step 1. In the braid group case, the construction of the large strong collapsing set used the well-known fact that for $i \neq j$ the quotient

$$B_n / \langle\!\langle \sigma_i \sigma_j^{-1} \rangle\!\rangle$$

is cyclic. To mimic this step, we will use the following fact:

> *Let $f \in \mathrm{Mod}(S_g^1)$ and suppose c is a curve with $i(c, f(c)) = 1$. Then the normal closure $\langle\!\langle f \rangle\!\rangle$ contains the commutator subgroup $\mathrm{Mod}(S_g^1)'$ and*
>
> $$\mathrm{Mod}(S_g^1)/\langle\!\langle f \rangle\!\rangle$$
>
> *is cyclic.*

This fact is an instance of the well-suited curve criterion of Lanier and the second author [**20**, Lemma 2.1]. While their argument is given explicitly for $\mathrm{Mod}(S_g)$, it applies verbatim for $\mathrm{Mod}(S_g^1)$. The key points are that—like $\mathrm{Mod}(S_g)$—the abelianization of $\mathrm{Mod}(S_g^1)$ is cyclic for $g \geq 1$ and—like $\mathrm{Mod}(S_g)$—the group $\mathrm{Mod}(S_g^1)$ has a generating set consisting of Dehn twists about curves that have pairwise intersection at most 1. (In the construction of the large strong collapsing set for B_n we noted that we could have used the well-suited arc criterion in the proof. Similarly, it is true here that we can give a proof that mimics the braid group case more closely. We leave it to the reader to decide which approach they prefer.)

The hyperelliptic involution and mod 2 homology. A hyperelliptic involution of S_g^1 is a homeomorphism ι of order two with $2g+1$ fixed points. The quotient $S_g^1/\langle\iota\rangle$ is D_{2g+1}, the disk with $2g+1$ marked points. These marked points are the images of the fixed points of ι. We denote the set of marked points by P.

Let D_{2g+1}° denote the disk with $2g+1$ punctures obtained by removing P. The homology group $H_1(D_{2g+1}^\circ) \cong (\mathbb{F}_2)^{2g+1}$ has a canonical generating set, namely, the classes represented by small loops around the punctures. This gives rise to a canonical homomorphism

$$H_1(D_{2g+1}^\circ) \to \mathbb{F}_2,$$

whereby each of these generators maps to 1. We denote the kernel of this map by $H_1(D_{2g+1}^\circ; \mathbb{F}_2)^{\text{even}}$. The elements of this kernel are exactly the ones represented by simple closed curves in D_{2g+1} surrounding an even number of marked points.

We will define a map

$$\Psi : H_1(S_g^1; \mathbb{F}_2) \to H_1(D_{2g+1}^\circ; \mathbb{F}_2)$$

as follows. Given $v \in H_1(S_g^1; \mathbb{F}_2)$, we may represent v by a simple closed curve c that avoids the fixed points of ι. The image of c in D_{2g+1} represents an element of $H_1(D_{2g+1}^\circ; \mathbb{F}_2)$.

Arnol'd [**2**] gave the following (easy-to-prove but) remarkable fact:

The map Ψ is an isomorphism

$$\Psi : H_1(S_g^1; \mathbb{F}_2) \xrightarrow{\cong} H_1(D_{2g+1}^\circ; \mathbb{F}_2)^{even}$$

The map Ψ^{-1} can be described as follows. Given an element v of $H_1(D_{2g+1}^\circ; \mathbb{F}_2)^{\text{even}}$ we represent it by a simple closed curve c, and $\Psi^{-1}(v)$ is the class represented by one component of the preimage of c.

Say that a subset of P is even if it has an even number of elements. By the above discussion we have natural bijections

$$H_1(D_{2g+1}^\circ; \mathbb{F}_2)^{\text{even}} \leftrightarrow H_1(S_g^1; \mathbb{F}_2) \leftrightarrow \{\text{even subsets of } P\}$$

Further, the nonzero elements of $H_1(S_g^1; \mathbb{F}_2)$ correspond to the nonempty even subsets of P.

The large collapsing set. Let us represent D_{2g+1} as a disk with the points of P lying on a circle. For each nonempty subset $A \subseteq P$ there is, up to isotopy, a unique curve c_A in D_{2g+1} that bounds a convex disk containing exactly the points of P contained in A.

If A is even then the preimage of c_A in S_g^1 has exactly two components. We choose one of these (arbitrarily) and call it $\tilde{c}_A$.

We will show that the set of Dehn twists

$$X_g = \{T_{\tilde{c}_A} \mid A \subseteq P \text{ even}, \quad A \neq \varnothing\}$$

is a strong collapsing set in $\text{Mod}(S_g^1)$.

In order to prove this, we take A and B to be distinct nonempty even subsets of P, and we assume that

$$f : \text{Mod}(S_g^1) \to G$$

is a homomorphism with

$$f(T_{\tilde{c}_A}) = f(T_{\tilde{c}_B}).$$

For any choices of A and B, there exists an arc c in D_{2g+1} that connects two marked points, that intersects c_A in one point, and that is disjoint from c_B. To check this, we consider two cases, according to whether or not $A \cup B$ is a proper subset of P or not. In the first case, let $p \in P \smallsetminus (A \cup B)$; we take c to connect p to a point $q \in A$. If $A \cup B = P$, then since A and B are even there is a point p in $A \cap B$, and c connects any such p to a point $q \in A \smallsetminus B$.

The preimage of c in S_g^1 is a simple closed curve $\tilde{c}$ with $i(\tilde{c}, \tilde{c}_A) = 1$ and $i(\tilde{c}, \tilde{c}_B) = 0$. It follows that

$$i(\tilde{c}, T_{\tilde{c}_A} T_{\tilde{c}_B}^{-1}(\tilde{c})) = i(\tilde{c}, T_{\tilde{c}_A}(\tilde{c})) = 1.$$

By the above well-suited curve criterion, we conclude that X_g is a strong collapsing set, as desired.

A derivative collapsing set. Say that two elements of the strong collapsing set X_g are dual if the corresponding curves have intersection number 1 (equivalently if they have algebraic intersection number 1). We define

$$X_g' = \{(x, y) \in X_g \times X_g \mid x \text{ is dual to } y\}.$$

The set X_g' is a collapsing set in the following sense: if $f : \text{Mod}(S_g^1) \to G$ is a non-cyclic homomorphism, then each $f(x, y)$ is an ordered pair of distinct elements, and if the f-image of any two elements of X'_g coincide, they all coincide. Both of these statements follow from the fact that X_g is a collapsing set.

Since X_g is in bijection with the nonzero elements of $H_1(S_g^1;\mathbb{F}_2)$, it has $2^{2g}-1$ elements. Given one nonzero element of $H_1(S_g^1;\mathbb{F}_2)$, there is a codimension-1 affine subspace of $H_1(S_g^1;\mathbb{F}_2)$ corresponding to dual elements in $H_1(S_g^1;\mathbb{F}_2)$. Thus we have

$$|X'_g| = (2^{2g}-1)\cdot 2^{2g-1}$$

We are finally ready for the proof of the Kielak–Pierro theorem.

Proof of Theorem 5.2. We proceed by induction on g. We already checked the base case $g=1$. Assume then that the theorem holds for $\operatorname{Mod}(S_g^1)$ with $g \geq 1$. We will show that it holds for $\operatorname{Mod}(S_{g+1}^1)$.

Let $f : \operatorname{Mod}(S_{g+1}^1) \to G$ be a non-cyclic quotient. We use essentially the same orbit-stabilizer argument that was used in the proof of Theorem 5.1. The group G acts by conjugation on the set of ordered pairs of elements of G. Because X'_{g+1} is a collapsing set in the sense described above, the orbit is bounded below by

$$|f(X'_{g+1})| = |X'_{g+1}| = (2^{2(g+1)}-1)\cdot 2^{2(g+1)-1} = (2^{2(g+1)}-1)\cdot 2^{2g+1}$$

The stabilizer contains a copy of the image of $\operatorname{Mod}(S_g^1)$. By induction this gives a lower bound of

$$|f(\operatorname{Mod}(S_g^1)| \geq 2^{g^2}\prod_{i=1}^{g}(2^{2i}-1).$$

Multiplying these together gives the desired bound

$$2^{(g+1)^2}\prod_{i=1}^{g+1}(2^{2i}-1).$$

For this bound to be realized, the elements of X_g must map to elements of order 2. From there it follows that the quotient is $\operatorname{Sp}_{2(g+1)}(\mathbb{F}_2)$. This completes the proof.

A final lament. Kolay's proof of Theorem 5.2 goes through the derived collapsing set X'_g. An analogous argument can be used to give an unnecessarily complicated proof of Theorem 5.1 (about braid groups). The situation suggests to the authors that there should be a proof of Theorem 5.2 that uses X_g directly, and decreases genus in two inductive steps. We were not able to find such a proof. We implore the reader to find one.

6. Speculations and representations

As suggested to us by Kordek, there is a strong analogy between the collision-implies-collapse property and Schur's lemma from representation theory. We can give weight to this analogy as follows.

A homomorphism $f : G \to H$ induces a linear map $f_* : \mathbb{C}[G] \to \mathbb{C}[H]$. The vector spaces $\mathbb{C}[G]$ and $\mathbb{C}[H]$ come equipped with a G-action and an H-action, respectively, where both groups act by conjugation on the basis elements. If $X \subseteq G$ is a totally symmetric set with $|X| = k$, and G_X is the stabilizer of X in G, then $\mathbb{C}[X]$ is a representation of G_X. By total symmetry G_X surjects onto Σ_k, and so $\mathbb{C}[X]$ is a representation of G_X; in fact this representation factors through the permutation representation of Σ_k on $\mathbb{C}[X]$. On the other hand, the vector space $\mathbb{C}[f(X)]$ is a representation of $f(G_X) \subseteq H_{f(X)}$. Lemma 2.1 implies that the latter representation has either the same dimension as $\mathbb{C}[X]$ or it has dimension 1. This statement can be derived from Schur's lemma using the following three facts: (1) the first representation factors through Σ_k, (2) f_* intertwines the two representations, and (3) the permutation representation of Σ_k is the direct product of two irreducible representations of Σ_k, namely, the standard representation and the trivial one.

Other groups. Because total symmetry can be understood within representation theory as above, we are led to speculate on which aspects of representation theory can be brought to bear in the theory of totally symmetric sets. To begin, we know that, while the representation theory of the symmetric group is rich in and of itself, there is a broad landscape of representations of various groups.

QUESTION 1. To what extent, and to what end, can the theory of totally symmetric sets be generalized to arbitrary groups besides Σ_k?

As an example of what we have in mind, we note that there is no lift of the totally symmetric set $\{(1\ i)\} \subseteq \Sigma_n$ to a totally symmetric set in B_n. On the other hand, there is a lift to a cyclically symmetric set, that is, a set with an action of $\mathbb{Z}/(n-1)$. Similarly, there are large sets of Dehn twists in $\mathrm{Mod}(S_g)$ that carry an action by the dihedral group D_{2g}. How can these sets be used in the classification of homomorphisms between braid groups and mapping class groups?

Extending the analogy to representation theory. Because of the connection between totally symmetric sets and representation theory described above, it is natural to ask which notions from representation theory have analogues for totally symmetric sets.

QUESTION 2. Which of the concepts in representation theory—direct sum, direct product, tensor product, etc.—have analogues in the theory of totally symmetric sets?

Already in their work, Salter and the first author give versions of subrepresentations and induced representations for totally symmetric sets [**9**].

Multiple totally symmetric sets. The arguments presented in this paper are carried out by analyzing the action of a homomorphism on a single totally symmetric set or collapsing set. But many groups, such as the braid group, contain totally symmetric sets that are compatible in some sense (for instance, elements either commute or braid).

QUESTION 3. How can multiple totally symmetric sets in a group be used to give stronger constraints on homomorphisms than can be obtained with a single totally symmetric sets?

One step in this direction is taken in the work of Scherich–Verberne, where they study homomorphisms of virtual, welded, and classical braid groups by considering multiple totally symmetric sets at once [**22**].

Bounds on representations. As we have seen, the fact that Σ_n has a totally symmetric set of cardinality $n-1$ can be used, along with the work of the first author and Salter, to give a lower bound on the dimension of a non-cyclic linear representation of Σ_n. We are curious to what extent this line of reasoning holds for other groups.

As a first test case, we might consider the monster group M. The smallest nontrivial representation of M has dimension $47 \cdot 59 \cdot 71 = 196,883$. Based on the case of the symmetric group, one might hope that this is because M contains a totally symmetric set of cardinality 196,883. But we already showed in Section 2.2 that M cannot contain a totally symmetric set whose cardinality is greater than 43.

QUESTION 4. What are the largest totally symmetric sets of the monster group? Can they give insight into the 196883-dimensional representation?

It is tantalizing that totally symmetric sets might give new insights into the notoriously mysterious monster group. And similarly for other groups, discovered and not.

Acknowledgments

We are grateful to Michael Griffin, Koichi Oyakawa, Larry Rolen, Shuxian Song, Isabelle Steinmann, and an anonymous referee for helpful comments and conversations. We would also like to thank Lei Chen, Kevin Kordek, Justin Lanier, Dan Minahan, and Nick Salter for enlightening conversations about totally symmetric sets.

References

[1] Javier Aramayona and Juan Souto, *Homomorphisms between mapping class groups*, Geom. Topol. **16** (2012), no. 4, 2285–2341, DOI 10.2140/gt.2012.16.2285. MR3033518

[2] V. I. Arnol′d, *A remark on the branching of hyperelliptic integrals as functions of the parameters* (Russian), Funkcional. Anal. i Priložen. **2** (1968), no. 3, 1–3. MR244266

[3] E. Artin, *Braids and permutations*, Ann. of Math. (2) **48** (1947), 643–649, DOI 10.2307/1969131. MR20989

[4] Barbara Baumeister, Dawid Kielak, and Emilio Pierro, *On the smallest non-abelian quotient of* $\mathrm{Aut}(F_n)$, Proc. Lond. Math. Soc. (3) **118** (2019), no. 6, 1547–1591, DOI 10.1112/plms.12232. MR3957829

[5] Robert W. Bell and Dan Margalit, *Braid groups and the co-Hopfian property*, J. Algebra **303** (2006), no. 1, 275–294, DOI 10.1016/j.jalgebra.2005.10.038. MR2253663

[6] A. J. Berrick, V. Gebhardt, and L. Paris, *Finite index subgroups of mapping class groups*, Proc. Lond. Math. Soc. (3) **108** (2014), no. 3, 575–599, DOI 10.1112/plms/pdt022. MR3180590

[7] Noah Caplinger, *Large totally symmetric sets*, New York J. Math. **29** (2023), 931–938. MR4627651

[8] Noah Caplinger and Kevin Kordek, *Small Quotients of Braid Groups*, E-prints, `arXiv:2009.10139`, 2020.

[9] Noah Caplinger, *Large totally symmetric sets*, New York J. Math. **29** (2023), 931–938. MR4627651

[10] Lei Chen, Kevin Kordek, and Dan Margalit, *Homomorphisms between braid groups*, `arXiv:1910.00712`, Oct 2019.

[11] Lei Chen and Aru Mukherjea, *From braid groups to mapping class groups*, Math. Z. **303** (2023), no. 2, Paper No. 27, 26, DOI 10.1007/s00209-022-03167-5. MR4527842

[12] Alice Chudnovsky, Kevin Kordek, Qiao Li, and Caleb Partin, *Finite quotients of braid groups*, Geom. Dedicata **207** (2020), 409–416, DOI 10.1007/s10711-019-00505-6. MR4117579

[13] M. Dehn, *Die Gruppe der Abbildungsklassen* (German), Acta Math. **69** (1938), no. 1, 135–206, DOI 10.1007/BF02547712. Das arithmetische Feld auf Flächen. MR1555438

[14] Joan L. Dyer and Edna K. Grossman, *The automorphism groups of the braid groups*, Amer. J. Math. **103** (1981), no. 6, 1151–1169, DOI 10.2307/2374228. MR636956

[15] Benson Farb and Dan Margalit, *A primer on mapping class groups*, Princeton Mathematical Series, vol. 49, Princeton University Press, Princeton, NJ, 2012. MR2850125

[16] Dawid Kielak and Emilio Pierro, *On the smallest non-trivial quotients of mapping class groups*, Groups Geom. Dyn. **14** (2020), no. 2, 489–512, DOI 10.4171/GGD/552. MR4118626

[17] Sudipta Kolay, *Smallest noncyclic quotients of braid and mapping class groups*, Geom. Topol. **27** (2023), no. 6, 2479–2496, DOI 10.2140/gt.2023.27.2479. MR4634752

[18] Kevin Kordek, Lily Qiao Li, and Caleb Partin, *Upper bounds for totally symmetric sets*, Involve **14** (2021), no. 5, 853–870, DOI 10.2140/involve.2021.14.853. MR4384026

[19] Kevin Kordek and Dan Margalit, *Homomorphisms of commutator subgroups of braid groups*, Bull. Lond. Math. Soc. **54** (2022), no. 1, 95–111, DOI 10.1112/blms.12560. MR4396925

[20] Justin Lanier and Dan Margalit, *Normal generators for mapping class groups are abundant*, Comment. Math. Helv. **97** (2022), no. 1, 1–59, DOI 10.4171/cmh/526. MR4410724

[21] Michael Lönne, *Fundamental groups of projective discriminant complements*, Duke Math. J. **150** (2009), no. 2, 357–405, DOI 10.1215/00127094-2009-055. MR2569617

[22] Nancy Scherich and Yvon Verberne, *Finite image homomorphisms of the braid group and its generalizations*, Glasg. Math. J. **65** (2023), no. 2, 430–445, DOI 10.1017/S0017089523000022. MR4625994

[23] Bruno P. Zimmermann, *On minimal finite quotients of mapping class groups*, Rocky Mountain J. Math. **42** (2012), no. 4, 1411–1420, DOI 10.1216/RMJ-2012-42-4-1411. MR2981051

Department of Mathematics, University of Chicago, 5734 University Ave., Chicago, Illinois 60637

Email address: nccaplinger@gmail.com

Department of Mathematics, Vanderbilt University, 1326 Stevenson Center Ln, Nashville, Tennessee 37240

Email address: dan.margalit@vanderbilt.edu

Contemporary Mathematics
Volume **812**, 2025
https://doi.org/10.1090/conm/812/16265

Proper 2-equivalences and amalgams over virtually free groups

M. Cárdenas, F. F. Lasheras, and A. Quintero

Dedicated to Mike Mihalik on his 70th birthday

Abstract. We recall that two finitely presented groups are said to be *proper* 2*-equivalent* if they can be realized as fundamental groups of finite 2-dimensional CW-complexes whose universal covers are proper homotopy equivalent. This yields a topological equivalence relation which is coarser than the quasi-isometry relation. For a 1-ended and semistable at infinity group, its proper 2-equivalence class is determined by the pro-isomorphism type of its fundamental pro-group. This paper is intended to be part of a program for a better understanding of the behaviour of proper 2-equivalences with respect to the basic constructions in combinatorial group theory (i.e., amalgamated free products and HNN-extensions). Previous joint work with R. Roy shows some results in the case the amalgamation subgroups are finite. Here we show that if a 1-ended finitely presented group G splits non-trivially as an amalgamated product $G_1 *_F G_2$ (or an HNN-extension $G_1 *_F$) over a finitely generated virtually free group F where the groups G_i have a fundamental pro-group of a special type (namely, semistable and pro-(finitely generated free)) at each end, then so does G at infinity. As a consequence, it follows from some previous work and recent results by M. Mihalik and S. Shepherd that such a group G as above is proper 2-equivalent to either $\mathbb{Z}\times\mathbb{Z}$ (and hence G is a virtually surface group) or $F_2\times\mathbb{Z}$, where F_2 denotes the free group on two generators.

1. Introduction

We will be working within the category of locally finite CW-complexes and proper maps. A *proper* map is a map with the property that the inverse image of every compact subset is compact. Thus, two locally finite CW-complexes are said to be proper homotopy equivalent if they are homotopy equivalent and all homotopies involved are proper.

Given an infinite (strongly) locally finite CW-complex Y, a *proper ray* in Y is a proper map $\omega : [0,\infty) \longrightarrow Y$. We say that two proper rays ω, ω' *define the same end* if their restriction to the natural numbers $\omega|_{\mathbb{N}}, \omega'|_{\mathbb{N}}$ are properly homotopic. This equivalence relation gives rise to the notion of *end determined by* ω as the corresponding equivalence class, as well as the space of ends $\mathcal{E}(Y)$ of Y as a compact totally disconnected metrizable space (see [**15**]). Moreover, we say that two proper

2020 *Mathematics Subject Classification.* Primary 20F69, 20F65.

Key words and phrases. Amalgamated products, virtually free groups, proper 2-equivalence, properly 3-realizable groups.

rays ω, ω' *define the same strong end* if they are in fact properly homotopic. The CW-complex Y is *semistable* at the end determined by ω if any other proper ray defining the same end in fact defines the same strong end.

Sometimes it is convenient to treat proper rays in Y in a more combinatorial way. In this respect, we observe that any proper ray $\omega : [0, \infty) \longrightarrow Y$ is properly homotopic to a proper parametrization of some *proper edge path ray* in Y, namely, an infinite sequence $(\sigma_1, \sigma_2, \dots)$ of edges in Y (which we will still denote by r) so that each subsequence $\sigma_1, \dots, \sigma_k$ describes a finite edge path in Y and no edge σ_i appears infinitely many times in the entire sequence. Moreover, any two of such proper parametrizations are properly homotopic, see [**15**, §16.1].

The number of ends of a finitely generated group G represents the number of ends of the (strongly) locally finite CW-complex $\widetilde{X}^1$ for some (equivalently any) CW-complex X with $\pi_1(X) \cong G$ and with finite 1-skeleton. This number equals $0, 1, 2$ or ∞ (finite groups have 0 ends, see [**15, 25**]). Here, $\widetilde{X}$ denotes the universal cover of X. We say that a finitely presented group G is *semistable at each end* (resp. *at infinity*, if G is 1-ended) if the (strongly) locally finite CW-complex $\widetilde{X}^2$ is so, for some (equivalently any) CW-complex X with $\pi_1(X) \cong G$ and with finite 2-skeleton. Observe that any finite-dimensional locally finite CW-complex is strongly locally finite, see [**15**, Prop. 10.1.12].

We recall that a proper cellular map $f : X \longrightarrow Y$ between finite-dimensional locally finite CW-complexes is a proper n-equivalence if there is another proper cellular map $g : Y \longrightarrow X$ such that the restrictions $g \circ f/X^{n-1}$ and $f \circ g/Y^{n-1}$ are proper homotopic to the inclusion maps $X^{n-1} \subseteq X$ and $Y^{n-1} \subseteq Y$. Observe that if two finite-dimensional locally finite CW-complexes are proper homotopy equivalent the they are proper n-equivalent for all n. In this respect, a topological equivalence relation for finitely presented groups was introduced in [**8**] which is coarser than the quasi-isometry relation. Namely, two finitely presented groups are said to be *proper* 2*-equivalent* if they can be realized as fundamental groups of finite 2-dimensional CW-complexes whose universal covers are proper 2-equivalent as (strongly) locally finite CW-complexes; in fact, one can always arrange for an actual proper homotopy equivalence between the universal covers, see [**8**] for details. For a 1-ended and semistable at infinity group, its proper 2-equivalence class is completely determined by the pro-isomorphism type of its fundamental pro-group. See also [**9**] for some specific results on proper 2 equivalences between infinite ended groups.

Also, we recall that a finitely presented group G is *properly* 3*-realizable* (abbreviated to P3R) if for some finite 2-dimensional CW-complex X with $\pi_1(X) \cong G$, the universal cover $\widetilde{X}$ of X has the proper homotopy type of a 3-manifold M (with boundary). Notice that it follows from [**2**] that $\widetilde{X}$ always has the proper homotopy type of a 4-manifold. This concept was originally motivated by the fact that Lefschetz duality then yields the freeness of the second cohomology $H^2(G; \mathbb{Z}G)$, as it is a direct summand of $H^2_c(\widetilde{X}; \mathbb{Z}) \cong H^2_c(M; \mathbb{Z}) \cong H_1(M, \partial M; \mathbb{Z}) \cong \widetilde{H}_0(\partial M; \mathbb{Z})$, see [**17**]. This P3R property is a virtual property which does not depend on the choice of X after taking wedge with a single 2-sphere S^2, see [**1**]. It is worth mentioning that direct products and ascending HNN-extensions of infinite finitely presented groups are P3R, see [**10, 18**]. In fact, any extension of an infinite finitely presented group by another infinite finitely presented group is P3R, see [**6**]. One-relator groups are also P3R, see [**5**]. Also, it has been proved in [**7**] that this property is a quasi-isometry invariant, and the class of P3R groups is closed under amalgamated products over finite groups, see [**4**]; moreover, this class contains the class of all finitely presented groups whose fundamental pro-group is of *telescopic type at each end*, see [**18, 19**], i.e., semistable and pro-(finitely generated free), see §2 below for details; in fact, it is conjectured that both classes coincide. Also, in the context of P3R groups one can also give a geometric interpretation to $H^2(G;\mathbb{Z}G)$, see [**8**, Prop. 5.10 and Cor. 5.11]. Examples of non P3R groups are given in [**14**].

It is worth noticing that certain type of groups actions already ensure a fundamental pro-group of telescopic type at infinity; for instance, those admitting the existence of a coaxial homeomorphism, see [**16**].

Splittings of 1-ended finitely presented groups over certain classes of subgroups have been of interest since the work of Rips-Sela, see [**24**]; in fact, all elementary splittings over $\mathbb{Z}$ can be described from their canonical JSJ decomposition. The behavior at infinity of these splittings are also of interest from the proper homotopy point of view. In this respect, this paper is intended to be part of a program to understand the behaviour of proper 2-equivalences with respect to the basic constructions in combinatorial group theory (i.e., amalgamated free products and HNN-extensions). In the case the amalgamation subgroups are finite, some results have already been obtained in [**9**] more generally for graphs of groups of that kind. Here we study, up to proper 2-equivalences, those 1-ended finitely presented groups G which split in a special way over a non-trivial finitely generated virtually free group F. More precisely, our main results are:

THEOREM 1.1. *If a 1-ended finitely presented group G splits non-trivially as an amalgamated product $G_1 *_F G_2$ (or an HNN-extension $G_1 *_F$) over a (non-trivial) finitely generated virtually free group F where the groups G_i have a fundamental pro-group of telescopic type at each end, then so does G at infinity. In particular, G is P3R.*

M. Mihalik and S. Shepherd have shown very recently that a group G as in Theorem 1.1 can not be simply connected at infinity, see [**21**, Cor. 1.5]; in fact, they have obtained a cohomological result implying non 1-acyclicity at infinity for 1-ended groups in a more general setting up to "artificial" splittings. Notice that for a 1-ended and semistable at infinity P3R group G, G not being simply connected at infinity amounts to $H^2(G;\mathbb{Z}G) \neq 0$ (i.e., not 1-acyclic at infinity, see [**17**]) by some previous results in [**8**], where the authors gave some geometric interpretation to that cohomology group via the so called ∂-number of G, which measures the minimum number of plane boundary components among all 3-manifolds realizing (up to proper homotopy) a universal cover for G.

It is worth mentioning that a particular subclass of such groups G as in Theorem 1.1 has already been studied in [**3**] regarding some boundary matters (see also [**19**]). On the other hand, observe that if a P3R group is 1-ended and semistable at infinity then it is proper 2-equivalent to either $\mathbb{Z}\times\mathbb{Z}\times\mathbb{Z}, \mathbb{Z}\times\mathbb{Z}$ or $F_2\times\mathbb{Z}$ (here F_2 denotes the free group on two generators), see [**8**, Thm. 5.1]. In this setting, we have:

COROLLARY 1.2. *Let G be as Theorem 1.1. Then G is proper 2-equivalent to either $\mathbb{Z}\times\mathbb{Z}$ (and hence it is a virtually surface group) or $F_2\times\mathbb{Z}$. Moreover, in the case one of the groups G_i is not virtually free then G is proper 2-equivalent to $F_2\times\mathbb{Z}$.*

PROOF. Clearly, G is a 1-ended and semistable at infinity P3R group and hence proper 2-equivalent to either $\mathbb{Z} \times \mathbb{Z}$ (and hence it is a virtually surface group, by [**8**, Thm. 5.17]) or $F_2 \times \mathbb{Z}$, since the case $\mathbb{Z} \times \mathbb{Z} \times \mathbb{Z}$ is being ruled out by [**21**, Cor. 1.5].

For the last part, we recall that a non-compact surface has free fundamental group, and hence any subgroup of it of infinite index must be free as well. Assume G_1 is not virtually free, and suppose $G = G_1 *_F G_2$ is proper 2-equivalent to $\mathbb{Z} \times \mathbb{Z}$. Then G has a surface subgroup $S \subseteq G$ of finite index and hence S is 1-ended as well. It follows that $G_1 \cap S$ is also of finite index in G_1 and hence it is not free (as G_1 is not virtually free). Since $G_1 \cap S \subseteq S$ is a subgroup of the surface group S, it follows that $G_1 \cap S$ must be of finite index in S (by the previous remark) and hence of finite index in G. But then G_1 has finite index in G. Let $a \in G_1 - F$ and $b \in G_2 - F$. The element ab has infinite order and $(ab)^k$ is never in G_1 for $k > 0$, contradicting the fact that G_1 has finite index in G. The argument is similar in the case of an HNN-extension $G = G_1 *_F$. □

COROLLARY 1.3. *The class of those finitely presented P3R groups which are* 1*-ended and semistable at infinity is closed under amalgamated products (resp. HNN-extensions) over (non-trivial) finitely generated virtually free groups; in fact, such amalgams are all proper* 2*-equivalent to* $F_2 \times \mathbb{Z}$.

PROOF. Notice that if a finitely presented group is 1-ended and semistable at infinity, then being P3R amounts to having a fundamental pro-group of telescopic type at infinity, by [**18**, Thm. 1.2]) and [**8**, Thm. 5.22]. □

REMARK 1.4. Given some fixed finitely presented groups G_1, G_2 and a finitely generated free group F, the various amalgamated products of the form $G_1 *_F G_2$ might behave in a quite different way, even if the factor groups are free groups. See [**23**] for instance, where $G_i = F_9$ and $F = F_{81}$ are all free groups of the corresponding rank and F has finite index in both factors. Note that finitely generated free groups trivially have a fundamental pro-group of telescopic type at each end; in fact, Theorem 1.1 above applies to the "exotic" examples in [**23**], since those amalgamated products are indeed 1-ended and semistable at infinity, by construction. This also follows from by [**11**, Lemma 3.14] together with [**12**, Thm. 1.2]. Furthermore, by [**8**, Remark 2.8], these particular examples are all proper 2-equivalent to $F_2 \times \mathbb{Z}$, as they are normal subgroups of fundamental groups of finite complexes which have the product $T_{10} \times T_{10}$ (of two 10-regular trees) as universal cover.

EXAMPLE 1.5. It is also worth mentioning that one can easily exhibit examples of different HNN-extensions of the form $\mathbb{Z}*_{\mathbb{Z}}$ which are 1-ended (and semistable at infinity) but not proper 2-equivalent to each other. For this, we may just consider the groups $G = \langle x, t; t^{-1}xt = x \rangle$ and $H = \langle x, t; t^{-1}xt = x^2 \rangle$. Clearly, $G = \mathbb{Z} \times \mathbb{Z}$ whereas the fundamental pro-group of H corresponds to a strictly telescopic tower (i.e., a tower of finitely generated free groups of strictly increasing rank) and hence H is proper 2-equivalent to $F_2 \times \mathbb{Z}$.

2. Some Preliminaries

This section is intended to provided the background and notation needed in the previous section, as well as to indicate what is behind those results. In what follows, we will be working within the category $tow - Gr$ of *towers of groups* whose objects are inverse sequences of groups

$$\underline{G} = \{G_0 \stackrel{\phi_1}{\longleftarrow} G_1 \stackrel{\phi_2}{\longleftarrow} G_2 \longleftarrow \cdots\}$$

A morphism in this category will be called a *pro-morphism.* See [**20**] for a general reference. More precisely, we recall that two towers (inverse sequences) of groups $G_0 \stackrel{\lambda_1}{\longleftarrow} G_1 \stackrel{\lambda_2}{\longleftarrow} G_2 \longleftarrow \cdots$ and $H_0 \stackrel{\mu_1}{\longleftarrow} H_1 \stackrel{\mu_2}{\longleftarrow} H_2 \longleftarrow \cdots$ are pro-isomorphic if after passing to subsequences there exists a commutative diagram:

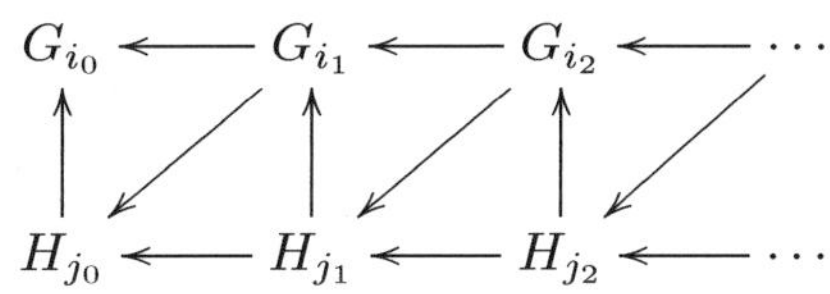

where the horizontal arrows are the obvious compositions of the corresponding λ's or μ's.

We will say that a tower is *pro-(finitely generated free)* if it is pro-isomorphic to a tower of finitely generated free groups. Moreover, a tower is said to be of *telescopic type* if it is pro-isomorphic to a a tower $\underline{P}$ of the form

$$\underline{P} = \{P_0 \stackrel{p_1}{\longleftarrow} P_1 \stackrel{p_2}{\longleftarrow} P_2 \longleftarrow \cdots\}$$

where each $P_i = \langle D_i \rangle$ is a finitely generated free group with basis D_i such that $D_{i-1} \subseteq D_i$ and the bonding homomorphisms p_k are the obvious projections.

LEMMA 2.1. *A tower of groups $\underline{H}$ is of telescopic type if and only if it is both pro-(finitely generated free) and semistable (or Mittag-Leffler).*

The proof of this Lemma is basically that in [**18**, Thm. 1.2]; nevertheless, we include it here for the sake of completeness.

PROOF. Assume $\underline{H}$ is pro-isomorphic to a tower $\underline{F} = \{F_0 \stackrel{\phi_1}{\longleftarrow} F_1 \stackrel{\phi_2}{\longleftarrow} F_2 \longleftarrow \cdots\}$ of finitely generated free groups which is semistable. Recall that $\underline{F}$ is said to be semistable (or Mittag-Leffler) if for each k there is $n \geq k$ so that for every $m \geq n$ we have $Im(\phi_{k+1} \cdots \phi_m) = Im(\phi_{k+1} \cdots \phi_n) \subset F_k$. It is well known that if $\underline{F}$ has this property then it is pro-isomorphic to a tower of groups $\underline{P} = \{P_0 \stackrel{\psi_1}{\longleftarrow} P_1 \stackrel{\psi_2}{\longleftarrow} P_2 \longleftarrow \cdots\}$ where the bonding maps ψ_n are all epimorphisms. Moreover, there is a level morphism $\{i_n\}$ inducing this pro-isomorphism such that every $i_n : P_n \longrightarrow F_n$ is a monomorphism between finitely generated groups, see [**20**, Ch. II, §6.2, Thm.7]. Thus, $\underline{P}$ is a tower of finitely generated free groups in which all bonding maps are epimorphisms. Finally, since $\psi_n : P_n \longrightarrow P_{n-1}$ is an epimorphism between free groups, it follows from [**13**] that P_n is a free product $P_n = P'_n * P''_n$, where $\psi_n(P''_n) = \{1\}$ and $\psi_n|P'_n$ is an isomorphism. Therefore, each ψ_n is a projection and hence $\underline{P}$ can be regarded as a telescopic tower, as each P''_n is finitely generated by the Grushko-Neumann theorem. □

We will also use the full subcategory $(Gr, tow - Gr)$ of $Mor(tow - Gr)$ whose objects are arrows $\underline{A} \longrightarrow G$, where $\underline{A}$ is an object in $tow - Gr$ and G is a group regarded as a constant tower whose bonding maps are the identity. Morphisms in $(Gr, tow - Gr)$ will also be called pro-morphisms.

Given a (strongly) locally finite CW-complex Y and a proper ray ω in Y together with an exhaustion by compact subsets $C_1 \subset C_2 \subset \cdots \subset Y$, so that $Y - C_i$ is a union of unbounded components, $C_i \subset int(C_{i+1})$ and $Y = \bigcup_{n=1}^{\infty} C_n$, the following tower

$$pro - \pi_1(Y, \omega) = \{\pi_1(Y, \omega(0)) \leftarrow \pi_1(Y - C_1, \omega(t_1)) \leftarrow \pi_1(Y - C_2, \omega(t_2)) \leftarrow \cdots\}$$

can be regarded as an object in $(Gr, tow - Gr)$ and it is called the *fundamental pro-group of* (Y, ω), where $\omega([t_i, \infty)) \subset Y - C_i$ and the bonding homomorphisms are induced by the inclusions (together with the corresponding change of base points along ω). This tower does not depend (up to pro-isomorphism) on the exhaustion $\{C_i\}_i$. From the inclusions $C_i \subset int(C_{i+1}) \subset C_{i+1}$ one can easily check that the above tower is also pro-isomorphic to the tower

$$\pi_1(Y, \omega(0)) \leftarrow \pi_1(Y - int(C_1), \omega(t_1)) \leftarrow \pi_1(Y - int(C_2), \omega(t_2)) \leftarrow \cdots$$

It is worth noticing that if ω and ω' define the same strong end of Y (i.e., they are in fact properly homotopic) then $pro - \pi_1(Y, \omega)$ and $pro - \pi_1(Y, \omega')$ are pro-isomorphic. Furthermore, it is worth mentioning that Y is semistable at the end determined by ω if the tower $pro - \pi_1(Y, \omega)$ is semistable (or Mittag-Leffler), and hence pro-isomorphic to a tower where all bonding maps are epimorphisms. Finally, given a finitely presented group G and a finite 2-dimensional CW-complex X with $\pi_1(X) \cong G$, we will refer to the fundamental pro-group of the universal cover $\widetilde{X}$ at each end (resp. at infinity, if 1-ended) as the fundamental pro-group of G at each end (resp. at infinity). Furthermore, we say that G is simply connected at infinity if for any compact subset $C \subset \widetilde{X}$ there is a compact subset $D \subset \widetilde{X}$ so that loops in $\widetilde{X} - D$ are homotopically trivial in $\widetilde{X} - C$; equivalently, the fundamental pro-group of G is pro-trivial. See also [**15**] as a general reference.

3. Main results

The purpose of this section is to prove Theorem 1.1.

3.1. Some geometric preliminaries. We start with some geometric considerations, following the description in [**22**] (compare with [**25**, §3]). Consider an amalgamated product $G = G_1 *_F G_2$ where G_1 and G_2 are finitely presented groups with presentations

$$\mathcal{P}_1 \equiv \langle g_1, \ldots, g_n, c_1, \ldots, c_k; r_1, \ldots, r_p \rangle,$$
$$\mathcal{P}_2 \equiv \langle h_1, \ldots, h_m, c_1, \ldots, c_k; s_1, \ldots, s_q \rangle,$$

respectively, where in both presentations $c_1, \ldots, c_k$ generate the subgroup F. Then, G has the following presentation

$$\mathcal{Q} \equiv \langle g_1, \ldots, g_n, h_1, \ldots, h_m, c_1, \ldots, c_k; r_1, \ldots, r_p, s_1, \ldots, s_q \rangle.$$

Let K_1, K_2 and X be the standard 2-dimensional CW-complexes associated to the above presentations $\mathcal{P}_1, \mathcal{P}_2$ and $\mathcal{Q}$ respectively, and let Γ be the Cayley graph of F with respect to the generating set $c_1, \ldots, c_k$. Thus, the universal cover $\widetilde{X}$ can be regarded as a push-out obtained from copies $g \cdot \widetilde{K}_1$ and $g \cdot \widetilde{K}_2$ $(g \in G)$ of the universal covers $\widetilde{K}_1$ and $\widetilde{K}_2$ respectively, such that $g \cdot \widetilde{K}_i \equiv g' \cdot \widetilde{K}_i$ if and only if $g^{-1}g' \in G_i$, $i = 1, 2$ (or disjoint otherwise), and glued together along copies $g \cdot \Gamma$

($g \in G$) of Γ within $g \cdot \widetilde{K}_i$ ($i = 1, 2$) such that $g \cdot \Gamma \equiv g' \cdot \Gamma$ if and only if $g^{-1}g' \in F$ (or disjoint otherwise). Also, $g_1 \cdot \widetilde{K}_1 \cap g_2 \cdot \widetilde{K}_2 = \emptyset$ unless there is $g \in G$ so that $g_1 \cdot \widetilde{K}_1 = g \cdot \widetilde{K}_1$ and $g_2 \cdot \widetilde{K}_2 = g \cdot \widetilde{K}_2$ and then $g_1 \cdot \widetilde{K}_1 \cap g_2 \cdot \widetilde{K}_2 = g \cdot \Gamma$. Observe that the universal cover $\widetilde{X}$ is modeled after the Bass-Serre tree for $G = G_1 *_F G_2$, and the notation above is compatible with the action of G on $\widetilde{X}$.

Consider root copies $1 \cdot \widetilde{K}_i \equiv \widetilde{K}_i \subset \widetilde{X}$ of the universal covers of K_i ($i = 1, 2$), and a root copy $1 \cdot \Gamma \equiv \Gamma$ common to both of them. One can define a "height function" as follows. Let $1 \cdot \widetilde{K}_1 \equiv \widetilde{K}_1 \subset \widetilde{X}$ have height 0 among all the copies $g \cdot \widetilde{K}_1$, and let all those copies $g \cdot \widetilde{K}_2 \subset \widetilde{X}$ intersecting $\widetilde{K}_1$ (including $1 \cdot \widetilde{K}_2 \equiv \widetilde{K}_2$) have height 0 as well. Next, a copy $g \cdot \widetilde{K}_1$ ($g \neq 1$) is said to have height 1 if it is glued through a copy of Γ to some copy $g' \cdot \widetilde{K}_2$ of height 0; similarly, a copy $g \cdot \widetilde{K}_2$ not of height 0 is said to have height 1 if it is glued through a copy of Γ to some copy $g' \cdot \widetilde{K}_1$ of height 1, and so on. This way one can define the height of any copy $g \cdot \widetilde{K}_i$ inside $\widetilde{X}$.

3.2. Generalized van Kampen's theorem: a special case. We will review a generalized van Kampen's theorem, which will be used later on in a particular form for the proof of Theorem 1.1. Here we follow the notation from [**15**, §6.2]. See also [**25**].

Let Γ be an oriented connected graph, and denote by E_0 and E_1 the sets of its vertices and (oriented) edges respectively. A generalized graph of groups $(\mathcal{G}, \Gamma)$ consists of a vertex group $G(w)$ for each $w \in E_0$, an edge group $G(e)$ for each $e \in E_1$ and homomorphisms $G(t(e)) \xleftarrow{\phi_e^+} G(e) \xrightarrow{\phi_e^-} G(o(e))$, where $o(e)$ and $t(e)$ are the initial and final vertices of e. Given a maximal tree $T \subseteq \Gamma$, the fundamental group of $(\mathcal{G}, \Gamma)$ based at T is defined as the quotient $(*_{w \in E_0} G(w)) * F(E_1)$ by the normal subgroup generated by (i) elements of the form $e^{-1}\phi_e^-(x)e(\phi_e^+(x))^{-1}$ where $e \in E_1$ and $x \in G(e)$, and (ii) elements e where e is a 1-cell of T.

Also, a graph of pointed CW-complexes $(\mathcal{X}, \Gamma)$ consists of a pointed CW-complex $(X(w), x(w))$ for each $w \in E_0$, a pointed CW-complex $(X(e), x(e))$ for each $e \in E_1$ and cellular maps $(X(o(e)), x(o(e))) \xleftarrow{p_e^-} (X(e), x(e)) \xrightarrow{p_e^+} (X(t(e)), x(t(e)))$. There is an associated total complex $Tot(\mathcal{X}, \Gamma)$ defined as the adjunction complex obtained from $\bigsqcup_{w \in E_0} X(w)$ and $\bigsqcup_{e \in E_1} X(e) \times [-1, 1]$ via the maps p_e defined as $p_e(y, \pm 1) = p_e^{\pm}(y)$ for all $y \in X(e)$. The map $s : \Gamma \longrightarrow Tot(\mathcal{X}, \Gamma)$, which takes every vertex $w \in E_0$ to $x(w)$ and every edge $e \in E_1$ homeomorphically onto $x(e) \times B^1$, sends a maximal tree $T \subseteq \Gamma$ isomorphically onto a tree $s(T) \subseteq Tot(\mathcal{X}, \Gamma)$.

On the other hand, given a connected CW-complex X and a tree $T \subseteq X$, the fundamental group of X based at T, $\pi_1(X, T)$, is defined as $\pi_1(X/T, \hat{v})$ where $\hat{v}$ is the vertex of X/T given by the image of T. For any vertex $v \in T$, there is a canonical isomorphism $\pi(X, v) \longrightarrow \pi_1(X, T)$. Moreover, for any maximal tree $\widehat{T} \subseteq X$, $\pi_1(X, T)$ and $\pi_1(X, \widehat{T})$ are also canonically isomorphic.

Observe that a graph of pointed CW-complexes $(\mathcal{X}, \Gamma)$ yields the corresponding generalized graph of groups $(\mathcal{G}, \Gamma)$ by passing to the fundamental groups; moreover, there is an isomorphism $\pi_1(\mathcal{G}, \Gamma; T) \cong \pi_1(Tot(\mathcal{X}, \Gamma), s(T))$, see [**15**, §6.2].

3.2.1. *Back to the universal cover* $\widetilde{X}$. We return now to §3.1 and assume $\widetilde{X}$ is 1-ended and F is now a free group (freely generated by $c_1, \ldots, c_k$) with the tree Γ as its corresponding Cayley graph. For the sake of simplicity, in this subsection we will denote by $\{X_\alpha, \alpha \in \mathcal{A}\}$ the collection of subcomplexes $\{g \cdot \widetilde{K}_i; g \in G, i = 1, 2\}$ in $\widetilde{X}$. Choose a proper edge path ray ω in the root copy $1 \cdot \Gamma \subset \widetilde{X}$ of the Cayley graph Γ of F.

Let $\{C_n\}_{n\geq 0}$ be an exhaustion of $\widetilde{X}$ by compact subsets with $C_i \subseteq int(C_{i+1})$. For each $\alpha \in \mathcal{A}$, we choose an end $e_\alpha \in \mathcal{E}(X_\alpha)$ and consider a proper edge path ray ω_{e_α} in $X_\alpha \subset \widetilde{X}$ determining that end. Next, given a connected component $X'_{\alpha'}$ of $X_\alpha - int(C_n \cap X_\alpha)$, we stick ourselves to the subray of ω_{e_α} within $X'_{\alpha'}$ unless it no longer determines an end of $X'_{\alpha'}$ in which case we pick a new proper edge path ray $\omega_{e_{\alpha'}}$ inside $X'_{\alpha'}$, and repeat the process with respect to $X'_{\alpha'}$ and any connected component $X''_{\alpha''}$ of $X'_{\alpha'} - (int(C_{n+1}) \cap X'_{\alpha'})$ and so on.

Given $n \geq 0$, let J_n be the graph with as many vertices as components $X'_{\alpha'}$ ($\alpha' \in \mathcal{A}'$) of $X_\alpha - int(C_n \cap X_\alpha)$, for each $\alpha \in \mathcal{A}$, and an edge $e'_{\alpha'\beta'\gamma'}$ for each (non-empty) connected component $Y'_{\alpha'\beta'\gamma'}$ of $X'_{\alpha'} \cap X'_{\beta'}$ ($\alpha \neq \beta$), which is a subtree of the corresponding copy of Γ and hence simply connected. Pick base points $x'_{\alpha'} = \omega_{e_{\alpha'}}(N_{\alpha'(n)}) \in X'_{\alpha'}$, with $\omega_{e_{\alpha'}}[N_{\alpha'(n)}, \infty) \subset X'_{\alpha'}$, and $y'_{\alpha'\beta'\gamma'} \in Y'_{\alpha'\beta'\gamma'}$ as well as edge paths $\rho_{\alpha'\beta'\gamma'}$ in $X'_{\alpha'}$ joining $y'_{\alpha'\beta'\gamma'}$ to $x'_{\alpha'}$ and $\rho_{\beta'\alpha'\gamma'}$ in $X'_{\beta'}$ joining $y'_{\alpha'\beta'\gamma'}$ to $x'_{\beta'}$.

Thus, for each $n \geq 0$ we have a generalized graph of groups $(\mathcal{G}_n, J_n)$ with vertex groups $\pi_1(X'_{\alpha'}, \omega_{e_{\alpha'}}(N_{\alpha'(n)}))$ and trivial structural induced homomorphisms:

$$\phi_{\alpha'\beta'\gamma'} : \pi_1(Y'_{\alpha'\beta'\gamma'}, y'_{\alpha'\beta'\gamma'}) \longrightarrow \pi_1(X'_{\alpha'}, x'_{\alpha'})$$

and

$$\phi_{\beta'\alpha'\gamma'} : \pi_1(Y'_{\alpha'\beta'\gamma'}, y'_{\alpha'\beta'\gamma'}) \longrightarrow \pi_1(X'_{\beta'}, x'_{\beta'})$$

since the subcomplexes $Y'_{\alpha\beta'\gamma'}$ are simply connected. Pick a maximal tree T_n in J_n, and denote by E_n the set of those edges in J_n which are not in T_n. An argument similar to that in [**15**, Thm. 6.2.11] leads to an isomorphism between the fundamental group of $\widetilde{X} - int(C_n)$ and $\pi_1(\mathcal{G}_n, J_n; T)$ (see also [**25**]). This isomorphism together with the choice of edge paths $\rho_{\alpha'}$ in $\widetilde{X} - int(C_n)$ from $\omega(N_n)$ to $\omega_{e_{\alpha'}}(N_{\alpha'(n)})$ turns into an isomorphism:

$$\Phi_n : (*_{\alpha' \in \mathcal{A}'}\ \pi_1(X'_{\alpha'}; \omega_{e_{\alpha'}}(N_{\alpha'(n)})) * F(E_n) \longrightarrow \pi_1(\widetilde{X} - int(C_n); \omega(N_n)).$$

Now we will further assume that the factor groups G_1 and G_2 are both semistable at each end, and hence so is each X_α, $\alpha \in \mathcal{A}$. It follows from [**22**] that the 1-ended amalgamated product $G = G_1 *_F G_2$ is semistable at infinity as well. Therefore, for each $\alpha \in \mathcal{A}$, given a proper edge path ray ω_e in X_α determining an end $e \in \mathcal{E}(X_\alpha)$ there exist edge paths λ_e^n ($n \geq 0$) inside $\widetilde{X}$ from the nth vertex $\omega(n)$ of the fixed proper edge path ray ω to the nth vertex $\omega_e(n)$ of ω_e, with the property that for any compact subset $C \subset \widetilde{X}$ all but finitely many of the λ_e^n's avoid C, and

(3.2.1.1) all but finitely many loops $\omega_e[n, n+1] \cdot (\lambda_e^{n+1})^{-1} \cdot (\omega[n, n+1])^{-1} \cdot \lambda_e^n$ (as depicted below) are homotopically trivial in $\widetilde{X} - C$ (see [**15**, §16.1.19]).

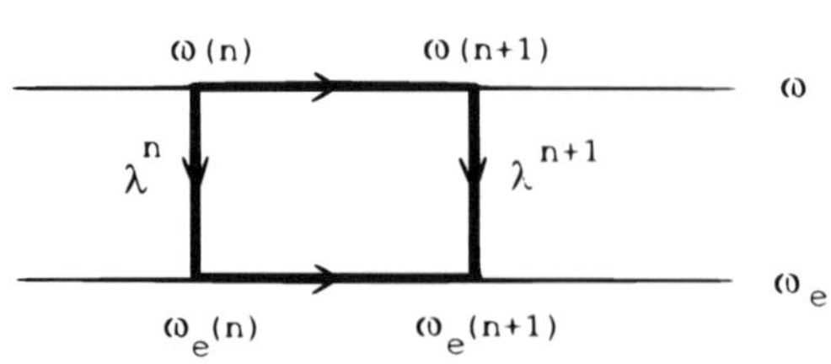

Let us denote $\Pi_n = *_{\alpha' \in \mathcal{A}'} \pi_1(X'_{\alpha'}; \omega_{e_{\alpha'}}(N_{\alpha'(n)}))$ and $H_n = F(E_n)$, $n \geq 0$. Now, the inclusion $C_n \subset C_{n+1}$ gives rise to a commutative diagram:

$$\begin{array}{ccc} \pi_1(\widetilde{X} - int(C_n); \omega(N_n)) & \xleftarrow{\omega_\#} & \pi_1(\widetilde{X} - int(C_{n+1}); \omega(N_{n+1})) \\ \Phi_n \uparrow \cong & & \Phi_{n+1} \uparrow \cong \\ \Pi_n * H_n & \xleftarrow{\iota_{n+1}} & \Pi_{n+1} * H_{n+1} \end{array}$$

where $\omega_\#$ is the composite of the inclusion-induced homomorphism followed by the base point change isomorphism along the proper edge path ray ω, and $\iota_{n+1} = \Phi_n^{-1} \circ \omega_\# \circ \Phi_{n+1}$. Observe that $\Pi_{n+1} = *_{\alpha'' \in \mathcal{A}''} \pi_1(X''_{\alpha''}; \omega_{e_{\alpha''}}(N_{\alpha''(n+1)}))$, where $X''_{\alpha''} \subset X'_{\alpha'}$ ranges over all connected components of $X'_{\alpha'} - int(C_{n+1} \cap X'_{\alpha'})$, and $\omega_{e_{\alpha''}}([N_{\alpha''(n+1)}, \infty)) \subset X''_{\alpha''}$.

LEMMA 3.1. *If $X''_{\alpha''} \subset X'_{\alpha'}$ is such that the corresponding chosen proper edge path ray $\omega_{e_{\alpha''}}$ is a subray of $\omega_{e_{\alpha'}}$, then there is an element $z_{\alpha''} \in \Pi_n * H_n$ such that on the factor $\pi_1(X''_{\alpha''}; \omega_{e_{\alpha''}}(N_{\alpha''(n+1)}))$ of Π_{n+1} the homomorphism ι_{n+1} can be expressed as*

$$x \longmapsto z_{\alpha''} \cdot \omega_{e_{\alpha'}\#}(x) \cdot z_{\alpha''}^{-1}$$

where $\omega_{e_{\alpha'}\#}$ is the composite of the inclusion-induced homomorphism $X''_{\alpha''} \subset X'_{\alpha}$ followed by the base point change isomorphism along the proper edge path ray $\omega_{e_{\alpha'}}$.

PROOF. Recall that Φ_{n+1} is defined on $\pi_1(X''_{\alpha''}; \omega_{e_{\alpha''}}(N_{\alpha''(n+1)}))$ by a change of base point along an edge path $\rho_{\alpha''}$ in $\widetilde{X} - int(C_{n+1})$ from $\omega(N_{n+1})$ to $\omega_{e_{\alpha''}}(N_{\alpha''(n+1)}) = \omega_{e_{\alpha'}}(N_{\alpha'(n+1)})$. Similarly, the isomorphism Φ_n is defined on $\pi_1(X'_{\alpha'}; \omega_{e_{\alpha'}}(N_{\alpha'(n)}))$ by a change of base point along an edge path $\rho_{\alpha'}$ in $\widetilde{X} - int(C_n)$ from $\omega(N_n)$ to $\omega_{e_{\alpha'}}(N_{\alpha'(n)})$. For the sake of simplicity, we will assume $N_n = N_{\alpha'(n)}$ and $N_{n+1} = N_{\alpha'(n+1)}$, and we take them large enough such that the corresponding λ-paths below as well as the corresponding homotopies from (3.2.1.1) avoid the compact subsets C_n and C_{n+1} from that point on.

Moreover, if we consider the elements $y_{n+1} \in \pi_1(\widetilde{X} - int(C_{n+1}); \omega(N_{n+1}))$ and $y_n \in \pi_1(\widetilde{X} - int(C_n); \omega(N_n))$ represented by the loops $\rho_{\alpha''} \cdot (\lambda_{e_{\alpha'}}^{N_{n+1}})^{-1}$ and $\lambda_{e_{\alpha'}}^{N_n} \cdot (\rho_{\alpha'})^{-1}$ respectively, we have the equalities for $x \in \pi_1(X''_{\alpha''}, \omega_{e_{\alpha'}}(N_{n+1})$:

$$\begin{aligned} \omega_\#(\Phi_{n+1}(x)) &= \omega_\#(\rho_{\alpha''\#}(x)) = \omega_\#(y_{n+1}) \cdot \omega_\#((\lambda_{e_{\alpha'}}^{N_{n+1}})_\#(x)) \cdot \omega_\#(y_{n+1}^{-1}) \overset{(*)}{=} \\ &= \omega_\#(y_{n+1}) \cdot (\lambda_{e_{\alpha'}}^{N_n})_\#(\omega_{e_{\alpha'}\#}(x)) \cdot \omega_\#(y_{n+1}^{-1}) = \\ &= \omega_\#(y_{n+1}) \cdot y_n \cdot \rho_{\alpha'\#}(\omega_{e_{\alpha'}\#}(x)) \cdot y_n^{-1} \cdot \omega_\#(y_{n+1}^{-1}) = \\ &= \omega_\#(y_{n+1}) \cdot y_n \cdot \Phi_n(\omega_{e_{\alpha'}\#}(x)) \cdot y_n^{-1} \cdot \omega_\#(y_{n+1}^{-1}) \end{aligned}$$

where $(*)$ follows from (3.2.1.1) above. Thus, $\iota_{n+1}(x) = \Phi_n^{-1} \circ \omega_\# \circ \Phi_{n+1}(x) = z_{\alpha''} \cdot \omega_{e_{\alpha'}\#}(x) \cdot z_{\alpha''}^{-1}$, with $z_{\alpha''} = \Phi_n^{-1}(\omega_\#(y_{n+1}) \cdot y_n)$. □

REMARK 3.2. The tower $\underline{\Pi H}$:

$$1 \longleftarrow \Pi_0 * H_0 \xleftarrow{\iota_1} \Pi_1 * H_1 \xleftarrow{\iota_2} \Pi_2 * H_2 \xleftarrow{\iota_3} \cdots$$

is then pro-isomorphic (by the level isomorphisms Φ_n) to $pro-\pi_1(\widetilde{X}, \omega)$ and hence semistable. Observe that if for a given component $X''_{\alpha''}$ of $\widetilde{X} - int(C_{n+1})$ there is no such component $X'_{\alpha'}$ of $\widetilde{X} - int(C_n)$ as in Lemma 3.1 (i.e., the former $X'_{\alpha'}$ splits producing a larger number of unbounded components) then we may as well consider

ι_{n+1} as the trivial homomorphism on the factor $\pi_1(X''_{\alpha''}, \omega_{e_{\alpha''}}(N_{\alpha''(n+1)})) \subset \Pi_{n+1}$, thus yielding a pro-isomorphic tower which we keep denoting by $\underline{\Pi H}$.

3.3. Proof of Theorem 1.1. First, we will make the following reduction. Observe that if $J \leq F$ is a finitely generated free subgroup with $[F : J] < \infty$ then, from the point of view of proper homotopy, the 2-skeleton of the universal cover of F behaves like a 2-*spherical object* under a tree T, i.e., it may be regarded as T together with finitely many 2-spheres attached to each vertex, where the tree T comes from the Cayley graph of the free subgroup $J \leq F$, as F and J are proper 2-equivalent (see [**8**, Remark 2.8]). Thus, for our purposes and for the sake of simplicity, from now on we will assume that the amalgamation subgroup F is already a finitely generated free group.

We keep the notation from §3.1 and consider such a 1-ended amalgamated product $G = G_1 *_F G_2$ with G_1 and G_2 finitely presented and having a fundamental pro-group of telescopic type at each end. In particular, G_1 and G_2 are both semistable at each end, and hence G is semistable at infinity as well (see [**22**]). As in §3.2.1, we consider a proper edge path ray ω in the root copy $1 \cdot \Gamma \equiv \Gamma \subset \widetilde{X}$, where the tree Γ is the Cayley graph of F with respect to the given generators. We wish to examine the pro-isomorphism type of the fundamental pro-group $pro-\pi_1(\widetilde{X}, \omega)$, in terms of the various copies $g \cdot \widetilde{K}_i \subset \widetilde{X}$, and show that it is of telescopic type as well. By Lemma 2.1, it suffices to show that $pro - \pi_1(\widetilde{X}, \omega)$ is pro-(finitely generated free) as it is already semistable at infinity because G is so, as stated lines above.

We will proceed as in §3.2.1 above. Consider an exhaustion $C_0 \subseteq C_1 \subseteq \cdots \subset \widetilde{X}$ by compact subsets with $C_n \subset int(C_{n+1})$, $n \geq 0$, and so that C_0 intersects the root copy $1 \cdot \Gamma \equiv \Gamma$. For every (strong) end $e \in \mathcal{E}(g \cdot \widetilde{K}_i)$ we take a proper edge path ray ω_e in $g \cdot \widetilde{K}_i$ defining the end e, together with edge paths $\lambda^n_{g,i,e}$ in $\widetilde{X}$, chosen as in §3.2.1, from the nth vertex $\omega(n)$ of ω to the nth vertex $\omega_e(n)$ of ω_e, and satisfying the property (3.2.1.1). Further, for each end $e \in \mathcal{E}(g \cdot \widetilde{K}_i)$ we choose pro-isomorphisms yielding commutative diagrams ($g \in G$, $i = 1, 2$), since by hypothesis the corresponding fundamental pro-group is of telescopic type

(3.3.1)

$$\begin{array}{ccccccc}
1 \leftarrow & \pi_1(g \cdot \widetilde{K}_i - int(D_{m_0(g,i,e)}), w_e(N_0)) & \leftarrow & \pi_1(g \cdot \widetilde{K}_i - int(D_{m_1(g,i,e)}), w_e(N_1)) & \leftarrow & \cdots \\
 & \downarrow \phi^0_{g,i,e} & \nwarrow \Psi^1_{g,i,e} & \downarrow \phi^1_{g,i,e} & \nwarrow \Psi^2_{g,i,e} & \\
1 \longleftarrow & F^0_{g,i,e} & \longleftarrow & F^1_{g,i,e} & \longleftarrow & \cdots
\end{array}$$

where, for each triple (g, i, e), (i) $m_{j+1}(g, i, e) > m_j(g, i, e) \geq j$, for each $j \geq 0$; (ii) the $D_{m_j(g,i,e)} = C_{m_j(g,i,e)} \cap g \cdot \widetilde{K}_i$ provide with an exhaustion of $g \cdot \widetilde{K}_i$ by compact subsets; (iii) the N_j-th vertex $\omega_e(N_j)$ of ω_e is some vertex such that $\omega_e([N_j, \infty)) \subset g \cdot \widetilde{K}_i - D_{m_j(g,i,e)}$; and (iv) the bonding maps in the upper row are induced by the corresponding inclusion and the obvious change of base point (along the base ray), whereas the lower row is a tower of finitely generated free groups of non-decreasing rank (with the obvious projections as bonding maps).

We will next proceed to replace the exhaustion $\{C_n\}_{n \geq 0}$ above by a new one $C'_0 \subset C'_1 \subset \cdots$ in order to be able to assemble all the above pro-isomorphisms and get the desired pro-(finitely generated free) fundamental pro-group. For this, let V^1_k, $1 \leq k \leq r_0$ be the unbounded connected components of $\widetilde{K}_1 - C_0$, and for each

$1 \leq k \leq r_0$ we choose an end $e_k \in \mathcal{E}(\widetilde{K}_1)$ such the corresponding proper edge path ray ω_{e_k} in $\widetilde{K}_1$ given above is eventually contained in V_k^1 and so it determines an end of V_k^1 (which we keep denoting by e_k).

Analogously, for those (finitely many) copies $g_j \cdot \widetilde{K}_2$ $(1 \leq j \leq q_0)$ of height 0 intersecting C_0, we consider the unbounded connected components $W_l^{g_j}$, $1 \leq l \leq s_0(g_j)$ of $g_j \cdot \widetilde{K}_2 - C_0$, together with ends $e_{(g_j,l)} \in \mathcal{E}(g_j \cdot \widetilde{K}_2)$ such that the corresponding proper edge path ray $\omega_{e_{(g_j,l)}}$ in $g_j \cdot \widetilde{K}_2$ given above is eventually contained in $W_l^{g_j}$ and so it determines an end of $W_l^{g_j}$ (which we keep denoting by $e_{(g_j,l)}$). Let $C'_0 \subset \widetilde{X}$ be the compact subset

$$C'_0 = \left(\bigcup_{k=1}^{r_0} D_{m_0(1,1,e_k)} \right) \cup \left(\bigcup_{j=1}^{q_0} \bigcup_{l=1}^{s_0(g_j)} D_{m_0(g_j,2,e_{(g_j,l)})} \right)$$

and take $N_0 \geq 1$ so that: (i) $\omega([N_0, \infty)) \subset \widetilde{X} - int(C'_0)$, and (ii) all the edge paths $\lambda^m_{1,1,e_k}$ $(1 \leq k \leq r_0)$ and $\lambda^m_{g_j,2,e_{(g_j,l)}}$ $(1 \leq j \leq l$ and $1 \leq l \leq s_0(g_j))$ lie outside the compact subset $C'_0 \subset \widetilde{X}$, for all $m \geq N_0$; moreover, (iii) the corresponding edge loops as in (3.2.1.1) obtained from the edge paths in (ii) (and the proper edge path rays ω, ω_{e_k} and $\omega_{e_{(g_j,l)}}$) are all homotopically trivial in $\widetilde{X} - C'_0$, for all $m \geq N_0$ (see (3.2.1.1)).

First, observe that any copy $g \cdot \widetilde{K}_i \subset \widetilde{X}$ $(i = 1, 2)$ not intersecting C'_0 does not contribute to $\pi_1(\widetilde{X} - int(C'_0), \omega(N_0))$, by construction, as it is simply connected and intersects the other copies of $\widetilde{K}_i$ along an entire copy of the tree Γ inside $\widetilde{X} - int(C'_0)$. Second, any copy $g \cdot \widetilde{K}_i \subset \widetilde{X}$ $(i = 1, 2)$ only intersecting $C'_0 - C_0$ contributes with a (possibly trivial) free factor $H^0_{g,i}$ to $\pi_1(\widetilde{X} - int(C'_0), \omega(N_0))$ as follows. Observe that such copy $g \cdot \widetilde{K}_i$ remains entirely in $\widetilde{X} - int(C'_0)$ and satisfies: (a) it intersects any other copy of $\widetilde{K}_j$ not involved in the definition of C'_0 along an entire copy of the tree Γ inside $\widetilde{X} - int(C'_0)$; and (b) for any copy $h \cdot \widetilde{K}_j$ of height 0 involved in the definition of C'_0, it intersects $h \cdot \widetilde{K}_j - int(C'_0 \cap h \cdot \widetilde{K}_j)$ and each of the (finitely many) components of such intersection is a subtree of a copy of Γ and hence contractible, thus contributing with a free factor $H^0_{g,i}$ of rank the number of those components minus one, by a generalized van Kampen's argument (see §3.2).

On the other hand, if some of the above components $W_l^{g_j}, V_k^1$ have non-empty intersection $W_l^{g_j} \cap V_k^1 \neq \emptyset$ then each of the connected components of such intersection is in turn a subtree of a copy of Γ and hence simply connected, thus contributing to $\pi_1(\widetilde{X} - int(C'_0), \omega(N_0))$ with another (possibly trivial) free factor $H^0_{g_j,l,k}$ of rank the number of connected components of $W_l^{g_j} \cap V_k^1$ minus one, again by a generalized van Kampen's argument as in §3.2.

Putting together all of the above, one can check that a global generalized van Kampen's argument yields the isomorphism Φ_0 from §3.2.1 corresponding to $\widetilde{X} - int(C'_0)$.

$$\Phi_0 : \Pi_0 * H_0 \longrightarrow \pi_1(\widetilde{X} - int(C'_0), \omega(N_0))$$

which allows us to express that fundamental group as a free product, where

$$\begin{aligned} \Pi_0 = \; & *_{k=1}^{r_0} \pi_1 \left(\widetilde{K}_1 - int(D_{m_0(1,1,e_k)}), \; \omega_{e_i}(N_0) \right) * \\ & *_{j=1}^{q_0} *_{l=1}^{s_0(g_j)} \; \pi_1 \left(g_j \cdot \widetilde{K}_2 - int(D_{m_0(g_j,2,e_{(g_j,l)})}), \; \omega_{e_{(g_j,l)}}(N_0) \right) \end{aligned}$$

and H_0 is the (abstract) finitely generated free group $H_0 = (*_{g,i} H^0_{g,i}) * (*_{g_j,l,k} H^0_{g_j,l,k})$ obtained as the free product of the various free groups mentioned in the discussion above (see §3.2.1).

Assume the compact subset $C'_{n-1} \subset \widetilde{X}$ $(n \geq 1)$ is already constructed. We proceed to construct a compact subset $C'_{n-1} \subset C'_n \subset \widetilde{X}$ and describe the fundamental group $\pi_1(\widetilde{X} - int(C'_n), \omega(N_n))$ as follows. For those (finitely many) copies $g_i \cdot \widetilde{K}_1$ $(1 \leq i \leq p_n)$ of height $\leq n$ intersecting $C_n \subset \widetilde{X}$, we consider the unbounded connected components $V^{g_i}_k$, $1 \leq k \leq r_n(g_i)$, of $g_i \cdot \widetilde{K}_1 - C_n$, together with ends $e_{(g_i,k)} \in \mathcal{E}(g_i \cdot \widetilde{K}_1)$ such that the corresponding proper edge path ray $\omega_{e_{(g_i,k)}}$ in $g_i \cdot \widetilde{K}_1$ given above is eventually contained in $V^{g_i}_k$ and so it determines an end of $V^{g_i}_k$ (which we keep denoting by $e_{(g_i,k)}$). On the other hand, for those (finitely many) copies $g_j \cdot \widetilde{K}_2$ $(1 \leq j \leq q_n)$ of height $\leq n$ intersecting C_n, we consider the unbounded connected components $W^{g_j}_l$, $1 \leq l \leq s_n(g_j)$, of $g_j \cdot \widetilde{K}_2 - C_n$, together with ends $e_{(g_j,l)} \in \mathcal{E}(g_j \cdot \widetilde{K}_2)$ such that the corresponding proper edge path ray $\omega_{e_{(g_j,l)}}$ in $g_j \cdot \widetilde{K}_2$ above is eventually contained in $W^{g_j}_l$ and so it determines an end of $W^{g_j}_l$ (which we keep denoting by $e_{(g_j,l)}$). We set

$$C'_n = \left(\bigcup_{i=1}^{p_n} \bigcup_{k=1}^{r_n(g_i)} D_{m_n(g_i,1,e_{(g_i,k)})} \right) \cup \left(\bigcup_{j=1}^{q_n} \bigcup_{l=1}^{s_n(g_j)} D_{m_n(g_j,2,e_{(g_j,l)})} \right)$$

and take $N_n \geq N_{n-1}$ so that: (i) $\omega([N_n, \infty)) \subset \widetilde{X} - int(C'_n)$, and (ii) all the edge paths $\lambda^m_{g_i,1,e_{(g_i,k)}}$ and $\lambda^m_{g_j,2,e_{(g_j,l)}}$ lie outside the compact subset $C'_n \subset \widetilde{X}$, for all $m \geq N_n$; moreover, (iii) the corresponding edge loops as in (3.2.1.1) obtained from the edge paths in (ii) (and the proper edge path rays $\omega, \omega_{e_{(g_i,k)}}$ and $\omega_{e_{(g_j,l)}}$) are all homotopically trivial in $\widetilde{X} - C'_n$, for all $m \geq N_n$. One can readily check that $C'_{n-1} \subseteq int(C'_n)$ and $\widetilde{X} = \bigcup_{n \geq 0} C'_n$.

Again, a global generalized van Kampen's argument yields the isomorphism Φ_n from §3.2.1 corresponding to $\widetilde{X} - int(C'_n)$

$$\Phi_n : \Pi_n * H_n \longrightarrow \pi_1(\widetilde{X} - int(C'_n), \omega(N_n))$$

where

$$\begin{aligned} \Pi_n = \; & *_{i=1}^{p_n} *_{k=1}^{r_n(g_i)} \pi_1\left(g_i \cdot \widetilde{K}_1 - int(D_{m_n(g_i,1,e_{(g_i,k)})}),\, \omega_{e_{(g_i,k)}}(N_n)\right) * \\ & *_{j=1}^{q_n} *_{l=1}^{s_n(g_j)} \pi_1\left(g_j \cdot \widetilde{K}_2 - int(D_{m_n(g_j,2,e_{(g_j,l)})}),\, \omega_{e_{(g_j,l)}}(N_n)\right) \end{aligned}$$

and H_n is an (abstract) finitely generated free group obtained in a similar fashion to the one described above for the case $n = 0$ (see §3.2.1). Moreover, each inclusion $\widetilde{X} - int(C'_n) \subset \widetilde{X} - int(C'_{n-1})$ (together with the obvious change of base points along the corresponding proper edge path ray) yields, as described previously in §3.2.1, a commutative diagram:

$$(3.3.2) \qquad \begin{array}{ccc} \pi_1(\widetilde{X} - int(C'_{n-1}), \omega(N_{n-1})) & \longleftarrow & \pi_1(\widetilde{X} - int(C'_n), \omega(N_n)) \\ \Phi_{n-1} \uparrow \cong & & \Phi_n \uparrow \cong \\ \Pi_{n-1} * H_{n-1} & \overset{\iota_n}{\dashleftarrow} & \Pi_n * H_n \end{array}$$

This describes a semistable tower of groups $\underline{H\Pi}$ as in §3.2.1:

$$1 \longleftarrow \Pi_0 * H_0 \xleftarrow{\iota_1} \Pi_1 * H_1 \xleftarrow{\iota_2} \Pi_2 * H_2 \xleftarrow{\iota_3} \cdots$$

since $\widetilde{X}$ is semistable at infinity. Now, for the finitely generated free groups

$$F_n = \left(*_{i=1}^{p_n} *_{k=1}^{r_n(g_i)} F^n_{g_i,1,e_{(g_i,k)}} \right) * \left(*_{j=1}^{q_n} *_{l=1}^{s_n(g_j)} F^n_{g_j,2,e_{(g_j,l)}} \right),$$

the combination of the various corresponding diagrams from (3.3.1), induced by the fact that the ends of G_1 and G_2 are of telescopic type, yields a pro-isomorphism given by a commutative diagram:

$$\begin{array}{ccccccccc}
1 & \longleftarrow & \Pi_0 & \xleftarrow{p_1} & \Pi_1 & \xleftarrow{p_2} & \Pi_2 & \xleftarrow{p_3} & \cdots \\
 & & {\scriptstyle\phi_0}\downarrow & \nwarrow^{\psi_1} & {\scriptstyle\phi_1}\downarrow & \nwarrow^{\psi_2} & {\scriptstyle\phi_2}\downarrow & \nwarrow^{\psi_3} & \\
1 & \longleftarrow & F_0 & \xleftarrow[f_1]{} & F_1 & \xleftarrow[f_2]{} & F_2 & \xleftarrow[f_3]{} & \cdots
\end{array}$$

where the groups and homomorphisms are the corresponding free products, and the homomorphisms p_j in the upper row are induced by the inclusion of the factors followed by a change of base point along the corresponding base ray, as in (3.3.1).

For the sake of notation and simplicity, we will simply write $\Pi_j = *_i \, \Pi^i_j$ and $F_j = *_i \, F^i_j$. By Lemma 3.1, for each pair i, j there is an element $z_{ij} \in \Pi_j * H_j$ so that $\iota_{j+1}(x) = z_{ij} \cdot p_{j+1}(x) \cdot z_{ij}^{-1}$ for each $x \in \Pi^i_{j+1}$ (or $\iota_{j+1}(x) = 1$, here we work with the pro-isomorphic tower from Remark 3.2).

We consider now the following diagram:

$$\begin{array}{ccccccccc}
1 & \longleftarrow & \Pi_0 * H_0 & \xleftarrow{\iota_1} & \Pi_1 * H_1 & \xleftarrow{\iota_2} & \Pi_2 * H_2 & \xleftarrow{\iota_3} & \cdots \\
 & & {\scriptstyle\widetilde{\phi}_0}\downarrow & \nwarrow^{\widetilde{\psi}_1} & {\scriptstyle\widetilde{\phi}_1}\downarrow & \nwarrow^{\widetilde{\psi}_2} & {\scriptstyle\widetilde{\phi}_2}\downarrow & \nwarrow^{\widetilde{\psi}_3} & \\
1 & \longleftarrow & F_0 * H_0 & \xleftarrow[\widetilde{f}_1]{} & F_1 * H_1 & \xleftarrow[\widetilde{f}_2]{} & F_2 * H_2 & \xleftarrow[\widetilde{f}_3]{} & \cdots
\end{array}$$

where $\widetilde{\phi}_j$ is defined as ϕ_j on Π_j and the identity on H_j. Moreover, $\widetilde{\psi}_j$ is defined as $\widetilde{\psi}_j(x) = z_{ij} \cdot \psi_j(x) \cdot z_{ij}^{-1}$ if $x \in F^i_j$, or $\widetilde{\psi}_j(x) = \iota_j(x)$ if $x \in H_j$. Finally, $\widetilde{f}_j$ is defined as the composite $\widetilde{f}_j = \widetilde{\phi}_{j-1} \circ \widetilde{\psi}_j$. Notice that for any $x \in \Pi_{j+1}$ we have $\widetilde{\psi}_{j+1} \circ \widetilde{\phi}_{j+1}(x) = z_{ij} \cdot \psi_{j+1}(\phi_{j+1}(x)) \cdot z_{ij}^{-1} = z_{ij} \cdot p_{j+1}(x) \cdot z_{ij}^{-1} = \iota_{j+1}(x)$. Similarly, for any element $t \in H_j$ we have $\widetilde{\psi}_j \circ \widetilde{\phi}_j(t) = \widetilde{\psi}_j(t) = \iota_j(t)$. Thus, the diagram above shows that the tower $\underline{\Pi H}$ (and hence $pro - \pi_1(\widetilde{X}, \omega)$ by (3.3.2)) is pro-(finitely generated free) and hence of telescopic type, by Lemma 2.1.

In the case of an HNN-extension $G = G_1 *_F$, given a presentation of G_1

$$\mathcal{P} \equiv \langle g_1, \ldots, g_n, c_1, \ldots, c_k, c'_1, \ldots, c'_k; r_1, \ldots, r_m \rangle$$

where $c_1, \ldots, c_k$ and $c'_1, \ldots, c'_k$ freely generate isomorphic subgroups F and F' of G_1, then $G = G_1 *_F$ admits the following presentation

$$\mathcal{Q} \equiv \langle t, g_1, \ldots, g_n, c_1, \ldots, c_k, c'_1, \ldots, c'_k; r_1, \ldots, r_m, t^{-1} c_i t = c'_i, 1 \leq i \leq k \rangle$$

Let K_1 and X be the standard 2-dimensional CW-complexes associated to the above presentations $\mathcal{P}$ and $\mathcal{Q}$ respectively, and let the trees Γ and Γ' be the Cayley graphs of F and F' with respect to the generating sets $c_1, \ldots, c_k$ and $c'_1, \ldots, c'_k$ respectively. Following the description in [**22**], let K be the 2-complex consisting of two vertices $*$ and $*'$, one loop at $*$ for each generator g_i, c_i and c_i', 2-cells attached to them according to the relations r_i, an edge labelled t from $*'$ to $*$, and for each $1 \leq i \leq k$ a loop d_i at $*'$ and a 2-cell attached to the loop $t^{-1}d_i^{-1}tc_i'$. The universal cover $\widetilde{K}$ of K consists of the universal cover $\widetilde{K}_1$ with a copy of $\Gamma' \times I$ attached to each copy of Γ'. The 2-complex X is now K with the edges c_i and d_i identified, and the universal cover $\widetilde{X}$ of X is a union of copies of $\widetilde{K}$ glued together appropriately along copies of Γ. Thus, an argument similar to the above shows that $\widetilde{X}$ also has a (non pro-trivial) fundamental pro-group of telescopic type.

References

[1] R. Ayala, M. Cárdenas, F. F. Lasheras, and A. Quintero, *Properly 3-realizable groups*, Proc. Amer. Math. Soc. **133** (2005), no. 5, 1527–1535, DOI 10.1090/S0002-9939-04-07628-2. MR2111954

[2] M. Cárdenas, T. Fernández, F. F. Lasheras, and A. Quintero, *Embedding proper homotopy types*, Colloq. Math. **95** (2003), no. 1, 1–20, DOI 10.4064/cm95-1-1. MR1967550

[3] M. Cárdenas, F. F. Lasheras, and A. Quintero, *Weak $\mathcal{Z}$-structures and one-relator groups*, J. Pure Appl. Algebra **227** (2023), no. 1, Paper No. 107177, 8, DOI 10.1016/j.jpaa.2022.107177. MR4457412

[4] M. Cardenas, F. F. Lasheras, A. Quintero, and D. Repovš, *Amalgamated products and properly 3-realizable groups*, J. Pure Appl. Algebra **208** (2007), no. 1, 293–296, DOI 10.1016/j.jpaa.2005.12.006. MR2269844

[5] M. Cárdenas, F. F. Lasheras, A. Quintero, and D. Repovš, *One-relator groups and proper 3-realizability*, Rev. Mat. Iberoam. **25** (2009), no. 2, 739–756, DOI 10.4171/RMI/581. MR2569552

[6] M. Cárdenas, F. F. Lasheras, A. Quintero, and R. Roy, *A note on group extensions and proper 3-realizability*, Mediterr. J. Math. **13** (2016), no. 5, 3303–3309, DOI 10.1007/s00009-016-0686-8. MR3554309

[7] M. Cárdenas, F. F. Lasheras, A. Quintero, and R. Roy, *Quasi-isometries and proper homotopy: the quasi-isometry invariance of proper 3-realizability of groups*, Proc. Amer. Math. Soc. **147** (2019), no. 4, 1797–1804, DOI 10.1090/proc/14373. MR3910444

[8] M. Cárdenas, F. F. Lasheras, A. Quintero, and R. Roy, *A topological equivalence relation for finitely presented groups*, J. Pure Appl. Algebra **224** (2020), no. 7, 106300, 25, DOI 10.1016/j.jpaa.2019.106300. MR4058235

[9] M. Cárdenas, F. F. Lasheras, A. Quintero, and R. Roy, *Proper 2-equivalences between infinite ended finitely presented groups*, Algebr. Geom. Topol. **23** (2023), no. 1, 1–11, DOI 10.2140/agt.2023.23.1. MR4567997

[10] M. Cárdenas, F. F. Lasheras, and R. Roy, *Direct products and properly 3-realisable groups*, Bull. Austral. Math. Soc. **70** (2004), no. 2, 199–205, DOI 10.1017/S0004972700034419. MR2094287

[11] G. Conner and M. Mihalik, *Commensurated subgroups and ends of groups*, J. Group Theory **16** (2013), no. 1, 107–139, DOI 10.1515/jgt-2012-0035. MR3008315

[12] G. R. Conner and M. L. Mihalik, *Commensurated subgroups, semistability and simple connectivity at infinity*, Algebr. Geom. Topol. **14** (2014), no. 6, 3509–3532, DOI 10.2140/agt.2014.14.3509. MR3302969

[13] H. Federer and B. Jónsson, *Some properties of free groups*, Trans. Amer. Math. Soc. **68** (1950), 1–27, DOI 10.2307/1990535. MR32643

[14] L. Funar, F. F. Lasheras, and D. Repovš, *Groups which are not properly 3-realizable*, Rev. Mat. Iberoam. **28** (2012), no. 2, 401–414. MR2916965

[15] R. Geoghegan, *Topological methods in group theory*, Graduate Texts in Mathematics, vol. 243, Springer, New York, 2008, DOI 10.1007/978-0-387-74614-2. MR2365352

[16] R. Geoghegan, C. Guilbault, and M. Mihalik, *Topological properties of spaces admitting a coaxial homeomorphism*, Algebr. Geom. Topol. **20** (2020), no. 2, 601–642, DOI 10.2140/agt.2020.20.601. MR4092308

[17] R. Geoghegan and M. L. Mihalik, *Free abelian cohomology of groups and ends of universal covers*, J. Pure Appl. Algebra **36** (1985), no. 2, 123–137, DOI 10.1016/0022-4049(85)90065-9. MR787167

[18] F. F. Lasheras, *Ascending HNN-extensions and properly 3-realisable groups*, Bull. Austral. Math. Soc. **72** (2005), no. 2, 187–196, DOI 10.1017/S0004972700035000. MR2183402

[19] F. F. Lasheras and R. Roy, *Relating the Freiheitssatz to the asymptotic behavior of a group*, Rev. Mat. Iberoam. **29** (2013), no. 1, 75–89, DOI 10.4171/RMI/713. MR3010122

[20] S. Mardešić and J. Segal, *Shape theory: The inverse system approach*, North-Holland Mathematical Library, vol. 26, North-Holland Publishing Co., Amsterdam-New York, 1982. MR0676973

[21] M. Mihalik and S. Shepherd, *Splittings of one-ended groups with one-ended halfspaces and non-1-acyclicity at infinity*, Preprint, 2023, arXiv: 2308.06218.

[22] M. L. Mihalik and S. T. Tschantz, *Semistability of amalgamated products and HNN-extensions*, Mem. Amer. Math. Soc. **98** (1992), no. 471, vi+86, DOI 10.1090/memo/0471. MR1110521

[23] D. Rattaggi, *Three amalgams with remarkable normal subgroup structures*, J. Pure Appl. Algebra **210** (2007), no. 2, 537–541, DOI 10.1016/j.jpaa.2006.10.008. MR2320016

[24] E. Rips and Z. Sela, *Cyclic splittings of finitely presented groups and the canonical JSJ decomposition*, Ann. of Math. (2) **146** (1997), no. 1, 53–109, DOI 10.2307/2951832. MR1469317

[25] P. Scott and T. Wall, *Topological methods in group theory*, Homological group theory (Proc. Sympos., Durham, 1977), London Math. Soc. Lecture Note Ser., vol. 36, Cambridge Univ. Press, Cambridge-New York, 1979, pp. 137–203. MR564422

Departamento de Geometría y Topología, Fac. Matemáticas, Universidad de Sevilla, C/. Tarfia s/n 41012-Sevilla, Spain
Email address: mcard@us.es

Departamento de Geometría y Topología, Fac. Matemáticas, Universidad de Sevilla, C/. Tarfia s/n 41012-Sevilla, Spain
Email address: lasheras@us.es

Departamento de Geometría y Topología, Fac. Matemáticas, Universidad de Sevilla, C/. Tarfia s/n 41012-Sevilla, Spain
Email address: quintero@us.es

Contemporary Mathematics
Volume **812**, 2025
https://doi.org/10.1090/conm/812/16266

On Gromov's conjecture for right-angled Artin groups

Alexander Dranishnikov and Satyanath Howladar

ABSTRACT. We explore in terms of KO-homology the 2nd obstruction for the universal covering $\widetilde{M}$ of a closed spin n-manifold M with positive scalar curvature (PSC) to have macroscopic dimension $\leq n-2$. As a corollary we prove that the universal covering $\widetilde{M}$ of an n-dimensional closed spin PSC manifold M with $\pi_1(M) \in$ RAAG has macroscopic dimension $\dim_{mc} \widetilde{M} \leq n-2$. This confirms Gromov's conjecture in the case of RAAG (right-angled Artin group) fundamental groups.

1. Introduction

The notion of macroscopic dimension was introduced by M. Gromov [**G2**] to study topology of manifolds with a positive scalar curvature metric.

DEFINITION 1.1. A metric space X has the macroscopic dimension $\dim_{\mathrm{mc}} X \leq k$ if there is a uniformly cobounded proper map $f : X \to K$ to a k-dimensional simplicial complex. Then $\dim_{mc} X = m$ where m is minimal among k with $\dim_{\mathrm{mc}} X \leq k$.

A map of a metric space $f : X \to Y$ is uniformly cobounded if there is a uniform upper bound on the diameter of preimages $f^{-1}(y)$, $y \in Y$.

GROMOV'S CONJECTURE. *The macroscopic dimension of the universal covering $\widetilde{M}$ of a closed positive scalar curvature n-manifold M satisfies the inequality $\dim_{mc} \widetilde{M} \leq n-2$ for the metric on $\widetilde{M}$ lifted from M.*

The main examples supporting Gromov's Conjecture are n-manifolds of the form $M = N \times S^2$. They admit metrics with PSC in view of the formula $Sc_{x_1,x_2} = Sc_{x_1} + Sc_{x_2}$ for the cartesian product $(X_1 \times X_2, \mathcal{G}_1 \oplus \mathcal{G}_2)$ of two Riemannian manifolds $(X_1, \mathcal{G}_1)$ and $(X_2, \mathcal{G}_2)$ and the fact that while Sc_N is bounded Sc_{S^2} can be chosen to be arbitrary large. Clearly, the projection $p : \widetilde{M} = \widetilde{N} \times S^2 \to \widetilde{N}$ is a proper uniformly cobounded map to a $(n-2)$-dimensional manifold which can be triangulated. Hence, $\dim_{mc} \widetilde{M} \leq n-2$.

2020 *Mathematics Subject Classification.* Primary 53C23; Secondary 57N65, 19L41, 19M05, 20F36.

Key words and phrases. Positive scalar curvature, inessential manifold, macroscopic dimension, right-angled Artin groups.

The first author was supported by Simons Foundation.

Since $\dim_{mc} X = 0$ for every compact metric space, the Gromov Conjecture holds trivially for manifolds with finite fundamental groups. Therefore, Gromov's conjecture is about manifolds with infinite fundamental groups. We note that even a weaker version of the Gromov Conjecture, that is $\dim_{mc} \widetilde{M} \leq n-1$, for positive scalar curvature manifolds looks out of reach. This is because it implies the Gromov-Lawson's Conjecture: *A closed aspherical manifold cannot carry a metric of positive scalar curvature.* The latter is known as a twin sister of the famous Novikov Higher Signature conjecture. The real analytic version of the Novikov conjecture is called the Strong Novikov Conjecture. It is a conjecture for discrete groups π: *The analytic assembly map* $\alpha : KO_*(B\pi) \to KO_*(C^*(\pi))$ *is a monomorphism, where* $C^*(\pi)$ *is reduced* C^**-algebra of* π. We note that the Strong Novikov conjecture for a group π implies the Higher Signature conjecture for all manifolds with the fundamental group π.

The (Strong) Novikov's conjecture is proven for large classes of groups (see the survey [**Yu**]). The Gromov Conjecture is proven in far fewer cases [**Bo1**],[**Bo3**], [**Dr1**],[**Dr2**], [**BD1**],[**BD2**],[**DD**]. In view of the above connection between the Gromov and the Novikov conjectures, it makes sense to state the Gromov Conjecture for groups instead of manifolds as the Gromov Conjecture for all positive scalar curvature manifolds with a given fundamental group π. Additionally, to make Gromov's conjecture more accessible we consider it in this paper for spin manifolds with the fundamental group π satisfying the Strong Novikov Conjecture. In this paper we proved the inequality $\dim_{mc} \widetilde{M} \leq n-1$ for spin manifolds with the fundamental group Γ that satisfies the Strong Novikov Conjecture and has its integral homology $H_*(\Gamma)$ torsion free (Corollary 2.11).

Gromov defined *inessential manifolds* M as those for which a classifying map $u : M \to B\Gamma$ of the universal covering $\widetilde{M}$ can be deformed to the $(n-1)$-skeleton $B\Gamma^{(n-1)}$ where $n = \dim M$. Clearly, for an inessential n-manifold M we have $\dim_{mc} \widetilde{M} \leq n-1$.

We call an n-manifold *strongly inessential* if a classifying map of its universal covering $u : M \to B\Gamma$ can be deformed to the $(n-2)$-skeleton. Since for strongly inessential n-manifolds $\dim_{mc} \widetilde{M} \leq n-2$, the following conjecture implies the Gromov's conjecture.

STRONG GROMOV'S CONJECTURE. *A closed positive scalar curvature manifold* M *is strongly inessential.*

We note that for a finite cover $M' \to M$ of a PSC manifold M the manifold M' has PSC and the universal cover of M' coincides with the universal cover of M, Therefore, the Strong Gromov Conjecture for a group Γ' implies the original Gromov Conjecture for all groups Γ containing Γ' as a finite index subgroup.

Dmitry Bolotov showed [**Bo2**] that there are inessential manifolds which are not strongly inessential.

In this paper we prove the Strong Gromov Conjecture for spin n-manifolds with $n > 4$, for some classes of groups, in particular, for right-angled Artin groups (RAAGs).

2. Preliminaries

We recall that for every spectrum E there is a connective cover $e \to E$, that is a spectrum e with the morphism $e \to E$ that induces isomorphisms of homotopy groups $\pi_i(e) \to \pi_i(E)$ for $i \geq 0$ and with $\pi_i(e) = 0$ for $i < 0$. By KO we denote the spectrum for real K-theory, by ko its connective cover, and by $per : ko \to KO$ the corresponding morphism of spectra. For more details we refer to [**Ru**]. We use the standard notation π_*^s for the stable homotopy groups. We use both notations for an E-homology of a space X: old-fashioned $E_*(X)$ and modern $H_*(X; E)$. We recall that $KO_n(pt) = \mathbb{Z}$ if $n = 0$ or $n = 4$ mod 8, $KO_n(pt) = \mathbb{Z}_2$ if $n = 1$ or $n = 2$ mod 8, and $KO_n(pt) = 0$ for all other values of n. By $\mathbb{S}$ we denote the spherical spectrum. Note that for any ring spectrum E there is a natural morphism $\mathbb{S} \to E$ which leads to the natural transformation of the stable homotopy to E-homology $\pi_*^s(X) \to H_*(X; E)$.

Let $M(\mathbb{Z}_m, n) = S^n \cup_\phi D^{n+1}$ denote the Moore space where $\phi : S^n \to S^n$ is a map of degree m.

PROPOSITION 2.1. *The homomorphism*

$$per : ko_i(M(\mathbb{Z}_m, n)) \to KO_i(M(\mathbb{Z}_m, n))$$

is an isomorphism for $i \geq n$.

PROOF. We apply the Five Lemma to the diagram with $M = M(\mathbb{Z}_m, n)$

$$\begin{array}{ccccccccc}
ko_i(S^n) & \xrightarrow{m} & ko_i(S^n) & \longrightarrow & ko_i(M) & \longrightarrow & ko_{i-1}(S^n) & \xrightarrow{m} & ko_{i-1}(S^n) \\
\cong\big\downarrow & & \cong\big\downarrow & & \big\downarrow & & \cong\big\downarrow & & \cong\big\downarrow \\
KO_i(S^n) & \xrightarrow{m} & KO_i(S^n) & \longrightarrow & KO_i(M) & \longrightarrow & KO_{i-1}(S^n) & \xrightarrow{m} & KO_{i-1}(S^n)
\end{array}$$

to get the result. Note that for $i = n$ on the right there will be isomorphisms, since $KO_{-1}(pt) = 0$. $\square$

The following proposition is taken from [**BD1**].

PROPOSITION 2.2. *The natural transformation $\pi_*^s(pt) \to ko_*(pt)$ induces an isomorphism $\pi_n^s(K/K^{(n-2)}) \to ko_n(K/K^{(n-2)})$ for any CW complex K.*

We use the following well-known fact (see 4C, [**Ha**]).

PROPOSITION 2.3. *Let X be an $(n-1)$-connected $(n+1)$-dimensional CW complex. Then X is homotopy equivalent to the wedge of spheres of dimensions n and $n+1$ together with the Moore spaces $M(\mathbb{Z}_m, n)$.*

Next, we recall the following classical result (see Corollary 6.10.3, [**TtD**]):

THEOREM 2.4. *Suppose that a CW-complex pair (X, A) satisfies the conditions $\pi_i(X) = 0$ for $i \leq m$ and $\pi_i(A) = 0$ for $i \leq m - 1$ with $m \geq 2$. Then the quotient map $q : (X, A) \to (X/A, *)$ induces isomorphisms $q_* : \pi_i(X, A) \to \pi_i(X/A)$ for $i \leq 2m - 1$.*

2.1. Inessential manifolds. Note that for an inessential n-manifold M we have $\dim_{mc} \widetilde{M} \leq n - 1$. Indeed, a lift $\widetilde{u_M} : \widetilde{M} \to E\Gamma^{(n-1)}$ of a classifying map is a uniformly cobounded proper map to an $(n-1)$-complex.

The first step in any proof of the Strong Gromov Conjecture would be establishing the inessentiality of a positive scalar curvature manifold. We recall that the inessentiality of a manifold can be characterized as follows [**Ba**] (see also [**BD1**], Proposition 3.2).

THEOREM 2.5. *Let M be a closed oriented n-manifold. Then the following are equivalent:*

1. M is inessential;

2. $(u_M)_*([M]) = 0$ in $H_n(B\Gamma)$ where $[M]$ is the fundamental class of M and $\Gamma = \pi_1(M)$.

Gromov calls an orientable closed n-manifold M *rationally inessential* if $(u_M)_*([M]) = 0$ in $H_n(B\Gamma;\mathbb{Q})$.

It is known that for a manifold to be spin is equivalent to be orientable for any of the K-theories: complex KU, real KO, or their connective covers ku or ko [**Ru**]. Suppose that a closed n-manifold M is orientable with respect to a generalized homology theory E. We call the manifold M to be *E-inessential* if it is E-orientable and $u_*([M]_E) = 0$ where $[M]_E \in H_n(M;E)$ is the E-fundamental class. Note that this property does not depend on the choice of E-orientation.

We recall the Rosenberg index theorem [**R1**]:

THEOREM 2.6. *Suppose that the fundamental group Γ of a positive scalar curvature spin manifold M satisfies the Strong Novikov Conjecture. Then M is KO-inessential.*

Note that in view of the Chern character isomorphism a KO-inessential manifold is rationally inessential. As Marcinkowski's example shows [**M**], the rational inessentiality of an n-manifold does not imply the inequality $\dim_{mc}\widetilde{M} \le n-1$. This makes difficult the first step in the Gromov's conjecture even in the case of groups satisfying the Strong Novikov Conjecture.

PROPOSITION 2.7 ([**BD1**, Proposition 3.3]). *A ko-inessential manifold is inessential.*

QUESTION 2.8. When a KO-inessential manifold is inessential?

In view of Proposition 2.7 this is true for manifolds with the fundamental group Γ for which the map $per : ko_n(B\Gamma) \to KO_n(B\Gamma)$ is injective.

In view of the Rosenberg Index theorem, we obtain the following

PROPOSITION 2.9. *Suppose that the fundamental group Γ of a positive scalar curvature closed spin n-manifold M has the following properties:*

(1) It satisfies the Strong Novikov Conjecture;

(2) The map $per : ko_n(B\Gamma) \to KO_n(B\Gamma)$ is injective.

Then M is inessential.

The conditions (1) and (2 for all n) are known as the Rosenberg-Stolz conditions. For closed spin manifolds with the fundamental group satisfying Rosenberg-Stoltz conditions there is a vanishing index criterion when they could carry a metric with positive scalar curvature [**R2**]. In other words the Gromov-Lawson-Rosenberg conjecture holds true for such groups.

PROPOSITION 2.10. *Suppose that a group Γ has torsion free integral homology. Then every KO-inessential manifold with the fundamental group Γ is inessential.*

PROOF. Since the KO-fundamental class $[M^n]_{KO}$ maps under the natural transformation $KO \to KU$ to a KU-fundamental class, every KO-inessential manifold is KU-inessential. In view of the diagram of the homology Chern character isomorphisms generated by a degree one map $f : M^n \to S^n$

$$\begin{array}{ccc} K_n(M^n)\otimes\mathbb{Q} & \xrightarrow[\cong]{ch} & \bigoplus_{i\in\mathbb{Z}} H_{n+2i}(M^n)\otimes\mathbb{Q} \\ {\scriptstyle f_*\otimes 1_{\mathbb{Q}}}\downarrow & & {\scriptstyle f_*\otimes 1_{\mathbb{Q}}}\downarrow \\ K_n(S^n)\otimes\mathbb{Q} & \xrightarrow[\cong]{ch} & \bigoplus_{i\in\mathbb{Z}} H_{n+2i}(S^n)\otimes\mathbb{Q} \end{array}$$

the image of a KU-fundamental class of M^n under the Chern character has nontrivial $H_n(M^n)$ component. Then the bottom arrow in the diagram

$$\begin{array}{ccc} H_n(M^n) & \xrightarrow{u_*} & H_n(B\Gamma) \\ \downarrow & & \downarrow \\ H_n(M^n)\otimes\mathbb{Q} & \xrightarrow{u_*\otimes 1_{\mathbb{Q}}} & H_n(B\Gamma)\otimes\mathbb{Q} \end{array}$$

is a zero homomorphism. The right vertical homomorphism is injective, since $H_n(B\Gamma)$ is torsion free. Therefore, the top arrow is zero. □

COROLLARY 2.11. *Suppose that the fundamental group Γ of a positive scalar curvature spin manifold M satisfies the Strong Novikov conjecture and all homology groups $H_*(\Gamma)$ are torsion free. Then*

$$\dim_{mc} \widetilde{M} \le n-1.$$

PROOF. We apply Rosenberg's theorem (Theorem 2.6) and Proposition 2.10. □

In [**BD1**] we proved the following addendum to Theorem 2.5.

PROPOSITION 2.12 ([**BD1**, Lemma 3.5]). *For an inessential manifold M with a CW complex structure a classifying map $u : M \to B\Gamma$ can be chosen such that*

$$u(M, M^{(n-1)}) \subset (B\Gamma^{(n-1)}, B\Gamma^{(n-2)}).$$

3. The second obstruction

Suppose that Γ acts on an abelian group π. By $\pi_\Gamma = \pi \otimes_{\mathbb{Z}\Gamma} \mathbb{Z}$ we denote the group of coinvariants. Since every ko-inessential manifold is inessential (Proposition 2.7) and for every inessential n-manifold there is a classifying map $u : M \to B\Gamma$ with $u(M, M^{(n-1)}) \subset (B\Gamma^{(n-1)}, B\Gamma^{(n-2)})$ (Proposition 2.12), for ko-inessential manifolds we always will consider such classifying maps u.

The following lemma shows that the second step in the Strong Gromov's conjecture can be also reduced to a question about ko-inessentiality.

LEMMA 3.1. *Suppose that a closed spin n-manifold M, $n > 4$, with $\Gamma = \pi_1(M)$ is ko-inessential in a way that $u_*([M]_{ko}) = 0$ in $ko_n(B\Gamma^{(n)})$ for a cellular classifying map $u : M \to B\Gamma$. Then M is strongly inessential.*

Moreover, M is strongly inessential if $j_*u_*([M]_{ko}) = 0$ where $j : B\Gamma^{(n)} \to B\Gamma^{(n)}/B\Gamma^{(n-2)}$ is the quotient map.

PROOF. We may assume that M has a CW complex structure with one n-dimensional cell. Let $\psi : D^n \to M$ be its characteristic map.

We show that the lifting problem

$$\begin{array}{ccc} M^{(n-1)} & \longrightarrow & B\Gamma^{(n-2)} \\ \downarrow \subset & \nearrow & \downarrow \subset \\ M & \xrightarrow{iu} & B\Gamma^{(n)}. \end{array}$$

has a solution. Here $i : B\Gamma^{(n-1)} \to B\Gamma^{(n)}$ is the inclusion. It would mean that there is a homotopy lift $\hat{u} : M \to B\Gamma^{(n-2)}$ of $i \circ u$ which agrees with u on $M^{(n-2)}$. Since $n \geq 4$, the map $\hat{u}$ induces an isomorphism of the fundamental groups and, hence, is classifying map.

Note that the $(n-1)$-homotopy groups of the homotopy fiber F of the inclusion $B\Gamma^{(n-2)} \to B\Gamma^{(n)}$ equals the relative n-homotopy group, $\pi_{n-1}(F) = \pi_n(B\Gamma^{(n)}, B\Gamma^{(n-2)})$. Then the first and the only obstruction to deform u to $B\Gamma^{(n-2)}$ is defined by the cocycle $c_u : C_n(M) \to \pi_{n-1}(F)$ represented by the composition

$$\pi_n(D^n, \partial D^n) \stackrel{\psi_*}{\to} \pi_n(M, M^{(n-1)}) \stackrel{u_*}{\to} \pi_n(B\Gamma^{(n-1)}, B\,\Gamma^{(n-2)}) \stackrel{i_*}{\to} \pi_n(B\Gamma^{(n)}, B\Gamma^{(n-2)})$$

with the cohomology class $o_u = [c_u] \in H^n(M; \pi_n(B\Gamma^{(n)}, B\Gamma^{(n-2)}))$. By the Poincare Duality with local coefficients, the cohomology class o_u is dual to the homology class

$$PD(o_u) \in H_0(M; \pi_n(B\Gamma^{(n)}, B\Gamma^{(n-2)})) = \pi_n(B\Gamma^{(n)}, B\Gamma^{(n-2)})_\Gamma$$

represented by $q_* i_* u_* \psi_*(1)$ where

$$q_* : \pi_n(B\Gamma^{(n)}, B\Gamma^{(n-2)}) \to \pi_n(B\Gamma^{(n)}, B\Gamma^{(n-2)})_\Gamma$$

is the projection onto the group of coinvariants.

Note that $\pi_n(B\Gamma^{(n-1)}, B\Gamma^{(n-2)}) = \pi_n(E\Gamma^{(n-1)}, E\Gamma^{(n-2)})$. Below we will identify these groups. Denote by

$$\tilde{i} : (E\Gamma^{(n-1)}, E\Gamma^{(n-2)}) \to (E\Gamma^{(n)}, E\Gamma^{(n-2)})$$

the inclusion induced by i.

Since $n \leq 2(n-2) - 1$, by Theorem 2.4,

$$\pi_n(E\Gamma^{(n-1)}, E\Gamma^{(n-2)}) = \pi_n(E\Gamma^{(n-1)}/E\Gamma^{(n-2)}).$$

It is easy to see that

$$\pi_n(E\Gamma^{(n-1)}/E\Gamma^{(n-2)})_\Gamma = \pi_n(B\Gamma^{(n-1)}/B\Gamma^{(n-2)}).$$

Similarly,

$$\pi_n(E\Gamma^{(n)}/E\Gamma^{(n-1)})_\Gamma = \pi_n(B\Gamma^{(n)}/B\Gamma^{(n-1)}).$$

The homotopy exact sequence of the triple $(E\Gamma^{(n)}, E\Gamma^{(n-1)}, E\Gamma^{(n-2)})$ produces the following commutative diagram

$$\begin{array}{ccccccc}
\pi_{n+1}(E\Gamma^{(n)}, E\Gamma^{(n-1)}) & \longrightarrow & \pi_n(E\Gamma^{(n-1)}, E\Gamma^{(n-2)}) & \xrightarrow{\tilde{i}_*} & im(\tilde{i}_*) & \longrightarrow & 0 \\
\downarrow & & \downarrow & & {\scriptstyle q_*}\downarrow & & \\
\pi_{n+1}(E\Gamma^{(n)}, E\Gamma^{(n-1)})_\Gamma & \longrightarrow & \pi_n(E\Gamma^{(n-1)}, E\Gamma^{(n-2)})_\Gamma & \xrightarrow{\tilde{i}_*\otimes_\Gamma 1_{\mathbb{Z}}} & im(\tilde{i}_*)_\Gamma & \longrightarrow & 0 \\
{\scriptstyle \cong}\downarrow & & {\scriptstyle \cong}\downarrow & & {\scriptstyle \xi}\downarrow & & \\
\pi_{n+1}(B\Gamma^{(n)}/B\Gamma^{(n-1)}) & \longrightarrow & \pi_n(B\Gamma^{(n-1)}/B\Gamma^{(n-2)}) & \xrightarrow{i_*} & im(i_*) & \longrightarrow & 0
\end{array}$$

where the middle row is exact being obtained by tensor product by $\mathbb{Z}$ over $\mathbb{Z}\Gamma$ of an exact row. By the Five Lemma the homomorphism ξ is an isomorphism.

Denote by $\bar{u} : M/M^{(n-1)} = S^n \to B\Gamma^{(n-1)}/B\Gamma^{(n-2)}$ the induced map. The following commutative diagram

$$\begin{array}{ccccccc}
\pi_n(M, M^{(n-1)}) & \xrightarrow{\tilde{i}_* u_*} & im(\tilde{i}_*) & \xrightarrow{q_*} & im(\tilde{i}_*)_\Gamma & & \\
{\scriptstyle \psi_*}\uparrow & & & & \downarrow{\scriptstyle \xi} & & \\
\pi_n(D^n, \partial D^n) & \xrightarrow{=} & \pi_n(M/M^{(n-1)}) & \xrightarrow{i_*\bar{u}_*} & im(i_*) & \xrightarrow{\subset} & \pi_n(B\Gamma^{(n)}/B\Gamma^{(n-2)})
\end{array}$$

implies that $i_*\bar{u}_*(1) = \xi q_* \tilde{i}_* u_* \psi_*(1)$. Thus, $i_*\bar{u}_*(1) = 0$ if and only if the obstruction o_u vanishes.

We show that $i_*\bar{u}_*(1) = 0$. The restriction $n > 4$ imply that $\bar{u}_*(1)$ survives to the ko-homology group:

$$\pi_n(B\Gamma^{(n)}/B\Gamma^{(n-2)}) \xrightarrow{\cong} \pi_n^s(B\Gamma^{(n)}/B\Gamma^{(n-2)}) \xrightarrow{\cong} ko_n(B\Gamma^{(n)}/B\Gamma^{(n-2)}).$$

The first isomorphism is obtained by taking n-times suspension and using Freudenthal Theorem, and the second isomorphism by Proposition 2.2. Then the commutative diagram

$$\begin{array}{ccc}
\pi_n(S^n) & \xrightarrow{\cong} & ko_n(S^n) \\
{\scriptstyle i_*\bar{u}_*}\downarrow & & {\scriptstyle i_*\bar{u}_*}\downarrow \\
\pi_n(B\Gamma^{(n)}/B\Gamma^{(n-2)}) & \xrightarrow{\cong} & ko_n(B\Gamma^{(n)}/B\Gamma^{(n-2)})
\end{array}$$

implies that $i_*\bar{u}_*(1) = 0$ for ko_n if and only if $i_*\bar{u}_*(1) = 0$ for π_n.

From the assumption and the diagram defined by the quotient maps $j' : M \to M/M^{(n-1)} = S^n$ and $j : B\Gamma^{(n)} \to B\Gamma^{(n)}/B\Gamma^{(n-2)}$

$$\begin{array}{ccc}
ko_n(M) & \xrightarrow{u_*} & ko_n(B\Gamma^{(n)}) \\
{\scriptstyle j'_*}\downarrow & & {\scriptstyle j_*}\downarrow \\
ko_n(S^n) & \xrightarrow{i_*\bar{u}_*} & ko_n(B\Gamma^{(n)}/B\Gamma^{(n-2)})
\end{array}$$

it follows that $i_*\bar{u}_*(1) = i_*\bar{u}_* j'_*([M]_{ko}) = j_* u_*([M]_{ko}) = 0$. □

REMARK 3.2. A slightly weaker lemma was proven in [**Dr1**]. Namely, a strong inessentiality of M was proven under assumption of $j_*u_*([M]_{ko}) = 0$ in $ko_n(B\Gamma^{(n-1)}/B\Gamma^{(n-2)})$ instead of $ko_n(B\Gamma^{(n)}/B\Gamma^{(n-2)})$. Since ko_* is a connective homology theory, for any ko-inessential n-manifold M with the fundamental group

Γ we have $u_*([M]_{ko}) = 0$ in $ko_n(B\Gamma^{(n+1)}/B\Gamma^{(n-2)})$. Thus, in the current version of lemma on the second obstruction we are one step closer to an affirmative answer to the following question.

QUESTION 3.3. Is every *ko*-inessential manifold strongly inessential ?

We note that Bolotov's example [**Bo2**] is *ko*-essential.

4. K-theory Stabilization Condition

The following condition on K-theory of a group Γ was introduced in [**Dr1**] In the proof of the Strong Gromov Conjecture for abelian fundamental groups in [**Dr1**] (see also [**Bo1**]) we considered the following property of groups Γ denoted by (*) : The inclusion homomorphism $KO_*(B\Gamma^{(m)}) \to KO_*(B\Gamma)$ is injective for all m.

This property depends on choice of $B\Gamma$. Here we introduce a weaker condition which is independent of the choice of $B\Gamma$ and sufficient for our purposes.

First, we define this condition for spaces. We state it for arbitrary generalized homology theory h_*.

1-Step Stabilization property for CW complex X: *For every m the image of the inclusion homomorphism $h_*(X^{(m)}) \to h_*(X)$ stabilizes at $X^{(m+1)}$.*

This property can be written as the formula

$$\mathrm{Ker}(\phi_{m,m+1})_* = \mathrm{Ker}(\phi_{m,\infty})_*$$

where $\phi_{m,m+\ell} : X^{(m)} \to X^{(m+\ell)}$, $\ell = 1, \dots, \infty$, denotes the inclusions.

We note that every CW complex has the 1-Step Stabilization property for the ordinary homology H_*.

By default the 1-Step Stabilization property is assumed for h_* in all dimensions, but often we use this property for fixed k or k depending on m. We denote the set of such k by $K(m) \subset \mathbb{N}$. In such situation we say that X has the 1-Step Stabilization property for h_* restricted to $K(m)$.

For example, for the stable homotopy groups π_*^s and, for that matter for any connective homology theory, the 1-Step Stabilization property restricted to $K(m) = \{k \in \mathbb{N} \mid k \leq m\}$ holds for any CW complex. In this paper we use the 1-Step Stabilization property for the real K-theory, KO_*, with $k = m, m+1$. Since the complex K-theory, KU_* is 2-periodic, it is not a restriction on k, but in the case of KO_* it is a potential restriction.

PROPOSITION 4.1. *The (restricted) 1-Step Stabilization property is a homotopy invariant.*

PROOF. Suppose that a CW complex X_1 has the 1-Step Stabilization property for h_*. Let $f : X_1 \to X_2$ be a homotopy equivalence with a homotopy inverse map g. We may assume that both maps f and g are cellular. Then in view of the cellular approximation theorem the composition

$$(f|_{X_1^{(m+1)}}) \circ \phi_{m,m+1} \circ g|_{X_2^{(m)}} : X_2^{(m)} \to X_2^{(m+1)}$$

is homotopic to the inclusion $\phi'_{m,m+1} : X_2^{(m)} \to X_2^{(m+1)}$. Thus, we have the following commutative diagram

$$\begin{array}{ccccc}
h_k(X_1^{(m)}) & \xrightarrow{(\phi_{m,m+1})_*} & h_k(X_1^{(m+1)}) & \xrightarrow{(\phi_{m+1,\infty})_*} & h_k(X_1) \\
{\scriptstyle g|_*}\uparrow & & {\scriptstyle f|_*}\downarrow & & {\scriptstyle g_*}\uparrow\cong \\
h_k(X_2^{(m)}) & \xrightarrow{(\phi'_{m,m+1})_*} & h_k(X_2^{(m+1)}) & \xrightarrow{(\phi'_{m+1,\infty})_*} & h_k(X_2).
\end{array}$$

Assume that $(\phi'_{m,m+1})_*(a) \neq 0$. Then $(f|)_*(\phi_{m,m+1})_*(g|)_*(a) \neq 0$ and, hence, $(\phi_{m,m+1})_*(g|)_*(a) \neq 0$. By the 1-step Stabilization property we obtain

$$0 \neq (\phi_{m+1,\infty})_*(\phi_{m,m+1})_* g|_*(a) = g_*(\phi'_{m+1,\infty})_*(f|)_*(\phi_{m,m+1})_*(g|)_*(a)$$
$$= g_*(\phi'_{m+1,\infty})_*(\phi'_{m,m+1})_*(a).$$

Therefore, $(\phi'_{m+1,\infty})_*(\phi'_{m,m+1})_*(a) \neq 0$. □

PROPOSITION 4.2. *If X has the (restricted) 1-Step Stabilization property for h_*, then so does the quotient complex $Y = X/X^{(n)}$.*

PROOF. We have to check the 1-Step Stabilization property for Y only for $m \geq n+1$. Let $a \in \operatorname{Ker}\{h_k(Y^{(m)}) \to h_k(Y)\}$ and let $\bar{a} \in h_k(Y^{(m+1)})$ be the image of a. We show that $\bar{a} = 0$. We note that $Y^{(m)} = X^{(m)}/X^{(n)}$. From the diagram generated by the exact sequence of pairs $(X^{(m)}, X^{(n)})$ and $(X, X^{(n)})$ it follows that $a \in \operatorname{Ker}\{\partial : h_k(Y^{(m)}) \to h_{k-1}(X^{(n)})\}$. Hence, by exactness a is the image of some $b \in h_k(X^{(m)})$. If the image of b in $h_k(X)$ is trivial, then by the 1-Step Stabilization property its image in $h_k(X^{(m+1)})$ is zero and, therefore, as one can see from the diagram below, $\bar{a} = 0$,

$$\begin{array}{ccccc}
h_k(X^{(n)}) & \longrightarrow & h_k(X^{(n)}) & \longrightarrow & h_k(X^{(n)}) \\
{\scriptstyle i_*}\downarrow & & {\scriptstyle i_*}\downarrow & & {\scriptstyle i_*}\downarrow \\
h_k(X^{(m)}) & \longrightarrow & h_k(X^{(m+1)}) & \longrightarrow & h_k(X) \\
{\scriptstyle j_*}\downarrow & & {\scriptstyle j_*}\downarrow & & {\scriptstyle j_*}\downarrow \\
h_k(Y^{(m)}) & \longrightarrow & h_k(Y^{(m+1)}) & \longrightarrow & h_k(Y).
\end{array}$$

Suppose that the image $b' \in h_k(X)$ of b is not zero. By exactness, there is $c \in h_k(X^{(n)})$ that maps by i_* onto b'. Then the image of $b - i_*(c)$ in $h_k(X)$ is zero. By the 1-Step Stabilization, $\bar{b} - i_*(c) = 0$ where $\bar{b} \in h_k(X^{(m+1)})$ is the image of b. Note that $\bar{a} = j_*(\bar{b}) = j_*(\bar{b} - i_*(c)) = 0$. □

Since all classifying CW complexes $B\Gamma$ for a group Γ are homotopy equivalent, we can define 1-Step Stabilization property for groups. Thus, a group Γ has the **1-step Stabilization property** for a generalized homology theory h_* if a classifying CW complex $B\Gamma$ has it.

We prove the following two propositions

PROPOSITION 4.3. *Suppose that a group Γ has the 1-Step Stabilization property for KO_* for all m, with $k = m$. Then every closed KO-inessential manifold M with the fundamental group $\pi_1(M) = \Gamma$ is inessential.*

PROPOSITION 4.4. *Suppose that an inessential and KO-inessential n-manifold M has the fundamental group Γ, which satisfies the 1-Step Stabilization property for KO_* for all m with $k = m+1$. Then M is strongly inessential.*

Proposition 4.3 and Proposition 4.4 lead to the main result of the paper.

THEOREM 4.5. *Suppose that a group Γ has a finite index subgroup Γ' with the 1-Step Stabilization property for KO_* for all m with $k \in K = \{m, m+1\}$ and suppose that Γ' satisfies the Strong Novikov Conjecture. Then the original Gromov conjecture holds for all spin n-manifolds, $n > 4$, with the fundamental group Γ.*

PROOF. Let M be a positive scalar curvature spin n-manifold with $\pi_1(M) = \Gamma$. Let $M' \to M$ be a finite covering manifold with the fundamental group Γ'. The Riemannian metric on M' lifted from M has a positive scalar curvature. By Rosenberg's theorem (Theorem 2.6) M' is KO-inessential. By Proposition 4.3 M' is inessential. By Proposition 4.4 M' is strongly inessential. This completes the proof since M and M' have same universal cover. □

PROOF OF PROPOSITION 4.3. Let $\dim M = n$. By Proposition 4.2 $B\Gamma/B\Gamma^{(n-1)}$ has the 1-Step Stabilization property property for KO_* with $m = n$ and $k = n$. We may assume that M has one n-dimensional cell and $u(M, M^{(n-1)}) \subset (B\Gamma^{(n)}, B\Gamma^{(n-1)})$. Since $u_*([M]_{KO}) = 0$ in $KO_n(B\Gamma)$, we obtain $q_*u_*([M]_{KO}) = 0$ in $KO_n(B\Gamma/B\Gamma^{(n-1)})$ where $q : B\Gamma \to B\Gamma/B\Gamma^{(n-1)}$ is the quotient map. Let q' denote the restriction of q to $B\Gamma^{(n+1)}$. Since $q'_*u_*([M]_{KO})$ lies in the image of the induced homomorphism for the inclusion

$$B\Gamma^{(n)}/B\Gamma^{(n-1)} \to B\Gamma^{(n+1)}/B\Gamma^{(n-1)},$$

by the 1-step Stabilization property applied for $m = n$ with $k = n$ we obtain $q'_*u_*([M]_{KO}) = 0$ in $KO_n(B\Gamma^{(n+1)}/B\Gamma^{(n-1)})$.

Let $\bar{u} : M/M^{(n-1)} = S^n \to B\Gamma^{(n+1)}/B\Gamma^{(n-1)}$ be the induced map by the classifying map u. In the commutative diagram

$$\begin{array}{ccccc} ko_n(M) & \xrightarrow{j'_*} & ko_n(S^n) & \xrightarrow{\bar{u}_*} & ko_n(B\Gamma^{(n+1)}/B\Gamma^{(n-1)}) \\ & & per \downarrow \cong & & \cong \downarrow per \\ KO_n(M) & \xrightarrow{j'_*} & KO_n(S^n) & \xrightarrow{\bar{u}_*} & KO_n(B\Gamma^{(n+1)}/B\Gamma^{(n-1)}) \end{array}$$

the *per* homomorphisms are isomorphisms in view of Proposition 2.3 and Proposition 2.1. Since $per \circ j'_*([M]_{ko}) = j'_*([M]_{KO})$, this implies that $\bar{u}_*j'_*([M]_{ko}) = 0$ in $ko_n(B\Gamma^{(n+1)}/B\Gamma^{(n-1)})$. In view of the natural transformation of homology theories $ko_* \to H_*(\ ;\mathbb{Z})$ it follows that $\bar{u}_*j'_*([M]) = 0$ in $H_n(B\Gamma^{(n+1)}/B\Gamma^{(n-1)})$. Since the homomorphism $q'_* : H_n(B\Gamma^{(n+1)}) \to H_n(B\Gamma^{(n+1)}/B\Gamma^{(n-1)})$ is injective, we obtain that $u_*([M]) = 0$. Theorem 2.5 completes the proof. □

PROOF OF PROPOSITION 4.4. Let $\dim M = n$. By Proposition 4.2 $B\Gamma$ has the 1-Step Stabilization property for KO_* with $k = m+1$. We may assume that M has a CW complex structure with one n-dimensional cell. Let $\psi : D^n \to M$ be its characteristic map. By Proposition 2.12 we may assume that the classifying map u satisfies the condition $u(M^{(n-1)}) \subset B\Gamma^{(n-2)}$ and $u(M) \subset B\Gamma^{(n-1)}$.

Note that in the diagram

$$\begin{array}{ccccc} ko_n(M/M^{(n-1)}) & \xrightarrow{\bar{u}_*} & ko_n(B\Gamma^{(n-1)}/B\Gamma^{(n-2)}) & \xrightarrow{(\bar{\phi}_{n-1,n})_*} & ko_n(B\Gamma^{(n)}/B\Gamma^{(n-2)}) \\ \cong\downarrow & & \cong\downarrow & & \cong\downarrow \\ KO_n(M/M^{(n-1)}) & \xrightarrow{\bar{u}_*} & KO_n(B\Gamma^{(n-1)}/B\Gamma^{(n-2)}) & \xrightarrow{(\bar{\phi}_{n-1,n})_*} & KO_n(B\Gamma^{(n)}/B\Gamma^{(n-2)}) \end{array}$$

the right vertical arrow is an isomorphism in view of Propositions 2.3 and 2.1. The 1-Step Stabilization property of $B\Gamma/B\Gamma^{(n-2)}$ applied with $m = n-1$ and $k = n$ implies that $(\bar{\phi}_{n-1,n})_*\bar{u}_*(1) = 0$ for KO_*. The above diagram implies that $i_*\bar{u}_*(1) = 0$ for ko_*. Lemma 3.1 completes the proof. □

5. Applications to the Gromov Conjecture

For applications of Theorem 4.5 to Gromov's conjecture we use the following stronger condition:

$(^*)_s$ *The classifying space $B\Gamma$ has the stable homotopy type of wedge of spheres.*

We recall that this means that the spectrum generated by $B\Gamma$ is the wedge sum of sphere spectra.

We note that in the case when $B\Gamma$ is a finite complex there is ℓ such that $\Sigma^\ell B\Gamma$ is homotopy equivalent to $\Sigma^\ell(\vee_j S^{n_j})$ [**BCM**].

PROPOSITION 5.1. *For geometrically finite groups property $(^*)_s$ implies the 1-Step Stabilization property for any homology theory h_*.*

PROOF. Let $f : \Sigma^\ell B\Gamma \to \Sigma^\ell(\vee S^{n_j})$ be a homotopy equivalence with a homotopy inverse map g. We may assume that f and g are cellular maps with respect to the natural CW structure on $\Sigma^\ell S^{n_j} = S^{n_j+\ell}$. Then the maps $\Sigma^\ell\phi_{m,m+1}$ and $\Sigma^\ell\phi_{m,m+1}(g|)(f|)$ are homotopic where $f|$ and $g|$ are the restrictions of f and g to the $(m+\ell)$-skeleton. Then for any nontrivial element $a \in h_{k+\ell}(\Sigma^\ell B\Gamma^{(m)})$ with $(\Sigma^\ell\phi_{m,m+1})_*(a) \neq 0$ we have $(f|)_*(a) \neq 0$. In this diagram

$$\begin{array}{ccccc} h_{k+\ell}(\Sigma^\ell B\Gamma^{(m)}) & \xrightarrow{(\Sigma^\ell\phi_{m,m+1})_*} & h_{k+\ell}(\Sigma^\ell B\Gamma^{(m+1)}) & \xrightarrow{(\Sigma^\ell\phi_{m+1,\infty})_*} & h_{k+\ell}(\Sigma^\ell B\Gamma) \\ f|_*\downarrow & & & & g_*\uparrow\cong \\ h_{k+\ell}(\bigvee_{n_j\le m} S^{n_j+\ell}) & \xrightarrow{\xi^0_*} & h_{k+\ell}(\bigvee_{n_j\le m+1} S^{n_j+\ell}) & \xrightarrow{\xi^1_*} & h_{k+\ell}(\bigvee S^{n_j+\ell}). \end{array}$$

Here the maps ξ^0 and ξ^1 are inclusions of a wedge in a large wedge. Since $\xi^1_*\xi^0_*$ is injective, it follows that

$$(\phi_{m+1,\infty})_*(\phi_{m,m+1})_*(a) \neq 0.$$

Thus, the 1-Step Stabilization property is verified. □

PROPOSITION 5.2. *Let Γ_1 and Γ_2 be geometrically finite groups satisfying the condition $(^*)_s$, then the free product $\Gamma_1 * \Gamma_2$ and the product $\Gamma_1 \times \Gamma_2$ satisfy $(^*)_s$.*

PROOF. Let $\Sigma^{\ell_1}B\Gamma_1$ and $\Sigma^{\ell_2}B\Gamma_2$ be homotopy equivalent to $\vee S^{n_j}$ and $\vee S^{m_i}$ respectively. We have $B(\Gamma_1 * \Gamma_2) = B\Gamma_1 \vee B\Gamma_2$, and thus for any k,

$$\Sigma^k B(\Gamma_1 * \Gamma_2) = \Sigma^k B\Gamma_1 \vee \Sigma^k B\Gamma_2,$$

Taking $k = max\{\ell_1, \ell_2\}$, we have the result for $\Gamma_1 * \Gamma_2$.

We note that $B(\Gamma_1 \times \Gamma_2) = B\Gamma_1 \times B\Gamma_2$. The suspension of product of CW complexes formula [**Ha**], $\Sigma(X \times Y) = \Sigma X \vee \Sigma Y \vee \Sigma(X \wedge Y)$ gives us

$$\Sigma^{\ell_1+\ell_2} B(\Gamma_1 \times \Gamma_2) = \Sigma^{\ell_1+\ell_2} B\Gamma_1 \vee \Sigma^{\ell_1+\ell_2} B\Gamma_2 \vee \Sigma^{\ell_1+\ell_2}(B\Gamma_1 \wedge B\Gamma_2).$$

The first two components on the right side are wedge sums of spheres. Since taking the reduced k-suspension is equivalent to the smash product with S^k, we obtain for the remaining component

$$\Sigma^{\ell_1+\ell_2}(B\Gamma_1 \wedge B\Gamma_2) = S^{\ell_1+\ell_2} \wedge (B\Gamma_1 \wedge B\Gamma_2) = (S^{\ell_1} \wedge B\Gamma_1) \wedge (S^{\ell_2} \wedge B\Gamma_2).$$

Here we used associativity and commutativity of smash product and the fact that $S^{\ell_1+\ell_2} = S^{\ell_1} \wedge S^{\ell_2}$. Thus, the right hand side is the same as $(\Sigma^{\ell_1} B\Gamma_1) \wedge (\Sigma^{\ell_2} B\Gamma_2)$, which is equivalent to

$$(\bigvee S^{n_j}) \wedge (\bigvee S^{m_i}) = \bigvee (S^{n_j} \wedge (\bigvee S^{m_i})) = \bigvee (S^{n_j} \wedge S^{m_i}) = \bigvee S^{n_j+m_i}.$$

□

PROPOSITION 5.3. *The surface groups $\pi_1(M_g)$ for all $g \geq 0$, satisfy the condition $(^*)_s$.*

PROOF. The surface M_g has one 0-cell, $2g$ many 1-cell, one 2-cell and the cell structure is obtained by attaching the 2-cell along $[a_1, b_1] \cdot [a_2, b_2] \cdots [a_g, b_g]$ to the 1-skeleton, which is wedge of $2g$ circles. Let $\varphi : S^1 \to \bigvee_{2g} S^1_i$ be the attaching map for the 2-cell. We claim that $\Sigma\varphi : S^2 \to \bigvee_{2g} S^2_i$ is nullhomotopic. Since $H_2(\bigvee_{2g} S^2_i) = \oplus_{i=1}^{2g} H_2(S^2_i)$, by Hurewicz Theorem $\pi_2(\bigvee_{2g} S^2_i) = \oplus_{i=1}^{2g} \pi_2(S^2_i)$. Thus,

$$[\Sigma\varphi] = ([\Sigma(p_1\varphi)], \dots, [\Sigma(p_{2g}\varphi]) \in \bigoplus_{i=1}^{2g} \pi_2(S^2_i)$$

where $p_i : \bigvee_{2g} S^1 \to S^1_i$ is the projection to i-th summand. Note that each $p_i \circ \varphi : S^1 \to S^1$ has degree 0. Hence $deg\Sigma(p_i\varphi) = deg(p_i\varphi) = 0$. Hence $[\Sigma\varphi] = 0$ and $\Sigma\varphi$ is null-homotopic. Therefore, ΣM_g is homotopy equivalent to $\left(\bigvee_{2g} S^2_i\right) \vee S^3$. □

Let X be a finite simplicial graph with the set of vertices $S = \{s_1, s_2, \dots, s_n\}$, and the set of edges E. We recall, that the right-angled Artin group defined by X is the group with following presentation

$$A_X = \langle S \mid [s_i, s_j] = 1 \text{ whenever } \{s_i, s_j\} \in E \rangle.$$

The Salvetti complex, $\mathcal{S}_X$, which is a classifying space, $\mathcal{S}_X := BA_X$ for the group A_X is constructed as follows: Begin with a wedge of circles attached to a point x_0 and labeled by the generators $s_1, \cdots, s_n$. For each edge, $\{s_i, s_j\} \in E$, consider the 2-torus obtained by attaching 2-cell to the wedge of circles corresponding to s_i and s_j along the relation $[s_i, s_j]$. For each triangle in X connecting three vertices s_i, s_j, s_k, consider the 3-torus defined by circles corresponding to them. Continue this process, having a k-torus for each set of k mutually commuting generators (i.e., generators spanning a complete subgraph in X). The resulting space is the Salvetti complex $\mathcal{S}_X$ for A_X.

PROPOSITION 5.4. *Suppose that $K = e^0 \cup (\cup_{i \in J} e_i^{n_i})$ is a subcomplex of an n-dimensional torus $T^n = S^1 \times \cdots \times S^1$ with the product CW complex structure and $S^1 = e^0 \cup e^1$. Then for sufficiently large ℓ the reduced ℓ-suspension $\Sigma^\ell K$ is homotopy equivalent to the wedge of spheres $\vee_{i \in J} S^{n_i+\ell}$.*

Moreover, there is a cellular homotopy equivalence $f : \Sigma^\ell K \to \vee_{i\in J} S^{n_i+\ell}$ which takes the cell $\Sigma^\ell e_i^{n_i}$ to $S^{n_i+\ell}$ for all $i \in J$.

PROOF. The proof is based on the formula

$$\Sigma(X \times Y) = \Sigma X \vee \Sigma Y \vee \Sigma(X \wedge Y).$$

We refer to the proof of Proposition 6.5 in [**Dr2**] for details. □

COROLLARY 5.5. *Any subcomplex of n-dimensional torus T^n satisfies $(^*)_s$.*

PROPOSITION 5.6. *RAAGs satisfy $(^*)_s$.*

PROOF. Let A_X be right-angled Artin group generated by n generators where X is the corresponding simple graph with n vertices $\{s_1, \cdots, s_n\}$. Since the Salvetti complex $\mathcal{S}_X$ is built as a union of k-tori, $T^k \subset T^n$ for each complete k-subgraph of X, clearly $\mathcal{S}_X$ is a subcomplex of T^n. Thus Lemma 5.5, completes the proof. □

In view of Proposition 5.1, the condition $(^*)_s$ implies the 1-Step Stabilization property for KO_*. Thus, Theorem 4.5 implies the following

COROLLARY 5.7. *Suppose that a group Γ satisfies $(^*)_s$ and the Strong Novikov conjecture holds true for Γ. Then the Strong Gromov conjecture holds for spin manifolds with the fundamental group Γ.*

COROLLARY 5.8. *The Strong Gromov Conjecture holds true for spin manifolds with the fundamental group Γ where*

- *Γ is a RAAG,*
- *Γ is a finite product of orientable surface groups.*

PROOF. Since these groups are CAT(0) groups, the Strong Novikov Conjecture holds for them and the result follows. □

In view of Proposition 4.5, if the Strong Gromov Conjecture holds for a group Γ' contained in Γ as a finite index subgroup, then the original Gromov's conjecture holds for Γ. Since the fundamental group of non-orientable surface contains as index 2, the fundamental group of an orientable surface, we obtain that the original Gromov Conjecture holds true for products of non-orientable surface groups.

REMARK 5.9. It is well known that each RAAG is contained in some right-angled Coxeter groups as finite index subgroup, [**DJ**], hence the Gromov Conjecture holds true for spin manifolds having fundamental groups those right-angled Coxeter groups.

QUESTION 5.10. Does the property $(^*)_s$ implies the (Strong) Novikov conjecture for a group Γ?

5.1. More groups satisfying 1-Step Stabilization Property. Given an abelian group G and $n \geq 1$, a CW complex X having trivial reduced homology $\tilde{H}_i(X) = 0$ for $i \neq n$ and $H_n(X) = G$ is called a Moore space and is denoted as $M(G, n)$. In the case of a finitely generated group G and $n \geq 2$ the Moore space can be chosen to be the finite wedge of n-sphere S^n and n-spheres with one $(n+1)$-dimensional cell attached $S^n \cup_\phi D^{n+1}$.

The condition $(^*)_s$ can be weakened to the following

$(^*)_m$ *The classifying space $B\Gamma$ has the stable homotopy type of a wedge of Moore spaces.*

PROPOSITION 5.11. *For geometrically finite groups property $(^*)_m$ implies the 1-Step Stabilization property for any homology theory h_*.*

PROOF. Let $f : \Sigma^\ell B\Gamma \to \Sigma^\ell(\vee M(G_j, n_j))$ be a homotopy equivalence with a homotopy inverse map g. Since $B\Gamma$ is geometrically finite, the groups G_j are finitely generated. We may assume that each G_j is isomorphic to $\mathbb{Z}$ or to $\mathbb{Z}_{m_j}$ with $m_j = p^{k_j}$ for a prime number p. Thus, $M(G_j, n_j)$ is either an n_j-sphere S^{n_j} or an n_j-sphere with (n_j+1)-cell attached, $S^{n_j} \cup_{\psi_j} D^{n_j+1}$. We may assume that f and g are cellular map with respect to the natural CW structure on $\Sigma^\ell M(G_j, n_j) = M(G_j, n_j + \ell)$. Let $m' = m + \ell$, $k' = k + \ell$, and $n'_j = n_j + \ell$. Let f^m and g^m denote the restrictions of f and g to the $(m + \ell)$-skeletons. Then for any nontrivial element $a \in h_{k+\ell}(\Sigma^\ell B\Gamma^{(m)})$ with $(\Sigma^\ell \phi_{m,m+1})_*(a) \neq 0$ we have $(f|)_*(a) \neq 0$:

Let $\phi^0 = \Sigma^\ell \phi_{m,m+1}$ and $\phi^1 = \Sigma^\ell \phi_{m+1,\infty}$. Consider the diagram

$$\begin{array}{ccccc}
h_{k'}(\Sigma^\ell B\Gamma^{(m)}) & \xrightarrow{\phi^0_*} & h_{k'}(\Sigma^\ell B\Gamma^{(m+1)}) & \xrightarrow{\phi^1_*} & h_{k'}(\Sigma^\ell B\Gamma) \\
g^m_* \uparrow & & f^{m+1}_* \downarrow & & g_* \uparrow \cong \\
h_{k'}(\bigvee M(G_j, n'_j)^{(m')}) & \xrightarrow{\xi^0_*} & h_{k'}(\bigvee M(G_j, n'_j)^{(m'+1)}) & \xrightarrow{\xi^1_*} & h_{k'}(\bigvee M(G_j, n'_j)).
\end{array}$$

where ξ^0 and ξ^1 are the inclusions of wedge sums. Let $a \in h_{k'}(\Sigma^\ell B\Gamma^{(m)})$ with $\phi^0_*(a) \neq 0$. We need to show that $\phi^1_*\phi^0_*(a) \neq 0$.

Since the maps ϕ^0 and $\phi^0 g^m f^m$ are homotopic, we obtain

$$g^{m+1}_* \xi^0(f^m_*(a)) = \phi^0_* g^m_*(f^m_*(a)) = \phi^0_*(a) \neq 0.$$

Therefore, $\xi^0_*(f^m_*(a)) \neq 0$. We note that

$$\bigvee M(G_j, n'_j)^{(m'+1)} = \bigvee_{n_j \leq m'} M(G_j, n'_j) \vee (\bigvee S^{m'+1})$$

and $\xi^0_*(f^m_*(a))$ lives in the summand $\bigvee_{n_j \leq m'} M(G_j, n'_j)$ which is a retract of $\bigvee M(G_j, n'_j)$. Hence, $\xi^1_*(\xi^0(f^m_*(a)) \neq 0$. Then

$$\phi^1_*\phi^0_*(a) = \phi^1_* g^{m+1}_* \xi^0_* f^m_*(a) = g_* \xi^1_* \xi^0_* f^m_*(a)\ ne0.$$

Thus, the 1-Step Stabilization property is verified. □

PROPOSITION 5.12. *All finitely generated groups with* $\dim B\Gamma \leq 2$ *have property* $(^*)_m$ *with a finite wedge of Moore spaces.*

PROOF. Apply Proposition 2.3. □

Suppose that a geometrically finite group Γ has property $(^*)_m$. Then for some ℓ the ℓ-suspension $\Sigma^\ell B\Gamma$ is homotopy equivalent to a wedge $\vee_{j=1}^s M(G_j, n_j)$ where $G_j = \mathbb{Z}$ or $\mathbb{Z}_{p^{k_j}}$ where p is a prime number and $n_j > \ell$. We call such Γ *2-avoiding* if $p^{k_j} \neq 2$ for all j.

PROPOSITION 5.13. *(* [**JN**] *Corollary 6.6) If $gcd(k, l) = d$ is odd or 4 divides k, then $M(\mathbb{Z}_k, m) \wedge M(\mathbb{Z}_l, n)$ is homotopy equivalent to $M(\mathbb{Z}_d, m+n) \vee M(\mathbb{Z}_d, m+n+1)$.*

PROPOSITION 5.14. *The product of geometrically finite groups $\Gamma_1 \times \Gamma_2$ has property $(^*)_m$, when each of Γ_1, Γ_2 has property $(^*)_m$, provided one of the groups is 2-avoiding.*

PROOF. Proceeding exactly similar to the proof of Proposition 5.2, this time $(\Sigma^{\ell_1}B\Gamma_1)\wedge(\Sigma^{\ell_2}B\Gamma_2)$ is homotopy equivalent to

$$(\vee_{j=1}^{s}M(G_j,n_j))\bigwedge(\vee_{i=1}^{t}M(G_i,m_i))=\bigvee_{i,j}M(G_j,n_j)\wedge M(G_i,m_i).$$

Since $M(\mathbb{Z},n)=S^n$, if any of G_j or G_i is $\mathbb{Z}$ we have $M(G_j,n_j)\wedge M(G_i,m_i)$ to be a suspension of one of the Moore spaces, which is again Moore space. It is enough to consider when we have same prime p for both groups G_j and G_i, since for $k=rs$ with r,s relatively prime we have $M(\mathbb{Z}_k,n)=M(\mathbb{Z}_r,n)\vee M(\mathbb{Z}_s,n)$. If $G_j=\mathbb{Z}_{p^j}$ and $G_i=\mathbb{Z}_{p^i}$, where p is some prime, then assuming $i\leq j$, we have $gcd(p^j,p^i)=p^i$, thus by Proposition 5.13 we have,

$$M(\mathbb{Z}_{p^j},n_j)\wedge M(\mathbb{Z}_{p^i},m_i)=M(\mathbb{Z}_{p^i},n_j+m_i)\vee M(\mathbb{Z}_{p^i},n_j+m_i+1).$$

Note that if $p=2$, $1\leq i$ and $2\leq j$ then Proposition 5.13 gives the above splitting. □

COROLLARY 5.15. *The Strong Gromov Conjecture holds for spin manifolds whose fundamental groups are finite products $\Gamma=\Gamma_1\times\cdots\times\Gamma_k$ of 2-dimensional finitely generated 2-avoiding groups such that each Γ_i satisfies the Strong Novikov Conjecture.*

We note that 2-avoiding torsion free one-relator groups are among examples of above Γ_i. Since they have finite asymptotic dimension [**BD**], [**Dr3**], [**T**], they satisfy the coarse Baum-Connes conjecture for both KU and KO. Therefore, they satisfy the Strong Gromov Conjecture [**HR**]. By the Lyndon-Cockeroft theorem [**L**],[**C**] they are 2-dimensional.

We recall that a 2-dimensional group is 2-avoiding if its abelianization does not contain $\mathbb{Z}_2$ as a direct summand. An affirmative answer to the following question would prove the original Gromov's conjecture for the products of all 2-dimensional groups.

QUESTION 5.16. Does every 2-dimensional group contain a 2-avoiding finite index subgroup?

We believe that the other required condition for 2-dimensional groups in Corollary 5.15, the Strong Novikov Conjecture, can be derived from results of Mathai [**Mat**] and Hanke-Schick [**HS**].

5.2. Bolotov's question. Dmitri Bolotov constructed an example M_b of a closed inessential spin 4-manifold which is not strongly inessential. At the end of his paper he asked if any product with a torus $M_b\times T^p$ can carry a metric of positive scalar curvature [**Bo2**]. Here we give a negative answer.

PROPOSITION 5.17. *The manifolds $M_b\times T^p$ for all $p>0$ do not admit a metric with positive scalar curvature.*

PROOF. The fundamental group of Bolotov's manifold is $\Gamma=\mathbb{Z}*\mathbb{Z}^3$. By Proposition 5.2 the group $\Gamma\times\mathbb{Z}^p$ satisfies the condition $(*)_s$. The Strong Novikov conjecture holds for the group $\Gamma\times\mathbb{Z}^p$ since it is constructed out of integers by taking operation of the product and the free product. Note that $M_b\times T^p$ is spin and inessential. By Theorem 4.5 the Strong Gromov conjecture holds for $M_b\times T^p$ for $p>0$. Therefore, $M_b\times T^p$ cannot carry a metric of positive scalar curvature. Hence M_b cannot carry a metric of positive scalar curvature as well. □

References

[Ba] I. K. Babenko, *Asymptotic invariants of smooth manifolds* (Russian, with Russian summary), Izv. Ross. Akad. Nauk Ser. Mat. **56** (1992), no. 4, 707–751, DOI 10.1070/IM1993v041n01ABEH002181; English transl., Russian Acad. Sci. Izv. Math. **41** (1993), no. 1, 1–38. MR1208148

[BD] G. Bell and A. Dranishnikov, *On asymptotic dimension of groups acting on trees*, Geom. Dedicata **103** (2004), 89–101, DOI 10.1023/B:GEOM.0000013843.53884.77. MR2034954

[Bo1] Dmitry V. Bolotov, *Macroscopic dimension of 3-manifolds*, Math. Phys. Anal. Geom. **6** (2003), no. 3, 291–299, DOI 10.1023/A:1024994930786. MR1997917

[Bo2] Dmitry V. Bolotov, *Gromov's macroscopic dimension conjecture*, Algebr. Geom. Topol. **6** (2006), 1669–1676, DOI 10.2140/agt.2006.6.1669. MR2253461

[Bo3] Dmitry Bolotov, *About the macroscopic dimension of certain PSC-manifolds*, Algebr. Geom. Topol. **9** (2009), no. 1, 21–27, DOI 10.2140/agt.2009.9.21. MR2471130

[BD1] Dmitry Bolotov and Alexander Dranishnikov, *On Gromov's scalar curvature conjecture*, Proc. Amer. Math. Soc. **138** (2010), no. 4, 1517–1524, DOI 10.1090/S0002-9939-09-10199-5. MR2578547

[BD2] Dmitry Bolotov and Alexander Dranishnikov, *On Gromov's conjecture for totally non-spin manifolds*, J. Topol. Anal. **8** (2016), no. 4, 571–587, DOI 10.1142/S1793525316500230. MR3545013

[Bou] A. K. Bousfield, *The localization of spectra with respect to homology*, Topology **18** (1979), no. 4, 257–281, DOI 10.1016/0040-9383(79)90018-1. MR551009

[BCM] R. R. Bruner, F. R. Cohen, and C. A. McGibbon, *On stable homotopy equivalences*, Quart. J. Math. Oxford Ser. (2) **46** (1995), no. 181, 11–20, DOI 10.1093/qmath/46.1.11. MR1326132

[Br] Kenneth S. Brown, *Cohomology of groups*, Graduate Texts in Mathematics, vol. 87, Springer-Verlag, New York-Berlin, 1982. MR672956

[C] W. H. Cockcroft, *On two-dimensional aspherical complexes*, Proc. London Math. Soc. (3) **4** (1954), 375–384, DOI 10.1112/plms/s3-4.1.375. MR63042

[DD] Michelle Daher and Alexander Dranishnikov, *On macroscopic dimension of non-spin 4-manifolds*, J. Topol. Anal. **14** (2022), no. 2, 343–352, DOI 10.1142/S1793525321500084. MR4446910

[DJ] Michael W. Davis and Tadeusz Januszkiewicz, *Right-angled Artin groups are commensurable with right-angled Coxeter groups*, J. Pure Appl. Algebra **153** (2000), no. 3, 229–235, DOI 10.1016/S0022-4049(99)00175-9. MR1783167

[Dr1] Alexander Dranishnikov, *On Gromov's positive scalar curvature conjecture for duality groups*, J. Topol. Anal. **6** (2014), no. 3, 397–419, DOI 10.1142/S1793525314500162. MR3217864

[Dr2] Alexander Dranishnikov, *Positive scalar curvature, macroscopic dimension, and inessential manifolds*, Perspectives in scalar curvature. Vol. 2, World Sci. Publ., Hackensack, NJ, 2023, pp. 231–248. MR4577916

[Dr3] Alexander Dranishnikov, *On asymptotic dimension of amalgamated products and right-angled Coxeter groups*, Algebr. Geom. Topol. **8** (2008), no. 3, 1281–1293, DOI 10.2140/agt.2008.8.1281. MR2443244

[G1] Mikhael Gromov, *Filling Riemannian manifolds*, J. Differential Geom. **18** (1983), no. 1, 1–147. MR697984

[G2] M. Gromov, *Positive curvature, macroscopic dimension, spectral gaps and higher signatures*, Functional analysis on the eve of the 21st century, Vol. II (New Brunswick, NJ, 1993), Progr. Math., vol. 132, Birkhäuser Boston, Boston, MA, 1996, pp. 1–213, DOI 10.1007/s10107-010-0354-x. MR1389019

[G3] Misha Gromov, *Metric structures for Riemannian and non-Riemannian spaces*, Progress in Mathematics, vol. 152, Birkhäuser Boston, Inc., Boston, MA, 1999. MR1699320

[GL] Mikhael Gromov and H. Blaine Lawson Jr., *Positive scalar curvature and the Dirac operator on complete Riemannian manifolds*, Inst. Hautes Études Sci. Publ. Math. **58** (1983), 83–196 (1984). MR720933

[Ha] Allen Hatcher, *Algebraic topology*, Cambridge University Press, Cambridge, 2002. MR1867354

[HS] Bernhard Hanke and Thomas Schick, *The strong Novikov conjecture for low degree cohomology*, Geom. Dedicata **135** (2008), 119–127, DOI 10.1007/s10711-008-9266-9. MR2413333

[HR] N. Higson, J. Roe, On the coarse Baum-Connes conjecture, London Math. Soc. Lecture Note Ser., 227 Cambridge University Press, Cambridge, 1995.

[JN] Joseph Neisendorfer, *Primary homotopy theory*, Mem. Amer. Math. Soc. **25** (1980), no. 232, iv+67, DOI 10.1090/memo/0232. MR567801

[L] Roger C. Lyndon, *Cohomology theory of groups with a single defining relation*, Ann. of Math. (2) **52** (1950), 650–665, DOI 10.2307/1969440. MR47046

[M] Michał Marcinkowski, *Gromov positive scalar curvature conjecture and rationally inessential macroscopically large manifolds*, J. Topol. **9** (2016), no. 1, 105–116, DOI 10.1112/jtopol/jtv037. MR3465842

[Mat] Varghese Mathai, *The Novikov conjecture for low degree cohomology classes*, Geom. Dedicata **99** (2003), 1–15, DOI 10.1023/A:1024941020306. MR1998926

[R1] Jonathan Rosenberg, *C^*-algebras, positive scalar curvature, and the Novikov conjecture*, Inst. Hautes Études Sci. Publ. Math. **58** (1983), 197–212 (1984). MR720934

[R2] Jonathan Rosenberg, *Manifolds of positive scalar curvature: a progress report*, Surveys in differential geometry. Vol. XI, Surv. Differ. Geom., vol. 11, Int. Press, Somerville, MA, 2007, pp. 259–294, DOI 10.4310/SDG.2006.v11.n1.a9. MR2408269

[Ru] Yuli B. Rudyak, *On Thom spectra, orientability, and cobordism*, Springer Monographs in Mathematics, Springer-Verlag, Berlin, 1998. MR1627486

[SY] Richard Schoen and Shing-Tung Yau, *Positive scalar curvature and minimal hypersurface singularities*, Surveys in differential geometry 2019. Differential geometry, Calabi-Yau theory, and general relativity. Part 2, Surv. Differ. Geom., vol. 24, Int. Press, Boston, MA, [2022] ©2022, pp. 441–480. MR4479726

[TtD] Tammo tom Dieck, *Algebraic topology*, EMS Textbooks in Mathematics, European Mathematical Society (EMS), Zürich, 2008, DOI 10.4171/048. MR2456045

[T] Panagiotis Tselekidis, *Asymptotic dimension of graphs of groups and one-relator groups*, Algebr. Geom. Topol. **23** (2023), no. 8, 3587–3613, DOI 10.2140/agt.2023.23.3587. MR4665278

[Yu] G. Yu, *The Novikov conjecture* (Russian, with Russian summary), Uspekhi Mat. Nauk **74** (2019), no. 3(447), 167–184, DOI 10.4213/rm9882; English transl., Russian Math. Surveys **74** (2019), no. 3, 525–541. MR3954330

Department of Mathematics, University of Florida, 358 Little Hall, Gainesville, Florida 32611-8105

Email address: dranish@ufl.edu

Department of Mathematics, University of Florida, 358 Little Hall, Gainesville, Florida 32611-8105

Email address: showladar@ufl.edu

Contemporary Mathematics
Volume **812**, 2025
https://doi.org/10.1090/conm/812/16267

Ends of large scale groups

Jerzy Dydak, Yuankui Ma, and Hussain Rashed

ABSTRACT. The aim of this paper is to create a theory of ends of large scale groups, a class containing both finitely generated groups and locally compact, metrizable and connected topological groups. Our generalization amounts to defining the space of ends of any coarse space and then applying it to large scale groups. We prove that, if the number of ends is finite, then it must be at most 2. Groups of two ends are characterized as having an infinite cyclic subgroup of bounded index.

Additionally, we prove a version of Svarc-Milnor Lemma for large scale groups and we prove that coarsely hyperbolic large scale groups have finite asymptotic dimension provided they have bounded geometry.

CONTENTS

1. Introduction

Historically, as noted in [**8**] on p.287, ends are the oldest coarse topological notion. Here is their internal description:

DEFINITION 1.1. A **Freudenthal end** is a decreasing sequence $(U_i)_{i\geq 1}$ of components of $X \setminus K_i$, where $(K_i)_{i\geq 1}$ is an **exhausting sequence**, i.e. K_i is compact,

---

2020 *Mathematics Subject Classification.* Primary 20F69; Secondary 54D35.

Key words and phrases. Dimension, coarse geometry, ends of groups, Freundenthal compactification, Higson corona.

The authors declare that all data supporting the findings of this study are available within the article.

$K_i \subset int(K_{i+1})$ for each $i \geq 1$, and $\bigcup_{i=1}^{\infty} K_i = X$. The set of ends of X is denoted by $Ends(X)$.

Ends were used by Freudenthal in 1930 in his famous compactification (see [**18**] for information about theorems in this section and see [**13**] for results related to the theory of dimension):

THEOREM 1.2. *Suppose X is a σ-compact, locally compact, connected and locally connected Hausdorff space. It has a compactification $\bar{X}$ such that $\bar{X} \setminus X$ is of dimension* 0 *and $\bar{X}$ dominates any compactification $\hat{X}$ of X whose corona is of dimension* 0.

REMARK 1.3. Given a σ-compact, locally compact, connected and locally connected Hausdorff space $(X, \mathcal{T})$. If $U \in \mathcal{T}$, we define the subset $U_{end} := \{(U_i) \in Ends(X) | U_i \subset U \text{ for some } i \geq 1\}$ and $\widetilde{U}_{end} := U_{end} \cup U$.

The family $\mathcal{T} \cup \{\widetilde{U}_{end} | U \in \mathcal{T}\}$ is a basis for a topology $\mathcal{T}_{end}$ on $X \cup Ends(X)$. The topological space $\bar{X} := X \cup Ends(X)$ is a compactification of X called the **Freudenthal compactification**. The space of ends $Ends(X) = \bar{X} \setminus X$ is of dimension 0, and $\bar{X}$ dominates any compactification $\hat{X}$ of X whose corona is of dimension 0. Moreover, the number of ends of X is the supremum of $n_i \geq 0$ where n_i is the number of all mutually disjoint unbounded components of $X \setminus K_i$, for all $i \geq 1$.

Initially, ends were useful as properties of topological groups:

THEOREM 1.4 (Freudenthal). *A path connected topological group has at most two ends.*

THEOREM 1.5 (Leo Zippin [**26**]). *If a locally compact, metrizable, connected topological group G is two-ended, then G contains a closed subgroup T isomorphic to the group of the reals such that the coset-space G/T is compact, moreover, the space G is the topological product of the axis of the reals by a compact connected set homeomorphic to the space G/T.*

THEOREM 1.6 (H. Hopf). *Let G be a finitely generated discrete group acting on a space X by covering transformations. Suppose the orbit space $B := X/G$ is compact. Then (i) and (ii), below, hold.*

(i) *The space of ends of X has 0, 1, 2 (discrete) elements or is a Cantor space.*
(ii) *If the space of ends of X is infinite, it is a Cantor space.*
(iii) *If G also acts on Y satisfying the hypotheses above, then X and Y have homeomorphic end spaces.*

Conclusion (iii) suggests to regard the space of ends of X as an invariant of the group G itself:

DEFINITION 1.7. Let $p : X \to B$ be a covering map with compact base B and the group of covering transformations G. The **end space** of G is

$$Ends(G) := Ends(X).$$

When applied to a Cayley graph of G, it gives the standard definition of ends of finitely generated groups (see [**8**], p.295). See [**14**] for basic results in this theory and see [**15**] for more general facts in coarse geometry related to groups. The paper [**17**] contains interesting results for ends of finitely generated groups.

Stone [**22**] assigns to each Boolean algebra B a totally disconnected compact Hausdorff space X_B called the Stone space associated to B; and conversely, every Stone space X (totally disconnected compact Hausdorff space) is assigned a Boolean algebra B_X consisting of all clopen subsets of X. Stone representation theorems assure us that two Boolean algebras are isomorphic if and only if their corresponding Stone spaces are homeomorphic and that two Stone spaces are homeomorphic if and only if their Boolean algebras of their clopen subsets are isomorphic. See [**23**] for basic results on Stone Duality.

Realizing that the space of ends of a finitely generated group is a Stone space, and that its Boolean algebra of all clopen subsets is isomorphic to the quotient Boolean algebra of the Boolean algebra consisting of all almost right invariant subsets by the ideal of finite subsets ($A \subset G$ is called almost right invariant if the symmetric difference $A \Delta A \cdot g$ is finite for all $g \in G$), E. Specker [**19**] defined the space of ends of an arbitrary group as the Stone space of the quotient Boolean algebra of the Boolean algebra of all almost right invariant subsets by the ideal of finite subsets. Specker partially generalized the first part of 1.6:

THEOREM 1.8 (Specker). *Let G be an arbitrary group. Then:*

1) $|ends(G)|$ *is either infinite or at most* 2.
2) $|ends(G)| = 0$ *if and only if G is finite.*
3) $|ends(G)| = 2$ *if and only if G contains an infinite cyclic subgroup of finite index.*

Adopting Specker's approach, W. Dicks and M. J. Dunwoody [**7**] considered ends of non-finitely generated groups. In particular, they proved the following result that is a generalization of the famous theorem of Stallings [**21**]:

THEOREM 1.9. *A group G has infinitely many ends if and only if one of the following conditions holds:*

(i) *G is countably infinite and locally finite,*
(ii) *G can be expressed as an amalgamated free product $A *_C B$ or an HNN extension $A*_C$, where C is a finite subgroup of A and B such that $[A : C] \geq 3$ and $[B : C] \geq 2$.*

Yves Cornulier [**5**] has also studied the space of ends of infinitely generated groups.

When a group G is equipped with a left-invariant proper metric, its almost right invariant subsets are exactly the coarsely clopen subsets of G equipped with the large scale structure induced by its metric. The concept of coarsely clopen subsets is defined for any large scale space X. Furthermore, the family of all coarsely clopen subsets $\mathcal{CC}(X)$ of a given large scale space X is a Boolean algebra. Consequently, it is natural to define the space of ends $Ends(X)$ of a coarsely connected large scale space X as the Stone space of the to the quotient Boolean algebra of $\mathcal{CC}(X)$ by the ideal of all bounded subsets.

When considering a group G as a large scale space with the large scale structure induced by the bornology of its finite subsets, its space of ends as a large scale space coincide with Specker's space of ends and therefore our construction of the space of ends of a large scale space naturally generalizes Specker's space of ends of a group.

The space of ends $Ends(X)$ of a large scale space X has the following properties:

1. It is a coarse invariant of large scale spaces.
2. $X \cup Ends(X)$ is a large scale compactification of X.
3. If X is a connected, locally connected, locally compact Hausdorff space, then there exists a large scale structure on X such that the coarse ends are in one-to-one correspondence with the Freudenthal ends of X.

Finiteness in large scale groups equipped with the large scale structure induced by the bornology of finite subsets (including countable groups equipped with a proper left-invariant metric, and a fortiori, finitely generated groups) corresponds to boundedness in large scale groups in general.

Many concepts involving finiteness can be naturally generalized to the setting of boundedness. For instance, we can define boundedly generated groups, subgroups of bounded index and locally bounded groups as counterparts of finitely generated groups, subgroups of finite index and locally finite groups, respectively. Consequently, many results involving finiteness hold, in general, in the realm of large scale groups. For instance:

1. In 4.6, we prove that a large scale group G is of asymptotic dimension 0 if and only if G is locally bounded (i.e. for every bounded subset B of G the subgroup $\langle B\rangle$ of G generated by B is bounded); which is a generalization of the fact that a countable group is of asymptotic dimension 0 if and only if it is locally finite (see Theorem 2, [**20**]).
2. In section 6 we give a coarse version of the Svarc-Milnor Lemma.
3. Let G be a large scale group that contains two infinite cyclic subgroups H and K. If H is of bounded index, then so is K.

Given a group G and a symmetric subset $K \subset G$ generating G; G can naturally be equipped with the norm $|.|_K : G \to [0,\infty)$ given by:

$$|x|_K = \begin{cases} 0 & \text{iff } x = 1_G \\ \inf\{n \in \mathbb{N} : x = x_1 \dots x_n \text{ and } x_1,\dots,x_n \in K \setminus \{1_G\}\} & \text{if } x \in G \setminus \{1_G\} \end{cases}$$

- The norm $|.|_K : G \to [0,\infty)$ on G induces a left-invariant metric $d_K : G \times G \to [0,\infty)$ given by $d_K(x,y) := |x^{-1}y|$ and it is called the **Cayley** metric with respect to the subset K.
- The Cayley metric d_K induces a graph $Cay(G,K) = (V,E)$ called the **Cayley graph of G with respect to the subset** K where $V = G$ and $\{x,y\} \in E$ if and only if $d_K(x,y) = 1$ if and only if $y = a \cdot x$ for some $a \in K \setminus \{1_G\}$.
- Passing to the geometric realization X_K, we can think of the Cayley graph as a geodesic metric space.

When K is finite then $d_K : G \times G \to [0,\infty)$ is a proper left-invariant metric that induces its large scale structure with G and the geodesic metric space X_K are coarsely equivalent. Moreover, for any bounded subset $B \subset G$ the following properties hold:

P_1) $G \setminus B$ contains only finitely many unbounded (infinite) K-components (or 1-components) and the union of all bounded (finite) K-components of $G \setminus B$ is bounded. By a K-component of $G \setminus B$ we mean the largest subset C of $G \setminus B$ such that for all $a, b \in C$, there exists a finite sequence

$g_1, \ldots, g_n$ of elements of G such that $a = g_1$ and $b = g_n$ with $g_i^{-1} \cdot g_{i+1} \in K$ for all $i < n$ (equivalently $d_K(g_i, g_{i+1}) = 1$ for all $i < n$).

P_2) B can be covered by finitely many subsets of the form $g \cdot K$, where $g \in G$.

Section 5 mainly gives an answer to when a large scale group is boundedly generated: it turns out that the sufficient condition for a large scale group to be boundedly generated is to satisfy the property P_1) for some symmetric bounded subset $K \subset G$.

Section 6 answers the same question using a coarse version of the Svarc-Milnor Lemma.

Section 7 concerns large scale groups that satisfies P_2) that we call large scale groups of bounded geometry. This class of groups includes large scale groups with large scale structure induced by the bornology of finite subsets as well as locally compact groups with the large scale structure induced by the family of all pre-compact subsets. boundedly generated large scale groups of bounded geometry are metrizable by a Cayley metric and their space of ends is metrizable.

Section 8 attempts to address the following questions:

Q_1) What are the conditions for a large scale structure of a large scale group to be induced by a metric?

Q_2) If the large scale structure of the boundedly generated group G is metrizable, when is it induced by a Cayley metric?

We conclude this section by given a coarse version and a simpler proof for the fact that hyperbolic groups have finite asymptotic dimension given in [**25**].

In **Section 9** we prove and generalize the following classical results:

1. Locally finite groups are infinitely-ended.
2. The number of ends of a group G is either infinite or at most 2.
3. A group G is 2-ended if and only if it contains an infinite cyclic group of finite index.

The coarse counterparts of these results are:

1. Locally bounded groups equipped with metrizable are infinitely-ended.
2. If G is a large scale group that contains a sequence $\{G_i\}_{i \geq 1}$ of coarsely geodesic subgroups such that for every boundedly generated subgroup H of G, there exists $n \in \mathbb{N}$ such that $H \subset G_n$; then the number of ends of G is either infinite or at most two.
3. Let G be a large scale group of bounded geometry that is σ-bounded and non-locally bounded. If G is 2-ended, then it contains an infinite cyclic group of bounded index.

At the end of section 9, we give the necessary and sufficient conditions for the space of coarse ends $Ends(G)$ of a coarsely geodesic large scale group G to be topologically metrizable. It turns out that the necessary and sufficient condition for $Ends(G)$ to be merizable is to satisfy the property P_1).

In the case when $Ends(G)$ is metrizable, its ends can be described in a similar manner as the Freudenthal ends. We construct a metric on $Ends(G)$ that induces its topology, and prove that if $|Ends(G)| = \infty$, then $Ends(G)$ is a Cantor space.

2. Ends of coarse spaces

In this section we generalize the concept of Freudenthal ends to arbitrary coarse spaces. See [**10**] for other ways to introduce ends in coarse spaces.

We follow a description of coarse spaces (quite often our terminology is that of **large scale spaces**) as in [**9**]. It is equivalent to Roe's definition of those spaces in [**24**].

Recall that a **star** $st(x,\mathcal{U})$ of $x \in X$ with respect to a family $\mathcal{U}$ of subsets of X is defined as the union of $U \in \mathcal{U}$ containing x. If $A \subset X$, then $st(A,\mathcal{U}) := \bigcup_{x\in A} st(x,\mathcal{U})$. Given two families $\mathcal{U}, \mathcal{V}$ of subsets of X, $st(\mathcal{U},\mathcal{V})$ is defined as the family $st(A,\mathcal{V})$, $A \in \mathcal{U}$.

DEFINITION 2.1. A **large scale space** is a set X equipped with a family $\mathbb{LSS}$ of covers (called **uniformly bounded** covers) satisfying the following two conditions:

1. $st(\mathcal{U},\mathcal{V}) \in \mathbb{LSS}$ if $\mathcal{U},\mathcal{V} \in \mathbb{LSS}$.
2. If $\mathcal{U} \in \mathbb{LSS}$ and every element of $\mathcal{V}$ is contained in some element of $\mathcal{U}$, then $\mathcal{V} \in \mathbb{LSS}$.

Sets which are contained in an element of $\mathcal{U} \in \mathbb{LSS}$ are called **bounded.**

In this paper we consider only large scale spaces that have the **lowest form of coarse connectivity**. Namely, the union of two bounded subsets of X is always bounded.

DEFINITION 2.2. The subsets A and C of a large scale space X are **coarsely disjoint** if for every uniformly bounded cover $\mathcal{U}$ of X the set $st(A,\mathcal{U}) \cap st(C,\mathcal{U})$ is bounded. A is **coarsely clopen** if A and A^c are coarsely disjoint.

A **non-trivial coarsely clopen** subset A of a large scale space X (an NCC-set for short) is one that is not bounded and A^c is not bounded.

LEMMA 2.3. $st(A_1 \cap A_2,\mathcal{U}) \cap st((A_1 \cap A_2)^c,\mathcal{U}) \subset st(A_1,\mathcal{U}) \cap st((A_1)^c,\mathcal{U}) \cup st(A_2,\mathcal{U}) \cap st((A_2)^c,\mathcal{U})$.

PROOF. Suppose $x \in st(A_1 \cap A_2,\mathcal{U}) \cap st((A_1 \cap A_2)^c,\mathcal{U})$. There is $y \in A_1 \cap A_2$ satisfying $x \in st(y,\mathcal{U})$ and there is $z \in A_1^c \cup A_2^c$ satisfying $x \in st(z,\mathcal{U})$. Thus either $x \in st((A_1)^c,\mathcal{U})$ or $x \in st((A_2)^c,\mathcal{U})$ and we are done. □

COROLLARY 2.4. *The intersection of two coarsely clopen subsets of X is coarsely clopen.*

PROOF. Apply 2.3. Notice that it is exactly here we need the union of two bounded subsets of X be always bounded. □

DEFINITION 2.5. A topology on X is **compatible** with the large scale structure on X if there is a uniformly bounded cover of X consisting of open subsets of X.

REMARK 2.6. The simplest non-trivial topology compatible with a large scale structure is the discrete topology.

DEFINITION 2.7. A **topological large scale space** is a set equipped with large scale structure and with a compatible topology. Additionally, we assume that the coarse structure is **coarsely connected**, i.e. the union of two bounded subsets of X is bounded.

EXAMPLE 2.8. Every metric space (X, d) has a natural topological large scale structure $\mathbb{LSS}_d = \{\mathcal{U}_r | r > 0\}$, where $\mathcal{U}_r$ is the family of all subsets of X whose diameter is at most r. Notice that for any subset A of X and any uniformly bounded cover $\mathcal{U}_r$, one has $st(A, \mathcal{U}_r) = B(A, r)$; in particular, a subset A of X is coarsely clopen if one of the following conditions holds:

1. For all $r > 0$, the subset $B(A, r) \cap A^c$ is bounded,
2. For all $r > 0$, the subset $A \cap B(A^c, r)$ is bounded,
3. For all $r > 0$, there is a bounded subset K_r of X such that $B(A \setminus K_r, r) \cap B(A^c, r)$ is empty,
4. For all $r > 0$, there is a bounded subset K_r of X such that $B(A, r) \cap B(A^c \setminus K_r, r)$ is empty.

LEMMA 2.9. *If A is a (non-trivial) coarsely clopen subset of X, then $st(A, \mathcal{U})$ is a (non-trivial) coarsely clopen subset of X for each uniformly bounded cover $\mathcal{U}$ of X.*

PROOF. Observe that if A is a non-trivial coarsely clopen subset of X, then $C := (st(A, \mathcal{U}))^c$ cannot be bounded. Indeed, in that case $A^c \subset st(st(C, \mathcal{U}), \mathcal{U})$ would be bounded.

Notice $st(st(A, \mathcal{U}), \mathcal{V}) \subset st(A, st(\mathcal{U}, \mathcal{V}))$ for any two covers $\mathcal{U}, \mathcal{V}$. Therefore $st(st(A, \mathcal{U}), \mathcal{V}) \cap st(st(A^c, \mathcal{U}), \mathcal{V}) \subset st(A, st(\mathcal{U}, \mathcal{V})) \cap st(A^c, st(\mathcal{U}, \mathcal{V}))$.
Since $st(A, \mathcal{U})^c \subset A^c \subset st(A^c, \mathcal{U})$ the proof is completed. $\square$

LEMMA 2.10. *If A is a coarsely clopen subset of X, then a subset C of A is coarsely clopen provided $A \subset st(C, \mathcal{V})$ for some uniformly bounded cover $\mathcal{V}$ of X.*

PROOF. Observe $C' := (st(st(A^c, \mathcal{V}), \mathcal{V}))^c \subset C$ is coarsely clopen by 2.9 and $B := C \setminus C' \subset A \cap st(st(A^c, \mathcal{V}), \mathcal{V}) \subset st(st(A, \mathcal{V}) \cap st(A^c, \mathcal{V}), \mathcal{V})$ is bounded as $st(A, \mathcal{V}) \cap st(A^c, \mathcal{V})$ is bounded. Adding a bounded set B to a coarsely clopen subset preserves being coarsely clopen as can be easily seen. $\square$

DEFINITION 2.11. An **end** of a large scale space X is a family E of unbounded and coarsely clopen subsets of X that is maximal with respect to the property of all finite intersections being unbounded. The set of all ends of X is denoted by $Ends(X)$. Equivalently, E is an end of the large scale space X if it is an ultrafilter of the quotient Boolean algebra of $\mathcal{CC}(X)$ by the ideal of all bounded subsets, where $\mathcal{CC}(X)$ is the family of all coarsely clopen subsets of X.

PROPOSITION 2.12. *If X is a topological large scale space and A belongs to an end E of X, then there is an open $V \in E$ contained in A. Consequently, two ends E and E' are equal if and only if $\{U \in E | U \text{ is open}\} = \{V \in E' | V \text{ is open}\}$.*

PROOF. Let $\mathcal{U}$ be a uniformly bounded cover of X consisting of open subsets. Notice $cl(B) \subset st(B, \mathcal{U})$ is bounded for every bounded subset of X.

Consider $B := st(A^c, \mathcal{U}) \cap st(A, \mathcal{U})$ and observe $V := A \setminus cl(B)$ is open. Indeed, if $x \in V$, then $st(x, \mathcal{U}) \setminus cl(B) \subset V$ as otherwise there is $y \in (st(x, \mathcal{U}) \setminus cl(B)) \setminus A$ resulting in $x \in st(A^c, \mathcal{U}) \cap A \subset B$, a contradiction.

Suppose $\{U \in E | U \text{ is open}\} = \{V \in E' | V \text{ is open}\}$ and $C \in E' \setminus E$. There is $A \in E$ such that $C \cap A$ is bounded. Choose $V \subset C$, $V \in E'$ and open. Now, $V \in E$ contradicting $A \cap V$ being bounded. $\square$

REMARK 2.13. Let X be a large scale space. If U is a coarsely clopen subset of X, we define $U_{end} := \{E \in Ends(X) | U \in E\}$ and $\widetilde{U}_{end} := U_{end} \cup U$.

PROPOSITION 2.14. *Let $\mathcal{T}$ be the topology of a topological large scale space X. The collection $\widetilde{\mathcal{B}} = \{\widetilde{U}_{end} | U$ open coarsely clopen subset of $X\}$ is a basis for a topology $\mathcal{T}_{end}$ on $X \cup Ends(X)$ that extends the topology $\mathcal{T}$.*

PROOF. Clearly $\widetilde{\mathcal{B}}$ covers $X \cup Ends(X)$ as either X is unbounded open coarsely clopen and $\widetilde{X}_{end} = X \cup Ends(X)$ or X is bounded and $Ends(X) = \emptyset$. Now, let $\widetilde{U}_{end}, \widetilde{V}_{end} \in \widetilde{\mathcal{B}}$. Notice $\widetilde{U}_{end} \cap \widetilde{V}_{end} = \widetilde{(U \cap V)}_{end} \in \widetilde{\mathcal{B}}$. Finally, every bounded element of $\mathcal{T}$ is contained in $\widetilde{\mathcal{B}}$ which implies that $\mathcal{T} \subset \mathcal{T}_{end}$. □

COROLLARY 2.15. *Let $\mathcal{T}$ be the topology of a topological large scale space X. The collection $\mathcal{B}_{\mathcal{T}} = \{U_{end} | U$ open coarsely clopen subset of $X\}$ is a basis for the topology of $Ends(X)$ as a subspace of $X \cup Ends(X)$.*

COROLLARY 2.16. *Given two compatible topologies on a large scale space X, the induced topologies on $Ends(X)$ coincide.*

PROOF. Use 2.12. □

DEFINITION 2.17. Let X be a large scale space. Two subsets $A, C \subset X$ are said to be **coarsely identical** if their symmetric difference $A \Delta C := (A \setminus C) \cup (C \setminus A)$ is a bounded subset of X.

EXAMPLE 2.18. Let A be a coarsely clopen subset of a large scale space X, and U be a uniformly bounded cover of X. Then A and $st(A, \mathcal{U})$ are coarsely identical.

LEMMA 2.19. *Let $E \in Ends(X)$ and $U \in E$, then $int(U), cl(U) \in E$.*

PROOF. Use 2.12 to see that $int(U) \in E$. It suffices to show that $int(U)$ and $cl(U)$ are coarsely identical. Let $\mathcal{U}$ be a uniformly bounded open cover of X. Notice that $cl(U) \setminus int(U) \subseteq st(U, \mathcal{U}) \cap st(U^c, \mathcal{U})$, and that $st(U, \mathcal{U}) \cap st(U^c, \mathcal{U})$ is bounded as U is coarsely clopen. □

PROPOSITION 2.20. *Let X be a topological large scale space. If X is Hausdorff, then the topological space $X \cup Ends(X)$ is Hausdorff.*

PROOF. We consider two cases:

CASE 1. Let $x \in X$ and $E \in Ends(X)$ and let $\mathcal{U}$ be a uniformly bounded open cover of X. Choose $U \in \mathcal{U}$ such that $x \in U$. Notice that $cl(U)$ is a bounded subset and hence $V := X \setminus cl(U)$ is open and coarsely clopen contained in E. Hence, U and $\widetilde{V}_{end}$ are disjoint open neighborhood of x and E, respectively.

CASE 2. If $E_1, E_2 \in Ends(X)$ are two distinct ends of X, then we can find $A \in E_1$ and $B \in E_2$ such that $A \cap B$ is a bounded subset of X. Without loss of generality, we may assume that $A \cap B = \emptyset$. Therefore, $U = int(A) \in E_1$ and $V = int(B) \in E_2$ are disjoint and hence $\widetilde{U}_{end}$ and $\widetilde{V}_{end}$ are disjoint neighborhoods of E_1 and E_2, respectively. □

X is an open dense subspace of $X \cup Ends(X)$. Moreover, $(X \cup Ends(X), \mathcal{T}_{end})$ is Hausdorff whence $(X, \mathcal{T})$ is Hausdorff.
The question then arises: is $(X \cup Ends(X), \mathcal{T}_{end})$ a compactification of $(X, \mathcal{T})$? The answer is positive in a coarse sense!

DEFINITION 2.21. Let Y be a topological space and $X \subseteq Y$ is a topological subspace equipped with a large scale structure compatible with its topology. Y is **large scale compact** if for any open cover $\{U_s\}_{s\in S}$ of Y, there is a finite subset F of S such that $Y \setminus \bigcup_{s\in F} U_s$ is a bounded subset of X. Y is a **large scale compactification** of X if in addition to being large scale compact, Y is Hausdorff and X is an open dense subspace of Y(see [**11**]).

LEMMA 2.22. *For any family $\{U^s\}_{s\in I}$ of coarsely clopen subsets of X such that $Ends(X) \subset \bigcup_{s\in I} U^s_{end}$, there is a finite subset F of I such that $X \setminus \bigcup_{s\in F} U^s$ is a bounded subset of X.*

PROOF. Let $\mathcal{F}$ be the collection of all finite subsets of I. Seeking contradiction assume that for any $F \in \mathcal{F}$, $A_F = X\setminus \bigcup_{s\in F} U^s$ is unbounded. The collection $\{A_F | F \in \mathcal{F}\}$ is contained in some end $E \in Ends(X)$. Hence, $\{A_F | F \in \mathcal{F}\} \subset E \in U^s_{end}$ for some $s \in I$ which implies that $U^s \in E$ and $X \setminus U^s \in E$, a contradiction. □

THEOREM 2.23. *Let X be a Hausdorff topological large scale space, the topological space $X \cup Ends(X)$ is a large scale compactification of X. Furthermore, $Ends(X)$ is a compact totally disconnected subspace of $X \cup Ends(X)$.*

PROOF. Let $\{O^s\}_{s\in I}$ be an open cover $X \cup Ends(X)$, we can find a subset $J \subset I$ such that $\{\widetilde{U^s}_{end}\}_{s\in J}$ of $\{O^s\}_{s\in I}$ that covers $Ends(X)$. In particular, $Ends(X) \subset \bigcup_{s\in J} U^s_{end}$. By the above lemma, there is a finite subset F of J such that $X \setminus \bigcup_{s\in F} U^s$ is a bounded subset of X. We claim that $Ends(X) \subset \bigcup_{s\in F} U^s_{end}$. To this end, notice that if $U \subset X$ is an arbitrary coarsely clopen subset of X and $V = X \setminus U$, then $Ends(X) = U_{end} \cup V_{end}$. Indeed, if $E \in Ends(X) \setminus (U_{end} \cup V_{end})$, then we can find $A, B \in E$ such that both $A \cap U$ and $B \cap V$ are bounded. Hence, there exist K, L bounded subsets of X such that $A \subset V \cup K$ and $B \subset U \cup L$. In particular, $A \cap B$ is bounded, a contradiction. Now, seeking contradiction assume that $E \in Ends(X) \setminus \bigcup_{s\in F} U^s_{end}$, then $E \notin U^s_{end}$ for all $s \in F$. Therefore, by above observation, $X\setminus U^s \in E$ for all $s \in F$ which implies that $\bigcap_{s\in F} (X\setminus U^s) = X\setminus \bigcup_{s\in F} U^s \in E$, a contradiction. This shows that $Ends(X)$ is compact and $X \cup Ends(X)$ is large scale compact as $(X \cup Ends(X)) \setminus \bigcup_{s\in F} \widetilde{U^s}_{end}$ is bounded.

Finally, we show that $Ends(X)$ is totally disconnected. Let $E_1, E_2 \in Ends(X)$ be two distinct ends of X, then we can find $A \in E_1$ and $B \in E_2$ such that $cl(A) \cap B$ is a bounded subset of X. Without loss of generality, we may assume that $cl(A) \cap B = \emptyset$. Clearly, $U = int(A) \in E_1$ and $V = X \setminus cl(A) \in E_2$. Moreover, $Ends(X) = U_{end} \cup V_{end}$ and $U_{end} \cap V_{end} = \emptyset$. □

Alternatively, one can show that the space of ends $Ends(X)$ is compact totally disconnected Hausdorff space by noticing that $Ends(X)$ is actually the Stone space (space of ultrafilters) of the quotient Boolean algebra of $\mathcal{CC}(X)$ by the ideal of all bounded subsets. Let us show that $Ends(X)$ is a coarse invariant of a space.

DEFINITION 2.24. Let $\varphi, \varphi' : X \to Y$ be maps between large scale spaces.

- φ and φ' are **close** if there is a uniformly bounded cover $\mathcal{U}$ of Y such that $\varphi(x) \in st(\varphi'(x), \mathcal{U})$ for each $x \in X$.

- φ is **coarsely proper** if $\varphi^{-1}(K)$ is bounded for each bounded subset K of Y.
- φ is **large scale continuous** if $\varphi(\mathcal{U})$ is a uniformly bounded cover of $\varphi(X)$ for each uniformly bounded cover $\mathcal{U}$ of X.
- φ is a **coarse equivalence** if there is a coarse, large scale continuous map $\psi : Y \to X$ such that $\psi \circ \varphi$ is close to id_X and $\varphi \circ \psi$ is close to id_Y. In such a case, φ and ψ are **coarse inverses** of each other, so X and Y are **coarsely equivalent**.

LEMMA 2.25. *Suppose $f : X \to Y$ is a coarsely proper and large scale continuous function of topological large scale spaces and $E \in Ends(X)$. Then, using the closure operation cl in $Y \cup Ends(Y)$, the following hold:*

a. *If $G \in \bigcap\limits_{A\in E} cl(f(A))$ is an end of Y and $V \in G$ is an open and coarsely clopen subset of Y, then $f^{-1}(V) \in E$.*
b. *$\bigcap\limits_{A\in E} cl(f(A))$ is a singleton belonging to $Ends(Y)$.*

PROOF. Observe $f^{-1}(D)$ is coarsely clopen in X if D is coarsely clopen in Y. Indeed, if $B := st(f^{-1}(D),\mathcal{U}) \cap st(f^{-1}(D)^c,\mathcal{U})$ is bounded for some uniformly bounded cover $\mathcal{U}$ of X, then $st(D,\mathcal{U}_f) \cap st(D^c,\mathcal{U}_f) \subset f(B)$ is bounded, a contradiction. Here, $\mathcal{U}_f$ is $f(\mathcal{U})$ union singletons outside of $f(X)$.

a. If $f^{-1}(V) \notin E$, then there is $A \in E$ such that $A \cap f^{-1}(V)$ is bounded, hence its image $B := f(A) \cap V$ is bounded. Put $A_1 = A \setminus f^{-1}(B)$ to get $f(A_1) \subset Y \setminus V$ contradicting $E \in cl(f(A_1))$. Thus $f^{-1}(V) \in E$.

b. $\bigcap\limits_{A\in E} cl(f(A))$ cannot be empty and it is contained in $Ends(Y)$. First of all, if $y \in Y$, then there is an open bounded neighborhood V of y in Y, so $f^{-1}(V)$ is bounded in X. Hence $A := X \setminus f^{-1}(V) \in E$, $f(A) \subset Y \setminus V$, and $cl(f(A))$ misses V. Secondly, assume that $\bigcap\limits_{A\in E} cl(f(A)) = \emptyset$. By 2.22, there exist finitely many $A_i \in E$, $i \leq n$, such that $\bigcap\limits_{i=1}^{n} cl(f(A_i))$ is bounded. However, $A := \bigcap\limits_{i=1}^{n} A_i \in E$, and $f(A)$ is bounded, a contradiction. Finally, if $E_1, E_2 \in \bigcap\limits_{A\in E} cl(f(A))$ and $E_1 \neq E_2$, then there exists two disjoint open and coarsely clopen subsets $V_1 \in E_1$ and $V_2 \in E_2$. By a) $f^{-1}(V_1) \in E$ and $f^{-1}(V_2) \in E$, a contradiction. □

COROLLARY 2.26. *If $f : X \to Y$ is a coarsely proper large scale continuous function of topological large scale spaces and $E \in Ends(X)$, then it induces a continuous function $f_{end} : Ends(X) \to Ends(Y)$ defined by $f_{end}(E) \in \bigcap\limits_{A\in E} cl(f(A))$.*

a. *If $f, g : X \to Y$ are close, then $f_{end} = g_{end}$.*
b. *If f is continuous, then $f \cup f_{end} : X \cup Ends(X) \to Y \cup Ends(Y)$ is continuous.*

PROOF. Given an open, coarsely clopen subset V of Y, notice that $f^{-1}_{end}(V_{end}) \subset U_{end}$, where $U = f^{-1}(V)$. Indeed, if $f_{end}(E) \in V_{end}$ then $E \in U_{end}$ by a) of 2.25. That proves continuity of f_{end} and of $f \cup f_{end}$ if f is continuous.

Suppose there is an open uniformly bounded cover $\mathcal{U}$ of Y such that $f(x) \in st(g(x),\mathcal{U})$ for each $x \in X$. If there is $E \in Ends(X)$ such that $f_{end}(E) \neq g_{end}(E)$, then we can choose open subsets $U \in f_{end}(E), V \in g_{end}(E)$ so that $st(U,\mathcal{U}) \cap V = \emptyset$.

Notice $f^{-1}(U) \in E$ and $g^{-1}(V) \in E$ by a) of 2.25. Now, there is $x \in f^{-1}(U) \cap g^{-1}(V)$, so $f(x) \in U$, $g(x) \in V$ contradicting $st(U,\mathcal{U}) \cap V = \emptyset$. □

COROLLARY 2.27. *If two topological large scale spaces X and Y are coarsely equivalent, then $Ends(X)$ is homeomorphic to $Ends(Y)$.*

PROOF. It suffices to show that if $f : X \to Y$ and $g : Y \to Z$ are coarsely proper large scale continuous functions of topological large scale spaces, then $(g \circ f)_{end} = f_{end} \circ g_{end}$. Indeed, let $E \in Ends(X)$ such that $(g \circ f)_{end}(E) \neq f_{end} \circ g_{end}(E)$. There exist open disjoint subsets $U \in (g \circ f)_{end}(E)$ and $V \in f_{end} \circ g_{end}(E)$. Now, $(g\circ f)^{-1}(U) \in E$ and $f^{-1}(V) \in g_{end}(E)$ by a) of 2.25. In particular, $(g\circ f)^{-1}(U), (g\circ f)^{-1}(V) \in E$, a contradiction. □

3. Comparison of coarse ends and Freundenthal ends

In this section we are concentrating on a relationship between Freudenthal ends and coarse ends. More specifically, we are interested in cases where there is a one-to-one correspondence between those ends in the following sense: each Freudenthal end is contained in a unique coarse end and each coarse end contains a Freudenthal end. To accomplish it, we need a large scale analog of local connectedness. Also, we need to generalize Freundenthal ends to non-locally compact spaces. See **[6]** for a theory of Freundenthal compactifications for general topological spaces.

DEFINITION 3.1. Suppose X is a locally connected topological space that is a union of an increasing sequence of closed subspaces $\{K_i\}_{i\geq 1}$ such that $K_i \subset int(K_{i+1})$ for each $i \geq 1$. Let $\mathcal{K}$ be the bornology generated by $\{K_i\}_{i\geq 1}$, i.e. all subsets B of X such that $B \subset K_i$ for some $i \geq 1$. A **Freundenthal end** of X with respect to $\mathcal{K}$ is a decreasing sequence $\{C_i\}_{i\geq 1}$, where each C_i is a non-empty component of $X \setminus K_i$.

Notice each C_i is unbounded as otherwise it is contained in some K_k, $k > i$, and $C_k \subset C_i \cap (X \setminus K_k) = \emptyset$, a contradiction.

DEFINITION 3.2. A large scale space X is **large scale chain-connected** if there is a uniformly bounded cover $\mathcal{U}$ of X such that every uniformly bounded cover $\mathcal{V}$ of X is a refinement of a uniformly bounded cover $\mathcal{W}$ that consists of $\mathcal{U}$-connected sets (that means any two points $x, y \in W \in \mathcal{W}$ can be connected by a chain of points $x_1 = x, \ldots, x_n = y$ with the property that for any $i < n$ there is $U \in \mathcal{U}$ containing both x_i and x_{i+1}).

DEFINITION 3.3. Suppose X is a large scale space that is large scale chain-connected via a uniformly bounded cover $\mathcal{U}$ and the bornology $\mathcal{B}$ of X has an increasing sequence $\{B_i\}_{i\geq 1}$ of bounded subsets of X that serves as a basis of it. A **Freundenthal end** of X with respect to $\mathcal{B}$ is a decreasing sequence $\{C_i\}_{i\geq 1}$, where each C_i is a non-empty $\mathcal{U}$-component of $X \setminus B_i$.

THEOREM 3.4. *Suppose X is a large scale space that is large scale chain-connected via a uniformly bounded cover $\mathcal{U}$ and has the property that for any bounded subset B of X the union of all bounded $\mathcal{U}$-components of $X \setminus B$ is bounded and there are only finitely many unbounded $\mathcal{U}$-components of $X \setminus B$. If there is an increasing sequence $\{B_i\}_{i\geq 1}$ of bounded subsets of X that is a basis for all bounded subsets of X, then the coarse ends of X are in one-to-one correspondence with Freundenthal ends of X with respect to its bornology.*

PROOF.

CLAIM 1. If C is a union of $\mathcal{U}$-components of $X \setminus B$ for some bounded subset B of X, then C is coarsely clopen.

PROOF OF CLAIM 1. Given a uniformly bounded cover $\mathcal{V}$ of X consisting of $\mathcal{U}$-connected subsets of X, the set $st(C, \mathcal{V}) \cap st(C^c, \mathcal{V})$ is contained in $st(B, \mathcal{V})$. Indeed, if $x \notin B$ belongs to $st(C, \mathcal{V}) \cap st(C^c, \mathcal{V})$, then there is $V \in \mathcal{V}$ containing x and intersecting both C and C^c. Therefore $V \setminus B$ must be contained in either C or C^c, so $V \cap B \neq \emptyset$ and $x \in st(B, \mathcal{V})$. □

CLAIM 2. If C is coarsely clopen, then every union of $\mathcal{U}$-components of C is coarsely clopen.

PROOF OF CLAIM 2. $B := st(C,\mathcal{U}) \cap st(C^c,\mathcal{U})$ is bounded and $C \setminus B$ is a union of $\mathcal{U}$-components of $X \setminus B$. By Claim 1, $C \setminus B$ is coarsely clopen, so so is C. □

PROOF OF 3.4. Consider a Freundenthal end $\{C_i\}_{i\geq 1}$ of X with respect to its bornology. $\{C_i\}_{i\geq 1}$ is contained in some coarse end of X by Claim 1. $\{C_i\}_{i\geq 1}$ cannot be contained in two different coarse ends E_1, E_2. Indeed, choose $A_1 \in E_1$, $A_2 \in E_2$ whose stars with respect to $\mathcal{U}$ are disjoint. There is $k > 1$ such that $st(A_1,\mathcal{U}) \cap st(A_1^c,\mathcal{U}) \subset B_k$. Notice that the union D_1 of $\mathcal{U}$-components of $X \setminus B_k$ containing points from $C_k \cap A_1$ does not intersect $C_k \cap A_1^c$. That means C_k can be expressed as the union of D_1 and $C_k \setminus D_1$, both unbounded unions of $\mathcal{U}$-components of $X \setminus B_k$. There is $j > k$ such that either $C_j \subset D_1$ or $C_j \subset C_k \setminus D_1$. In the first case $C_j \notin E_2$ and in the second case $C_j \notin E_1$, a contradiction.

Given a coarse end E of X and given $A \in E$ there is $i \geq 1$ such that $st(A,\mathcal{U}) \cap st(A^c,\mathcal{U}) \subset B_i$. Consider $j > i$ satisfying $st(B_i,\mathcal{U}) \subset B_j$. Notice $A \setminus B_j$ is a union of $\mathcal{U}$-components of $X \setminus B_j$. Therefore there is an unbounded U-component of $X \setminus B_j$ belonging to E. That means, starting from some $k > i$, E contains exactly one unbounded component C_n of $X \setminus B_n$ for all $n \geq k$. Those components can be easily extended to a Freundenthal end of X contained in E. □

□

X admits an **exhaustion by compact sets** if there exists a sequence of compact subsets $(K_n)_{n\in\mathbb{N}}$ that covers X such that $K_n \subset int(K_{n+1})$ for all $n \in \mathbb{N}$. The sequence of compact subsets $(K_n)_{n\in\mathbb{N}}$ is called an **exhausting sequence**. X is **σ-compact** if it is a countable union of compact subsets of X. It is known that a locally compact σ-compact space admits an exhaustion by compact sets.

LEMMA 3.5. *Let X be a Freudenthal space. X admits an exhausting sequence $(K_n)_{n\in\mathbb{N}}$ of compact subsets such that both K_n and $int(K_n)$ are connected.*

PROOF. Since X is σ-compact, it can be written as a countable union of compact subsets, say $X := \bigcup_{n\in\mathbb{N}} L_n$. As X is locally compact and locally connected, each L_n can be covered by finitely many open connected and relatively compact subsets. Therefore X is a countable union of open connected relatively compact subsets, say $X := \bigcup_{n\in\mathbb{N}} O_n$. Now, we inductively construct an increasing sequence $(U_n)_{n\in\mathbb{N}\cup\{0\}}$ of open connected relatively compact subsets satisfying $cl_X(U_n) \subset U_{n+1}$ for all $n \in \mathbb{N}\cup\{0\}$, and $X := \bigcup_{n\in\mathbb{N}} U_n$. Put $U_0 := \emptyset$, and assume that $U_1, \ldots, U_n$ are open connected relatively compact non-empty subsets such that $cl_X(U_i) \subset U_{i+1}$ for all $1 \leq i < n$. Since $cl_X(U_n)$ is compact, there exist $O_{n_1}, \ldots, O_{n_m}$ open connected relatively compact subsets such that $cl_X(U_n) \subset \bigcup_{i=1}^{m} O_{n_i}$ and $cl_X(U_n) \cap O_{n_i} \neq \emptyset$ for all $1 \leq i \leq m$. Put $U_{n+1} := \bigcup_{i=1}^{m} O_{n_i}$; since X is connected, $cl_X(U_n)$ is properly contained in U_{n+1}. Moreover, U_{n+1} is open connected relatively compact subset. Finally, since $cl_X(\bigcup_{n\in\mathbb{N}} U_n) = \bigcup_{n\in\mathbb{N}} cl_X(U_n) = \bigcup_{n\in\mathbb{N}} U_n$, and X is connected, so $X = \bigcup_{n\in\mathbb{N}} U_n$.

Now, put $K_n := cl_X(U_n)$ for all $n \in \mathbb{N}$. Notice that for any open subset $U \subset X$, one has $U \subset int(cl_X(U))$. Notice that:

$$U_n \subset int(cl_X(U_n)) = int(K_n) \subset K_n = cl_X(U_n) \subset U_{n+1} \subset int(cl_X(U_{n+1})) = int(K_{n+1}),$$

and therefore, $int(K_n)$ is connected for all $n \in \mathbb{N}$, and $(K_n)_{n\in\mathbb{N}}$ is an exhausting sequence consisting of compact connected subsets. □

THEOREM 3.6. *Suppose X is a connected, locally compact, σ-compact space. There is a large scale structure on X such that coarse ends of X are in one-to-one correspondence with Freundenthal ends of X.*

PROOF. By 3.5, let $(K_n)_{n\in\mathbb{N}}$ be an exhausting sequence of compact subsets such that both K_n and $int(K_n)$ are connected. Consider all open covers $\mathcal{V}$ of X consisting of open connected subsets of X with the property that $st(K, \mathcal{V})$ is pre-compact for each compact subset K of X. The large scale structure on X consists of all refinements of such covers.

The basic cover $\mathcal{U}$ consists of components of sets $int(K_{i+2}) \setminus K_i$ for $i \geq 0$, where we put $K_0 = \emptyset$. That implies the cover of X consisting of singletons is uniformly bounded as needed.

Given two covers $\mathcal{V}, \mathcal{W}$ of X consisting of open connected subsets of X with the property that $st(K, \mathcal{V})$ is pre-compact for each compact subset K of X, elements of the cover $st(\mathcal{V}, \mathcal{W})$ are $\mathcal{W}$-connected, so the large scale is large scale chain-connected.

Given a bounded subset B of X, cover $cl(B)$ by finitely many elements $U_1, \ldots, U_k$ of $\mathcal{U}$. Any $\mathcal{U}$-chain joining $x \in X \setminus B$ to $b \in B$ must intersect one of U_i. Therefore $X \setminus B$ has only finitely many $st(\mathcal{U}, \mathcal{U})$-components. Apply 3.4. □

4. Large scale groups

In this section we introduce the concept of a large scale group that generalizes the following classes of groups:

1. Finitely generated groups with word metrics,
2. Countable groups with proper left-invariant metrics,
3. Locally compact topological groups.

Ideally, the name of those groups should be coarse groups but it has been already used in literature for similar but different objects (see [**16**]).

See [**4**] for a discussion of large scale structures on a group induced by right-invariant metrics versus left-invariant metrics.

Recall that a **bornology** on a set X is a cover of X that is stable under inclusion and is stable under finite unions.

DEFINITION 4.1. A **large scale group** is a group G equipped with a large scale structure $\mathbb{LSS}$ induced by a bornology $\mathcal{B}$. That means every uniformly bounded cover in $\mathbb{LSS}$ is a refinement of $\{g \cdot B\}_{g\in G}$ for some $B \in \mathcal{B}$.

LEMMA 4.2. *A bornology $\mathcal{B}$ on a group G induces a large scale structure on G if and only if $\mathcal{B}$ is stable under inverses and products.*

PROOF. Notice $\mathcal{B}$ is stable under inverses and products if and only if for all non-empty $B_1, B_2 \in \mathcal{B}$ the set $(B_1 \cdot B_2^{-1}) \cdot B_2$ belongs to $\mathcal{B}$.

Let $\mathcal{U} := \{g \cdot B_2\}_{g\in G}$. Notice $(B_1 \cdot B_2^{-1}) \cdot B_2 = \bigcup\limits_{g\in G} B_1 \cap (g \cdot B_2) = st(B_1,\mathcal{U})$. Conversely, if $B_3 := \bigcup\limits_{g\in G} B_1 \cap (g \cdot B_2) \in \mathcal{B}$, then $st(h \cdot B_1,\mathcal{U}) \subset h \cdot B_3$ for all $h \in G$. □

COROLLARY 4.3. *Large scale groups include the following classes of groups:*

1. *Finitely generated groups with word metrics,*
2. *Countable groups with proper left-invariant metrics,*
3. *Locally compact topological groups.*

PROOF. In cases 1) and 2) the bornologies consist of all finite subsets of G.

In case 3) the bornology $\mathcal{B}$ consist of all subsets of compact sets in G. Indeed, given $B_1, B_2 \in \mathcal{B}$ the set $(B_1 \cdot B_2^{-1}) \cdot B_2$ is pre-compact (its closure is compact). □

LEMMA 4.4. *Given a subgroup H of a group G the restriction of a cover $\{g \cdot B\}_{g\in G}$ to H is a refinement of the cover $\{h \cdot ((B^{-1} \cdot B) \cap H)\}_{h\in H}$ of H.*

PROOF. Since B is non-empty, so is $B^{-1} \cdot B$. For each $g \in G$ such that $H \cap (g \cdot B) \neq \emptyset$ pick $h_g \in H \cap (g \cdot B)$. If $h \in (g \cdot B) \cap H$, then $b := g^{-1} \cdot h \in B$ and $b_g := g^{-1} \cdot h_g \in B$, so $h = g \cdot b = h_g \cdot b_g^{-1} \cdot b \in h_g \cdot ((B^{-1} \cdot B) \cap H)$. □

COROLLARY 4.5. *Given a subgroup H of a large scale group G the induced large scale on H equals the large scale generated by the restriction of the bornology of G to H.*

THEOREM 4.6. *A large scale group G is of asymptotic dimension 0 if and only if for every bounded subset B of G the subgroup $< B >$ of G generated by B is bounded.*

PROOF. See [**2**] for a discussion of asymptotic dimension 0. Given a bounded subset B of G consider a set of elements $\{g_i\}_{i\in J}$ representing all cosets $g\cdot < B >$. Notice $\{g_i\cdot < B >\}_{i\in J}$ is a uniformly bounded of G consisting of mutually disjoint sets for which $\{g \cdot B\}_{g\in G}$ is a refinement.

Suppose G is of asymptotic dimension 0 and $B \subset G$ is bounded and symmetric. Choose a uniformly bounded cover $\{U_i\}_{i\in J}$ consisting of mutually disjoint sets for which $\{g \cdot B\}_{g\in G}$ is a refinement. Let $1_G \in U_0$. If $h \in U_0$, then $h \cdot B \subset U_0$ as otherwise $h \cdot B \subset U_i$ for some $i \neq 0$ and $h \in U_0 \cap U_i$, a contradiction. Consequently, any finite product of elements of B belongs to U_0. Hence $< B > \subset U_0$ is bounded. □

DEFINITION 4.7. A subgroup H of a large scale group G is of **bounded index** in G if there is a bounded subset B of G such that $B \cdot H = G$.

PROPOSITION 4.8. *A subgroup H of a large scale group G is of bounded index in G if and only if the inclusion $H \to G$ is a coarse equivalence.*

PROOF. If $B \cdot H = G$, then $H \cdot B^{-1} = G$, so the star of H with respect to $\{g \cdot B^{-1}\}_{g \in G}$ equals G and the inclusion $H \to G$ is a coarse equivalence.

Conversely, if the star of H with respect to $\{g \cdot B\}_{g \in G}$ equals G, then for each $g \in G$ there is $f \in G$ such that $g \in f \cdot B$ and there is $h \in H \cap (f \cdot B)$. Since $f^{-1} \cdot g \in B$ and $f^{-1} \cdot h \in B$, so $h^{-1} \cdot g \in B \cdot B^{-1}$ and $g \in H \cdot (B \cdot B^{-1})$. Thus $G = (B \cdot B^{-1}) \cdot H$ and H is of bounded index in G. □

5. Connectivity in large scale groups

In this section we introduce concepts needed to generalize being finitely generated to being boundedly generated.

DEFINITION 5.1. Suppose K is a symmetric subset of a group G (that means $K^{-1} = K$). A **K-chain** is a finite sequence $g_1, \ldots, g_k$ of elements of G such that $g_i^{-1} \cdot g_{i+1} \in K$ for each $i < k$. A subset C of G is **K-connected** if every two elements of C can be connected by a K-chain. C is a **K-component** of $A \subset G$ if C is an equivalence class of the equivalence relation $\sim$ on A defined as follows: $g \sim h$ if g and h can be connected by a K-chain in A. We always assume $1_G \in K$ as that does not change connectivity.

If G is K-connected, then the **K-norm** on G of $g \in G$ is the length of the shortest K-chain joining 1_G and $g \in G$.

DEFINITION 5.2. A large scale group G is **boundedly generated** if there is a symmetric bounded set K such that every element g of G is a finite product of elements of K. Equivalently, G is K-connected. In this case we say G is **K-generated**.

PROPOSITION 5.3. *Suppose K is a symmetric bounded subset of a group large scale G. If, for some bounded subset B of G, $G \setminus B$ has finitely many unbounded K-components and the union of all bounded K-components is bounded, then G is boundedly generated.*

PROOF. Let L be the union of $K \cup B$ and of the following:

1. The union of all bounded K-components of $G \setminus B$,
2. One point from each non-empty unbounded K-component of $G \setminus B$.

Put $M = L \cup L^{-1}$ and notice G is M-connected. Indeed, the M-component of 1_G contains B and all K-components of $G \setminus B$. □

EXAMPLE 5.4. Consider the topological group of the rational numbers $(\mathbb{Q}, +)$ with the bornology $\mathcal{B}$ consisting of all pre-compact subsets of G. $\mathbb{Q}$ is boundedly generated by the compact $K := \{\frac{1}{n} : n \in \mathbb{N}\} \cup \{0\}$.

EXAMPLE 5.5. Let $M_n(\mathbb{R})$ be the set of $n \times n$ matrices over $\mathbb{R}$. Topologically, $M_n(\mathbb{R})$ is homeomorphic to $\mathbb{R}^{n^2}$. The special linear group:

$$SL_2(\mathbb{R}) := \{A \in M_2(\mathbb{R}) : \det(A) = 1\},$$

is a topological subgroup of $M_2(\mathbb{R})$ and it is compactly generated by:

$$K := \left\{ \begin{pmatrix} 1 & r \\ 0 & 1 \end{pmatrix} : r \in \mathbb{R}, |r| \leq 1 \right\} \cup \left\{ \begin{pmatrix} 0 & 1 \\ -1 & 0 \end{pmatrix} \right\}.$$

Therefore, $SL_2(\mathbb{R})$, with the bornology $\mathcal{B}$ consisting of all pre-compact subsets of G, is boundedly generated.

6. Svarc-Milnor Lemma for large scale groups

Geometric group theorists traditionally restrict their attention to finitely generated groups equipped with a word metric. A typical proof of Švarc-Milnor Lemma (see [**24**] or [**1**], p.140) involves such metrics.

THEOREM 6.1. *A group G acting properly and cocompactly via isometries on a length space X is finitely generated and induces a quasi-isometry equivalence $g \to g \cdot x_0$ for any $x_0 \in X$.*

THEOREM 6.2 ([**3**]). *If a group G acts cocompactly and properly via isometries on a proper metric space X, then $g \to g \cdot x_0$ induces a coarse equivalence between G and X for all $x_0 \in X$.*

DEFINITION 6.3. A group G acts on a large scale space X by **uniform coarse equivalences** if for every uniformly bounded cover $\{U_s\}_{s\in S}$ of X the cover $\{g \cdot U_s\}_{s\in S, g\in G}$ is uniformly bounded.

The action is **cobounded** if there is a bounded subset B of X so that $G \cdot B = X$.

If G is a large scale group, then the action is **proper** if for every bounded subset B of X the set $\{g \in G | (g \cdot B) \cap B \neq \emptyset\}$ is bounded in G and for every bounded subset K of G the set $K \cdot x$ is bounded in X for each $x \in X$.

THEOREM 6.4. *Suppose a large scale group G acts by uniform coarse equivalences on a large scale space X. If the action is proper and cobounded, then for each $x_0 \in X$ the map $g \to g \cdot x_0$ is a coarse equivalence from G to X.*

PROOF. The map $g \to g \cdot x_0$ is large scale continuous as for each bounded subset K of G the family $\{g \cdot K\}_{g\in G}$ is sent to the family $\{g \cdot (K \cdot x_0)\}_{g\in G}$ which is uniformly bounded as $K \cdot x_0$ is bounded in X.

Given a uniformly bounded cover $\mathcal{U} = \{U_s\}_{s\in S}$ of X let $B := st(x_0, g \cdot \mathcal{U})$. Notice B is bounded. The set $K := \{g \in G | (g \cdot B) \cap B \neq \emptyset\}$ is bounded in G. Now, given $s \in S$ such that $g_s \cdot x_0 \in U_s$ for some $g_s \in G$, then for any $g \in G$ so that $g \cdot x_0 \in U_s$ one has $x_0 \in g^{-1} \cdot U_s$. Therefore $g^{-1} \cdot U_s \subset B$ resulting in $x_0 \in ((g_s^{-1} \cdot g) \cdot B) \cap B$. Hence $g \in g_s \cdot K$ and the inverse of the cover U under the map $g \to g \cdot x_0$ is a refinement of the cover $\{g \cdot K\}_{g\in G}$. Thus the map is a coarse embedding.

If $G \cdot B = X$ for some bounded B in X, then $st(G \cdot x_0, \{g \cdot B\}_{g\in G}) = X$, so the inclusion $G \cdot x_0 \to X$ is a coarse equivalence. □

7. Large scale groups of bounded geometry

In this section we introduce the concept of bounded geometry for large scale groups. Typically, bounded geometry is defined for discrete metric spaces by requiring that for each $r > 0$ there is $N_r \in \mathbb{N}$ such that every r-ball contains at most N_r elements. We want a coarse invariant, so we extend this definition to arbitrary large scale spaces as follows:

DEFINITION 7.1. A large scale space X has **bounded geometry** if it is coarsely equivalent to a large scale space Y with the property that for each uniformly bounded cover $\mathcal{U}$ of Y there is $N(\mathcal{U}) \in \mathbb{N}$ such that every element of $\mathcal{U}$ contains at most $N(\mathcal{U})$ elements.

COROLLARY 7.2. *Suppose X and Y are large scale spaces and Y has bounded geometry. If there is a coarse embedding $f : X \to Y$, then X has bounded geometry.*

PROOF. Pick a coarse equivalence $g : Y \to Z$, where Z has the property that for each uniformly bounded cover $\mathcal{U}$ of Z there is $N(\mathcal{U}) \in \mathbb{N}$ such that every element of $\mathcal{U}$ contains at most $N(\mathcal{U})$ elements. Observe that X is coarsely equivalent to $g(f(X))$. □

PROPOSITION 7.3. *A large scale group G is of bounded geometry if and only if there is a bounded set K such that for every bounded subset B of G there are elements $g_i \in G$, $i \leq k$, so that $B \subset \bigcup_{i=1}^{k} g_i \cdot K$.*

PROOF. Suppose $f : G \to Y$ is a coarse embedding, where Y has the property that for each uniformly bounded cover $\mathcal{U}$ of Y there is $N(\mathcal{U}) \in \mathbb{N}$ such that every element of $\mathcal{U}$ contains at most $N(\mathcal{U})$ elements. The family $\{f^{-1}(y)\}_{y \in Y}$ is a uniformly bounded cover of G, so there is a bounded subset K of G with the property that for every $y \in Y$ there is $g(y) \in G$ such that $f^{-1}(y) \subset g(y) \cdot K$. Given a bounded subset B of G, $f(B)$ is bounded, hence it is finite and $B \subset \bigcup_{y \in f(B)} f^{-1}(y) \subset \bigcup_{y \in f(B)} g(y) \cdot K$.

Suppose there is a bounded set K such that for every bounded subset B of G there are elements $g_i \in G$, $i \leq k$, so that $B \subset \bigcup_{i=1}^{k} g_i \cdot K$. We may assume K is symmetric by switching to $K \cup K^{-1}$. Also, add 1_G to K. Choose a maximal subset Y of G with the property that $g \neq h \in Y$ implies $g^{-1} \cdot h \notin K \cdot K^{-1}$.

Notice that Y is of bounded geometry. Indeed, given a uniformly bounded cover $\mathcal{U}$ of Y there is a bounded subset B of G such that U is a refinement of $\{g \cdot B\}_{g \in G}$. Suppose $B \subset \bigcup_{i=1}^{k} g_i \cdot K$ and $y, z \in Y \cap (g \cdot B)$ are two different elements of Y. If $y, z \in g \cdot g_i \cdot K$, then $y^{-1} \cdot z \in K^{-1} \cdot K$, a contradiction.

To conclude that the inclusion $Y \to G$ is a coarse equivalence, it is sufficient to show $G = \bigcup_{y \in Y} y \cdot K$. Suppose $x \in G \setminus Y$. There is $y \in Y$ such that $k := x^{-1} \cdot y \in K$. Now, $x = y \cdot k^{-1}$ and we are done. □

COROLLARY 7.4. *The following large scale groups are of bounded geometry:*

1. *Groups with bornology consisting of finite sets,*
2. *Locally compact topological groups.*

PROOF. In case 2) any pre-compact neighborhood K of 1_G works. □

EXAMPLE 7.5. Consider the special linear group $SL_n(\mathbb{R})$ and the general linear group $GL_n(\mathbb{R})$. They are both locally compact σ-compact groups as the determinant function $det : M_n(\mathbb{R}) \to \mathbb{R}$ is continuous with $det^{-1}(\mathbb{R} \setminus \{0\}) = GL_n(\mathbb{R})$ and $det^{-1}(\{1\}) = SL_n(\mathbb{R})$. Therefore, both $SL_n(\mathbb{R})$ and $GL_n(\mathbb{R})$, with the bornology consisting of all pre-compact subsets, are of bounded geometry.

PROPOSITION 7.6. *Let G be boundedly generated large scale group of bounded geometry. There exist a symmetric bounded subset K of G such that for every bounded subset L of G its complement $G \setminus L$ has finitely many K-components.*

PROOF. Choose a bounded symmetric subset $M \subset G$ such that $1_G \in M$, $G =< M >$ and every bounded subset of G is covered by finitely many subsets of the form $g \cdot M$. Let $\mathcal{U} := \{g \cdot M\}_{g \in G}$ and $L \subset G$ be a bounded subset. For the bounded subset $st(L,\mathcal{U})$, choose $m \in \mathbb{N}$ such that $st(L,\mathcal{U}) \subset \bigcup_{i=1}^{m} g_i \cdot M$. If $L = \emptyset$, then $G \setminus L$ has exactly one M-component. Fix $a \in L$ and for each $x \in G \setminus L$ choose a K-chain c_x joining x to a. Let $i_x \le m$ be the first index of c_x such that $x_{i_x} \in \bigcup_{i=1}^{m} g_i \cdot M \setminus L$. For $y \in G \setminus L$, if $x_{i_x}, y_{i_y} \in g_i \cdot M$, then $x_{i_x}^{-1} \cdot y_{i_y} \in M^2$. Since x and x_{i_x} are M-connected, as well as y and y_{i_y} are M-connected, so x and y are M^2-connected. Let $K = M^2$, then $G \setminus L$ has at most m K-components. □

John Roe [**25**] defines a geodesic metric space X to be of **bounded growth** if for each $s > 0$ there is a number N_s such that each ball of radius $S + s$ in X can be covered by at most N_s balls of radius S. We consider this definition excessively restrictive as it is of an all-scale character instead of being of a large scale character. It would seem that changing Roe's definition to require that there is $G > 0$ such that for each $s > G$ there are numbers N_s and D_s such that each ball of radius $r + s$, $r \ge D_s$, in X can be covered by at most N_s balls of radius r, would be an improvement but it is not clear if the new concept is an invariant of quasi-isometries.

Our next result shows that metrizable large scale groups of bounded geometry have a property resembling bounded growth.

PROPOSITION 7.7. *If a large scale group G is metrizable by the K-norm, then the following conditions are equivalent:*

1. *For every bounded subset B of G there are elements $g_i \in G$, $i \le k$, so that $B \subset \bigcup_{i=1}^{k} g_i \cdot K$.*
2. *There are elements $g_i \in G$, $i \le k$, so that $K \cdot K \subset \bigcup_{i=1}^{k} g_i \cdot K$.*
3. *For each $s \ge 1$ there is $N_s \ge 1$ such that for each $n \ge 1$ the set K^{n+s} can be covered by at most N_s sets of the form $g \cdot K^n$.*

PROOF. 1) $\Longrightarrow$ 2) and 3) $\Longrightarrow$ 2) are trivial.

2) $\Longrightarrow$ 1). Notice $K \cdot K \cdot K \subset \bigcup_{i=1}^{k} g_i \cdot K \cdot K \subset \bigcup_{i,j=1}^{k} g_i \cdot g_j \cdot K$. Apply induction to get that for every $n \ge 1$ there are elements $h_i \in G$, $i \le k$, so that $K^n \subset \bigcup_{i=1}^{k} h_i \cdot K$. Since every bounded subset B of G is contained in some K^n, 1) follows.

2) $\Longrightarrow$ 3). As in the above there is $N_s \ge 1$ such that K^{s+1} can be covered by at most N_s sets of the form $g \cdot K$. Therefore $K^{n+s} = K^{s+1} \cdot K^{n-1}$ can be covered by at most N_s sets of the form $g \cdot K \cdot K^{n-1}$. □

REMARK 7.8. Notice that the reals form a large scale group of bounded geometry but the Cayley graph of the reals constructed using $K = [-1, 1]$ is not of bounded growth in the sense of Roe's definition. Indeed, for $s < 1/2$ the ball $B(r, 2s)$ at any vertex r of the graph cannot be covered by finitely many balls of radius s.

8. Metrizable large scale groups

In this section we discuss large scale groups whose large scale structure is metrizable or coarsely equivalent to a geodesic space.

DEFINITION 8.1. A **basis** $\mathcal{A}$ for a bornology $\mathcal{B}$ on a group G is a collection of subsets of G such that for every $B \in \mathcal{B}$ there exists $A \in \mathcal{A}$ such that $B \subset A$.

PROPOSITION 8.2. *Suppose G is a large scale group. The following conditions are equivalent:*

1. *G is metrizable (i.e. its large scale structure is generated by a metric).*
2. *The bornology of G has a countable basis.*
3. *There is a left-invariant metric d on G inducing the large scale structure on G.*

PROOF. 1) $\Longrightarrow$ 2) and 3) $\Longrightarrow$ 1) are obvious.

2) $\Longrightarrow$ 3) Choose an increasing sequence $\{B_n\}_{n\geq 1}$ of symmetric bounded subsets of G containing 1_G that serves as a basis of the bornology of G. We may assume $B_n * B_n \subset B_{n+1}$ for each $n \geq 1$. Define the norm $|g|$ on G as follows:

1. $|1_G| = 0$.
2. If $g \neq 1_G$, then $|g|$ is the smallest n such that $g \in B_n$.

Define the the metric d on G via $d(g,h) = |g^{-1}\cdot h|$. Notice the cover $\{B(g,n+1)\}_{g\in G}$ equals $\{g\cdot B_n\}_{g\in G}$, so the large scale structure of G equals the large scale structure induced from (G,d). □

DEFINITION 8.3. A large scale group G is **σ-bounded** if it is a countable union of bounded subsets.

When G is equipped with a metrizable large scale structure, it is σ-bounded. The converse is true when G is of bounded geometry.

PROPOSITION 8.4. *If $(G,\mathbb{LSS})$ is a σ-bounded large scale group of bounded geometry, then $\mathbb{LSS}$ is metrizable. Consequently, G has a countable basis of boundedly generated subgroups.*

PROOF. Pick a bounded symmetric set K containing 1_G such that for every bounded subset B of G there are elements $g_i \in G$, $i \leq k$, so that $B \subset \bigcup\limits_{i=1}^{k} g_i \cdot K$. Suppose $G = \bigcup\limits_{i=1}^{\infty} B_i$, where each B_i is bounded and $B_i \subset B_j$ if $i < j$. Each $g \in G$ has an index $n(g)$ such that $g \in B_{n(g)}$. Now, if B is bounded and $B \subset \bigcup\limits_{i=1}^{k} g_i \cdot K$, then for $m \geq n(g_i)$ for all $i \leq n$ one has $B \subset B_m \cdot K$. Therefore, $\{B_n \cdot K : n \in \mathbb{N}\}$ is a countable basis for the bornology of G. □

PROPOSITION 8.5. *Suppose G is a large scale group. The following conditions are equivalent:*

1. *G is generated by a symmetric bounded set K and its bornology has $\{K^n\}_{n\geq 1}$ as a countable basis.*
2. *G is coarsely equivalent to a connected graph Γ whose set of vertices is equal to G and the graph metric is left-invariant.*

3. *G is coarsely dominated by a geodesic space, i.e. there is a geodesic space X and large scale continuous functions $f : G \to X$, $g : X \to G$ such that $g \circ f$ is close to id_G.*

PROOF. 1) $\Longrightarrow$ 2). Extend G to a connected graph Γ by requiring that g and h form an edge if and only if $g^{-1} \cdot h \in K$. Let d be the graph metric on Γ. Notice the cover $\{B(g, n+1)\}_{g \in G}$ equals $\{g \cdot K^n\}_{g \in G}$, so the large scale structure of G equals the large scale structure induced from (G, d).

2) $\Longrightarrow$ 3) is obvious.

3) $\Longrightarrow$ 1). Let $\alpha : G \to X$ and $\beta : X \to G$ be two large scale continuous functions such that $\beta \circ \alpha$ is close to id_G. Choose a symmetric bounded subset K of G such that $\beta \circ \alpha(g) \in g \cdot K$ for each $g \in G$ and the cover $\{\beta(B(x, 2)\}_{x \in X}$ is a refinement of the cover $\{g \cdot K\}_{g \in G}$. Given a bounded subset L of G containing 1_G there is $m > 0$ such that $\alpha(L) \subset B(x_0, m)$, $x_0 = \alpha(1_G)$. If $g \in L$, then one can connect x_0 to $\alpha(g)$ via a chain $x_0, \ldots, x_n$ so that $x_{i+1} \in B(x_i, 1)$ for each $i < n$. That means $\beta(x_{i+1})$ and $\beta(x_i)$ belong to a set of the form $h \cdot K$ resulting in $\beta(x_{i+1}) \cdot \beta(x_i)^{-1}$ belonging to $K \cdot K$. That implies $\beta(\alpha(g)) \in (K \cdot K)^n$. Hence $\beta(\alpha(g)) \in (K \cdot K)^n \cap (g \cdot K)$ and $g \in K^{2n+1}$. As n can be chosen uniformly for all $g \in L$, $L \subset K^{2n+1}$. $\square$

A large scale group G that satisfies one of the equivalent properties of 8.5 is referred to as a **coarsely geodesic group**.

DEFINITION 8.6. Let G be a coarsely geodesic large scale group. A **Cayley graph of** G is a connected graph Γ whose set of vertices equals G and such that the inclusion $G \to \Gamma$ is a coarse equivalence, where Γ is equipped with its graph metric which is left-invariant. The metric on G induced from a Cayley graph will be called a **Cayley metric**.

REMARK 8.7. The set K of all vertices of G at distance 1 from 1_G is symmetric and generates G.

PROPOSITION 8.8. *If G is a boundedly generated large scale group of bounded geometry, then it is coarsely geodesic.*

PROOF. Pick a bounded symmetric set K containing 1_G such that for every bounded subset B of G there are elements $g_i \in G$, $i \le k$, so that $B \subset \bigcup_{i=1}^{k} g_i \cdot K$. We may assume K generates G. Each $g \in G$ has an index $n(g)$ such that $g \in K^{n(g)}$. Now, if B is bounded and $B \subset \bigcup_{i=1}^{k} g_i \cdot K$, then for $m \ge n(g_i)$ for all $i \le n$ one has $B \subset K^{m+1}$. By 8.5, G is coarsely geodesic. $\square$

EXAMPLE 8.9. $SL_2(\mathbb{R})$, with the bornology consisting of all pre-compact subsets, is coarsely geodesic; whereas, the large scale structure of $GL_n(\mathbb{R})$ induced from the bornology consisting of all pre-compact subsets is metrizable.

DEFINITION 8.10. A sequence of subgroups $\{G_n\}_{n \in \mathbb{N}}$ of a large scale group G is a **basis for boundedly generated subgroups** of G if every boundedly generated subgroup H of G is contained in some G_n. Equivalently, for any bounded subset B of G there is $n \in \mathbb{N}$ such that $B \subset G_n$.

Clearly, if a large scale group G has a metrizable large scale structure, then it can be expressed as the union of an increasing sequence of boundedly generated subgroups $\{G_n\}_{n\in\mathbb{N}}$ such that $\{G_n\}_{n\in\mathbb{N}}$ is a basis for boundedly generated subgroups. The converse is partially true; that is if $\{G_n\}_{n\in\mathbb{N}}$ is a basis for boundedly generated subgroups such that each G_n is coarsely geodesic then the large scale structure of G is metrizable.

PROPOSITION 8.11. *Let G be a large scale group that has a basis $\{G_n\}_{n\in\mathbb{N}}$ for boundedly generated subgroups consisting of unbounded subgroups that are coarsely geodesic; then its large scale structure is metrizable.*

PROOF. Notice for each G_n there exists a symmetric bounded subset $K_n \subset G$ containing 1_G such that $G_n = \langle K_n\rangle$ and that $(K_n^m)_{m\geq 1}$ is a basis for G_n. In particular, $\{K_n^m : n, m \in \mathbb{N}\}$ is a countable basis for G. □

DEFINITION 8.12. A coarse group G is **locally bounded** if for every bounded subset B of G the subgroup $\langle B\rangle$ of G generated by B is bounded.

LEMMA 8.13. *Let $\varphi : G \to H$ be a group homomorphism between large scale groups. If $\varphi : G \to H$ is a coarse equivalence and H is locally bounded, then G is locally bounded.*

PROOF. It straightforward to see that, as $\varphi : G \to H$ is a group homomorphism, for every subset $A \subset G$, $\varphi(\langle A\rangle) = \langle\varphi(A)\rangle$. Notice that as $\varphi : G \to H$ is coarsely proper, if $A \subset G$ is unbounded, then $\varphi(A)$ is an unbounded subset of H. Seeking contradiction, assume that $K \subset G$ is coarsely bounded such that $\langle K\rangle$ is unbounded subset of G. On one hand, $\varphi(\langle K\rangle) = \langle\varphi(K)\rangle$ in unbounded. On the other hand, since $\varphi : G \to H$ is controlled, $\varphi(K)$ is a coarsely bounded subset of H, and hence, as H is locally bounded, $\langle\varphi(K)\rangle$ is coarsely bounded; a contradiction. □

PROPOSITION 8.14. *An unbounded large scale group G is locally bounded and its large scale structure is metrizable if and only if it admits a strictly increasing sequence $\{G_n\}_{n\in\mathbb{N}}$ of bounded subgroups that is a basis for boundedly generated subgroups.*

PROOF. Assume that the large scale group G is locally bounded and its large scale structure is metrizable. For each $n \in \mathbb{N}$, put $B_n := B(1_G, n)$. We inductively define the strictly increasing sequence of bounded subgroups as follows:

$G_1 := \langle B_1\rangle$, and $G_n := \langle B_n \cup \{x_n\}\rangle$ for all $n \geq 2$, where $x_n \in G \setminus G_{n-1}$. Then, $\{G_n\}_{n\in\mathbb{N}}$ is the desired sequence.

Conversely, assume that $\{G_n\}_{n\in\mathbb{N}}$ is strictly increasing sequence of bounded subgroups that is a basis for boundedly generated subgroups. It is a basis for its bounded sets, so the large scale structure of G is metrizable. Moreover, every boundedly generated subgroup is contained in some G_n, and hence bounded, so G is locally bounded. □

PROPOSITION 8.15. *Let G be a large scale group G that is non-locally bounded and is not boundedly generated. Its large scale structure is metrizable if and only if G admits a strictly increasing sequence $\{G_n\}_{n\in\mathbb{N}}$ of unbounded boundedly generated subgroups whose large scale structures as large scale subspace of G are metrizable, and that $\{G_n\}_{n\in\mathbb{N}}$ is a basis for boundedly generated subgroups.*

PROOF. Assume that G is non-locally bounded and its large scale structure is metrizable. For each $n \in \mathbb{N}$, put $B_n := B(1_G, n)$. We inductively define the strictly increasing sequence of the subgroups as follows:

$G_1 := \langle B_1 \rangle$, and $G_n := \langle B_n \cup \{x_n\} \rangle$ for all $n \geq 2$, where $x_n \in G \setminus G_{n-1}$. Notice for any bounded subset $B \subset G$, there exists $n \in \mathbb{N}$ such that $B \subset B_n$ and hence $\langle B \rangle \subset G_n$. If each G_n is bounded then G is locally bounded, a contradiction. Therefor, there exists $N \geq 1$ such that G_n is unbounded for all $n \geq N$. Without loss of generality, we may assume that $N = 1$. It is clear that for each $n \geq N$, the large scale structure of G_n is metrizable large scale structure as a large scale subspace of G.

Conversely, assume that G admits a strictly increasing sequence $\{G_n\}_{n \in \mathbb{N}}$ of unbounded boundedly generated subgroups whose coarse structures as large scale subspace of G are metrizable, and that $\{G_n\}_{n \in \mathbb{N}}$ is a basis for boundedly generated subgroups. Let $\mathcal{B}_n$ be a countable basis for the bounded subsets of G_n, for all $n \in \mathbb{N}$. Then $\mathcal{B} := \bigcup_{n \in \mathbb{N}} \mathcal{B}_n$ is a countable basis for the bounded subsets of G. Clearly, G is non-locally bounded. $\square$

8.1. Coarse hyperbolicity.

DEFINITION 8.16. A large scale group G is **coarsely hyperbolic** if it is large scale equivalent to a geodesic space that is hyperbolic in the sense of Gromov.

Notice that G has Cayley graphs if it is coarsely hyperbolic and, since being hyperbolic is a coarse invariant of geodesic spaces, G is coarsely hyperbolic if and only if one (hence every) of its Cayley graphs is hyperbolic. Therefore, if G is metrizable via a K-norm, it is coarsely hyperbolic if there is $\delta > 0$ such that for every two K-geodesics c and d on G emanating from 1_G (those are K-chains of length equal to the K-norm of the terminal elements) the distance between ith elements of the chains is less than δ if $i \leq (|c| + |d| - d(g_c, g_d))/2$, where g_c is the terminal point of c and g_d is the terminal point of d.

The proof of the theorem below is a simplification of the one in **[25]**. Also, we fix a gap in the original proof by showing that any ball $B(x, ps)$ intersects at most $N_{2\delta}$ elements of the cover of A, not only those with $x \in A$.

THEOREM 8.17. *Suppose G is a large scale group of bounded geometry. If G is coarsely hyperbolic, then the asymptotic dimension of G is finite.*

PROOF. If G is bounded, it is of asymptotic dimension at most 0, so assume G is unbounded.

Choose K satisfying two conditions:

1. G is metrizable via the K-norm,
2. There is $m \geq 1$ such that $K \cdot K$ is covered by m balls of radius 2 in G, i.e. sets of the form $g \cdot K$, $g \in G$.
3. There is $\delta > 1$ such that for every two K-geodesics c and d on G emanating from 1_G the distance between ith elements of the chains is less than δ if $i \leq (|c| + |d| - d(g_c, g_d))/2$, where g_c is the terminal point of c and g_d is the terminal point of d.

By 7.7 for each $t \geq 1$ there is N_t such that for each $S \geq 1$ the set K^{S+t} can be covered by at most N_t sets of the form $g \cdot K^S$.

Consider $A := K^{2n+2p \cdot s} \setminus K^{2n}$ for some $n, p \geq 1$ so that $n > ps$. Notice $K^{2n} \setminus K^{2n-1} \neq \emptyset$ as otherwise $K^q = K^{2n}$ for all $g \geq 2n$ and G is bounded.

Similarly, $A \neq \emptyset$. In $K^{2n} \setminus K^{2n-1}$ choose a maximal set of points $\{x_i\}_{i\in J}$ that are separated by at least $2p \cdot s$ in terms of the K-metric. For each $i \in J$ choose a K-geodesic c_i from 1_G to x_i.

For each $g \in G$ choose a K-geodesic c_g from 1_G to g. Given $g \in A$ let $r(g)$ be the last term of c_g belonging to $K^{2n} \setminus K^{2n-1}$. Let U_i, $i \in J$, be the set of all $g \in A$ so that $d(r(g), x_i) \leq 2ps$. Obviously, U_i, $i \in J$, form a cover of A. If $g \in U_i$, then $d(g, x_i) \leq d(g, r(g)) + d(r(g), x_i) < 2ps + 2ps = 4ps$. That means each U_i is of diameter at most $8ps$.

Suppose $x \in G$ and $B(x, ps)$ intersects U_i for some $i \in J$ and $y \in B(x, ps) \cap U_i$. Since $|x| \geq |y| - ps > 2n - ps$, $z := c_x(2n-ps)$ exists. Moreover, $d(z, c_y(2n-ps)) < \delta$ as $|x| + |y| - d(x, y) \geq 2n - ps + 2n - ps = 2(2n - ps)$. Also, $|r(y)| + |x_i| - d(r(y), x_i) \geq 2n + 2n - 2ps = 2(2n - ps)$, so $d(c_y(2n - ps), c_i(2n - ps)) < \delta$. Thus $d(z, c_i(2n-ps)) < 2\delta$ and $d(z, x_i) < ps + 2\delta$. However, $B(z, ps + 2\delta)$ can be covered by $N_{2\delta}$ balls of radius ps and each of them can contain at most one x_i.

Now we are ready to show that the asymptotic dimension of G is at most $2 \cdot N_{2\delta} - 1$. Indeed, given $r > 0$ we choose $p \geq 1$ such that $ps > r$. Now, each $A_n := K^{2n+2p\cdot s} \setminus K^{2n}$, $n > ps$ is covered by at most $N_{2\delta}$ sets of diameter at most $8ps$. Add $A_{ps} := K^{2ps}$ to obtain a cover of G uniformly bounded by $8ps$ such that any r-ball intersects at most $2 \cdot N_{2\delta}$ of its elements. It is so because any r-ball can intersect at most two annuli A_n. □

9. Ends of large scale groups

In this section we apply the general theory of ends of large scale spaces to large scale groups.

LEMMA 9.1. *Suppose G is a large scale group, B containing 1_G is a bounded subset of G, and $\mathcal{U} = \{g \cdot B\}_{g\in G}$. If $A \subset G$, then $A \cdot B \subset st(A, \mathcal{U}) \subset A \cdot B^{-1} \cdot B$.*

PROOF. If $a \in A$ and $b \in B$, then $a, a \cdot b \in a \cdot B$ so $A \cdot B \subset st(A, U)$. If $a \in A \cap (g \cdot B)$, then $g \in a \cdot B^{-1}$ and $g \cdot B \subset A \cdot B^{-1} \cdot B$. □

LEMMA 9.2. *A subset A of a large scale group is coarsely clopen if and only if for each bounded subset B of G the set $(A \cdot B) \cap (A^c \cdot B)$ is bounded.*

PROOF. Apply 9.1. □

We begin with a large scale version of the first part of 1.9.

PROPOSITION 9.3. *Let G be a large scale group equipped with a metrizable large scale structure. If G is an unbounded and locally bounded group, then its number of coarse ends is infinite.*

PROOF. It suffices to show that G has infinitely many unbounded coarsely clopen subsets of G that are mutually disjoint. By 8.14, G can be expressed as a union of a strictly increasing sequence $\{G_n\}_{n\in\mathbb{N}}$ of bounded subgroups that is a basis for boundedly generated subgroups. Choose $g_n \in G_{n+1} \setminus G_n$ for each $n \in \mathbb{N}$. Given an infinite subset I of natural numbers define A_I as $\bigcup\limits_{n\in I} g_n \cdot G_n$. We claim that A_I is unbounded coarsely clopen. Indeed, A_I is unbounded as for any $N \in \mathbb{N}$, there exists $n \in I$ such that $n > N$ and hence $g_n \in A_I \setminus G_N$. Since, $\{G_n\}_{n\in\mathbb{N}}$ is a basis for the bounded sets of G, A_I must be unbounded. To show A_I is coarsely clopen assume $B \subset G$ is bounded and choose $n \in I$ so that $B \subset G_n$, and therefore $\{G_n : n \in I\}$ is a basis for the bounded sets of G. Notice for any $i \in I$, if $i < n$, then $(g_i \cdot G_i) \cdot G_n = G_n$; and if $i \geq n$, then $(g_i \cdot G_i) \cdot G_n = g_i \cdot G_i$. In particular, $A_I \cdot G_n = G_n \cup \bigcup_{i \in I, i \geq n} g_i \cdot G_i$, so $(A_I \cdot G_n)\Delta A_I$ is bounded, and hence A_I is coarsely clopen. To complete the proof notice $A_I \cap A_J = \emptyset$ if $I \cap J = \emptyset$. Indeed, if $g \in A_I \cap A_J$, then there exist $i < j$ such that $g_i \cdot h = g_j \cdot h'$, where $h \in G_i$ and $h' \in G_j$, with $i \in I$ and $j \in J$. Therefore $g_j \in G_j$, a contradiction. Finally, since $\mathbb{N}$ can be expressed as an infinite union of mutually disjoint infinite subsets of $\mathbb{N}$, G has infinitely many mutually disjoint unbounded coarsely clopen subsets and hence infinitely many coarse ends. □

DEFINITION 9.4. Let G be a large scale group and A be a subset of G. We say that G **acts coarsely trivially on** A if the subsets A and $x \cdot A$ are coarsely identical for all $x \in G$; that is, if $x \cdot A\Delta A$ is coarsely bounded for all $x \in G$.

LEMMA 9.5. *Let G be a coarsely geodesic large scale group. If G contains three mutually disjoint unbounded coarsely clopen subsets, then it acts coarsely trivially on at most one of them.*

PROOF. Let $K \subset G$ be symmetric bounded subset containing 1_G and generating G such that the Clayey metric d_K induces its large scale structure. Let A_1, A_2 and A_3 be mutually disjoint unbounded coarsely clopen subsets. Seeking contradiction, assume that G acts trivially on A_1 and A_2. Put $A_4 := G \setminus \bigcup_{i=1}^{3} A_i$. Clearly that A_4 is coarsely clopen disjoint from A_1, A_2 and A_3. Choose $N \in \mathbb{N}$ large enough so that for all $x \in A_i \setminus B(1_G, N)$ and $y \in (G \setminus A_i) \setminus B(1_G, N)$ then $d(x, y) > 2$, for all $1 \leq i \leq 4$. If A_4 is coarsely bounded, we may assume that $A_4 \subset B(1_G, N)$. Choose $m \in \mathbb{N}$ such that $(A_i \setminus B(1_G, m)) \cap B(G \setminus A_i, 2 \cdot N + 2) = \emptyset$, for all $1 \leq i \leq 3$. In particular, $m > N$ and if $x \in A_i$ such that $d_K(x, 1_g) \geq m$, then $B(x, 2 \cdot N + 2) \subset A_i$, for all $1 \leq i \leq 3$. If A_4 is unbounded, we assume also that $(A_4 \setminus B(1_G, m)) \cap B(G \setminus A_4, 2 \cdot N + 2) = \emptyset$. Notice that if $x \in A_i$ such that $d_K(x, 1_G) \geq m$, then $x \cdot B(1_G, N) = B(x, N) \subset B(x, 2 \cdot N + 2) \subset A_i$, for all $1 \leq i \leq 3$.

Now, pick $x \in A_3$ such that $d_K(x, 1_G) \geq m$. On one hand, as $d_K(x^{-1}, 1_G) \geq m$, so $x^{-1} \in G \setminus B(1_G, N)$ and hence $B(x^{-1}, N)$ is contained in exactly one coarsely clopen A_i for some $1 \leq i \leq 3$, if A_4 is bounded; and $B(x^{-1}, N)$ is contained in exactly one coarsely clopen A_i for some $1 \leq i \leq 4$, if A_4 is unbounded.

As G acts coarsely trivially on A_1, $x \cdot A_1\Delta A_1$ is coarsely bounded, and thus we can choose $g_1 \in A_1$ such that $d_K(g_1, 1_G) \geq m$ and $x \cdot g_1 \in A_1$. Since $G =< K >$, G is K-connected, and hence we can find a K-chain $g_1 = a_0, a_1, \ldots, a_s = 1_G$ connecting g_1 and 1_G. Notice that $d_K(a_i, a_{i+1}) = 1$ for all $0 \leq i \leq s - 1$. Let a_s be the first element of the K-chain that hits $B(1_G, N)$. Then, as $d_K(a_{s-1}, a_s) = 1$, $a_s \in A_1 \cap B(1_G, N)$ and hence $x \cdot a_s \in A_3$. Since d_K is left-invariant, the sequence $x \cdot g_1 = x \cdot a_0, x \cdot a_1, \ldots, x \cdot a_s$ is still a K-chain connecting $x \cdot g_1 \in A_1$ to $x \cdot a_s \in A_3$ and $d_K(x \cdot a_i, x \cdot a_{i+1}) = 1$ for all $0 \leq i \leq s - 1$. Thus, $x \cdot g_1 = x \cdot a_0, x \cdot a_1, \ldots, x \cdot a_s$ must hit $B(1_G, N)$, as otherwise, we would have two consecutive terms $x \cdot a_i$ and $x \cdot a_{i+1}$ of $x \cdot g_1 = x \cdot a_0, x \cdot a_1, \ldots, x \cdot a_s$ such that $d_K(x \cdot a_i, x \cdot a_{i+1}) > 2$, a contradiction. Therefore, there exists $x_1 \in A_1$ such that $x \cdot x_1 \in B(1_G, N)$ which implies that $x_1 \in B(x^{-1}, N)$, and hence $B(x^{-1}, N)$ must be contained in A_1.

On the other hand, as G acts coarsely trivially on A_2, we can find $x_2 \in A_2$ such that $x \cdot x_2 \in B(1_G, N)$ which implies that $x_2 \in B(x^{-1}, N)$ and hence $B(x^{-1}, N)$

intersects A_2, which contradicts the fact that $B(x^{-1}, N) \subset A_1$. Consequently, G can act coarsely trivially on at most one of the coarsely clopen subsets A_1, A_2 and A_3. □

It can be seen that if A is a coarsely clopen subset of the large scale group G, then for any $x \in G$, $x \cdot A$ is coarsely clopen because if $K \subset G$ is bounded $((x \cdot A) \cdot K) \cap (x \cdot A)^c \cdot K) = x \cdot ((A \cdot K) \cap (A^c \cdot K))$ which is bounded. If, moreover, A is unbounded, then $x \cdot A$ is unbounded. In particular, if $E \in Ends(G)$, then $x \cdot E = \{x \cdot A : A \in E\}$ is also a coarse end; and therefore, the action of G on itself by left multiplication extends to $Ends(G)$. G **acts trivially on** $Ends(G)$ if $x \cdot E = E$ for all $x \in G$ and $E \in Ends(G)$.

LEMMA 9.6. *A large scale group G acts trivially on $Ends(G)$ if and only if for all $E \in Ends(G)$ and $A \in E$, G acts coarsely trivially on A.*

PROOF. Assume that G acts trivially on $Ends(G)$, $E \in Ends(G)$ and $A \in E$. For $x \in G$, we aim to show that A and $x \cdot A$ are coarsely identical. Put $Ends(G) = \{E_\alpha : \alpha \in I\}$, for some index set I. Assume that $E = E_\beta$ for some $\beta \in I$. As G acts trivially on $Ends(G)$, so both A and $x \cdot A$ are in E_β. For all $\alpha \neq \beta$, we can find $A^\alpha \in E_\alpha$ such that $x \cdot A \cap A^\alpha = \emptyset$. Since $Ends(G) = A_{end} \cup (\bigcup_{\alpha \in I \setminus \{\beta\}} A^\alpha_{end})$. By 2.22, there exists a coarsely bounded subset $K \subset G$ such that $G = A \cup (\bigcup_{\alpha \in I \setminus \{\beta\}} A^\alpha) \cup K$. Therefore, $(x \cdot A) \setminus A \subset (x \cdot A) \cap K$ which is coarsely bounded. Similarly, $A \setminus (x \cdot A)$ is coarsely bounded and therefore A and $x \cdot A$ are coarsely identical. Consequently, G acts trivially on A.

Conversely, assume that for all $E \in Ends(G)$ and $A \in E$, G acts trivially on A. It is obvious that $x \cdot E = E$, for all $x \in G$. □

THEOREM 9.7. *Let G be a coarsely geodesic large scale group. The number of the coarse ends of G is either infinite or at most 2.*

PROOF. If G is bounded, then $Ends(G)$ is empty. If G is locally bounded and unbounded, then $Ends(G)$ is infinite by 9.3.

Assume G is unbounded, non-locally bounded, and its number of the coarse ends $m \geq 3$ is finite. Notice that, by 2.27, for any $x \in G$, the map $\sigma^x : G \to G$ that maps each $g \in G$ to $x \cdot g$ is a coarse equivalence, it induces a homeomorphism $\sigma^x_{end} : Ends(G) \to Ends(G)$. Let $Homeo(Ends(G))$ be the finite group of all homeomorphisms from $Ends(G)$ to itself. The map $\rho : G \to Homeo(Ends(G))$ given by: $\rho(x) = \sigma^x_{end}$ is a group homomorphism with $H := Ker(\rho)$ is a subgroup of finite index in G. By 4.8, G and H are coarsely equivalent and hence $Ends(G)$ and $Ends(H)$ are homeomorphic by 2.27. Moreover, H acts trivially on $Ends(H)$, with $|Ends(H)| \geq 3$. Now, let $E_1, E_2, E_3 \in Ends(H)$ be three distinct ends, and choose $A_i \in E_i$ for all $1 \leq i \leq 3$ such that A_1, A_2 and A_3 are mutually disjoint. By 9.6, H acts trivially on A_1, A_2 and A_3. This contradicts Lemma 9.5. □

As a result, a locally compact compactly generated group G equipped with the large scale structure induced by the bornology of all pre-compact subsets of G has either infinitely many coarse ends or at most 2. For example, the topological group $SL_2(\mathbb{R})$ has either infinitely many coarse ends or at most 2 coarse ends. We aim to generalize 9.7 to a larger class of large scale groups that includes:

- Locally compact σ-compact groups equipped with the large scale structure induced by bornology of all pre-compact subsets.
- Infinite countable groups equipped with the bornology of all finite subsets.

Although these two classes of large scale groups are not necessarily coarsely geodesic, they can be expressed as a union of a sequence of coarsely geodesic unbounded subgroups where that sequence is a basis for boundedly generated subgroups.

LEMMA 9.8. *Let G be a large scale group that has a basis $\{G_n\}_{n\in\mathbb{N}}$ for boundedly generated subgroups consisting of unbounded subgroups. If A is an unbounded coarsely clopen subset of G, then there is $n \in \mathbb{N}$ such that $A \cap G_n$ is unbounded.*

PROOF. Let G_1 be generated by a symmetric bounded subset B of G. Since $(A \cdot B)\Delta A$ is bounded, there is $m \in \mathbb{N}$ such that $(A \cdot B)\Delta A \subset G_m$. Notice that $(A \cdot B) \setminus G_m = A \setminus G_m$. If $A \setminus G_m = \emptyset$, we are done so assume $A \setminus G_m \neq \emptyset$. Now, for each $g \in A \setminus G_m$ and $b \in B$, we have $g \cdot b \in (A \cdot B) \setminus G_m = A \setminus G_m$. By induction, that implies $g \cdot G_1 \subset A \setminus G_m$. Choose $n \in \mathbb{N}$ such that $< B \cup g > \subset G_n$, then $g \cdot G_1 \subset A \cap G_n$ and hence $A \cap G_n$ is unbounded. $\square$

As a consequence, a large scale group G that can be expressed as a union of a sequence of boundedly generated subgroups where that sequence is a basis for boundedly generated unbounded subgroups, then $|Ends(G)| \leq \sup_{n\in\mathbb{N}} |Ends(G_n)|$.

COROLLARY 9.9. *Let G be a large scale group that has a basis $\{G_n\}_{n\in\mathbb{N}}$ for boundedly generated subgroups consisting of unbounded subgroups. If $m \leq \infty$ and each G_n has at most m ends, then the number of ends of G is at most m.*

PROOF. The case $m = \infty$ is clear, so assume $m < \infty$.

If G has $(m + 1)$ mutually disjoint unbounded coarsely clopen subsets A_i, $1 \leq i \leq m + 1$, then by 9.8 we can find an index n such that each $A_i \cap G_n$ is an unbounded coarsely clopen subset of G_n. Hence $|Ends(G_n)| \geq m + 1$, a contradiction. $\square$

Combining 9.8 and 9.9, we generalize 9.7.

THEOREM 9.10. *Let G be a large scale group that admits a basis $\{G_n\}_{n\in\mathbb{N}}$ for boundedly generated subgroups consisting of coarsely geodesic subgroups. The number of the coarse ends of G is either infinite or at most 2.*

PROOF. Notice for each G_n there exists a symmetric bounded subset $K_i \subset G$ containing 1_G such that $G_i = \langle K_i \rangle$ and that $(K_i^n)_{n\in\mathbb{N}}$ is a basis for the bounded subsets G_i. In particular, $\{K_i^n : i, n \in \mathbb{N}\}$ is a countable basis for the bounded subsets of G, and hence its large scale structure is metrizable. If G is bounded, then $Ends(G)$ is empty. If G is locally bounded and unbounded, then $Ends(G)$ is infinite by 9.3.

Assume G is unbounded, non-locally bounded and its number of coarse ends is finite. Notice that the boundedly generated groups cannot be all bounded. Choose $N \in \mathbb{N}$ such that G_n unbounded for all $n \geq N$. Let $H_1 :=< \bigcup_{i=1}^{N} K_i >$, and $H_i := G_{N+i}$, for all $i \in \mathbb{N}$. Clearly, $\{H_n\}_{n\in\mathbb{N}}$ is a basis for boundedly generated subgroups consisting of unbounded coarsely geodesic subgroups. Seeking contradiction, assume that G has $m \geq 3$ coarse ends. Without loss of generality, we may assume that G acts trivially on $Ends(G)$. Notice that G contains three disjoint unbounded coarsely clopen subsets A_1, A_2 and A_3, on which, by 9.6, G acts trivially. By 9.8, there exists $n \in \mathbb{N}$ such that $A_i \cap G_n$ is unbounded for all $1 \leq i \leq 3$. Moreover, G_n acts trivially on $A_i \cap G_n$, for all $1 \leq i \leq 3$, which contradicts 9.5. Hence, $|Ends(G)| \leq 2$. □

COROLLARY 9.11. *Let G be a locally compact σ-compact topological group equipped with the large scale structure induced by the bornology of all pre-compact subsets. The number of coarse ends of G is either infinite or at most 2.*

PROOF. Let $(K_n)_{n\in\mathbb{N}}$ be an exhausting sequence of G. The sequence $\{G_n\}_{n\in\mathbb{N}}$, where $G_n := \langle K_n \rangle$ is a basis for boundedly generated subgroups consisting of subgroups that are coarsely geodesic. □

EXAMPLE 9.12. Consider the topological groups $G = SL_n(\mathbb{R})$ and $H = GL_n(\mathbb{R})$ equipped with the large scale structure induced the bornology of all pre-compact subsets. Then, G and H have either infinitely many coarse ends or at most 2.

COROLLARY 9.13. *Let G be a countable large scale group equipped with the large scale structure induced by the bornology of all finite subsets. The number of coarse ends of G is either infinite or at most 2.*

PROOF. G can be expressed as an increasing sequence of finitely generated subgroups. □

9.13 can be generalized to arbitrary groups equipped with the large scale structure induced by the bornology of all finite subsets. We will first prove that if an uncountable locally finite group has finitely many coarse ends with respect to the large scale structure induced by the bornology of all finite subsets, then it has exactly one coarse end.

LEMMA 9.14. *Let G be a locally finite uncountable group equipped with the large scale structure induced by the bornology of finite subsets. For every infinitely countable $C \subset G$ and non-trivial coarsely clopen subset $A \subset G$, there exists a countable subgroup H containing C such that*

$$A \cdot H \setminus A \subset H$$

and

$$A^c \cdot H \setminus A^c \subset H.$$

PROOF. Inductively, one can construct a strictly increasing sequence of finite subgroups $(H_n)_{n\in\mathbb{N}\cup\{0\}}$ such that:

1. H_n contains at least n elements of C.
2. $A \cdot H_n \setminus A \subset H_{n+1}$.
3. $A^c \cdot H_n \setminus A^c \subset H_{n+1}$.

Indeed, put $H_0 := \{1_G\}$, and $H_1 :=< a_1 >$ where $a_1 \in C \setminus \{1_G\}$. Now, for $n \in \mathbb{N}$, assume that H_n is strictly in the finite subgroup H_{n+1} and that:

1. H_n contains at least n elements of C, and H_{n+1} contains at least $n+1$ elements of C.
2. $A \cdot H_n \setminus A \subset H_{n+1}$.
3. $A^c \cdot H_n \setminus A^c \subset H_{n+1}$.

Let H_{n+2} be the subgroup of G generated by $(A\cdot H_{n+1}\setminus A)\cup(A^c\cdot H_{n+1}\setminus A^c)\cup\{a_{n+2}\}$, where $a_{n+2}\in C\setminus H_{n+1}$. Since A is coarsely clopen and G is locally finite, H_{n+2} is finite. Let $H := \bigcup_{n\in\mathbb{N}} H_n$. □

THEOREM 9.15. *Let G be a locally finite uncountable group equipped with the large scale structure induced by the bornology of finite subsets. If G has finitely many coarse ends, then it has exactly one coarse end.*

PROOF. By 8.13, there is no loss of generality in assuming that G acts trivially on its coarse ends. Seeking contradiction assume that G has more than one coarse end. Then, G has a non-trivial coarsely clopen subset $A\subset G$ such that G acts trivially on both A and A^c, by 9.6. Let C be a countably infinite subset that contains countably infinite elements of A and countably infinite elements A^c. By 9.14, we can choose a countably infinite subgroup H that contains C such that:

$$A\cdot H\setminus A\subset H$$

and

$$A^c\cdot H\setminus A^c\subset H.$$

Now, let $g\in G\setminus H$, then either $g\in A$ or $g\in A^c$. Without loss of generality assume that $g\in A$. Then, $g\cdot H\subset A$. Indeed, if $g\cdot h\in A^c$ for some $h\in H$, then $g=(g\cdot h)\cdot h^{-1}\in A^c\cdot H\setminus A^c\subset H$, a contradiction. Since $g\cdot H\subset A$, so $H\subset g^{-1}\cdot A$ and therefore $H\cap A^c\subset g^{-1}\cdot A\cap A^c$. But, $H\cap A^c$ is infinite, and $g^{-1}\cdot A\cap A^c$ is finite because G acts trivially on A, a contradiction. Hence, either A is infinite and A^c is finite or A^c is infinite and A is finite, which shows that G has exactly one coarse end, as desired. □

THEOREM 9.16. *Let G be group equipped with the large scale structure induced by the bornology of all finite subsets. If G has finitely many ends, then it has at most 2 ends.*

PROOF. Without loss of generality, assume that G is non-locally finite uncountable group that has finitely many coarse ends. G has a subgroup H of finite index that has the same number of coarse ends as G; and H acts on its coarse ends trivially. Suppose H has at least 3 coarse ends. In that case we can find three mutually disjoint coarsely clopen subsets A_i of H, $1\le i\le 3$. Let $B_i := A_i^c$, and choose countable unbounded subsets C_i and D_i of each A_i and B_i, respectively. Since G is a non-locally finite group; H is a non-locally finite group, by 8.13. Choose a finite subset K of H such that $\langle K\rangle$ is infinite. Let H' be the subgroup of H generated by $K\cup\bigcup_{i=1}^{3}(C_i\cup D_i)$. Notice that H' is countably generated and hence it is countable. Moreover, each $A_i' = A_i\cap H'$ is an unbounded coarsely clopen subset of H' on which H' acts trivially. Now, since H' is non-locally finite, H' can be expressed as an increasing sequence of infinite finitely generated subgroups $\{H_n\}_{n\in\mathbb{N}}$. By 9.8, there exists $n\in\mathbb{N}$ such that each $A_i'\cap H_n$, $1\le i\le 3$, is unbounded coarsely clopen subset of H_n on which H_n acts trivially, contradicting 9.5. □

Now we generalize the Specker's theorem 1.8 that says that $|Ends(G)|=2$ if and only if G contains an infinite cyclic subgroup of finite index; by showing that any group of bounded geometry of two ends contains an unbounded cyclic subgroup of bounded index.

LEMMA 9.17. *Suppose K is a symmetric bounded subset of a large scale group G containing 1_G, B is a bounded subset of G contained in a bounded K-connected subset Q. If $G \setminus B$ has two unbounded K-components L and R on which G acts trivially, then G has an infinite cyclic subgroup of bounded index.*

PROOF. By 5.3, G is boundedly generated. It is indeed generated by K. To this end, it suffices to show that G is K-connected. Choose $a_L \in L$ and $a_R \in R$. Put $g := a_L \cdot a_R^{-1}$. Since $g^{-1} \cdot R \Delta R$ is bounded, there is $x \in R$ such that $g^{-1} \cdot x \in R$. Choose an K-chain c in R connecting a_R and $g^{-1} \cdot x$. Notice $g \cdot c$ is an K-chain connecting a_L and $x \in R$. Therefore G is K-connected and $B \neq \emptyset$. By switching to $b^{-1} \cdot B$, $b^{-1} \cdot L$, and $b^{-1} \cdot R$ for some $b \in B$, we may assume $1_G \in B \subset Q$. □

CLAIM 1. There exists $g \in R$ such that $g \cdot B \subset R$, $L \subset g \cdot L$, and $g \cdot R \subset R$.

PROOF OF THE CLAIM 1. Choose $g \in R \setminus (Q \cdot Q^{-1} \cup Q^{-1} \cdot Q)$, then $g \cdot Q \cap Q = \emptyset$, and hence $g \cdot Q \cap B = \emptyset$. In particular, $g \cdot Q$ is an K-connected subset of $G \setminus B$ with $g \in R \cap g \cdot Q$ which implies that $g \cdot Q \subset R$. To see that $L \subset g \cdot L$, notice that $L \cap g \cdot B = \emptyset$, and thus L is an unbounded K-connected subset of $G \setminus g \cdot B$ and therefore either $L \subset g \cdot L$ or $L \subset g \cdot R$. Since G acts trivially on L, we have $L \subset g \cdot L$. Finally, to prove that $g \cdot R \subset R$. Indeed, notice that $g^{-1} \cdot Q \cap Q = \emptyset$ and $g^{-1} \cdot B \subset g^{-1} \cdot Q$. So either $g^{-1} \cdot B \subset L$, or $g^{-1} \cdot B \subset R$. If $g^{-1} \cdot B \subset L$, then $B \subset g \cdot L$ which implies that $g \cdot R \subset G \setminus g \cdot L \subset G \setminus (L \cup B) = R$, and we are done. Seeking contradiction, assume that $g^{-1} \cdot B \subset R$. Then, $B \subset g \cdot R$ and therefore $g \cdot L \cap B = \emptyset$. Now, $g \cdot L$ is an K-connected subset of $G \setminus B$, and $g \cdot L \cap L \neq \emptyset$, so $g \cdot L \subset L$ leading to $g \cdot L = L$. Since G is K-connected, we can find a K-chain $a_R = a_0, a_1, \ldots, a_k = a_L$ in G connecting $a_R \in R$ to $a_L \in L$. There exists $1 \leq i \leq k-1$ such that $a_i \in B$ and $a_{i+1} \in L$. Since $(g \cdot a_i, g \cdot a_{i+1}) \in K$, and $g \cdot a_{i+1} \in L$, then $g \cdot a_i \in L$, as $g \cdot a_i \in G \setminus B$ and L is a K-component of $G \setminus B$. In particular, $g \cdot Q \cap L \neq \emptyset$, and hence $g \in g \cdot Q \subset L$, which contradicts the fact that $g \in R$. Therefore, $g \cdot R \subset R$ as claimed.

Inductively, we have:

- $g^n \cdot L \subset g^{n+1} \cdot L$,
- $g^n \cdot B \subset g^{n+1} \cdot L$,
- $g^{n+1} \cdot R \subset g^n \cdot R$,
- $g^{n+1} \cdot B \subset g^n \cdot R$,
- $g^n \cdot B \cap g^{n+1} \cdot B = \emptyset$, for all $n \in \mathbb{Z}$.

The last equality tells us that g is of infinite order. □

CLAIM 2. G has a cyclic subgroup of bounded index.

PROOF OF THE CLAIM 2. Put $H := \langle g \rangle$, and $K_0 := (g \cdot L \setminus L) \cup g \cdot B$. Clearly, K_0 is a coarsely bounded subset of G. We claim that $G = H \cdot K_0$. Given $x \in G$ find $m \in \mathbb{N} \cup \{0\}$ such that $d_K(x, g^m \cdot B) \leq d_K(x, g^n \cdot B)$, for all $n \in \mathbb{N}$. If $d_K(x, g^m \cdot B) = 0$, then $x \in g^m \cdot B \subset H \cdot K_0$. Assume that $d_K(x, g^m \cdot B) > 0$, then $x \notin g^m \cdot B$, and hence either $x \in g^m \cdot R$ or $x \in g^m \cdot L$. We consider two cases:

CASE 1. If $x \in g^m \cdot R$, then $x \in g^m \cdot R \setminus g^{m+1} \cdot R$. Indeed, seeking contradiction, assume that $x \in g^{m+1} \cdot R$, then as $g^m \cdot B \subset g^{m+1} \cdot L$, any K-chain connecting $x \in g^{m+1} \cdot R$ to $x \in g^m \cdot B$ must pass through $g^{m+1} \cdot B$ and hence $d_K(x, g^{m+1} \cdot B) \leq d_K(x, g^m \cdot B)$. Moreover, as $g^m \cdot B \cap g^{m+1} \cdot B = \emptyset$, $d_K(x, g^{m+1} \cdot B) < d_K(x, g^m \cdot B)$, a contradiction. Thus, $x \in g^m \cdot R \setminus g^{m+1} \cdot R \subset (g^{m+1} \cdot L \setminus g^m \cdot L) \cup g^{m+1} \cdot B = g^m \cdot (g \cdot L \setminus L \cup g \cdot B) \subset H \cdot K_0$.

CASE 2. If $x \in g^m \cdot L$, then $x \in g^m \cdot L \setminus g^{m-1} \cdot L$. Indeed, seeking contradiction, assume that $x \in g^{m-1} \cdot L$, then as $g^m \cdot B \subset g^{m-1} \cdot R$, any K-chain connecting $x \in g^{m-1} \cdot L$ to $x \in g^m \cdot B$ must pass through $g^{m-1} \cdot B$ and hence $d_K(x, g^{m-1} \cdot B) \leq d_K(x, g^m \cdot B)$. Moreover, as $g^{m-1} \cdot B \cap g^m \cdot B = \emptyset$, $d_K(x, g^{m-1} \cdot B) < d_K(x, g^m \cdot B)$, a contradiction. Thus, $x \in g^m \cdot L \setminus g^{m-1} \cdot L = g^{m-1} \cdot (g \cdot L \setminus L) \subset H \cdot K_0$. Therefore, $G = H \cdot K_0$ as desired. □

LEMMA 9.18. *If A is a coarsely clopen subset of a coarsely geodesic large scale group G with a large scale structure metrizable by a Cayley metric d_K, where K is a symmetric bounded subset of G containing 1_G and generates G; then there exists a bounded subset L of G such that every 1-component of $G \setminus L$ intersecting $A \setminus L$ is completely contained in $A \setminus L$. In particular, A is a union of 1-components of $G \setminus L$.*

PROOF. Let $L := (A \cdot K) \cap (A^c \cdot K)$. If $a \in A \setminus L$, then $a^{-1} \cdot g \notin K$ for all $g \in G \setminus A$. Now, let $g \in G$ such that $a^{-1} \cdot g \in K$, then $g \in A$. □

LEMMA 9.19. *Let G be a boundedly generated large scale group of bounded geometry. G acts coarsely trivially on a subset A and its complement A^c, for some non-trivial coarsely clopen subset $A \subset G$, if and only if it contains an unbounded cyclic subgroup of bounded index that has two coarse ends.*

PROOF. It is obvious that G is unbounded. If G is cyclic, then we have nothing to prove. Without loss of generality, we may assume that G is non-cyclic. By 8.8, G is coarsely geodesic group. Let $d_K : G \times G \to [0, \infty)$ be a Cayley metric induces its large scale structure, where K is a coarsely bounded symmetric subset of G containing 1_G. Let $A \subset G$ be a non-trivial coarsely clopen subset on which G acts coarsely trivially. By 7.6 and 9.18, we can find a coarsely bounded subset $L \subset G$ such that both $A \setminus L$ and $A^c \setminus L$ are union of unbounded 1-components (which are K-components) on which G acts coarsely trivially. Notice that K^n is K-connected for all $n \in \mathbb{N}$. Choose $r \in \mathbb{N}$ such that $r > 1$ and $L \subset K^r$. Pick $g \in A \setminus K^{2r}$. Assume that $d_K(g, 1_G) = m$. Notice that $m > 2r$ and for any $a \in K$, then $d_K(g \cdot a, 1_G) \geq 2r - 1$ and hence $g \cdot a \in A \setminus L$. Put $g := a_1 \cdot \ldots \cdot a_m$, where $a_1, \ldots, a_m \in K$, and let $F(K)$ be the free group generated by K; we discuss two cases:

CASE 1. If g is cyclically reduced in the sense that $a_1 \cdot a_m \neq 1_G$, then $(a_1 \cdot \ldots \cdot a_m)^n = (a_1 \cdot \ldots \cdot a_m) \cdot \ldots \cdot (a_1 \cdot \ldots \cdot a_m)$ is a reduced word, for all $n \in \mathbb{N}$. Since every element of G is uniquely represented as a reduced word in $F(K)$, we have $d_K(g^n, 1_G) = m \cdot n$. Put $H :=< g >$, then as seen in 9.17, $g^n \in A$, for all $n \in \mathbb{N}$, and H is infinite cyclic subgroup of bounded index in G. Furthermore, $g^n \notin K^n$, for all $n \in \mathbb{N}$, which proves that H is unbounded.

CASE 2. If $g = a_1 \cdot \ldots \cdot a_m$ is not cyclically reduced; that is $a_1 \cdot a_m = 1_G$; then as $G =< K >$ is not cyclic, there exists $a \in K$ such that $a_1 \cdot a \neq 1_G$. In particular, $g \cdot a \in A \setminus K^r$ and $g \cdot a$ is cyclically reduced. Hence, $H :=< g \cdot a >$ is unbounded cyclic subgroup of bounded index in G.

Finally, we prove that $H = \langle g \rangle$ has exactly two coarse ends. Without loss of generality, assume that g cyclically reduced and $d_K(g, 1_G) = m$. Then, $d_K(g^n, 1_G) = n \cdot m$, for all $n \in \mathbb{N}$. Clearly, $A_H = A \cap H$ is non-trivial coarsely clopen. Without loss of generality, assume that $A_H = \{g^n : n \in \mathbb{N}\}$ and $A_H^c = \{g^n : n \in \mathbb{Z} \setminus \mathbb{N}\}$. Choose $E_1, E_2 \in Ends(H)$ such that $A_H \in E_1$ and $A_H^c \in E_2$. We claim that

$Ends(H) = \{E_1, E_2\}$. To this end, we prove that for arbitrary unbounded coarsely clopen subset $C \subset H$ of H, then either $C \in E_1$ or $C \in E_2$. We may assume that C is non-trivial coarsely clopen.

CLAIM. If $C \cap A_H$ is unbounded, then $C \cap A_H$ and A_H are coarsely identical. Similarly, if $C \cap A_H^c$ is unbounded, then $C \cap A_H^c$ and A_H^c are coarsely identical.

PROOF OF THE CLAIM. Assume that $C \cap A_H$ is unbounded. Seeking contradiction, assume that there exists an infinite subset $I \subset \mathbb{N}$ such that $\mathbb{N} \setminus I$ is infinite and $C \cap A_H = \{g^n : n \in I\}$. Clearly, we can write $C \cap A_H = \{g^{n_i} : n_i \in \mathbb{N}\}$ as I is infinite countable subset of $\mathbb{N}$. Notice that there are infinitely many non-consecutive pairs (n_i, n_{i+1}) in the sense that $n_i < n_i + 1 < n_{i+1}$. Take $N > m + 1$, then $B(C, N) \cap C^c$ is infinite and hence unbounded, as K^r contains only finitely many elements of H for all $r \in \mathbb{N}$; a contradiction. Hence, $C \cap A_H$ and A_H are coarsely identical. Similarly, if $C \cap A_H^c$ is unbounded, then $C \cap A_H^c$ and A_H^c are coarsely identical.

If both $C \cap A_H$ and $C \cap A_H^c$ are unbounded, then $H \setminus C$ is bounded which contradicts the fact that C is non-trivial coarsely clopen. Thus, either $C \in E_1$ or $C \in E_2$ which implies that H has exactly two coarse ends. □

The converse is obvious. □

COROLLARY 9.20. *Let G be a σ-bounded large scale group of bounded geometry. G has two coarse ends if and only if it acts coarsely trivially on A and A^c for some non-trivial coarsely clopen subset $A \subset G$.*

PROOF. If G has two coarse ends, then it acts coarsely trivially A and A^c for some non-trivial coarsely clopen subset $A \subset G$, by 9.6. Assume that G acts coarsely trivially on A and A^c for some non-trivial coarsely clopen subset $A \subset G$. We consider two cases:

CASE 1. If G is boundedly generated, then G has two coarse ends by and 9.19.

CASE 2. Assume that G is not boundedly generated. By 8.15, G can be expressed as the union of a strictly increasing sequence $\{G_n\}_{n\in\mathbb{N}}$ of unbounded boundedly generated subgroups and that $\{G_n\}_{n\in\mathbb{N}}$ is a basis for boundedly generated subgroups. By 9.8, we may assume that $G_n \cap A$ and $G_n \cap A^c$ are unbounded for all $n \in \mathbb{N}$. By **Case 1**, each G_n has exactly two coarse ends. Applying 9.9, we obtain that $|Ends(G)| = 2$.

□

THEOREM 9.21. *Let G be a boundedly generated large scale group of bounded geometry. If G has two ends, then it contains an unbounded cyclic subgroup of bounded index.*

PROOF. Apply 9.6 and 9.19. □

LEMMA 9.22. *Let G be a coarse group containing two unbounded cyclic subgroups H and K. If H is of bounded index in G, then so is K.*

PROOF. Since there is a coarse equivalence $f : G \to H$ inverse to the inclusion $i : H \to G$, $f|K : K \to H$ is a coarse embedding, hence a coarse equivalence. Consequently, the inclusion $K \to G$ is a coarse equivalence and K is of bounded index in G. □

THEOREM 9.23. *Let G be a σ-bounded large scale group of bounded geometry. If G has two coarse ends, then it is boundedly generated.*

PROOF. Without loss of generality, assume that G acts trivially on its ends. By 8.4, its large scale structure is metrizable. Since G has finitely many ends, it is non-locally bounded by 9.3. Seeking contradiction, assume that G is not boundedly generated, then by 8.15 and 8.8, G admits a strictly increasing sequence $(G_n)_{n\in\mathbb{N}}$ of unbounded coarsely geodesic subgroups that is a basis for boundedly generated subgroups. Let A be a non-trivial coarsely clopen subset of G. By 9.8, there exists $n_0 \in \mathbb{N}$ such that $A \cap G_{n_0}$ and $A^c \cap G_{n_0}$ are unbounded. Without loss of generality, assume that $n_0 = 1$. Consequently, $A \cap G_n$ and $A^c \cap G_n$ are unbounded, for all $n \in \mathbb{N}$. By 9.17, each G_n has an unbounded cyclic subgroup H_n of bounded index.

CLAIM. For all $n \in \mathbb{N}$, there exists $N_n \in \mathbb{N}$ such that $N_n > n$, $A \cap G_{N_n} \setminus G_n \neq \emptyset$ and $A^c \cap G_{N_n} \setminus G_n \neq \emptyset$.

PROOF OF THE CLAIM. It suffices to prove that A and A^c cannot be contained in any G_n. Seeking contradiction, assume that $A \subset G_n$ for some $n \in \mathbb{N}$. Choose $z \in G_{n+1} \setminus G_n$. On one hand, if $A \cap A \cdot z \neq \emptyset$, then we can find $a, b \in A$ such that $z = a^{-1} \cdot b \in A^{-1} \cdot A \subset G_n$, a contradiction. On the other hand, if $A \cap A \cdot z = \emptyset$, then that contradicts the fact that A is unbounded coarsely clopen. Therefore, for all $n \in \mathbb{N}$, there exists $N_n \in \mathbb{N}$ such that $N_n > n$ and $A \cap G_{N_n} \setminus G_n \neq \emptyset$. Similarly, for all $n \in \mathbb{N}$, there exists $N_n \in \mathbb{N}$ such that $N_n > n$ and $A^c \cap G_{N_n} \setminus G_n \neq \emptyset$.

Now, put $H_1 :=< t >$; as $A\Delta A \cdot t$ and $A^c \Delta A^c \cdot t$ are bounded subsets, there exists $m \in \mathbb{N}$ such that $A\Delta A \cdot t, A^c \Delta A^c \cdot t \subset G_m$. Choose $N_m \in \mathbb{N}$ large enough such that $A \cap G_{N_m} \setminus G_m \neq \emptyset$ and $A^c \cap G_{N_m} \setminus G_m \neq \emptyset$; and pick $x \in A \cap G_{N_m} \setminus G_m$. Notice that $x \cdot t^n \in (G_{N_m} \setminus G_m) \cdot t^n = G_{N_m} \setminus G_m$, for all $n \in \mathbb{Z}$, as $< t > \subset G_m$. We claim that $x \cdot t^n \in A \cap G_{N_m} \setminus G_m$, for all $n \in \mathbb{Z}$. If $x \cdot t \notin A$, then $x \cdot t \in A \cdot t \setminus A \subset G_m$, a contradiction. Assume that $x \cdot t^n \in A \cap G_{N_m} \setminus G_m$, then $x \cdot t^{n+1} \in A \cdot t$. If $x \cdot t^{n+1} \notin A$, then $x \cdot t^{n+1} \in A \cdot t \setminus A \subset G_m$, a contradiction. Therefore, $x \cdot t^n \in A \cap G_{N_m} \setminus G_m$, for all $n \in \mathbb{Z}$ as claimed.

Put $B := \{x \cdot t^n : n \in \mathbb{Z}\}$; then $B \subset A \cap G_{N_m}$ which implies that $H_1 = x^{-1} \cdot B \subset x^{-1} \cdot A \cap G_{N_m}$. Since G acts trivially on A, $x^{-1} \cdot A \setminus A$ is bounded, and hence $H_1 \cap A$ is unbounded and $H_1 \cap A^c$ is bounded. One other hand, we can find $y \in A^c \cap G_{N_m} \setminus G_m$ such that $y \cdot t^n \in A^c \cap G_{N_m} \setminus G_N$. Let $C := \{y \cdot t^n : n \in \mathbb{Z}\}$; then $C \subset A^c \cap G_{N_m}$ which implies that $H_1 = y^{-1} \cdot C \subset y^{-1} \cdot A^c \cap G_{N_m}$. Since G acts trivially on A^c, $y^{-1} \cdot A^c \setminus A^c$ is bounded, and hence $H_1 \cap A^c$ is unbounded and $H_1 \cap A$ is bounded; a contradiction. □

□

THEOREM 9.24. *If $(G, \mathbb{LSS})$ is a large scale group of bounded geometry. Then, the following conditions are equivalent:*

(a) *G is boundedly generated with 2 coarse ends.*
(b) *G is coarsely geodesic with 2 coarse ends.*
(c) *$\mathbb{LSS}$ is metrizable and G has 2 coarse ends.*
(d) *G is σ-bounded with 2 coarse ends.*
(e) *G contains an unbounded cyclic subgroup of bounded index that has two coarse ends.*

PROOF. $(a) \implies (b)$ follows from 8.8.

$(d) \implies (e)$ follows from 9.21.

The other implications straightforward. □

COROLLARY 9.25. *Let G be a locally compact σ-compact topological group equipped with the large scale structure induced by the bornology of all pre-compact subsets. If G is not compactly generated, then either G is 1-ended or it has infinitely many ends.*

PROOF. Apply 9.24. □

COROLLARY 9.26. *Let G be a countable large scale group equipped with the large scale structure induced by the the bornology of all finite subsets. If G has two ends, then it is finitely generated and contains an infinite cyclic subgroup of finite index.*

PROOF. Apply 9.24. □

EXAMPLE 9.27. The group of the rationals $(\mathbb{Q}, +)$ has exactly one coarse end.

PROOF. Every finitely generated subgroup of $\mathbb{Q}$ is cyclic, and $\mathbb{Q}$ is not finitely generated. Therefore, $\mathbb{Q}$ can be written as a union of strictly increasing sequence of cyclic subgroups. From 9.9, and 9.26, $|Ends(\mathbb{Q})| = 1$. □

THEOREM 9.28. *Let G be a group equipped with the large scale structure induced by the bornology of all finite subsets. G has two coarse ends if and only if G is finitely generated and contains an infinite cyclic subgroup of finite index.*

PROOF. If G is countable, then we are done. Seeking contradiction assume that G is uncountable. If G is locally finite, then, by 9.16, it has one coarse end, a contradiction. Without loss of generality, we may assume that G acts on its coarse ends trivially. Since G is non-locally finite, there exists a finite subset $K \subset G$ such that $< K >$ is infinite subgroup. Let A be a non-trivial coarsely clopen subset of G. Let B and C be countably infinite subsets of A and A^c, respectively. Let H be arbitrary countable subgroup of G; then $H' :=< K \cup B \cup C \cup H >$ is a countable non-locally finite subgroup of G. Moreover, H' acts trivially on $A \cap H'$ and $A^c \cap H'$, with are unbounded coarsely clopen subsets of H'. By ??, H' has exactly two coarse ends. By 9.24, H' is a finitely generated subgroup of G containing H. In particular, every countable subgroup H of G is contained in a finitely generated subgroup H' that has two coarse ends.

Let $H_1 = H'$, and choose $x_1 \in G \setminus H_1$, and a finitely generated subgroup H_2 of two coarse ends containing $< H_1 \cup \{x_1\} >$. Inductively, we construct a strictly increasing sequence H_n of finitely generated subgroups of G each having two coarse ends. Put $G' := \bigcup_{n \in \mathbb{N}} H_n$. Clearly, G' is a countable subgroup of G. Moreover, G' has two coarse ends as $|Ends(G')| \leq 2$, by 9.9; and hence $|Ends(G')| = 2$ as $A \cap G'$ and $A^c \cap G'$ are unbounded coarsely clopen subsets of G' that belong to two distinct coarse ends of G'. Therefore, by 9.24, G' is a finitely generated subgroup of G. However, any finite subset $K' \subset G'$ is contained in H_n, for some $n \in \mathbb{N}$, which contradicts the fact that G' is finitely generated. Hence, G must be countable. □

10. Metrizablity of the space of coarse ends of coarsely geodesic large scale groups

We give the necessary and sufficient conditions for the space of coarse ends $Ends(G)$ of a coarsely geodesic large scale group G to be topologically metrizable. Then, we construct a metric on $Ends(G)$ that induces its topology. When $|Ends(G)| = \infty$, we show that the space of coarse ends is a Cantor space which generalizes part(ii) of 1.6.

LEMMA 10.1. *If A is a coarsely clopen subset of a coarsely geodesic coarse group G with a coarse structure metrizable by a Cayley metric d. Then:*

(a) *Every union of 1-components of A is coarsely clopen.*
(b) *There is a bounded subset L of G such that every 1-component of $G \setminus L$ intersecting $A \setminus L$ is completely contained in $A \setminus L$.*

PROOF.

(a) Let $K = \{g \in G : d(g, 1_G) \leq 1\}$. Notice that $\{E_{K^n}\}_{n \in \mathbb{N}}$ is a basis for $\mathcal{E}_d$, so it suffices to show that $E_{K^n}[C] \cap G \setminus C$ is bounded for all $n \in \mathbb{N}$. Since E_K is an entourage containing the diagonal, and A is coarsely clopen, then A and $E_K[A]$ are coarsely identical. Without loss of generality, we may assume that $E_K[A] = A$. Since $C \subset A$ is a union of 1-components of A, $E_K[C] = C \cdot K = C$. Therefore $E_{K^n}[C] = C \cdot K^n = C$ for all $n \geq 1$ and hence $E_{K^n}[C] \cap G \setminus C = \emptyset$. Thus, C is coarsely clopen.

(b) Let $L := (A \cdot K) \cap (A^c \cdot K)$. If $g \in A \setminus L$, then $g \notin E_K[G \setminus A]$. Now, let $h \in G$ such that $(g, h) \in E_K$, then $h \in A$. Indeed, if $h \in G \setminus A$, then $g \in E_K[G \setminus A]$, a contradiction. □

PROPOSITION 10.2. *Suppose G is a coarsely geodesic large scale group metrizable by a Cayley metric d. $Ends(G)$ is metrizable if both of the following conditions are satisfied:*

1. *For each bounded subset B of G the union of bounded 1-components of $G \setminus B$ is bounded.*
2. *For each bounded subset B of G its complement $G \setminus B$ has finitely many unbounded 1-components.*

Moreover, $Ends(G)$ can be described as the family of decreasing sequences $\{A_n\}_{n \in \mathbb{N}}$ of unbounded 1-components of $G \setminus K_i$, where $\{K_n\}_{n \in \mathbb{N}}$ is an increasing sequence of bounded subsets of G that is a basis of bounded subsets of G.

PROOF. Suppose $\{K_n\}_{n \in \mathbb{N}}$ is an increasing sequence of bounded subsets of G that is a basis of bounded subsets of G. By 10.1 every unbounded 1-component A_n of $G \setminus K_n$ is coarsely clopen. Therefore each decreasing sequence $\{A_n\}_{n \in \mathbb{N}}$ of unbounded 1-components of $G \setminus K_n$ is contained in an end E of G. It cannot be contained in two different ends E and E'. Indeed, in that case we can pick disjoint coarsely clopen subsets $C \in E$ and $D \in E'$. By 9.18 there are bounded subsets B_C and B_D of G such that $C \setminus B_C$ is a union of 1-components of $G \setminus B_C$ and $D \setminus B_D$ is a union of 1-components of $G \setminus B_D$. Find $n \in \mathbb{N}$ such that $B_C \cup B_D \subset K_n$. Notice A_n does not intersect at least one of $C \setminus B_C$ or $D \setminus B_D$, a contradiction.

Suppose $E \in Ends(G)$. By 9.18 there is a bounded subset B of G such that $A \setminus B$ is a union of 1-components of $G \setminus B$. Find $n \in \mathbb{N}$ such that $B \subset K_n$. Now, $A \setminus K_n$ is the union of 1-components of $G \setminus K_n$, so A must contain exactly one unbounded 1-component of $G \setminus K_n$. That shows $Ends(G)$ has a countable basis, hence it is metrizable. $\square$

In particular, the space of coarse ends of a group that belongs to the class of coarsely geodesic large scale groups of bounded geometry is topologically metrizable.

COROLLARY 10.3. *The space of coarse ends of a boundedly generated large scale group of bounded geometry is topologically metrizable. In particular, The space of coarse ends of a coarsely geodesic large scale group of bounded geometry is topologically metrizable*

PROOF. Apply 10.2 and 7.6. $\square$

Our next goal is to prove that the converse of 10.2 holds. To that end, we use the well-known facts that the number of ultrafilters of an infinite set X is $2^{|\mathcal{P}(X)|}$, and that a first countable separable Hausdorff space X has cardinality at most $2^{|\mathbb{N}|}$.

LEMMA 10.4. *Let X be a large scale space and $\{A_n\}_{n\in\mathbb{N}}$ be a family of mutually disjoint coarsely clopen subsets of X. If for each infinite subset P of naturals $\mathbb{N}$ the union $\bigcup_{n\in P} A_n$ is an unbounded coarsely clopen subset of X, then the space of ends $Ends(X)$ of X is not metrizable.*

PROOF. Let $\mathcal{B}$ be the Boolean quotient algebra of $\mathcal{P}(\mathbb{N})$ by the ideal of all finite subsets of $\mathbb{N}$. Given an ultrafilter $\mathcal{F}$ of $\mathcal{B}$ consisting of infinite sets, notice that as $\mathcal{F}$ satisfies the Finite Intersection Property, the family $\{Ends(\bigcup_{n\in P} A_n) : P \in \mathcal{F}\}$ of closed subsets of the compact space $Ends(X)$ satisfies the Finite Intersection Property, and hence $\bigcap_{P\in\mathcal{F}} Ends(\bigcup_{n\in P} A_n) \neq \emptyset$. For each ultrafilter $\mathcal{F}$ of $\mathcal{B}$, choose a coarse end $E_{\mathcal{F}} \in \bigcap_{P\in\mathcal{F}} Ends(\bigcup_{n\in P} A_n)$, and define the map $\varphi : \mathcal{U}(\mathcal{B}) \to Ends(X)$ sending each ultrafilter $\mathcal{F}$ of $\mathcal{B}$ to $E_{\mathcal{F}}$. Given two different ultrafilters $\mathcal{F}_1$ and $\mathcal{F}_2$ there are disjoint $P \in \mathcal{F}_1$ and $Q \in \mathcal{F}_2$ resulting in $E_{\mathcal{F}_1} \neq E_{\mathcal{F}_2}$. That means $Ends(X)$ contains at least $2^{|\mathcal{P}(\mathcal{B})|}$ points. Since $2^{|\mathcal{P}(\mathcal{B})|} > 2^{|\mathbb{N}|}$, $Ends(X)$ cannot be metrizable. $\square$

PROPOSITION 10.5. *Suppose G is a coarse group metrizable by a Cayley metric d. $Ends(G)$ is non-metrizable if and only if one of the following conditions is satisfied:*

1. *There is a bounded subset B of G such that the union of bounded 1-components of $G \setminus B$ is unbounded.*
2. *There is a bounded subset B of G such that of $G \setminus B$ has infinitely many unbounded 1-components.*

PROOF. In case of 1) choose a sequence A_n of bounded 1-components of $G \setminus B$ such that $g_n \in A_n$ and $d(g_n, 1_G) \to \infty$. Notice that for each infinite subset P of naturals $\mathbb{N}$ the union $\bigcup_{n\in P} A_n$ is an unbounded coarsely clopen subset of G by 10.1. Therefore, by 10.4, $Ends(G)$ is non-metrizable. In case of 2) choose a sequence A_n of unbounded 1-components of $G \setminus B$ that are mutually disjoint. Apply 10.1 and 10.4. $\square$

COROLLARY 10.6. *Suppose G is a coarsely geodesic large scale group metrizable by a Cayley metric d. The space of ends $Ends(G)$ is topologically metrizable if and only if both of the following conditions are satisfied:*

1. *For each bounded subset B of G the union of bounded 1-components of $G \setminus B$ is bounded.*
2. *For each bounded subset B of G its complement $G \setminus B$ has finitely many unbounded 1-components.*

Moreover, $Ends(G)$ can be described as the family of decreasing sequences $\{A_n\}_{n\in\mathbb{N}}$ of unbounded 1-components of $G \setminus K_n$, where $\{K_n\}_{n\in\mathbb{N}}$ is an increasing sequence of bounded subsets of G that is a basis of bounded subsets of G.

PROOF. Apply 10.2 and 10.5. □

Given a coarsely geodesic large scale group G metrizable by a Cayley metric d. If $Ends(G)$ is topologically metrizable, then there exists an increasing sequence $\mathcal{K} = (K_n)_{n\in\mathbb{N}}$ of bounded subsets of G that is a basis of bounded subsets of G, such that every decreasing sequences $(A_n)_{n\in\mathbb{N}}$ of unbounded 1-components of $G \setminus K_n$ determines exactly one coarse end, and that every coarse end is determined by a such sequence. We utilize these facts to construct an explicit metric on $Ends(G)$ that induces its topology. For the coarse ends $E, E' \in Ends(G)$ determined by the decreasing sequences $(A_n)_{n\in\mathbb{N}}$ and $(B_n)_{n\in\mathbb{N}}$ of of unbounded 1-components of $G \setminus K_n$, respectively; we define:

$$n_{\mathcal{K}}(E, E') = \begin{cases} \min\{n \in \mathbb{N} : A_n \cap B_n = \emptyset\} & \text{if } E \neq E' \\ \infty & \text{if } E = E' \end{cases}$$

LEMMA 10.7. *Let G be a coarsely geodesic large scale group metrizable by a Cayley metric d. Assume that $Ends(G)$ is topologically metrizable. Fix an increasing sequence $\mathcal{K} = (K_n)_{n\in\mathbb{N}}$ of bounded subsets of G that is a basis of bounded subsets of G, such that every decreasing sequences $(A_n)_{n\in\mathbb{N}}$ of unbounded 1-components of $G \setminus K_n$ determines exactly one coarse end, and that every coarse end is determined by a such sequence. Define the function $f : \mathbb{N} \to [0, \infty)$ by:*

$$f(n) = \exp(n)$$

and the function $d_{\mathcal{K}} : Ends(G) \times Ends(G) \to [0, \infty)$ by:

$$d_{\mathcal{K}}(E, E') := f(-n_{\mathcal{K}}(E, E')).$$

Then $(Ends(G), d_{\mathcal{K}})$ is an ultrametric space.

PROOF. Clearly the function $d_{\mathcal{K}} : Ends(G) \times Ends(G) \to [0, \infty)$ is non-negative symmetric and $d_{\mathcal{K}}(E, E') = 0$ if and only if $E = E'$. Let $E_1, E_2, E_3 \in Ends(G)$ be coarse ends determined by the decreasing sequences $(A_n)_{n\in\mathbb{N}}$, $(B_n)_{n\in\mathbb{N}}$ and $(C_n)_{n\in\mathbb{N}}$, respectively, where each A_n, B_n and C_n is an unbounded 1-component of $G \setminus K_n$. We claim that:

$$d_{\mathcal{K}}(E_1, E_3) \leq \max\{d_{\mathcal{K}}(E_1, E_2), d_{\mathcal{K}}(E_2, E_3)\}.$$

Without loss of generality, assume that $e_1 \neq e_3$, $\max\{d_{\mathcal{K}}(E_1, E_2), d(E_2, E_3)\} \leq d_{\mathcal{K}}(E_1, E_3)$ and $d_{\mathcal{K}}(E_1, E_2) < d_{\mathcal{K}}(E_1, E_3)$. We aim to show that $d_{\mathcal{K}}(E_2, E_3) = d_{\mathcal{K}}(E_1, E_3)$. To this end, it suffices to show that $d_{\mathcal{K}}(E_1, E_3) \leq d_{\mathcal{K}}(E_2, E_3)$ which is satisfied if and only if $n_{\mathcal{K}}(E_2, E_3) \leq n_{\mathcal{K}}(E_1, E_3)$. Since $d_{\mathcal{K}}(E_1, E_2) < d_{\mathcal{K}}(E_1, E_3)$, one has $n_{\mathcal{K}}(E_1, E_3) < n_{\mathcal{K}}(E_1, E_2)$ and hence $A_{n_{\mathcal{K}}(E_1,E_3)} = B_{n_{\mathcal{K}}(E_1,E_3)}$. However,

$A_{n_{\mathcal{K}}(E_1,E_3)} \cap C_{n_{\mathcal{K}}(E_1,E_3)} = \emptyset$ and therefore $B_{n_{\mathcal{K}}(E_1,E_3)} \cap C_{n_{\mathcal{K}}(E_1,E_3)} = \emptyset$ which implies that $n_{\mathcal{K}}(E_2,E_3) \leq n_{\mathcal{K}}(E_1,E_3)$. $\square$

THEOREM 10.8. *Let G be a coarsely geodesic large scale group metrizable by a Cayley metric d. Assume that $Ends(G)$ is topologically metrizable. Fix an increasing sequence $\mathcal{K} = (K_n)_{n\in\mathbb{N}}$ of bounded subsets of G as in 10.7. The ultrametric $d_{\mathcal{K}} : Ends(G) \times Ends(G) \to [0,\infty)$ given by $d_{\mathcal{K}}(E,E') := f(n_{\mathcal{K}}(E,E')) = \exp(-n_{\mathcal{K}}(E,E'))$ induces the topology of $Ends(G)$ as a subspace of $G \cup Ends(G)$.*

PROOF. Notice that the family:
$\{U_{end} : U$ is an unbounded 1-components of $G\backslash K_n$ for some $n \in \mathbb{N}\}$ is a basis for $Ends(G)$. Let $E \in Ends(G)$ be a coarse end determined by the decreasing sequence $(A_n)_{n\in\mathbb{N}}$, where each A_n is an unbounded 1-component of $G \setminus K_n$. For $r > 0$, we claim that the open ball $B_{d_{\mathcal{K}}}(E,r) = \{E' \in Ends(G) : d_{\mathcal{K}}(E,E') < r\}$ is open subset of $Ends(G)$ as a subspace of $G \cup Ends(G)$. Choose $n \in \mathbb{N}$ such that $f(n) < r$. Notice that U_{end}, where $U = A_n$, is an open set containing E and if $E'' \in U_{end}$ is a coarse end determined by the decreasing sequence $(B_n)_{n\in\mathbb{N}}$, where each B_n is an unbounded 1-component of $G \setminus K_n$, then $n_{\mathcal{K}}(E,E'') > n$ as $A_n = B_n$; and therefore, $d_{\mathcal{K}}(E,E'') = f(n_{\mathcal{K}}(E,E'')) < f(n) < r$. Thus, $E \in U_{end} \subset B_{d_{\mathcal{K}}}(E,r)$ as desired.

Conversely, pick a basis element U_{end} of $Ends(G)$. Then, $U = A_i$ for some an unbounded 1-component A_i of $G\backslash K_i$. Let $E \in U_{end}$ be a coarse end determined by the decreasing sequence $(A_n)_{n\in\mathbb{N}}$, where each A_n is an unbounded 1-component of $G \setminus K_n$, and choose $m \in \mathbb{N}$ such that $m > i$. Put $r := f(m)$ and let $E' \in B_{d_{\mathcal{K}}}(E,r)$ be a coarse end determined by the decreasing sequence $(B_n)_{n\in\mathbb{N}}$, where each B_n is an unbounded 1-component of $G \setminus K_n$, then $f(n_{\mathcal{K}}(E,E')) = d_{\mathcal{K}}(E,E') < r = f(m)$ which implies that $m < n_{\mathcal{K}}(E,E')$. Consequently, $A_m = B_m$ and as $i < m$, we must have $A_i = B_i$ and hence $B_{d_{\mathcal{K}}}(E,r) \subset U_{end}$. $\square$

Using the Brouwer's Theorem which states that a second countable totally disconnected compact Hausdorff space with no isolated points is homeomorphic to the Cantor middle-third set, we will show that if $|Ends(G)| = \infty$, then $Ends(G)$ is homeomorphic to the Cantor middle-third set.

COROLLARY 10.9. *Let G be a coarsely geodesic large scale group metrizable by a Cayley metric d. Assume that $Ends(G)$ is topologically metrizable. Fix an increasing sequence $\mathcal{K} = (K_n)_{n\in\mathbb{N}}$ of bounded subsets of G as in 10.7. If $|Ends(G)| = \infty$, then $Ends(G)$ is homeomorphic to the Cantor middle-third set.*

PROOF. We aim to show that every $e \in Ends(G)$ is an accumulation point. Indeed, let $E \in Ends(G)$ be a coarse end determined by the decreasing sequence $(A_n)_{n\in\mathbb{N}}$, where each A_n is an unbounded 1-component of $G\backslash K_n$. For $r > 0$ choose $n_r \in \mathbb{N}$ such that $\exp(-n_r) < r$. Since $Ends(G)$ is infinite, we can find $N \in \mathbb{N}$ such that $n_r \leq N$, and a coarse end $E' \in Ends(G)$ determined by the decreasing sequence $(B_n)_{n\in\mathbb{N}}$, where each B_n is an unbounded 1-component of $G \setminus K_n$ such that $A_N \cap B_N = \emptyset$ and $A_{N-1} \cap B_{N-1} \neq \emptyset$. Now, as $n_{\mathcal{K}}(E,E') = N \geq n_r$, one has $d_{\mathcal{K}}(E,E') \leq \exp(-n_r) < r$ and therefore $E' \in B_{d_{\mathcal{K}}}(E,r) \cap (Ends(G) \setminus \{E\})$. $\square$

Acknowledgements

The authors are grateful to Ross Geoghegan and Mike Mihalik for their help in understanding classical theory of ends of finitely generated groups.

References

[1] M. R. Bridson and A. Haefliger, *Metric spaces of non-positive curvature*, Grundlehren der mathematischen Wissenschaften [Fundamental Principles of Mathematical Sciences], vol. 319, Springer-Verlag, Berlin, 1999, DOI 10.1007/978-3-662-12494-9. MR1744486

[2] N. Brodskiy, J. Dydak, J. Higes, and A. Mitra, *Dimension zero at all scales*, Topology Appl. **154** (2007), no. 14, 2729–2740, DOI 10.1016/j.topol.2007.05.006. MR2340955

[3] N. Brodskiy, J. Dydak, and A. Mitra, *Švarc-Milnor lemma: a proof by definition*, Topology Proc. **31** (2007), no. 1, 31–36. MR2363149

[4] N. Brodskiy, J. Dydak, and A. Mitra, *Coarse structures and group actions*, Colloq. Math. **111** (2008), no. 1, 149–158, DOI 10.4064/cm111-1-13. MR2353938

[5] Y. Cornulier, *On the space of ends of infinitely generated groups*, Topology Appl. **263** (2019), 279–298, DOI 10.1016/j.topol.2019.05.013. MR3969224

[6] R. F. Dickman and R. A. McCoy, *The Freudenthal compactification*, Dissertationes Math. (Rozprawy Mat.) **262** (1988), 35. MR931673

[7] W. Dicks and M. J. Dunwoody, *Groups acting on graphs*, Cambridge Studies in Advanced Mathematics, vol. 17, Cambridge University Press, Cambridge, 1989. MR1001965

[8] C. Druţu and M. Kapovich, *Geometric group theory*, American Mathematical Society Colloquium Publications, vol. 63, American Mathematical Society, Providence, RI, 2018, DOI 10.1090/coll/063. MR3753580

[9] J. Dydak and C. S. Hoffland, *An alternative definition of coarse structures*, Topology Appl. **155** (2008), no. 9, 1013–1021, DOI 10.1016/j.topol.2008.01.002. MR2401211

[10] J. Dydak, *Ends and simple coarse structures*, Mediterr. J. Math. **17** (2020), no. 1, Paper No. 4, 22, DOI 10.1007/s00009-019-1437-4. MR4046844

[11] J erzy Dydak, *Unifying large scale and small scale geometry*, `arXiv:1803.09154` [math.MG]

[12] Jerzy Dydak and Hussain Rashed, *Ends of spaces via linear algebra*, in preparation

[13] R. Engelking, *Theory of dimensions finite and infinite*, Sigma Series in Pure Mathematics, vol. 10, Heldermann Verlag, Lemgo, 1995. MR1363947

[14] R. Geoghegan, *Topological methods in group theory*, Graduate Texts in Mathematics, vol. 243, Springer, New York, 2008, DOI 10.1007/978-0-387-74614-2. MR2365352

[15] M. Gromov, *Asymptotic invariants of infinite groups*, Geometric group theory, Vol. 2 (Sussex, 1991), London Math. Soc. Lecture Note Ser., vol. 182, Cambridge Univ. Press, Cambridge, 1993, pp. 1–295. MR1253544

[16] A. Leitner and F. Vigolo, *An invitation to coarse groups*, Lecture Notes in Mathematics, vol. 2339, Springer, Cham, 2023, DOI 10.1007/978-3-031-42760-2. MR4696828

[17] Mike Mihalik, *A Manual for Ends, Semistability and Simple Connectivity at Infinity for Groups and Spaces*, to appear.

[18] G. Peschke, *The theory of ends*, Nieuw Arch. Wisk. (4) **8** (1990), no. 1, 1–12. MR1056654

[19] E. Specker, *Endenverbände von Räumen und Gruppen* (German), Math. Ann. **122** (1950), 167–174, DOI 10.1007/BF01342894. MR38984

[20] J. Smith, *On asymptotic dimension of countable abelian groups*, Topology Appl. **153** (2006), no. 12, 2047–2054, DOI 10.1016/j.topol.2005.07.011. MR2237596

[21] J. R. Stallings, *On torsion-free groups with infinitely many ends*, Ann. of Math. (2) **88** (1968), 312–334, DOI 10.2307/1970577. MR228573

[22] M. H. Stone, *The theory of representations for Boolean algebras*, Trans. Amer. Math. Soc. **40** (1936), no. 1, 37–111, DOI 10.2307/1989664. MR1501865

[23] Hussain Rashed, *The Stone's Representation Theorems and compactifications of a discrete space*, `arXiv:2112.07821` [math.GN].

[24] J. Roe, *Lectures on coarse geometry*, University Lecture Series, vol. 31, American Mathematical Society, Providence, RI, 2003, DOI 10.1090/ulect/031. MR2007488

[25] J. Roe, *Hyperbolic groups have finite asymptotic dimension*, Proc. Amer. Math. Soc. **133** (2005), no. 9, 2489–2490, DOI 10.1090/S0002-9939-05-08138-4. MR2146189

[26] L. Zippin, *Two-ended topological groups*, Proc. Amer. Math. Soc. **1** (1950), 309–315, DOI 10.2307/2032374. MR36241

University of Tennessee, Knoxville, Tennessee 37996
Email address: jdydak@utk.edu

Xi'an Technological University, No.2 Xuefu zhong lu, Weiyang district, Xi'an, People's Republic of China 710021
Email address: mayuankui@xatu.edu.cn

Jubail Industrial College, jubail, KSA
Email address: rashedhs@rcjy.edu.sa, hrashed1@vols.utk.edu

Contemporary Mathematics
Volume **812**, 2025
https://doi.org/10.1090/conm/812/16268

The expansion set construction of classifying spaces for generalized Thompson groups

Daniel S. Farley

Abstract. We summarize the theory of *expansion sets*, which offer a convenient setting for establishing finiteness properties of generalized Thompson groups.

1. Introduction

In this note, we will describe a general way to build classifying spaces for generalized Thompson groups. For us, a *generalized Thompson group* has a *piecewise* definition. The basic example is Thompson's group F, which is a group of homeomorphisms of $[0,1]$ with a "local" definition: each $f \in F$ sends any sufficiently small standard dyadic interval $[a/2^n, (a+1)/2^n]$ to another by an increasing linear homeomorphism. In [**4**, **6**], the piecewise character of such definitions is expressed using inverse semigroups. An *inverse semigroup* (for the purposes of [**4**, **6**]) is a collection S of partial bijections of a given set X, such that S is closed under inverses and compositions. We let Γ_S denote the collection of bijections of X that locally agree with members of S, in a suitable sense (see Example 2.6). The groups Γ_S are the main object of interest. The class of such groups includes Thompson's groups F, T, and V (for which [**3**] is a standard reference), among many others.

More specifically, we are interested in the topological finiteness properties F_n of generalized Thompson groups, especially the F_∞ property. Recall that a group G has *type* F_∞ if there is a $K(G,1)$ complex having finitely many cells in each dimension.

We will describe a method that builds classifying spaces for generalized Thompson groups. The central idea is that of an *expansion set*, which offers a "piecewise" construction of such classifying spaces. The members of an expansion set $\mathcal{B}$ play something like the role of atoms in the construction of a simplicial complex $\Delta^f_{\mathcal{B}}$. Each $b \in \mathcal{B}$ has a *support*, denoted *supp*(b), and a partially ordered set $\mathcal{E}(b)$ of *expansions*. Vertices in $\Delta^f_{\mathcal{B}}$ are finite subsets of $\mathcal{B}$ with disjoint supports. The local topology of $\Delta^f_{\mathcal{B}}$ is entirely encoded by the sets $\mathcal{E}(b)$, and the complex $\Delta^f_{\mathcal{B}}$ has a natural height function. Thus, the theory of expansion sets facilitates the application of Brown's Finiteness Criterion [**2**], resulting in a proof that the given group has type F_∞ under favorable circumstances. Most of the required topological arguments are handled in [**4**], so the only input that is needed from the user is the choice of

---

2020 *Mathematics Subject Classification.* Primary 20F65, 20J05; Secondary 20M18.

expansion set, which is often not difficult to produce. As a special instance of our approach, we will sketch a proof that Thompson's group V has type F_∞ (a result first proved in [**2**]).

The theory of expansion sets, in effect, offers a generalized approach to the types of complexes considered by Stein [**12**]. The complex $\Delta^f_{\mathcal{B}}$ is frequently (but not always) locally finite, for instance.

This note is intended only as a guide and a summary. The interested reader should see [**4, 6**] for proofs, a more extensive introduction, and for more examples. (The main body of this paper is primarily a summary of [**4**], although [**6**] and [**4**] overlap somewhat. The similarities and differences are discussed in the introduction to [**4**].)

2. Expansion Sets

2.1. The definition, and an example. Let X be a (non-empty) set. By an *expansion set over* X, we mean a set $\mathcal{B}$ with certain additional properties. First, each $b \in \mathcal{B}$ has a *support*, denoted *supp*(b), which is a non-empty subset of X. A *vertex* is a finite non-empty subset $v = \{b_1, \ldots, b_n\}$ of $\mathcal{B}$ such that the supports of the b_i are pairwise disjoint. Let $\mathcal{V}$ denote the set of all vertices. For a given $v = \{b_1, \ldots, b_n\} \in \mathcal{V}$, we let

$$supp(v) = \bigcup_{i=1}^{n} supp(b_i); \quad P(v) = \{supp(b_1), \ldots, supp(b_n)\}.$$

The latter are the *support* of v and the *partition determined by* v, respectively.

Each $b \in \mathcal{B}$ is further assigned a partially ordered set $\mathcal{E}(b)$ of *expansions* from b. Each $\mathcal{E}(b)$ is a subset of $\mathcal{V}$, and collectively the $\mathcal{E}(b)$ are required to satisfy the following additional properties:

(1) the vertex $\{b\}$ is the least member of $\mathcal{E}(b)$. That is, if $v \in \mathcal{E}(b)$, then $\{b\} \leqslant v$;
(2) if $v_1, v_2 \in \mathcal{E}(b)$ and $v_1 \lneqq v_2$, then $P(v_2)$ properly refines $P(v_1)$;
(3) if $v_1, v_2 \in \mathcal{E}(b)$ and $v_1 = \{b'_1, \ldots, b'_m\}$, then $v_1 \leqslant v_2$ if and only if

$$v_2 = \bigcup_{i=1}^{m} v'_i$$

for some choices of $v'_i \in \mathcal{E}(b'_i)$ $(i = 1, \ldots, m)$.

This completes the definition of expansion set (see [**4**], Definition 2.1).

Example 2.1 (A class of "toy" examples). Recall that an *ultrametric on* X is a metric d such that, for all $x, y, z \in X$,

$$d(x, z) \leqslant max\{d(x, y), d(y, z)\}.$$

Let (X, d) be a compact ultrametric space. Each metric ball $B_\epsilon(x)$ $(x \in X;$ $\epsilon > 0)$ admits a unique maximal proper partition $P(B_\epsilon(x))$ into finitely many balls. Thus, if P' is any other proper partition of $B_\epsilon(x)$ into metric balls, then P' refines $P(B_\epsilon(x))$.

We can define an expansion set over X as follows. We let $\mathcal{B}$ be the set of all metric balls: $\mathcal{B} = \{B_\epsilon(x) \mid \epsilon > 0; x \in X\}$. The support of each metric ball is the ball itself: *supp*$(B_\epsilon(x)) = B_\epsilon(x)$. For each $B_\epsilon(x) \in \mathcal{B}$, we define $\mathcal{E}(B_\epsilon(x))$ as follows:

$$\mathcal{E}(B_\epsilon(x)) = \{\{B_\epsilon(x)\}, P(B_\epsilon(x))\}.$$

Here $\{B_\epsilon(x)\} \leqslant P(B_\epsilon(x))$. (Indeed, this is the only partial order on the set $\mathcal{E}(B_\epsilon(x))$ that is consistent with the definition of an expansion set.)

It is easy to verify that the assignment $(\mathcal{B}, X, supp, \mathcal{E})$ is an expansion set over X.

REMARK 2.2 (A clarification of the definition). We note, first, that (2) implies $supp(v) = supp(b)$, for each $v \in \mathcal{E}(b)$.

Condition (3) should be viewed as a "coherence" property of $\mathcal{E}(b)$, rather than a "closure" property. If $v_1 = \{b'_1, \ldots, b'_m\} \in \mathcal{E}(b)$, then it is generally *not* the case that a vertex of the form

$$v_2 = \bigcup_{i=1}^{m} v'_i$$

is in $\mathcal{E}(b)$ (where $v'_i \in \mathcal{E}(b'_i)$ for each i). (I.e., $\mathcal{E}(b)$ is not "closed under expansion", in the appropriate sense.) However, if one *assumes* that $v_1, v_2 \in \mathcal{E}(b)$ and $v_1 \leqslant v_2$, then v_2 does indeed admit such a description, as (3) asserts.

2.2. The simplicial complexes $\Delta_{\mathcal{B}}$ and $\Delta^f_{\mathcal{B}}$. Every expansion set $\mathcal{B}$ determines a simplicial complex $\Delta_{\mathcal{B}}$ and (under mild hypotheses) a certain subcomplex $\Delta^f_{\mathcal{B}}$. The set $\mathcal{V}$ from Subsection 2.1 is the vertex set of $\Delta_{\mathcal{B}}$. Simplices are determined by the partially ordered sets $\mathcal{E}(b)$ in the following way. Let $\{v_0, \ldots, v_k\} \subseteq \mathcal{V}$ be non-empty. Assume that

$$|v_0| < |v_1| < |v_2| < \ldots < |v_k|,$$

where the vertical bars denote cardinality. (If $|v_i| = |v_j|$ for some $i \neq j$, then $\{v_0, \ldots, v_k\}$ cannot be a simplex.) Assume that $v_0 = \{b_1, \ldots, b_\ell\}$. Here it is convenient to define restriction functions r_{b_i}:

$$r_{b_i}(v) = \{b \in v \mid supp(b) \subseteq supp(b_i)\}.$$

(In colloquial terms, $r_{b_i}(v)$ is the portion of v whose support lies in that of b_i.) The collection $\{v_0, \ldots, v_k\}$ is then a simplex if and only if

$$\{b_i\} = r_{b_i}(v_0) \leqslant r_{b_i}(v_1) \leqslant \ldots \leqslant r_{b_i}(v_k)$$

in the partially ordered set $\mathcal{E}(b_i)$, for each $b_i \in v_0$. We let $\mathcal{S}$ denote the set of simplices. Then $\Delta_{\mathcal{B}} = (\mathcal{V}, \mathcal{S})$ is a simplicial complex. (See [**4**], Subsection 2.1 for a more extended discussion, and for detailed proof.)

We note that $\Delta_{\mathcal{B}}$ has a natural height function, namely the assignment $v \mapsto |v|$. This enables us to speak of ascending and descending links and stars of vertices. The ascending star of a vertex $v_0 = \{b_1, \ldots, b_\ell\}$ is then very simple to describe: it is the simplicial product

$$\prod_{i=1}^{\ell} \mathcal{E}(b_i),$$

where we regard each $\mathcal{E}(b_i)$ as a simplicial complex in the usual way (i.e., members of $\mathcal{E}(b_i)$ are vertices, and non-empty finite ascending chains are the simplices). (See Proposition 2.16 from [**4**].) Moreover, each simplex $\sigma = \{v_0, \ldots, v_k\} \in \mathcal{S}$ is a simplex in the ascending star of its "lowest" vertex v_0 (assuming that $|v_0|$ is the vertex of minimal height in σ). In this sense, the above description of the ascending star entirely determines the simplicial structure of $\Delta_{\mathcal{B}}$.

We let $\Delta^f_{\mathcal{B}}$ denote the subcomplex of $\Delta_{\mathcal{B}}$ spanned by *full-support vertices*; i.e., vertices v such that $supp(v) = X$. We will generally want to assume that the set of full-support vertices is non-empty, and there is no real loss in assuming this.

Note that any two vertices of $\Delta_{\mathcal{B}}$ that are connected by an edge must necessarily have the same support, by Remark 2.2. Thus, $\Delta^f_{\mathcal{B}}$, if non-empty, is a connected component of $\Delta_{\mathcal{B}}$, and it is in fact the complex of interest to us. The remaining vertices of $\Delta_{\mathcal{B}}$ play supporting roles; in [**6**] such vertices were called *pseudovertices.* In [**4**], the latter term was dropped, but it remains useful to treat such vertices on an equal basis, since doing so allows us to conduct a "piecewise" analysis of the topology of $\Delta^f_{\mathcal{B}}$. We will consider the results of such an analysis in the next subsection, where we will summarize various facts about the local topology of $\Delta^f_{\mathcal{B}}$ (and of $\Delta_{\mathcal{B}}$).

Example 2.3. (The simplicial complex $\Delta_{\mathcal{B}}$ in the case of the toy example) In the class of examples considered in Example 2.1, each of the partially ordered sets $\mathcal{E}(b)$ has just two members: $\{b\}$ and $P(b)$. Here b is an ultrametric ball and $P(b)$ is the maximal proper partition of b into metric subballs. Since $\{b\} \leqslant P(b)$, the simplicial realization of $\mathcal{E}(b)$ is a line segment. It then follows from the above discussion that the ascending star of each vertex $v \in \Delta_{\mathcal{B}}$ is a cube. Thus, $\Delta_{\mathcal{B}}$ is the union of such cubes.

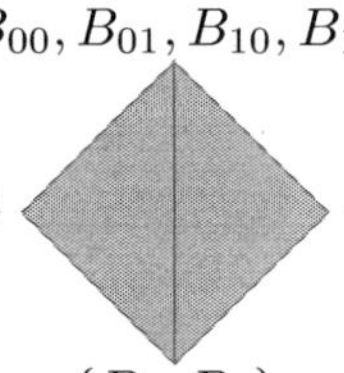

Figure 1. The ascending star of the vertex $\{B_0, B_1\}$ in $\Delta^f_{\mathcal{B}}$.

For the sake of specificity, consider the case in which X is the set of infinite binary strings $a_1a_2\dots$ ($a_i \in \{0,1\}$). Given two such strings ω_1 and ω_2, we define

$$d(\omega_1, \omega_2) = e^{-\ell(\omega_1,\omega_2)},$$

where $\ell(\omega_1, \omega_2)$ is the length of the longest prefix common to ω_1 and ω_2. The assignment d is an ultrametric, and (X, d) is a compact ultrametric space. The metric balls in (X, d) each take the form

$$B_\alpha = \{\omega \in X \mid \alpha \text{ is a prefix of } \omega\},$$

for some finite (possibly empty) binary string α. For a given ball $B_\alpha \in \mathcal{B}$, the maximal partition $P(B_\alpha)$ is $\{B_{\alpha 0}, B_{\alpha 1}\}$. Thus, we have the equality

$$\mathcal{E}(B_\alpha) = \{\{B_\alpha\}, \{B_{\alpha 0}, B_{\alpha 1}\}\}.$$

Figure 1 depicts the ascending star of $\{B_0, B_1\}$ in $\Delta^f_{\mathcal{B}}$.

Example 2.4 (The expansion set construction for groups defined by finite similarity structures). We next consider a class of groups that were first defined by Hughes [**9**]. Let (X, d) be a compact ultrametric space. A *finite similarity structure* on X, denoted Sim_X, assigns to each pair of metric balls $B_1, B_2 \subseteq X$ a finite (possibly empty) set $\mathrm{Sim}_X(B_1, B_2)$ of bijective similarities $h : B_1 \to B_2$. A function $h : (A, d_1) \to (B, d_2)$ between metric spaces is a *similarity* if there is a constant $\lambda > 0$ such that $d_2(h(a_1), h(a_2)) = \lambda d_1(a_1, a_2)$, for each $a_1, a_2 \in A$.

The sets $\mathrm{Sim}_X(B_1, B_2)$ are required to have various closure properties:

(1) (Compositions) If $h_1 \in \mathrm{Sim}_X(B_1, B_2)$ and $h_2 \in \mathrm{Sim}_X(B_2, B_3)$, then $h_2 \circ h_1 \in \mathrm{Sim}_X(B_1, B_3)$;
(2) (Inverses) If $h \in \mathrm{Sim}_X(B_1, B_2)$, then $h^{-1} \in \mathrm{Sim}_X(B_2, B_1)$;
(3) (Identities) The set $\mathrm{Sim}_X(B, B)$ contains the identity function id_B;
(4) (Restrictions) If $h \in \mathrm{Sim}_X(B_1, B_2)$ and B' is a subball of B_1, then $h_{|B'} \in \mathrm{Sim}_X(B', h(B'))$.

A function $\gamma : X \to X$ is *locally determined by* Sim_X if there are partitions $\mathcal{P} = \{B_1, \ldots, B_m\}$ and $\mathcal{P}' = \{B'_1, \ldots, B'_m\}$ of X into balls such that each restriction $\gamma_{|B_i}$ is a member of $\mathrm{Sim}_X(B_i, B'_i)$. Any such function γ is necessarily a homeomorphism. The collection of all such homeomorphisms forms a group under composition. This group, denoted $\Gamma(\mathrm{Sim}_X)$ (or Γ if Sim_X is understood) is the *group locally determined by* Sim_X.

A function $f : B \to X$ is a *local similarity embedding* if B is a metric ball in X and there is a partition $\mathcal{P} = \{B_1, \ldots, B_m\}$ of B into balls such that each image $f(B_i)$ is a ball and $f_{|B_i} \in \mathrm{Sim}_X(B_i, f(B_i))$ for $i = 1, \ldots, m$.

With the above background, we can now describe an expansion set $\mathcal{B} = \mathcal{B}(\mathrm{Sim}_X)$ for each finite similarity structure Sim_X. We first define

$$\mathcal{A} = \{(f, B) \mid f : B \to X \text{ is a local similarity embedding}\}.$$

We write $(f, B) \sim (g, B')$ if there is $h \in \mathrm{Sim}_X(B, B')$ such that $g \circ h = f$. The relation $\sim$ is an equivalence relation on $\mathcal{A}$ by the closure properties of Sim_X. The equivalence classes are denoted $[f, B]$, and we let $\mathcal{B}$ denote the set of all equivalence classes. For a given pair $[f, B] \in \mathcal{B}$, define $supp([f, B]) = f(B)$. Finally, letting $P(B)$ denote the maximal partition of the metric ball B, we set

$$\mathcal{E}([f, B]) = \{\{[f, B]\}, \{[f_{|B'}, B'] \mid B' \in P(B)\}\}.$$

(Note that $\mathcal{E}([f, B])$ is again a line segment.) It is straightforward to check that these assignments are well-defined, and make $\mathcal{B}$ an expansion set.

The resulting complexes $\Delta_{\mathcal{B}}$ and $\Delta^f_{\mathcal{B}}$ admit natural actions by the group Γ: for a given vertex $\{[f_1, B_1], \ldots, [f_m, B_m]\}$ and $\gamma \in \Gamma$, we define

$$\gamma \cdot \{[f_1, B_1], \ldots, [f_m, B_m]\} = \{[\gamma f_1, B_1], \ldots, [\gamma f_m, B_m]\}$$

It can be shown that the above action is simplicial and has finite vertex stabilizers.

REMARK 2.5 (A note on the pair notation $[f, B]$). Hughes [**9**] proved that the groups $\Gamma(\mathrm{Sim}_X)$ have the Haagerup property. The pairs $[f, B]$, exactly as defined above, were essential in that proof. Hughes and the author [**7**] showed that many of the groups $\Gamma(\mathrm{Sim}_X)$ have type F_∞. although the methods of [**7**] are somewhat different from those outlined here. (The main argument of [**7**] involves an application of Brown's Finiteness Criterion [**2**], but our work there owes more to Brown than to Stein [**12**].)

This pair notation plays a prominent role in both [**4**] and [**6**]. Such pairs almost always make up the expansion set $\mathcal{B}$ in practice, although the arguments in [**4**] are based on an abstract definition of expansion set (as outlined in Subsection 2.1).

EXAMPLE 2.6 (Groups that are locally determined by inverse semigroups). In [**6**], Hughes and the author generalized the basic construction of Example 2.4. A *partial bijection* of a set X is a bijection $s : A \to B$ between subsets of X. An *inverse semigroup acting on* X is a collection S of partial bijections of a set X that is closed under compositions and inverses. (Here composition is defined on overlaps: if s_1, s_2 are partial bijections of X, then $s_1 \circ s_2$ is defined at all points in the domain of s_2 whose s_2-images are members of the domain of s_1.)

Let X be a set and let S be an inverse semigroup acting on X. An injective function $f : Y \to X$ is *locally determined by* S if there is a finite subset $\{s_1, \dots, s_m\} \subseteq S$ such that f is a disjoint union of the s_i; i.e., if we let A_i denote the domains of the s_i, then $\{A_1, \dots, A_m\}$ is a partition of Y, and $f_{|A_i} = s_i$ for $i = 1, \dots, m$.

We let Γ_S denote the set of all bijections of X that are locally determined by S. It is straightforward to check that Γ_S is a group. This group is the main object of study in [**4**] and [**6**].

We now give a partial description of an expansion set $\mathcal{B}$ associated to Γ_S. First, we let $\mathcal{D}_S^+$ be the set of non-empty *domains*; i.e.

$$\mathcal{D}_S^+ = \{D \subseteq X \mid D \text{ is the domain of some } s \in S; D \neq \varnothing\}.$$

We let

$$\mathcal{A} = \{(f, D) \mid f \text{ is locally determined by } S \text{ with domain } D \subseteq X\}.$$

We write $(f_1, D_1) \sim (f_2, D_2)$ if there is a partial bijection $h : D_1 \to D_2$ in S such that $f_2 \circ h = f_1$. The relation $\sim$ is an equivalence relation; we let $[f, D]$ denote the equivalence class of (f, D). Now let

$$\mathcal{B} = \{[f, D] \mid (f, D) \in \mathcal{A}\}.$$

We define *supp*$([f, D]) = f(D)$ (exactly as in Example 2.4). The assignment *supp* is well-defined.

The final step is to define the expansions $\mathcal{E}(b)$ for each $b \in \mathcal{B}$. Here the specific definition of $\mathcal{E}$ will depend on the inverse semigroup, and even on the desired properties of the complex $\Delta_{\mathcal{B}}^f$. We must therefore refer the interested reader to [**4**] for a wider discussion, which includes the Brin-Thompson groups nV [**8**], Röver's group [**1**], the Lodha-Moore group [**10**], and generalizations of the Lodha-Moore group [**5**]. (Citations indicate the first proofs that the groups in question have type F_∞; see also [**11**], which proves the F_∞ property for a "T-like" version of the Lodha-Moore group.)

2.3. The topology of the complexes $\Delta_{\mathcal{B}}^f$ and the F_∞ property. This subsection is devoted to the topological properties of the complexes $\Delta_{\mathcal{B}}^f$, specifically in connection with Brown's Finiteness Criterion [**2**]. The approach developed in [**4**], and to be summarized here, reduces all of the relevant topology to that of the sets $\mathcal{E}(b)$, where the latter are to be considered as simplicial complexes.

2.3.1. *A sufficient condition for contractibility of* $\Delta_{\mathcal{B}}^f$. An edge-path in $\Delta_{\mathcal{B}}$ is *ascending* if the heights of the vertices along the path strictly increase. We write $v_1 \preceq v_2$ if there is an ascending edge-path from v_1 to v_2. It is clear that $\preceq$ is a partial order on the vertices of $\Delta_{\mathcal{B}}$.

For any $b \in \mathcal{B}$, the simplicial complex $\mathcal{E}(b)$ is a subcomplex of $\Delta_{\mathcal{B}}$. For a vertex $v \in \Delta_{\mathcal{B}}$ such that $\{b\} \precneq v$, we can therefore consider the subcomplex $\mathcal{E}^v(b) \subseteq \mathcal{E}(b)$ spanned by all vertices $u \in \mathcal{E}(b)$ such that $\{b\} \preceq u \preceq v$. We refer to the link of

$\{b\}$ in $\mathcal{E}^v(b)$ as the *ascending link of* $\{b\}$ *relative to* v. (Recall that $\{b\}$ has minimal height within $\mathcal{E}(b)$, so that the entire link of $\{b\}$ in $\mathcal{E}^v(b)$ is an ascending link.)

Recall that a partially ordered set is *directed* if any two of its members have a common upper bound.

We can now state a sufficient condition for the complex $\Delta^f_{\mathcal{B}}$ to be contractible:

THEOREM 2.7 (see Theorem 2.32 from [**4**]). *If*

(1) *the vertices of* $\Delta^f_{\mathcal{B}}$ *are a directed set with respect to* $\preceq$, *and*
(2) *for each* $b \in \mathcal{B}$ *and* v *such that* $\{b\} \precneqq v$, *the ascending link of* $\{b\}$ *relative to* v *is contractible,*

then $\Delta^f_{\mathcal{B}}$ *is contractible.*

The proof uses ideas similar to those from Stein [**12**]; the author learned the basic ideas from [**8**]. One argues that any map of a sphere $f : S^n \to \Delta^f_{\mathcal{B}}$ can be homotoped by local "pushes" to any vertex $\tilde{v}$ that is a common upper bound for the carrier of $f(S^n)$. (See Subsection 2.3 from [**4**].)

EXAMPLE 2.8 (Contractibility of the complexes from Examples 2.3 and 2.4). Let $\mathcal{B}$ be the expansion set from Example 2.3. We note that a vertex in $\Delta^f_{\mathcal{B}}$ is simply a partition of X into finitely many metric balls. We check the hypotheses of Theorem 2.7.

We first argue that the vertices of $\Delta^f_{\mathcal{B}}$ are a directed set with respect to $\preceq$. If $v_1 = \{b_1, \ldots, b_m\}$ and v_2 are connected by an edge, and $v_1 \preceq v_2$, it follows from the definitions of the sets $\mathcal{E}(b)$ and from the description of the ascending star of v_1 that v_2 is obtained by replacing some or all of the $b_i \in v_1$ with the members of their maximal partitions. It follows that, more generally, $v_1 \preceq v_2$ if and only if v_2 refines v_1. Thus, the statement that two vertices $v', v'' \in \Delta^f_{\mathcal{B}}$ have a common upper bound is equivalent to the statement that the partitions v', v'' have a common (finite) refinement by a collection of metric balls. Since the latter is clear, any two vertices have a common upper bound, proving (1).

Now suppose that $\{b\} \preceq v$, where $v \in \Delta_{\mathcal{B}}$. It follows that there is a sequence

$$\{b\} = v_0, v_1, v_2, \ldots, v_k = v,$$

where $|v_i| < |v_{i+1}|$ and v_{i+1} is a member of the ascending star of v_i, for $i = 0, \ldots, k-1$. Note that the above description entirely determines v_1: it must be that $v_1 = P(b)$, the maximal partition of the ultrametric ball b. (This is because the ascending star of $\{b\}$ is $\mathcal{E}(b)$, and $P(b)$ is the only vertex of $\mathcal{E}(b)$ having height greater than $|\{b\}| = 1$.) It follows that the ascending link of $\{b\}$ relative to v is the vertex $v_1 = P(b)$, which is contractible. It now follows from Theorem 2.7 that $\Delta^f_{\mathcal{B}}$ is contractible.

Next, we consider the case of Example 2.4. For a given pair $[f, B] \in \mathcal{B}$, there is a partition $\mathcal{P} = \{A_1, \ldots, A_\ell\}$ of B into ultrametric balls such that $f_{|A_i} \in \mathrm{Sim}_X(A_i, f(A_i))$, for $i = 1, \ldots, \ell$. The partition $\mathcal{P}$ can be obtained from B by a process of repeatedly replacing a given ball B' by the balls from its maximal partition. Such a process of subdivision can be mirrored by a sequence of expansions from $\{[f, B]\}$, which results in an ascending path from $\{[f, B]\}$ to

$$\{[f_{|A_1}, A_1], \ldots, [f_{|A_\ell}, A_\ell]\}.$$

The definition of the equivalence relation $\sim$ now implies that, for all i,

$$[f_{|A_i}, A_i] \sim [id, f(A_i)],$$

since $f_{|A_i} \in \mathrm{Sim}_X(A_i, f(A_i))$ by hypothesis.

It follows that there is an ascending path from $\{[f, B]\}$ to a vertex of the general form $v' = \{[id, B'_1], \ldots, [id, B'_m]\}$, for certain metric balls B_i. More generally, for a vertex of the form $v = \{[f_1, B_1], \ldots, [f_k, B_n]\} \in \Delta^f_{\mathcal{B}}$, we can apply the above reasoning entry-by-entry to produce an ascending edge-path from v to a vertex v' with the above form (i.e., one with "*id*" in the first coordinate of each pair). Let us call such vertices *positive*. A positive vertex v' can effectively be identified with a partition of X into balls. Any two positive vertices have a common upper bound, by essentially the reasoning given for the case of Example 2.3.

Thus, given any two vertices v_1, v_2 in $\Delta^f_{\mathcal{B}}$, we can find positive vertices v'_1, v'_2 such that $v_i \leq v'_i$ for $i = 1, 2$. We can, further, find a common upper bound v' for v'_1 and v'_2. Thus, v' is a common upper bound for v_1 and v_2, proving that the vertex set of $\Delta^f_{\mathcal{B}}$ is directed.

The proof of (2) from Theorem 2.7 is essentially identical to the one for the case of Example 2.3. It therefore follows that the complexes from Example 2.4 are contractible.

REMARK 2.9 (A remark about the general case). Theorem 2.7 facilitates proofs of contractibility by focusing effort on the choice of $\mathcal{E}(b)$, which is often a finite simplicial complex in practice. (It is not necessarily so; the Lodha-Moore group is among the exceptions – see Subsection 4.6 from [**4**].) The challenge is to make $\mathcal{E}(b)$ large enough that the vertices of $\Delta^f_{\mathcal{B}}$ are a directed set under $\leq$, while still keeping $\mathcal{E}(b)$ small enough that it is possible to show that the relative ascending links are contractible. For lack of space, we must refer the reader to Section 4 of [**4**] for a discussion of additional examples.

2.3.2. *The F_∞ property.* Finally, we state the main result of [**4**] in a somewhat adapted form (see Theorem 3.16 of [**4**]):

THEOREM 2.10 (A template for proving that Γ_S has type F_∞). *Assume that the hypotheses of Theorem 2.7 are satisfied. Let $\widehat{S}$ be the inverse semigroup of injections that are locally determined by S (See Example 2.6). If*

(1) *the support of each $b \in \mathcal{B}$ is the domain of some member of $\widehat{S}$;*
(2) *each stabilizer group $\widehat{S}_b$ ($b \in \mathcal{B}$) has type F_∞, and each set $\mathcal{E}(b)$ is finite;*
(3) *the action of Γ_S on $\mathcal{B}$ has finitely many orbits;*
(4) *$\mathcal{B}$ is rich in contractions and has the bounded contractions property,*

then Γ has type F_∞.

The proof is essentially an application of Brown's Finiteness Criterion [**2**].

REMARK 2.11 (Discussion of Theorem 2.10).

- The action of $\widehat{S}$ on $\mathcal{B}$ is the natural one: if $[f, B] \in \mathcal{B}$, $\gamma \in \widehat{S}$, and *supp*$([f, B])$ is contained in the domain of γ, then we define $\gamma \cdot [f, B] = [\gamma f, B]$. (In [**4**], the action in question is treated in a more abstract setting. See Subsection 3.1-3.3 of [**4**] for details.)
- (2) guarantees that the cell stabilizers have type F_∞, while (3) guarantees that Γ_S acts cocompactly on each finite-height subcomplex $\Delta^f_{\mathcal{B},n}$ (i.e., the subcomplex of $\Delta^f_{\mathcal{B}}$ spanned by vertices of height n or less). Here, the stabilizer group $\widehat{S}_b$ has a specialized definition:

$$\widehat{S}_b = \{\hat{s} \in \widehat{S} \mid dom(\hat{s}) = supp(b);\ \hat{s} \cdot b = b\}.$$

- The expansion set $\mathcal{B}$ is said to have the *bounded contractions property* if there is a global bound on the height of vertices v that appear in the sets $\mathcal{E}(b)$; i.e., there is some fixed constant C_0 such that, if $b \in \mathcal{B}$ and $v \in \mathcal{E}(b)$, then $|v| \leqslant C_0$. We say that $\mathcal{B}$ has the *rich-in-contractions property* if there is some fixed constant C_1 such that, if $|v| \geqslant C_1$, then there is some vertex $v' \subseteq v$ such that $v' \in \mathcal{E}(b)$, for some $b \in \mathcal{B}$.

  These two conditions are used to analyze the descending links in $\Delta^f_{\mathcal{B}}$. In the presence of these two properties, proving that the descending link is highly connected becomes a simple combinatorial problem. The details of the analysis are in Subsection 2.5 of [**4**].
- Theorem 3.16 from [**4**] covers groups of "F", "T", and "V" types. The version given here covers only the "V" groups.

EXAMPLE 2.12 (A proof that Thompson's group V has type F_∞). Let X be the set of infinite binary strings and, for a finite binary string α, let B_α denote the set of all infinite binary strings having α as a prefix.

If α and β are finite (possibly empty) binary strings, we define $\sigma^\beta_\alpha : B_\alpha \to B_\beta$ to be the transformation that removes the prefix α from a string in the domain and replaces it with β. It is a straightforward exercise to check that

$$S = \{\sigma^\beta_\alpha \mid \alpha, \beta \text{ are finite binary strings } \} \cup \{0\}$$

is an inverse semigroup acting on X. (The 0 element is the empty function.)

We can now define $\mathcal{B}$ exactly as in Example 2.6. For each $[f, B_\alpha] \in \mathcal{B}$, we define

$$\mathcal{E}([f, B_\alpha]) = \{\{[f, B_\alpha]\}, \{[f_|, B_{\alpha 0}], [f_|, B_{\alpha 1}]\}\}.$$

(Thus, $\mathcal{E}([f, B_\alpha])$ is a line segment, exactly as in Example 2.4.)

We now check the hypotheses of Theorem 2.10. We note, first, that the hypotheses of Theorem 2.7 are satisfied, by the argument from Example 2.8. Next, the support of $[f, B_\alpha]$ is $f(B_\alpha)$, which is the domain of $f^{-1} \in \widehat{S}$, so (1) is satisfied. For a given $b = [f, B_\alpha]$ and $\gamma \in \widehat{S}_b$, we have the equation $[f, B_\alpha] = [\gamma f, B_\alpha]$, which easily implies that $\gamma = id_{supp(b)}$ by the definition of $\sim$. Thus $\widehat{S}_b$ is trivial, so (2) is satisfied. It can be shown (see the proof of Theorem 4.8 from [**4**]) that there are two orbits under the action of Γ_S on $\mathcal{B}$ (one consisting of all b such that *supp*$(b) = X$, the other consisting of the remaining b). Thus, (3) is satisfied. Finally, we note that the bounded contractions property is satisfied with constant $C_0 = 2$ by the definition of $\mathcal{E}$, and the rich-in-contractions property is also satisfied with constant $C_1 = 2$.

To see the last point, consider any two-element vertex $v = \{[f, B_\alpha], [g, B_\beta]\}$. Up to equivalence, we can write

$$[f, B_\alpha] = [f\sigma^\alpha_0, B_0] \quad \text{and} \quad [g, B_\beta] = [g\sigma^\beta_1, B_1].$$

We then define $\hat{g} : X \to X$ to be the member of $\widehat{S}$ that agrees with $f\sigma^\alpha_0$ on B_0, and with $g\sigma^\beta_1$ on B_1. It follows that

$$\mathcal{E}([\hat{g}, X]) = \{\{[\hat{g}, X]\}, \{[f, B_\alpha], [g, B_\beta]\}\},$$

so $v \in \mathcal{E}([\hat{g}, X])$, as required.

Acknowledgements

I would like to thank the organizers of Mike Mihalik's 70th birthday conference for the suggestion that a short paper of this kind could be suitable for the conference proceedings.

References

[1] J. Belk and F. Matucci, *Röver's simple group is of type F_∞*, Publ. Mat. **60** (2016), no. 2, 501–524, DOI 10.5565/PUBLMAT_60216_07. MR3521498

[2] K. S. Brown, *Finiteness properties of groups*, Proceedings of the Northwestern conference on cohomology of groups (Evanston, Ill., 1985), J. Pure Appl. Algebra **44** (1987), no. 1-3, 45–75, DOI 10.1016/0022-4049(87)90015-6. MR885095

[3] J. W. Cannon, W. J. Floyd, and W. R. Parry, *Introductory notes on Richard Thompson's groups*, Enseign. Math. (2) **42** (1996), no. 3-4, 215–256. MR1426438

[4] Daniel Farley, *Finiteness properties of generalized Thompson groups via expansion sets*, arXiv:2312.10211.

[5] Daniel S. Farley, *Finiteness properties of some groups of piecewise projective homeomorphisms*, arXiv:2204.03278.

[6] Daniel S. Farley and Bruce Hughes, *Finiteness properties of locally defined groups*, arXiv:2010.08035.

[7] Daniel S. Farley and Bruce Hughes, *Finiteness properties of some groups of local similarities*, Proc. Edinb. Math. Soc. (2) **58** (2015), no. 2, 379–402. MR3341445

[8] M. G. Fluch, M. Marschler, S. Witzel, and M. C. B. Zaremsky, *The Brin-Thompson groups sV are of type* F_∞, Pacific J. Math. **266** (2013), no. 2, 283–295, DOI 10.2140/pjm.2013.266.283. MR3130623

[9] B. Hughes, *Local similarities and the Haagerup property*, Groups Geom. Dyn. **3** (2009), no. 2, 299–315, DOI 10.4171/GGD/58. With an appendix by Daniel S. Farley. MR2486801

[10] Y. Lodha, *A nonamenable type* F_∞ *group of piecewise projective homeomorphisms*, J. Topol. **13** (2020), no. 4, 1767–1838, DOI 10.1112/topo.12172. MR4186144

[11] Y. Lodha and M. C. B. Zaremsky, *The BNSR-invariants of the Lodha-Moore groups, and an exotic simple group of type* F_∞, Math. Proc. Cambridge Philos. Soc. **174** (2023), no. 1, 25–48, DOI 10.1017/S0305004122000111. MR4523108

[12] M. Stein, *Groups of piecewise linear homeomorphisms*, Trans. Amer. Math. Soc. **332** (1992), no. 2, 477–514, DOI 10.2307/2154179. MR1094555

Department of Mathematics, Miami University, Oxford, Ohio 45056
Email address: farleyds@miamioh.edu

Contemporary Mathematics
Volume **812**, 2025
https://doi.org/10.1090/conm/812/16269

Le Conte de la Mesure sur les Complexes Cubiques CAT(0)

Talia Fernós

In honor of Mike Mihalik's 70th birthday, happy birthday!

Abstract. We revisit the topic of probability measures on CAT(0) cube complexes and prove that an amenable group acting on a CAT(0) cube complex, regardless of dimension, necessarily preserves an interval in the Roller compactification. In the finite dimensional case, we prove that there must be an orbit of cardinality 2^N, where N is bounded by the dimension. This is a slight extension of the author's previuos Tits Alternative.

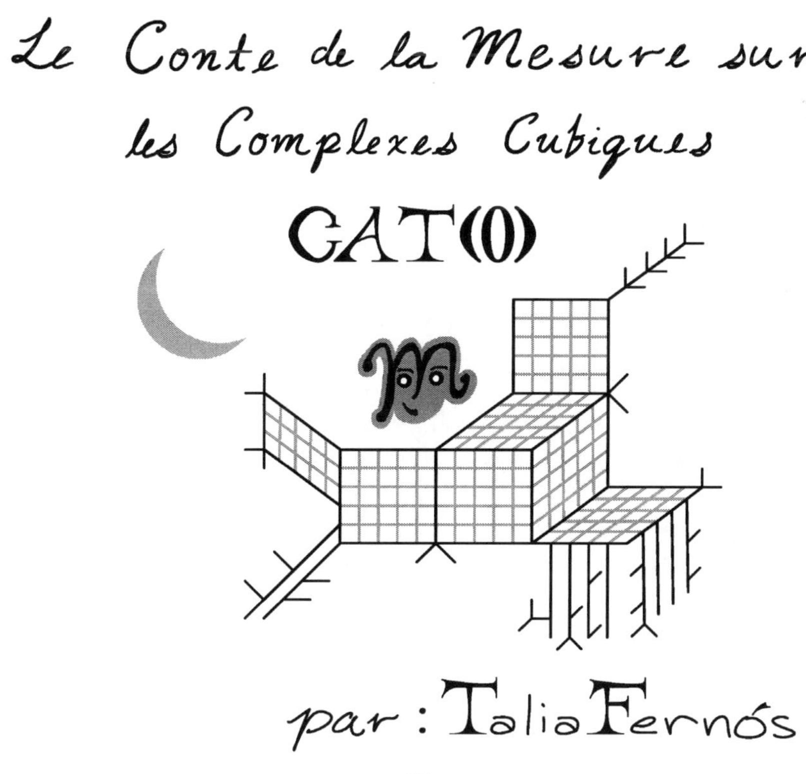

1. Introduction

The idea for this short paper came through preparations for a talk I gave in summer of 2023 at the Séminaire Virtuel Francophone – Groupes et Géométrie. The seminar is aimed to be expository in nature and so I thought I would revisit the topic of probability measures on CAT(0) cube complexes, and their associated intervals (see Section 3.1). This association was introduced by Monod and Shalom in [**MS04**, Proposition 4.5] for trees and developed further in my joint work with Chatterji and Iozzi [**CFI16**]. It was also used either implicitly or explicitly in [**Fer18**, **FLM18**, **FLM21**], and even [**FFH23**]. The key tool for this is the Lifting Decomposition which is discussed in Section 3.

As I was preparing my lecture, I realized that I could in fact make a small improvement on my previous contribution to the story of Tits Alternatives. Therefore, I take the opportunity to share that here. I also include some of the drawings I made (using notability on an ipad).

THEOREM. *Let X be a CAT(0) cube complex of finite dimension D, and suppose that $\Gamma \to \mathrm{Aut}\, X$ preserves an interval in the Roller compactification $\overline{X}$. Then Γ must have an orbit in $\overline{X}$ of cardinality 2^N, for some $0 \leqslant N \leqslant D$.*

If X is not finite dimensional and Γ is assumed to be amenable then it must preserve an interval $\mathcal{I} \subset \overline{X}$.

Combining with the Tits Alternative from [**Fer18**], we obtain the following (the reader should compare statements):

COROLLARY (Tits Alternative). *Suppose X is a finite dimensional CAT(0) cube complex. Given a group action $\Gamma \to \mathrm{Aut}\, X$ either Γ contains a freely acting free group on 2 generators, or Γ has an orbit of cardinality 2^N in $\overline{X}$, for some $0 \leqslant N \leqslant D$, where D is the dimension of X.*

Examples. The following examples show that the bounds above are optimal. We note that all the groups are amenable.

- ⋆ Consider the standard action of the infinite dihedral group D_∞ on $\mathbb{Z}$. Of course, there is no fixed point but there is an orbit of cardinality 2 in the Roller compactification, namely $\{\pm\infty\}$.
- ⋆ Similarly, D_∞^N acts on $\mathbb{Z}^N$ and has a unique finite orbit which has cardinality 2^N.
- ⋆ Finally, $\bigoplus\limits_{n\in\mathbb{N}} \mathrm{D}_\infty$ acts on $\bigoplus\limits_{n\in\mathbb{N}} \mathbb{Z}$ which is of course infinite dimensional, and does not have a finite orbit in the Roller compactification, but it is an interval.

We note that there are *many* Tits Alternatives available in the literature [**Tit72**, **BFH04**, **SW05**, **CS11**, **Fio18**, **MP22**, **OP22**, **GJN23**, **Gen21**]. However, in this brief note, I will not endeavor to give a comprehensive overview of the Tits Alternatives, nor to give a thorough account of CAT(0) cube complexes. The reader can see the references provided for further details.

2020 *Mathematics Subject Classification.* Primary 20F65, 20E05, 60B05; Secondary 05C12, 05E18, 06A07, 20F29.

Key words and phrases. CAT(0), Cube Complexes, Tits Alternative.

The author was partially supported by NSF grant DMS-2005640.

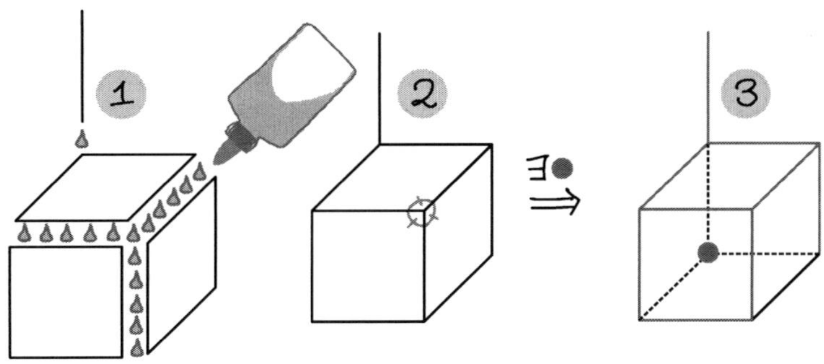

FIGURE 1. Making a cube complex CAT(0)

Acknowledgments

I thank Dafne Sanchez for help with the coloring scheme, inspired by Matisse. I also thank Indira Chatterji, François Dahmani, Anne Lonjou, and Yves Stalder for the invitation to speak at the Séminaire Virtuel Francophone – Groupes et Géométrie which inspired me to write this article. I thank all of my collaborators and mentors with whom I have discussed CAT(0) cube complexes for helping me explore how to understand and communicate their beauty. Finally, I thank the National Science Foundation for their generous support of my work, particularly through NSF grant DMS-2005640.

2. Basics

A cube complex is a space obtained by gluing unit cubes isometrically along their faces. Moreover, it is said to be CAT(0) if it is nonpositively curved and simply connected.

You may think of the process of creating a CAT(0) cube complex as gluing together cubes in a particular order. If at some point there is positive curvature that is created locally, it is due to something as in Figure 1. There, we have created part of the boundary of a 3-dimensional cube by gluing 3 unit squares together. That creates a cone angle of $3\pi/2$. Since $3\pi/2 < 2\pi$, it is positively curved. However, we may annul the positive curvature by "filling" in the other side to create a 3-dimensional cube. Nonpositive curvature of a CAT(0) cube complex is equivalent to the Gromov Link Condition. We refer the reader to [**Sag95**, **CN05**, **Nic04**, **Rol16**] for more details.

2.1. Functoriality: Walled Spaces. A useful way to think of CAT(0) cube complexes is via this functorial construction. Begin with a set Y together with a collection of (nonempty) two-sided (walls) partitions $\mathcal{P} \subset 2^Y$. Two sided here means that if $h \in \mathcal{P}$ then $Y \setminus h \in \mathcal{P}$. The wall may be thought of as the pair $\{h, Y \setminus h\}$.

Next, create a graph by declaring that each maximal non-empty intersection of sets from $\mathcal{P}$ is a vertex. We connect two vertices if their corresponding defining intersections differ by the choice of exactly one side of one partition from $\mathcal{P}$. In Figure 2, you can see a choice of 5 partitions. Each region is not empty. The regions labeled A and U_1 differ by the choice of one partition and therefore, their associated vertices a and u_1 respectively are connected by an edge.

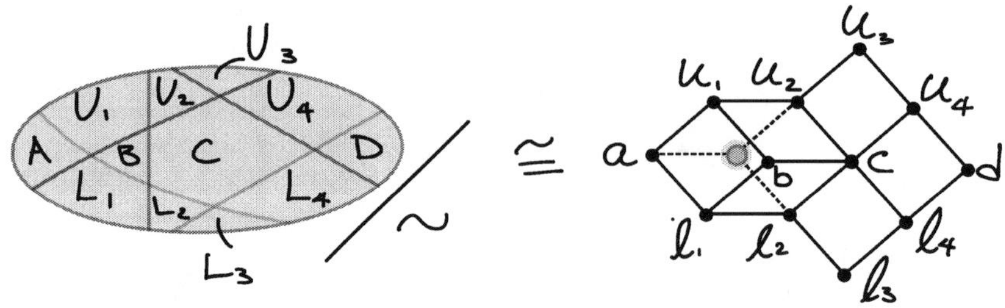

FIGURE 2. Functorial Construction

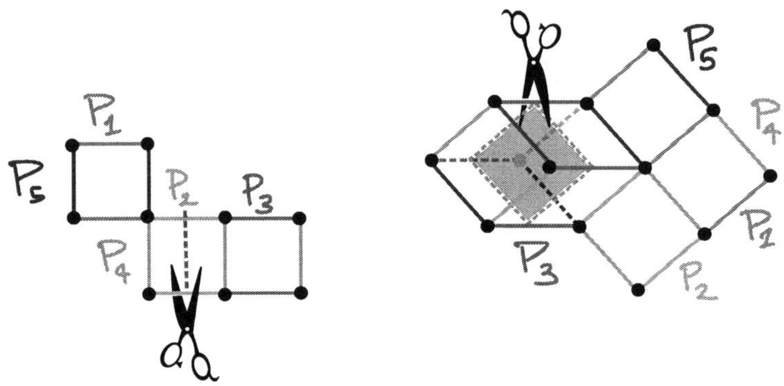

FIGURE 3. Two CAT(0) Cube Complexes with 10 halfspaces

Once we have done this, we may not have a CAT(0) cube complex. However, we may take the cubical completion (as in Figure 2, where we must add the unlabled "back" vertex to creat a 3-cube) to annul positive curvature. We note that this can always be done, except that distances may become infinite (if there are infinitely many partitions separating two regions, i.e. the finite interval condition is not satisfied) or the dimension may become infinite (if there are families of pairwise transverse partitions of unbounded cardinality, see Section 2.3 for the definition of transverse).

Conversely, start with (the vertex set of) a CAT(0) cube complex X. Each edge in X belongs to an equivalence class generated by "being parallel across a square". The compliment of that parallelism class has two sides, and the vertices that belong to each side give the two-sided partition of the vertex set (see Figure 3).

Finally, if we apply the previous construction to this collection of two-sided partitions, we get a CAT(0) cube complex, which is cannonically isomorphic to (the vertex set of) X.

2.2. The Roller Compactification. A *halfspace* of a CAT(0) cube complex is one side of a two-sided partition, as discussed above. We have established that the vertex set, denoted by X, is cannonically determined by the halfspace structure $\mathfrak{H} \subset 2^X$. By mapping a vertex in X to the collection of halfspaces that contain it, we obtain an isometric injection:

$$X \hookrightarrow 2^{\mathfrak{H}}.$$

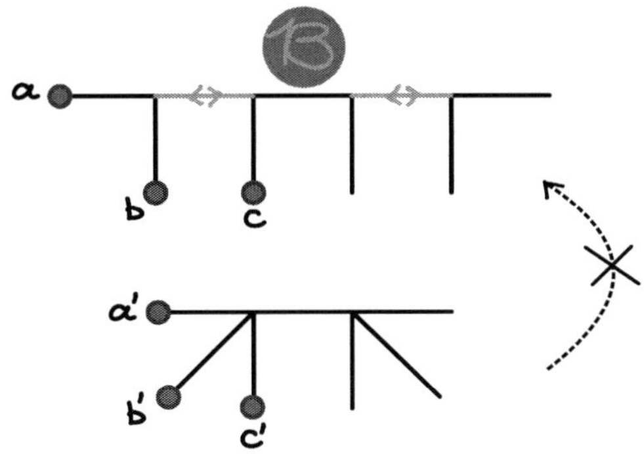

FIGURE 4. An example where a section to the projection is impossible

Here the extended metric on $2^{\mathfrak{H}}$ is given by half the cardinality of the symmetric difference $d(S,T) = \frac{1}{2}(S\triangle T) \in [0,\infty]$. While this metric is useful, we will rely on the standard topology on $2^{\mathfrak{H}} \cong \mathrm{Map}(\mathfrak{H} \to \{0,1\})$. This is given by declaring the basic open sets to be cylinder sets, which are themeselves determined by specifying values in *finitely many coordinates.* Equivalently, this is the topology of pointwise convergence of maps $\mathfrak{H} \to \{0,1\}$. With this topology, $2^{\mathfrak{H}}$ is compact. We note that the induced topology may be different than the metric topology on X. This is directly comparable to the weak-$*$ topology and respectively the metric topology on a Hilbert space.

Having found an injection of X into the compact space $2^{\mathfrak{H}}$, we may take the closure in the image and this defines the *Roller compactification.* By removing the image of X inside this closure, we are left with the *Roller boundary.*

Note that once we obtain the closure of X in $2^{\mathfrak{H}}$, there is a cannonical extension of the halfspaces as partitions of X to $\overline{X}$. Those are the basic clopen sets for our totally disconnected topology on $\overline{X}$.

The following is then immediate from the functorial construction, and can be thought of as forgetfulness.

COROLLARY. *Let X be a CAT(0) cube complex with halfspace structure $\mathfrak{H}$. Let $\mathfrak{H}' \subset \mathfrak{H}$ be involution invariant and X' the associated CAT(0) cube complex. Then the map $2^{\mathfrak{H}} \twoheadrightarrow 2^{\mathfrak{H}'}$, given by $S \mapsto S \cap \mathfrak{H}'$ induces a 1-Lipschitz projection $\overline{X} \twoheadrightarrow \overline{X'}$.*

We shall see in Section 3 how and when we can find a cannonical section to this map. The example in Figure 4 is given by taking $\mathfrak{H}' = \mathfrak{H} \setminus \mathcal{B}$ and a valence consideration shows that an isometric section is impossible.

The Roller boundary of a tree is, as a set, the visual boundary, but the topology is different when it is not (large-scale) locally finite. Consider the tree given by identifying at 0 infinitely many copies of $[0,\infty]$, as in Figure 5. It is not difficult to see that $\infty_n \to 0$ since all halfspaces eventually contain 0.

The Roller compactification has an important multiplicative property:

$$\overline{X_1 \times X_2} = \overline{X}_1 \times \overline{X}_2.$$

Combining this, with the discussion for trees, we obtain the Roller compactification of $\mathbb{Z}^2$, as in Figure 6, or for $F_2 \times \mathbb{Z}$ as in Figure 7.

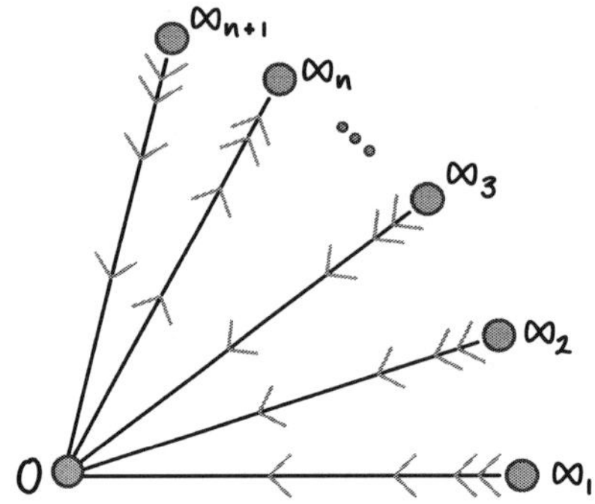

FIGURE 5. Halfspaces containing ∞_n.

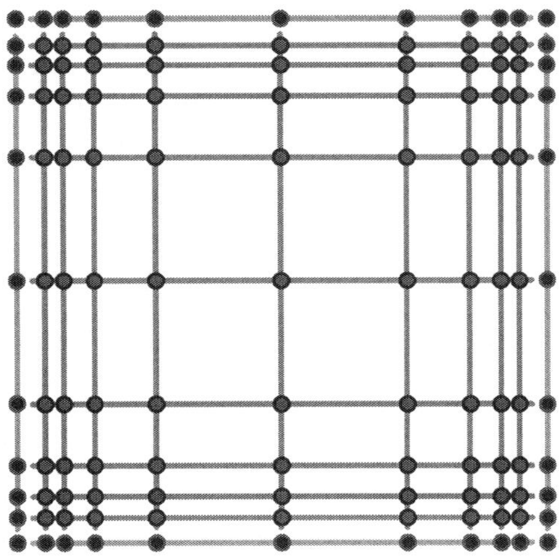

FIGURE 6. The Roller Compactification for $\mathbb{Z}^2$

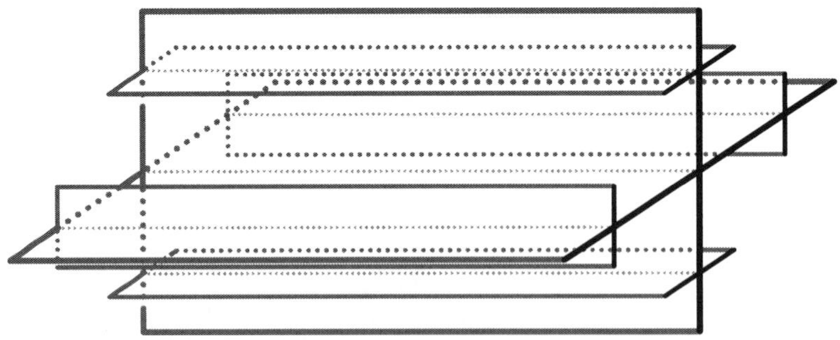

FIGURE 7. Roller Compactification for $F_2 \times \mathbb{Z}$

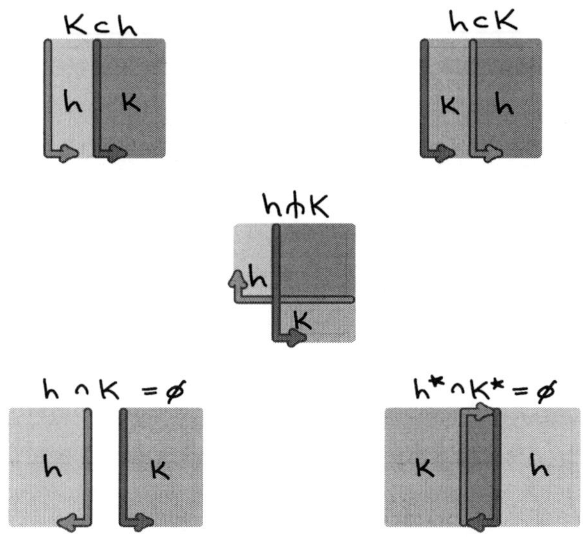

FIGURE 8. The possible relationships between halfspaces h, k, with $k \neq h, h^*$.

2.3. Comparing halfspaces. Given a pair of halfspaces $h, k \in \mathfrak{H}$ such that $k \neq h, h^*$ then either all of the pairwise intersections $h \cap k, h \cap k^*, h^* \cap k, h^* \cap k^*$ are not empty (and in this case, we say that h and k are *transverse*, and write $h \pitchfork k$) or one of the following other cases hold (see Figure 8):

$$\begin{array}{cc} k \subset h & h \subset k \\ h^* \cap k^* = \varnothing & h \cap k = \varnothing. \end{array}$$

LEMMA. *The dimension of X is the supremum cardinality of subsets $S \subset \mathfrak{H}$ of pairwise transverse halfspaces.*

It is straightforward to verify that if $\mathfrak{H} = \mathfrak{H}_1 \sqcup \mathfrak{H}_2$ is a nontrivial disjoint decomposition into nonempty involution invariant, pairwise transverse sets and X, X_1, and X_2 are the assocaited CAT(0) cube complexes respectively then there is a cannonical isomorphism $X \cong X_1 \times X_2$. We say that X is irreducible if it is not isomorphic to a product.

THEOREM ([**CS11**, Proposition 2.6]). *Let X be finite dimensional. Then, there is a cannonical decomposition of X into a product of irreducible CAT(0) cube complexes.*

A pair of halfspaces $h, k \in \mathfrak{H}$, with $h \neq k^*$ are said to be *facing* if $h^* \cap k^* = \varnothing$. This corresponds to the lower right corner in Figure 8. A triple of halfspaces $h, k, \ell \in \mathfrak{H}$ are said to be a *facing triple* if they are pairwise facing.

2.4. Intervals in $\overline{X}$. Let $\mathcal{I} \subset \overline{X}$ be an arbitrary subset. The (involution invariant) collection of halfspaces that separate points in $\mathcal{I}$ is

$$\mathfrak{H}_{\mathcal{I}} := \{h \in \mathfrak{H} : h \cap \mathcal{I}, h^* \cap \mathcal{I} \neq \varnothing\}.$$

Similarly, we denote the collection of halfspaces that contain $\mathcal{I}$ as $\mathfrak{H}^+_{\mathcal{I}}$, and the collection of compliments of halfspaces that contain $\mathcal{I}$, namely the halfspaces

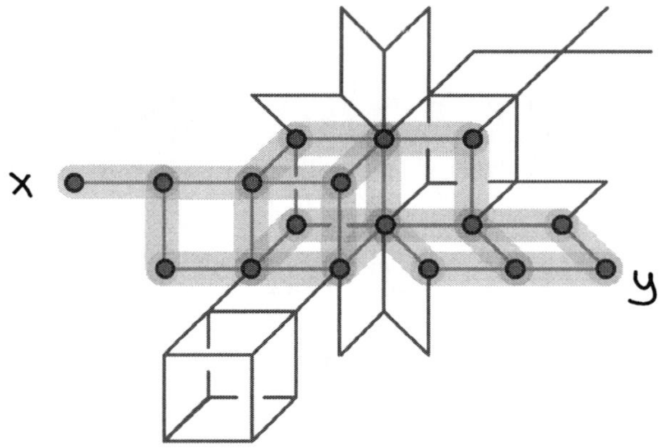

FIGURE 9. The shaded region is an interval in the ambient CAT(0) cube complex

that trivially intersect $\mathcal{I}$ is denoted by $\mathfrak{H}_{\mathcal{I}}^{-}$, meaning that $\mathfrak{H}_{\mathcal{I}}^{-} = (\mathfrak{H}_{\mathcal{I}}^{+})^{*}$. We have established the following decomposition:

$$\mathfrak{H} = \mathfrak{H}_{\mathcal{I}} \sqcup \mathfrak{H}_{\mathcal{I}}^{+} \sqcup \mathfrak{H}_{\mathcal{I}}^{-}.$$

We say that, $\mathcal{I}$ is an *interval* if there exists $x, y \in \overline{X}$ such that $\mathcal{I} = \bigcap_{h \in \mathfrak{H}_{\{x,y\}}^{+}} h$. In this case we write $\mathcal{I} = \mathcal{I}(x, y)$ and say that it is the *interval between* x *and* y. We also call x and y *endpoints* of $\mathcal{I}$. Clearly, $\mathfrak{H}_{\mathcal{I}(x,y)}$ does not contain a facing triple.

The structure of intervals in an arbitrary CAT(0) cube complex can be quite exotic; please see the examples in Figures 9 and 10 or come up with your own!

Nevertheless, the following theorem of [**BCG^{+}09**], which relies on Dilworth's Theorem for partially ordered sets, shows that intervals are not too wild.

THEOREM. *Let D be the dimension of X. If $\mathcal{I} \subset \overline{X}$ is an interval then there exists an isometric embedding $\mathcal{I} \hookrightarrow \overline{\mathbb{Z}^D}$.*

COROLLARY. *If $\mathcal{I} \subset \overline{X}$ is an interval then the set of end points of $\mathcal{I}$ has cardinality 2^N, for some $0 \leqslant N \leqslant D$.*

2.5. The Helly Property in $\overline{X}$.

The following is an important property of halfspaces in a CAT(0) cube complex. The first item is the standard Helly property for halfspaces. The second follows from the first by applying the finite intersection property to the collection of sets $\mathcal{S}$, each of which is clopen in the compact space $\overline{X}$. Comparing it with the classical Helly property for convex sets in euclidean space, we may consider it as telling us that, in a sense, CAT(0) cube complexes are of Helly-dimension 1.

LEMMA. *Suppose that $\mathcal{S} \subset \mathfrak{H}$ is such that $h \cap k \neq \varnothing$ for all $h, k \in \mathcal{S}$.*

- *If $\#\mathcal{S} < \infty$ then $\bigcap_{h \in \mathcal{S}} h \cap X \neq \varnothing$.*
- *Otherwise $\overline{X} \supseteq \bigcap_{h \in \mathcal{S}} h \neq \varnothing$.*

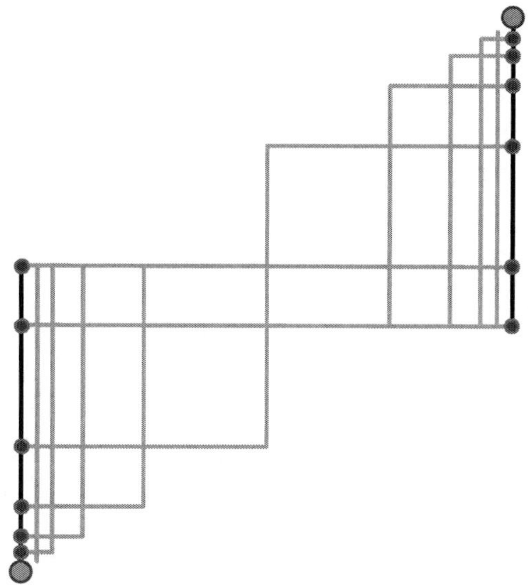

FIGURE 10. This Roller compactification is an interval

3. Lifting Decompositions

Let X be a CAT(0) cube complex with associated halfspace structure $\mathfrak{H}$. We shall say $\mathcal{S} \subset \mathfrak{H}$ is *consistent* if it satisfies the following two properties:

(1) If $h \in \mathcal{S}$ then $h^* \notin \mathcal{S}$.
(2) If $h, k \in \mathfrak{H}$, with $h \subset k$, and $h \in \mathcal{S}$ then $k \in \mathcal{S}$.

We note that, according to the functorial construction, if $\mathcal{S}$ is consistent and $\mathcal{S} \sqcup \mathcal{S}^* = \mathfrak{H}$ then $\mathcal{S}$ is the collection of halfspaces containing a single point.

PROPOSITION. *Let X be a CAT(0) cube complex with associated halfspaces $\mathfrak{H}$ and $\mathcal{S} \subset \mathfrak{H}$ consistent. Set $\mathfrak{H}_{\mathcal{S}} := \mathfrak{H} \setminus (\mathcal{S} \sqcup \mathcal{S}^*)$ and let $X(\mathfrak{H}_{\mathcal{S}})$ be the associated CAT(0) cube complex. Then the map $2^{\mathfrak{H}_{\mathcal{S}}} \to 2^{\mathfrak{H}}$ given by $E \mapsto E \sqcup \mathcal{S}$ induces an isometric injection $\overline{X}(\mathfrak{H}_{\mathcal{S}}) \hookrightarrow \overline{X}$, whose image is exactly*

$$\bigcap_{h \in \mathcal{S}} h \subset \overline{X}.$$

Furthermore, if $\mathcal{S}$ is Γ-invariant, for some action $\Gamma \to Aut(X)$ then with the restricted action on the image, the above natural injection is Γ-equivariant.

3.1. Probability Measures. Consider a probability measure $\mu \in \mathrm{Prob}(\overline{X})$ and the associated collection of halfspaces $\mathfrak{H}_\mu^+ := \{h \in \mathfrak{H} : \mu(h) > 1/2\}$. It is straightforward to verify that $\mathfrak{H}_\mu^+$ is consistent and that the collection of halfspaces $\mathfrak{H}_\mu = \{h \in \mathfrak{H} : \mu(h) = 1/2\}$ does not contain any facing triples. Applying the Lifting Decomposition to $\mathfrak{H}_\mu^+$, we get that $\bigcap_{h \in \mathfrak{H}_\mu^+} h$ is isomorphic to $\overline{X}(\mathfrak{H}_\mu)$.

LEMMA ([**CFI16**, Lemma 4.7]). *If $\mu \in Prob(\overline{X})$ then $\overline{X}(\mathfrak{H}_\mu)$ is an interval.*

3.2. Medians. Let $x, y, z \in \overline{X}$ and consider the associated probability measure $\mu_{(x,y,z)} = \frac{1}{3}(\delta_x + \delta_y + \delta_z)$. A simple parity argument shows that if $h \in \mathfrak{H}$ then

$\mu_{(x,y,z)}(h) \neq \frac{1}{2}$. Therefore, we have that $\bigcap_{h\in\mathfrak{H}_\mu^+} h$ is a single point, which we shall call the *median* of the triple and denote it by $m(x,y,z)$.

While this is not the standard definition of the median, it fits nicely within our context. We note that several of the natural properties of the median (e.g. invariance under permutation of the points) are immediate, including the property that

$$m(x,x,y) = x.$$

4. Proof of the Main Theorem

Suppose Γ is amenable acting on X. Since $\overline{X}$ is compact and metrizable, there must be a Γ-invariant probability measure $\mu \in \mathrm{Prob}(\overline{X})$. By the Lifting Decomposition and the previous lemma, it follows that $\bigcap_{h\in\mathfrak{H}_\mu^+} h$ is a Γ-invariant interval.

Suppose now that X is of finite dimension D and that Γ is not necessarily amenable but preserves an interval $\mathcal{I} \subset \overline{X}$. By Corollary 2.4 (see also [**Fer18**, Corollary 2.9]), the number of end points on which $\mathcal{I}$ is an interval is $2^{D'}$ for some $0 \leqslant D' \leqslant D$ and we identify these with $(\mathbb{Z}/2\mathbb{Z})^{D'} \cong \{0,1\}^{D'}$. By [**CS11**, Proposition 2.6] we have that $\mathrm{Aut}\left(\{0,1\}^{D'}\right) \cong \{0,1\}^{D'} \rtimes \mathrm{Sym}(D')$. We may therefore project $\Gamma \to \mathrm{Aut}\left(\{0,1\}^{D'}\right)$, and without loss of generality, assume $\Gamma \leqslant \mathrm{Aut}\left(\{0,1\}^{D'}\right)$.

Let $\Gamma_0 = \Gamma \cap \{0,1\}^{D'} \lhd \Gamma$ and note that $|\Gamma_0| = 2^N$ for some $0 \leqslant N \leqslant D'$. Fix a choice of right Γ_0-coset representatives $S \subset \Gamma$, with trivial $\{0,1\}^{D'}$-coordinate. Let $\mathcal{O} \in \{0,1\}^{D'}$ be defined component-wise by $\mathcal{O}_i \equiv 0$, for $i \in \{1,\dots,D'\}$. Note that $\mathcal{O}$ is fixed by $\mathrm{Sym}(D')$ and has trivial stabilizer in Γ_0. Let $(\mathcal{O},\alpha) \in S$. Then $\Gamma_0(\mathcal{O},\alpha).\mathcal{O} = \Gamma_0.\mathcal{O}$. We have shown that

$$\Gamma.\mathcal{O} = \bigsqcup_{(\mathcal{O},\alpha)\in S} \Gamma_0(\mathcal{O},\alpha).\mathcal{O} = \Gamma_0.\mathcal{O}$$

Therefore, by the orbit stabilizer theorem, we have

$$|\Gamma.\mathcal{O}| = |\Gamma_0.\mathcal{O}| = |\Gamma_0|/|\mathrm{stab}_{\Gamma_0}(\mathcal{O})| = 2^N. \quad \square$$

References

[BCG+09] J. Brodzki, S. J. Campbell, E. Guentner, G. A. Niblo, and N. J. Wright, *Property A and* CAT(0) *cube complexes*, J. Funct. Anal. **256** (2009), no. 5, 1408–1431, DOI 10.1016/j.jfa.2008.10.018. MR2490224

[BFH04] Mladen Bestvina, Mark Feighn, and Michael Handel, *Solvable subgroups of* Out(F_n) *are virtually Abelian*, Geom. Dedicata **104** (2004), 71–96, DOI 10.1023/B:GEOM.0000022864.30278.34. MR2043955

[CFI16] Indira Chatterji, Talia Fernós, and Alessandra Iozzi, *The median class and superrigidity of actions on* CAT(0) *cube complexes*, J. Topol. **9** (2016), no. 2, 349–400, DOI 10.1112/jtopol/jtu025. MR3509968

[CN05] Indira Chatterji and Graham Niblo, *From wall spaces to* CAT(0) *cube complexes*, Internat. J. Algebra Comput. **15** (2005), no. 5-6, 875–885, DOI 10.1142/S0218196705002669. MR2197811

[CS11] Pierre-Emmanuel Caprace and Michah Sageev, *Rank rigidity for CAT(0) cube complexes*, Geom. Funct. Anal. **21** (2011), no. 4, 851–891, DOI 10.1007/s00039-011-0126-7. MR2827012

[Fer18] Talia Fernós, *The Furstenberg-Poisson boundary and* CAT(0) *cube complexes*, Ergodic Theory Dynam. Systems **38** (2018), no. 6, 2180–2223, DOI 10.1017/etds.2016.124. MR3833346

[FFH23] Talia Fernós, David Futer, and Mark Hagen, *Homotopy equivalent boundaries of cube complexes*, Geom. Dedicata **218** (2024), no. 2, Paper No. 33, 83, DOI 10.1007/s10711-023-00877-w. MR4695822

[Fio18] Elia Fioravanti, *The Tits alternative for finite rank median spaces*, Enseign. Math. **64** (2018), no. 1-2, 89–126, DOI 10.4171/LEM/64-1/2-4. MR3959849

[FLM18] Talia Fernós, Jean Lécureux, and Frédéric Mathéus, *Random walks and boundaries of* CAT(0) *cubical complexes*, Comment. Math. Helv. **93** (2018), no. 2, 291–333, DOI 10.4171/CMH/435. MR3811753

[FLM21] Talia Fernós, Jean Lécureux, and Frédéric Mathéus, *Contact graphs, boundaries, and a central limit theorem for cat(0) cubical complexes*, 2021.

[Gen21] Anthony Genevois, *Cyclic hyperbolicity in cat(0) cube complexes*, 2021.

[GJN23] Radhika Gupta, Kasia Jankiewicz, and Thomas Ng, *Groups acting on CAT(0) cube complexes with uniform exponential growth*, Algebr. Geom. Topol. **23** (2023), no. 1, 13–42, DOI 10.2140/agt.2023.23.13. MR4567998

[MP22] Alexandre Martin and Piotr Przytycki, *Acylindrical actions for two-dimensional Artin groups of hyperbolic type*, Int. Math. Res. Not. IMRN **17** (2022), 13099–13127, DOI 10.1093/imrn/rnab068. MR4475273

[MS04] Nicolas Monod and Yehuda Shalom, *Cocycle superrigidity and bounded cohomology for negatively curved spaces*, J. Differential Geom. **67** (2004), no. 3, 395–455. MR2153026

[Nic04] Bogdan Nica, *Cubulating spaces with walls*, Algebr. Geom. Topol. **4** (2004), 297–309, DOI 10.2140/agt.2004.4.297. MR2059193

[OP22] Damian Osajda and Piotr Przytycki, *Tits alternative for 2-dimensional* CAT(0) *complexes*, Forum Math. Pi **10** (2022), Paper No. e25, 19, DOI 10.1017/fmp.2022.20. MR4522694

[Rol16] Martin Roller, *Poc sets, median algebras and group actions*, 2016.

[Sag95] Michah Sageev, *Ends of group pairs and non-positively curved cube complexes*, Proc. London Math. Soc. (3) **71** (1995), no. 3, 585–617, DOI 10.1112/plms/s3-71.3.585. MR1347406

[SW05] Michah Sageev and Daniel T. Wise, *The Tits alternative for* CAT(0) *cubical complexes*, Bull. London Math. Soc. **37** (2005), no. 5, 706–710, DOI 10.1112/S002460930500456X. MR2164832

[Tit72] J. Tits, *Free subgroups in linear groups*, J. Algebra **20** (1972), 250–270, DOI 10.1016/0021-8693(72)90058-0. MR286898

Department of Mathematics, University of North Carolina, Greensboro, North Carolina

Current address: Department of Mathematics, Vanderbilt University, 1326 Stevenson Center, Nashville, TN, USA

Email address: Talia.Fernos@Vanderbilt.edu

Contemporary Mathematics
Volume **812**, 2025
https://doi.org/10.1090/conm/812/16270

A generalized theory of expansions and collapses with applications to $\mathcal{Z}$-compactification

Craig R. Guilbault and Daniel Gulbrandsen

ABSTRACT. We generalize the dual notions of "expansion" and "collapse" so they can be applied to arbitrary metric spaces. We also expand the theory to allow for infinitely many such moves. Those tools are then employed to prove a variety of compactification theorems. We are particularly interested in $\mathcal{Z}$-set compactifications, which play an important role in geometric group theory and in algebraic and geometric topology.

In this paper, we expand upon the dual notions of *expansion* and *collapse* in ways that make them more applicable to spaces encountered in geometric group theory. In their original forms, these are combinatorial moves. An *elementary simplicial collapse* removes from a simplicial complex K a simplex and one of its faces in a way that results in a homotopy equivalent subcomplex L. The reverse is an *elementary simplicial expansion.* A *collapse* of K to L is a finite sequence of elementary collapses; an *expansion* is the same process, viewed in reverse. These definitions generalize to polyhedral and CW complexes where they form the foundation of *simple homotopy theory*, which plays a central role in algebraic and geometric topology. See, for example, [**Whi39**], [**Whi41**], [**Whi49**], [**Whi50**], [**Coh73**] and [**RS82**].

The primary objectives of this paper are as follows:

i) extend the classical notions of *expansion* and *collapse* to allow for *infinite* sequences of such moves,
ii) generalize the definition of an *elementary expansion/collapse* so that it can be applied to a broader class of topological spaces, and
iii) use i) and ii) to obtain compactification theorems which allow us to add "nice boundaries" to a variety of spaces.

We are particularly interested in obtaining $\mathcal{Z}$-compactifications, which play an important role in geometric and algebraic topology and geometric group theory. Here is a sample theorem.

THEOREM 0.1. *Every (infinitely) collapsible locally finite simplicial, cubical, or CW complex is $\mathcal{Z}$-compactifiable.*

2020 *Mathematics Subject Classification.* Primary 57Q10; Secondary 57Q15, 57Q70, 52B70.
Key words and phrases. collapse, expansion, $\mathcal{Z}$-set, homotopy negligible set, $\mathcal{Z}$-compactification, homotopy negligible compactification.
This research was supported in part by Simons Foundation Grant 427244, CRG.

A concrete application can be found in the second author's dissertation [**Gul23**], where it is shown that every locally finite CAT(0) cube complex X is collapsible. The resulting $\mathcal{Z}$-boundary (referred to as the *cubical boundary*) shares properties with the visual boundary, but retains some of the combinatorial structure of X. It can be viewed as a hybrid of the visual and Roller boundaries.

None of the goals listed above are entirely new. Allowing an *infinite* sequence expansions or collapses is a fairly obvious idea, but its appearance in the literature is surprisingly rare. One exception is a recent arXiv posting by Adiprasito and Funar and [**AF21**]. A different approach was developed by Siebenmann [**Sie70**], under which only finitely many moves are permitted, but a move can consist of an infinite discrete collection of *simultaneous* elementary expansions or collapses. The resulting theory, known as *infinite simple homotopy theory*, is important but different from what we do here. For example, the move just described can only be performed on a noncompact space. By contrast, our approach can *end* with a noncompact space, but at each step along the way the spaces are typically compact.

To accomplish the second goal, we leave the world of combinatorially constructed spaces to define a "compact topological collapse" (or simply *compact collapse*) of one space onto another. These compact collapses are the same as Ferry's *contractible retractions* [**Fer80**]. We will comment more on that work in Section 4.

Our main compactification theorems generalize some well known constructions. One of those is the $\mathcal{Z}$-compactification of the mapping telescope of an inverse sequence of finite polyhedra, which played a key role in work by Chapman and Siebenmann [**CS76**, §4]. Another is the addition of the visual boundary to a proper CAT(0) space X, as described by Bridson and Haefliger [**BH99**, §II.8.5]. They do not mention $\mathcal{Z}$-sets; that connection, and its usefulness, was noted by Bestvina [**Bes96**, Example 1.2(ii)]. Yet another is the proof, as presented by Jeremy Brazas in his blog *Wild Topology*, of a theorem by Borsuk, showing that all *dendrites* are contractible. Each of these is a special case of results presented in Section 5.

The existence of an infinite sequence of compact expansions $C_0 \swarrow C_1 \swarrow C_3 \swarrow C_4 \swarrow \cdots$, beginning with a compactum C_0, leads to a inverse sequence of retractions

$$C_0 \xleftarrow{r_1} C_1 \xleftarrow{r_2} C_2 \xleftarrow{r_3} \cdots$$

a direct sequence of inclusions

$$C_0 \xrightarrow{s_0} C_1 \xrightarrow{s_1} C_2 \xrightarrow{s_2} \cdots$$

and a set $X = \cup_{i=0}^{\infty} C_i$, whose topology is ambiguous. Our compactification strategy relies on the inverse limit construction, which determines the correct (but not always the obvious) topology for X. This realization led to an unexpected, but interesting, fourth major theme of this article:

iv) explore the topology of inverse sequences of retractions and corresponding direct sequences of inclusion maps.

This topic provides another connection to the literature. In [**MP15**], Marsh and Prajs studied inverse sequences of retractions from a different perspective. Whereas we use these sequences for constructing and compactifying noncompact spaces, they *begin* with a compact space Y and look for situations when Y can be realized as the inverse limit of compact subspaces $C_0, C_1, C_2, \cdots$ where the bonding maps are retractions.

The layout of this paper is as follows. Section 1 is general topology. We identify topologies $\mathcal{T}_I$ and $\mathcal{T}_D$ on X—one is strongly associated with the inverse sequence; the other with the direct sequence. We prove a useful theorem that shows when they agree; then we use the inverse limit construction to define a canonical compactification of $(X, \mathcal{T}_I)$ that is used throughout the paper. In Section 2, we bring these abstractions to life with a variety of examples. In Section 3 we review the classical theory of expansions and collapses, and develop terminology and notation that allows for infinitely many such moves. In Section 4 we develop generalizations of the concepts in Section 3. In Section 5 we apply all of this to prove a variety of compactification theorems; of particular interest are $\mathcal{Z}$-compactifications. We conclude the paper with a variety of applications.

1. The general setting

In the most general setting, we are interested in spaces that can be expressed as monotone unions of compact metric spaces $C_0 \subseteq C_1 \subseteq C_2 \subseteq \cdots$ where, for each $i > 0$, there is a retraction $r_i : C_i \to C_{i-1}$. This situation is concisely described with an an inverse sequence

$$C_0 \xleftarrow{r_1} C_1 \xleftarrow{r_2} C_2 \xleftarrow{r_3} \cdots \tag{1.1}$$

There is a corresponding direct sequence

$$C_0 \xrightarrow{s_0} C_1 \xrightarrow{s_1} C_2 \xrightarrow{s_2} \cdots \tag{1.2}$$

where the s_i are inclusions. There is also a set $X = \cup_{i=0}^{\infty} C_i$. Sequences (1.1) and (1.2) suggest preferred topologies for X, but those topologies may not agree. We will return to that issue after establishing some notation.

For $i \leq j$, let $r_{ij} : C_j \to C_i$ denote the obvious composition, with special cases $r_{ii} = \mathrm{id}_{C_i}$ and $r_{(i-1)i} = r_i$. Note that: each r_{ij} is a retraction; $r_{ij} \circ r_{jk} = r_{ik}$; and $r_{ik}|_{C_j} = r_{ij}$ whenever $i \leq j \leq k$. Thus, we may define retractions (on the level of sets) $r_{i\infty} : X \to C_i$ by letting $r_{i\infty}|_{C_j} = r_{ij}$. Similarly, let $s_{ij} : C_i \to C_j$ denote the inclusion map or, equivalently, the composition of maps from (1.2). Let $s_{i\infty} : C_i \to X$ also denote inclusion.

One natural way to topologize X is with the *coherent topology* with respect to the collection $\{C_i\}_{i=0}^{\infty}$. Under that topology, $U \subseteq X$ is open if and only if $U \cap C_i$ is open in C_i for all i. This is equivalent to the *final topology* on X with respect to the collection $\{s_{i\infty} : C_i \to X\}_{i=0}^{\infty}$ which, by definition, is the finest topology on X under which all of these functions are continuous. We denote that topology by $\mathcal{T}_D$ since it is closely related to the *direct* sequence (1.2).

An equally natural topology for X is the *initial topology* with respect to the set of functions $\{r_{i\infty} : X \to C_i\}_{i=0}^{\infty}$. By definition, this is the coarsest topology under which these functions are all continuous, i.e., it is the topology generated by the subbasis consisting of sets of the form $r_{i\infty}^{-1}(U)$, where U is an open subset of C_i for some i. We denote this topology by $\mathcal{T}_I$ since it is most naturally associated with the *inverse* sequence (1.1). In the special cases under consideration here, this subbasis is actually a basis.

LEMMA 1.1. *Given the above setup, the defining subbasis*

$$\mathcal{S}_I = \left\{ r_{i\infty}^{-1}(U) \mid U \text{ is an open subset of } C_i \text{ for some } i \right\}$$

for $\mathcal{T}_I$ is a basis for $\mathcal{T}_I$.

PROOF. It suffices to show that if U and V are open subsets of C_i and C_j respectively, then $r_{i\infty}^{-1}(U) \cap r_{j\infty}^{-1}(V) \in \mathcal{S}_I$. First note that, if $i = j$, then $r_{i\infty}^{-1}(U) \cap r_{i\infty}^{-1}(V) = r_{i\infty}^{-1}(U \cap V) \in \mathcal{S}$. If $i < j$, then $r_{i\infty}^{-1}(U) = r_{j\infty}^{-1}\left(r_{ij}^{-1}(U)\right)$, so the desired conclusion follows from the previous case. □

REMARK 1.2. Since the inclusion maps in sequence (1.2) do not depend upon the bonding maps in (1.1), $\mathcal{T}_D$ is essentially well-defined by the collection $\{C_i\}_{i=0}^{\infty}$. By contrast, the topology $\mathcal{T}_I$ depends on the bonding maps in (1.1). Example of this phenomenon can be found in Examples 2.1–2.3 of the following section..

A few more definitions and elementary observations will be useful as we proceed.

LEMMA 1.3. *Given the above setup,*

(1) *$\mathcal{T}_D$ is as fine as or finer than $\mathcal{T}_I$, i.e., $\mathcal{T}_I \subseteq \mathcal{T}_D$, and*
(2) *in both $(X, \mathcal{T}_D)$ and $(X, \mathcal{T}_I)$, each C_i is a subspace of X.*

PROOF. To verify 1), let U be an open subset of C_i and $r_{i\infty}^{-1}(U)$ be the corresponding basic open set in $\mathcal{T}_I$. Since $r_{i\infty}|_{C_i} = \mathrm{id}_{C_i}$, it follows that $r_{i\infty}^{-1}(U) \cap C_i = U$. So, for all $j \leq i$, $r_{i\infty}^{-1}(U) \cap C_j = U \cap C_j$ is open in C_j. For $j > i$, $r_{i\infty}^{-1}(U) \cap C_j = r_{ij}^{-1}(U)$, which is open in C_j since r_{ij} is continuous. Hence $r_{i\infty}^{-1}(U) \in \mathcal{T}_D$.

For 2), let $\mathcal{T}_{C_i}$ denote the original topology on C_i, $\mathcal{T}'_{C_i}$ the subspace topology on C_i induced by $\mathcal{T}_D$, and $\mathcal{T}''_{C_i}$ the subspace topology on C_i induced by $\mathcal{T}_I$.

If $U' \in \mathcal{T}'_{C_i}$, there exists $U \in \mathcal{T}_D$ such that $U' = U \cap C_i$, so by definition of $\mathcal{T}_D$, $U' \in \mathcal{T}_{C_i}$. Conversely, if $U' \in \mathcal{T}_{C_i}$, then $r_{i\infty}^{-1}(U') \in \mathcal{T}_I$ so, by 1), $r_{i\infty}^{-1}(U') \in \mathcal{T}_D$. Since $r_{i\infty}^{-1}(U') \cap C_i = U'$, $U' \in \mathcal{T}'_{C_i}$.

The latter half of the above argument assures that $\mathcal{T}_{C_i} \subseteq \mathcal{T}''_{C_i}$. For the reverse inclusion, let $r_{j\infty}^{-1}(U)$ be a basic open subset in $\mathcal{T}_I$ where $U \in \mathcal{T}_{C_j}$ and $U'' = r_{j\infty}^{-1}(U) \cap C_i \in \mathcal{T}''_{C_i}$. By again following the argument in 1), we know that $r_{j\infty}^{-1}(U) \cap C_i$ is either $U \cap C_i$ or $r_{ij}^{-1}(U)$, depending on whether $i \leq j$ or $i > j$. In both cases we have an element of $\mathcal{T}_{C_i}$. □

NOTATION 1. Going forward, we will often simplify notation by letting X_D and X_I denote $(X, \mathcal{T}_D)$ and $(X, \mathcal{T}_I)$. Similarly, when $A \subseteq X$, we will let $\mathrm{Int}_D\, A$ and $\mathrm{Int}_I\, A$ to denote the interiors of A in those spaces. Lemma 1.3 allows us to view C_i as a subspace of X_D or X_I, without need for specification.

We say that $A \subseteq X$ is *j-stationary* (with respect to the sequence $\{C_i, r_i\}$) if $A \subseteq C_j$ and $r_{j\infty}^{-1}(A) = A$. The following is immediate.

LEMMA 1.4. *Given the above setup,*

(1) *every subset of a j-stationary set is j-stationary,*
(2) *every union of j-stationary sets is j-stationary, and*
(3) *if $A \subseteq X$ is j-stationary, then $r_{jk}^{-1}(A) = A$ and $r_{k\infty}^{-1}(A) = A$ for all $k \geq j$. In particular, A is k-stationary, for all $k \geq j$.*

We say that $A \subseteq X$ is *j-insulated* if there exists a j-stationary open neighborhood of A in C_j. Lemma 1.4 gives the following.

LEMMA 1.5. *If $A \subseteq X$ is j-insulated, then A is k-insulated for all $k \geq j$.*

We will say that $A \subseteq X$ is *stationary* [*insulated*] if there is some $j \in \mathbb{N}$ for which A is j-stationary [j-insulated]. If a set $\{x\}$ has one of these properties, we attribute that property to the point x.

LEMMA 1.6. *If each $x \in C_i$ is insulated, then C_i is insulated.*

PROOF. By hypothesis, for each $x \in C_i$, there exists a $j_x \geq 0$ and a j_x-stable open neighborhood U_{j_x} of x in C_{j_x}. When $j_x \leq i$, Lemma 1.4 and continuity imply that U_{j_x} is open in C_i, whereas, when $j_x > i$, $U_{j_x} \cap C_i$ is open in C_i. So, by compactness, there exists a finite sequence $j_{x_1} \leq \cdots \leq j_{x_k}$ such that the sets $U_{j_{x_1}}, \cdots, U_{j_{x_k}}$ cover C_i. If $j_{x_k} \leq j$, then C_i is a union of j-stationary open subsets $U_{j_{x_1}}, \cdots, U_{j_{x_k}}$, and is therefore j-insulated. If $j_{x_k} > i$, then each of $U_{j_{x_1}}, \cdots, U_{j_{x_r}}$ is j_{x_k}-stationary and open in $C_{j_{x_k}}$. Since $C_i \subseteq \cup_{r=1}^{k} U_{j_{x_r}}$ it follows that C_i is j_{x_k}-insulated. □

THEOREM 1.7. *Given the above setup, the following are equivalent.*

(1) *Each point of X is insulated.*
(2) $X = \cup_{i=0}^{\infty} \operatorname{Int}_I C_i$.
(3) $\mathcal{T}_I = \mathcal{T}_D$.

PROOF. We begin by proving 1)⇔2). For the forward implication, note that Lemma 1.6 guarantees that, for each i, there exists an integer $k > i$ and an open subset $U_k \subseteq C_k$ that contains C_i and for which $r_{k\infty}^{-1}(U_k) = U_k$. This implies that U_k is a basic open set in $\mathcal{T}_I$, so $C_i \subseteq \operatorname{Int}_I C_k$. The desired conclusion follows.

For the reverse implication, let $x \in X$. Then there exists $j \geq 0$ such that $x \in \operatorname{Int}_I C_j$, so there exists a basic open neighborhood $r_{k\infty}^{-1}(U_k)$ of x contained in C_j. If $k \geq j$, then $r_{k\infty}^{-1}(U_k) \subseteq C_k$, so x is k-insulated. If $k < j$, then $r_{kj}^{-1}(U_k)$ is open in C_j and $r_{j\infty}^{-1}\left(r_{kj}^{-1}(U_k)\right) = r_{k\infty}^{-1}(U_k) \subseteq C_j$, so x is j-insulated.

Next we show 2)⇒3). We already know $\mathcal{T}_I \subseteq \mathcal{T}_D$, so let $V \in \mathcal{T}_D$ and $x \in V$. By hypothesis, there is a $j \geq 0$ and an open neighborhood U_j of x in C_j such that $r_{j\infty}^{-1}(U_j) = U_j$. Since $V \in \mathcal{T}_D$, $V \cap C_j$ is open in C_j, so $V_j := V \cap U_j$ is open in C_j. Moreover, $r_{j\infty}^{-1}(V_j) = V_j$, so V_j is a basic open set in $\mathcal{T}_I$. Since $x \in V_j \subseteq V$, $V \in \mathcal{T}_I$.

Lastly, we use a contrapositive argument to show 3)⇒1). Suppose $x \in X$ is not insulated, and let i_0 be the minimum integer for which $x \in C_{i_0}$. For each $k \geq i_0$, let $\mathcal{B}_k = \{B_{k,j}\}_{j=k}^{\infty}$ be a countable neighborhood basis at x of open subsets in C_k, and for each $k \geq i_0$ and $j \geq k$, let $B'_{k,j} = r_{k\infty}^{-1}(B_{k,j})$ be the corresponding basic open neighborhood of x in $(X, \mathcal{T}_I)$. If $\mathcal{B}'_k$ denotes $\{B'_{k,j}\}_{j=k}^{\infty}$, then $\cup_{k=i_0}^{\infty} \mathcal{B}'_k$ is a neighborhood basis at x in $(X, \mathcal{T}_I)$. Since x is not insulated, for each $k \geq i_0$ and $j \geq k$, we may choose $x_{kj} \in B'_{k,j} - C_j$. Let A be the set of all such $x_{k,j}$ and notice that $x \notin A$ and $A \cap C_i$ is finite, for all $i \geq 0$. As such, A is closed in $(X, \mathcal{T}_D)$. But, since every element of $\cup_{k=i_0}^{\infty} \mathcal{B}'_k$ intersects A, A is not closed in $(X, \mathcal{T}_I)$. □

COROLLARY 1.8. *If any of the conditions in Theorem 1.7 hold, then $X = \cup_{i=0}^{\infty} \operatorname{int}_D C_i$. The converse of this assertion is false.*

PROOF. The assertion is immediate from condition 3), while part iii) of Example 2.3 in Section 2 is a counterexample to the converse. □

When the conditions in Theorem 1.7 hold, we say that $\{C_i, r_i\}$ is *fully insulated.* Example 2.5 from Section 2 shows that, for this to be true, it is not enough to assume that each C_i is stationary.

1.1. Inverse and direct limits associated with X. We now provide further justification for associating the topologies $\mathcal{T}_D$ and $\mathcal{T}_I$ with the direct and inverse sequences (1.2) and (1.1). In particular, we look at the roles played by the limits of those sequences.

The *direct limit* of (1.2) can be defined by

$$\varinjlim \{C_i, s_i\} := (\sqcup_{i=0}^{\infty} C_i \times \{i\})/\sim$$

where $(x,i) \sim (y,j)$ if there exist $k \geq \max\{i,j\}$ such that $s_{ik}(x) = s_{jk}(y)$. The *direct limit topology* is the quotient topology induced by

$$q : \sqcup_{i=0}^{\infty} C_i \times \{i\} \to (\sqcup_{i=0}^{\infty} C_i \times \{i\})/\sim$$

where the left-hand side is given the disjoint union topology. In the case at hand, there is a bijection $f : X \to \varinjlim \{C_i, s_i\}$ taking $x \in X$ to the equivalence class $[(x,i)]$, where i is chosen sufficiently large that $x \in C_i$. If we use f to pull back the direct limit topology, we obtain a topology on X under which $U \subseteq X$ open if and only if $U \cap C_i$ is open in C_i for all i. In other words, we arrive at $\mathcal{T}_D$, which we think of, informally, as the direct limit topology on X.

There is a similar, but more subtle, relationship between $\mathcal{T}_I$ and the inverse limit of (1.1). Recall that

$$\varprojlim \{C_i, r_i\} := \{(x_0, x_1, x_2, \cdots) \mid r_i(x_i) = x_{i-1} \text{ for all } i > 0\}$$

is topologized as a subspace of $\prod_{i=0}^{\infty} C_i$, where the latter is given the product topology. There is a natural injection $e : X \to \varprojlim \{C_i, r_i\}$ defined by $e(x) = (r_{i\infty}(x))_{i=0}^{\infty}$. Under both $\mathcal{T}_D$ and $\mathcal{T}_I$, the coordinate function are continuous, so e is a continuous injection.

THEOREM 1.9. *The function $e : X \to \varprojlim \{C_i, r_i\}$ is an embedding if and only if X is given the topology $\mathcal{T}_I$; This function need not be surjective, but $e(X)$ is dense in $\varprojlim \{C_i, r_i\}$.*

PROOF. Clearly, there is a unique topology on X under which e is an embedding. We show that $\mathcal{T}_I$ is that topology.

Let $U = r_{j\infty}^{-1}(V_j)$ be a basic open set in $\mathcal{T}_I$, where V_j is an open subset of C_j, and let $\pi_j : \prod_{i=0}^{\infty} C_i \to C_j$ be the projection map. Then

$$\begin{aligned} e(U) &= \{e(x) \mid x \in X \text{ and } r_{j\infty}(x) \in V_j\} \\ &= e(X) \cap \pi_j^{-1}(V_j) \\ &= e(X) \cap \left(\varprojlim \{C_i, r_i\} \cap \pi_i^{-1}(V_i)\right) \end{aligned}$$

Since $\pi_i^{-1}(V_i)$ is open in the product topology, then $\varprojlim \{C_i, r_i\} \cap \pi_i^{-1}(V_i)$ is open in $\varprojlim \{C_i, r_i\}$, so $e(U)$ is open in $e(X)$. Therefore, the corestriction of e to $e(X)$ is an open map, hence a homeomorphism.

To see that $e(X)$ is dense in $\varprojlim \{C_i, r_i\}$, notice that if $x \in C_i$ then $r_{j\infty}(x) = x$ for all $j \geq i$. As such, $e(X)$ is the set of all eventually constant sequences $(x_0, \cdots, x_{k-1}, c, c, c, \cdots)$ such that $c \in C_k$, $x_{k-1} = r_k(c)$, and $x_i = r_{i+1}(x_{i+1})$ for all $i < k-1$. Clearly, every element of $\varprojlim \{C_i, r_i\}$ is the limit of a sequence of such points.

Non-surjectivity of e occurs in most of the examples presented in Section 2. $\square$

In most cases of interest to us, X_I (and therefore X_D) will be noncompact. We have assumed that each C_i is a compact metric space, so $\prod_{i=0}^{\infty} C_i$ is compact and metrizable and $\varprojlim \{C_i, r_i\}$ is a closed subset of $\prod_{i=0}^{\infty} C_i$. See, for example, [**Dug78**]. If we adopt a standard convention, that a *compactification* of a space S is a compact Hausdorff space that contains S as a dense subspace, then $\varprojlim \{C_i, r_i\}$ is a compactification of $e(X)$. Letting $R := \varprojlim \{C_i, r_i\} - e(X)$, we have

$$\varprojlim \{C_i, r_i\} = \overline{e(X)} = e(X) \sqcup R$$

By following a standard procedure we arrive at:

COROLLARY 1.10. *The space X_I is separable and metrizable and admits a compactification $\overline{X_I} = X_I \sqcup R$ which is homeomorphic to $\varprojlim \{C_i, r_i\}$.*

PROOF. As a subspace of the compact metrizable space $\prod_{i=0}^{\infty} C_i$, $e(X)$ is separable and metrizable, so X_I is separable and metrizable. To obtain $\overline{X_I} = X_I \sqcup R$, extend $e : X_I \to e(X)$ to the function $\overline{e} : \overline{X_I} \to \overline{e(X)}$ which is the identity on R. Then topologize $\overline{X}$ by declaring $U \subseteq \overline{X_I}$ to be open if and only if $\overline{e}(U)$ is open in $\overline{e(X)}$. □

Under the most general of the circumstances described so far, X_I and $\overline{X_I}$ can be quite unusual. Since we are primarily interested in locally compact metric spaces and their compactifications, the following observation is relevant.

PROPOSITION 1.11. *For an arbitrary compactification $\overline{X}$ of a Hausdorff space X, X is open in $\overline{X}$ if and only if it is locally compact. For the compactifications $\overline{e(X)} = e(X) \sqcup R$ and $\overline{X_I} = X_I \sqcup R$ described above, these conditions hold whenever the conditions in Theorem 1.7 are satisfied.*

PROOF. For the forward implication of the initial assertion, note that $\overline{X}$ is locally compact and Hausdorff. Hence, local compactness is truly a local property [**Mun00**, Th.29.2], so every open subset of $\overline{X}$ is locally compact. The reverse implication is well known but less obvious; see [**Dug78**, XI.8.3].

To prove the latter assertion, let $x \in X_I$ be arbitrary. By hypothesis, there exists a $j \geq 0$ and an open neighborhood U_j of x in C_j such that $r_{j\infty}^{-1}(U_j) = U_j$. Since C_j is compact, it is locally compact, so U_j contains a compact neighborhood V_j of x in the space C_j. But $r_{j\infty}^{-1}(V_j) = V_j$, so V_j is also a neighborhood of x in X_I. □

REMARK 1.12. Example 2.2 shows that the converse of the second sentence of Proposition 1.11 is false. One might ask whether local compactness of $(X, \mathcal{T}_I)$ implies that $\mathcal{T}_I = \mathcal{T}_D$. The same example shows that the answer is "no".

2. Examples

In this section, we present a collection examples that fit the above setup and exhibit a variety of properties. As additional topics are developed, we will often refer back to these examples.

EXAMPLE 2.1. Let $P = \{p_1, p_2, \cdots\}$ be a sequence of points with the discrete topology. For $i \geq 1$, let C_i be the cone on $\{p_1, \cdots, p_i\}$ with cone point q and let

$C_0 = \{q\}$. Then, as a set, $X = \cup_{i=0}^{\infty} C_i$ is the cone on P, and we have a direct system of inclusion maps

$$C_0 \xrightarrow{s_0} C_1 \xrightarrow{s_1} C_2 \xrightarrow{s_2} \cdots$$

For each $i > 0$, let $r_i : C_i \to C_{i-1}$ be the retraction which maps segment segment $\overline{qp}_i$ linearly onto $\overline{qp}_{i-1}$ by sending p_i to p_{i-1}. This gives an inverse sequence

$$C_0 \xleftarrow{r_1} C_1 \xleftarrow{r_2} C_2 \xleftarrow{r_3} \cdots$$

With a little effort, one can show that X_D does not have a countable local basis at q, and hence is nonmetrizable. By contrast, X_I is not only metrizable, but planar. In fact, $\overline{X_I} = X_I \sqcup R$ is homeomorphic to the left-hand compactum shown in Figure 1 with $R \approx (0, 1]$ corresponding to the limiting arc minus the point q.

EXAMPLE 2.2. Take the same collection $\{C_i\}$ as in the previous example, but now define $r'_i : C_i \to C_{i-1}$ to take the entire segment $\overline{qp}_i$ to the point q. The direct system and X_D remain the same, but with these bonding maps, X_I is homeomorphic to the planar continuum on the right in Figure 1. In this case, R is the empty set.

FIGURE 1. $\overline{X_I} = X_I \sqcup R$ in Examples 2.1 (left) and 2.2 (right)

EXAMPLE 2.3. For each $i \geq 0$, let C_i be the closed interval $[0, i]$, divided into subintervals of length 1. Define three different sequences of retractions of C_i onto C_{i-1}.

i) Let $r_i : C_i \to C_{i-1}$ be the retraction that maps $[i-1, i]$ to its left endpoint.
ii) Let $r'_i : C_i \to C_{i-1}$ be the retraction that maps $[i-1, i]$ linearly onto $[i-2, i-1]$ with $r_i(i) = i-1$.
iii) Let $r''_i : C_i \to C_{i-1}$ be the retraction that maps $[i-1, i]$ linearly onto $[0, i-1]$ with $r_i(i) = 0$.

In all three cases, X_D is homeomorphic to the standard ray $[0, \infty)$. In case i), X_I is also homeomorphic to $[0, \infty)$, and $\overline{X_I}$ is the one-point compactification $[0, \infty]$. In case ii), X_I is again homeomorphic to $[0, \infty)$, but $\overline{X_I}$ is a version of the "topologist's sin $(1/x)$ curve" with R corresponding to the limit arc. See Figure 2.

In case iii), $\overline{X_I}$ is homeomorphic to a famous indecomposable continuum B, known as the *bucket handle* or the *Brouwer-Janiszewski-Knaster continuum.* Here R and X_I are both dense in $\overline{X_I}$, and the topology $\mathcal{T}_I$ on $[0, \infty)$ is not the standard one. Open sets are most easily seen by taking preimages of open subsets of B under a rather exotic injection $e : [0, \infty) \to B$ under which the sequence

$\{e(n)\}_{n\in\mathbb{Z}^+}$ converges to $e(0)$.[1] As such, every neighborhood of 0 in $\mathcal{T}_I$ contains an unbounded sequence of open intervals. Other points in $([0,\infty),\mathcal{T}_I)$ have similar neighborhoods. See Figure 2. One takeaway from this example is that, even when X_D is metrizable—or even a locally finite simplicial complex—the space X_I can be quite different. The example also illustrates the significance of the insulation conditions; sequences i) and ii) are fully insulated while sequence iii) is not.

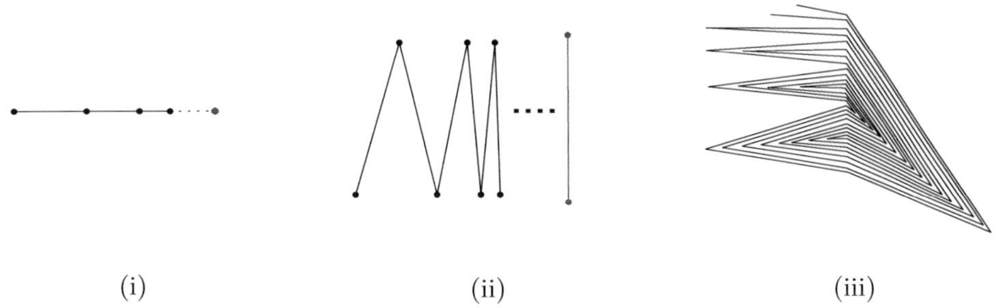

FIGURE 2. $\overline{X_I} = X_I \sqcup R$ in Example 2.3

EXAMPLE 2.4. For each integer $i \geq 0$, let I^i be the standard i-cube

$$\underbrace{[0,1]\times\cdots[0,1]}_{i\text{-times}} \quad\times\quad 0\times 0\times\cdots$$

where the additional 0-coordinates are added for notational convenience, and so that $\{I^i\}_{i=0}^{\infty}$ forms a nested sequence. For $i>0$, let $r_i : I^i \to I^{i-1}$ be the projection map. We may view $X = \cup_{i=0}^{\infty} I^i$ as a subset I_0^ω of the countable infinite product I^ω; in particular, I_0^ω consists of all sequences of elements of $[0,1]$ that end in a sequence of 0's. An argument similar to the one alluded to in Example 2.1 shows that, under the $\mathcal{T}_D$ (which is commonly used when I_0^ω is viewed as a CW complex), I_0^ω is nonmetrizable. By contrast, $\mathcal{T}_I$ is the same as the subspace topology induced by the product topology on I^ω. Additionally, $\overline{I_0^\omega}$ (as defined in Section 1.1) is equivalent to taking the closure of I_0^ω in I^ω. These observation require some work since, by definition, $e(X)$ and $\varprojlim\{I^i, r_i\}$ are somewhat thin subsets of $I^0\times I^1\times I^2\times\cdots$.

For readers interested in non-locally finite *cube complexes*, I_0^ω is a beautifully simple example. In that context, it is common to topologize I_0^ω using the path length metric d induced by the Euclidean metric on each I^i. Interestingly, that topology is neither $\mathcal{T}_D$ nor $\mathcal{T}_I$. Being metric, it it is clearly not $\mathcal{T}_D$. On the other hand, if $\mathbf{e}_i$ denotes the point with all coordinates 0 except for 1 as its $(i+1)^{\text{st}}$ coordinate, then $\{\mathbf{e}_i\}$ converges to $(0,0,0,\cdots)$ under $\mathcal{T}_I$, but has no limit points under d since $d(\mathbf{e}_i,\mathbf{e}_j)=\sqrt{2}$ for all $i\neq j$.

EXAMPLE 2.5. Let C_0 be the origin $\mathbf{0}$ in $\mathbb{R}^2$, C_j the closed circular disk of radius j centered at $(0,j)$, and $r_j : C_j \to C_{j-1}$ the map taking x to the first point where the ray $\overrightarrow{x\mathbf{0}}$ intersects C_{j-1}. Then X_I is the open upper half-plane, plus the point $\mathbf{0}$, endowed with the subspace topology; and $\overline{X_I}$ can be seen as the result

[1] We are abusing notion slightly by calling this map e.

of adding an open interval $(0,\pi)$ to X_I, with one point being added to the end of each ray in X_I emanating from $\mathbf{0}$. Another way to envision this example is to embed the construction in the unit disk with $(0,-1)$ as the point of tangency. Under that approach, $\overline{X_I}$ can be realized as the unit disk, and the remainder R as $\mathbb{S}^1 - \{(0,-1)\}$.

In this example, each C_i is stationary but not insulated. As a consequence, Theorem 1.7 cannot be applied, and we end up with a space X_I that is not locally compact and not an open subset of $\overline{X_I}$. The reader is invited to compare $\mathcal{T}_I$ and $\mathcal{T}_D$.

EXAMPLE 2.6. Let C_0 be origin and, for $j>0$, let C_j be the convex hull of

$$A_j = \left\{\exp\left(\frac{k\pi i}{2^{j-1}}\right) \mid k = 0,1,2,\cdots,2^j - 1\right\}$$

in $\mathbb{R}^2$, and $r_j : C_j \to C_{j-1}$ the radial projection of C_j onto C_{j-1}. Then X_I is the open unit disk, together with the countable dense set $\cup_{j\geq 1} A_j$ of points on the unit circle (with the subspace topology); and $\overline{X_I}$ is homeomorphic to the closed disk. As in the previous example, X_I is not locally compact and is not open in $\overline{X_I}$.

EXAMPLE 2.7. Let (Y,d) be a proper CAT(0) space and y_0 an arbitrary base point. Let $C_0 = \{y_0\}$ and, for each $i>0$, $C_i = B_d[y_0,i]$, the closed ball of radius i. Let $r_i : C_i \to C_{i-1}$ be the geodesic retraction taking $y \in C_i$ to the nearest point in C_{i-1}. As a set, $X = Y$, and since each C_i is insulated, $\mathcal{T}_I = \mathcal{T}_D$. With a little effort, one sees that these topologies agree with original metric topology and that $\overline{X_I}$ is homeomorphic to the compactification $\overline{Y} := Y \cup \partial_\infty Y$ obtained by adjoining the visual sphere at infinity to Y via the "cone topology".

Note also that the only points of C_i with (possibly) non-singleton preimages in C_{i+1} lie in the metric sphere $S_d[y_0,i]$. It follows that

$$R = \overline{X_I} - X_I = \varprojlim \{C_i, r_i\} - e(X)$$

is equal to the inverse limit of the sequence

$$\{y_0\} \xleftarrow{f_1} S_d[y_0,1] \xleftarrow{f_2} S_d[y_0,2] \xleftarrow{f_3} \cdots$$

where f_i is the restriction of r_i. See [**BH99**, §II.8] for more details related to this construction. We will revisit to this example in Section 6.2 and prove some additional properties.

EXAMPLE 2.8. Let

$$K_0 \xleftarrow{f_1} K_1 \xleftarrow{f_2} K_2 \xleftarrow{f_3} \cdots \tag{2.1}$$

be an inverse sequence of compact metric spaces and, for each integer $i>0$, let

$$\mathrm{Map}(f_i) := K_i \times [i-1,i] \sqcup K_{i-1} / \sim$$

(the *mapping cylinder*) be the quotient space, where $\sim$ is generated by the rule $(x,i-1) \sim f_i(x)$. By identifying K_{i-1} with its image in $\mathrm{Map}(f_i)$ and K_i with the image of $K_i \times \{i\}$, we view K_{i-1} and K_i as subsets of $\mathrm{Map}(f_i)$ and refer to them as the *range end* and the *domain end* of $\mathrm{Map}(f_i)$. There is a retraction of $\mathrm{Map}(f_i)$ onto K_{i-1} (often called the *mapping cylinder collapse*) which sends (x,t) to $f(x)$.

Let $C_0 = K_0$ and for each $i>0$, let

$$C_i = \mathrm{Map}(f_1) \cup_{K_1} \mathrm{Map}(f_2) \cup_{K_2} \cdots \cup_{K_{i-1}} \mathrm{Map}(f_i)$$

By applying the mapping cylinder collapses, we get retractions $r_i : C_{i-1} \leftarrow C_i$ which can be assembled into an inverse sequence of compact metrizable spaces

$$C_0 \xleftarrow{r_1} C_1 \xleftarrow{r_2} C_2 \xleftarrow{r_3} \cdots$$

In this case

$$X = \cup_{i=0}^{\infty} C_i = \operatorname{Map}(f_1) \cup_{K_1} \operatorname{Map}(f_2) \cup_{K_2} \cdots$$

is called the *inverse mapping telescope* of sequence (2.1).

By reasoning similar to the previous example, $\mathcal{T}_I = \mathcal{T}_D$ and $R = \varprojlim \{C_i, r_i\} - e(X)$ is the inverse limit of sequence (2.1). We will return to this compactification in Section 6.1.

3. Expansions and collapses in the simplicial category

For our favorite applications of the above setting, we will need to place conditions on the retraction maps in our inverse sequences. Necessity will direct us to the notion of a *compact collapse*—a generalization of a *simplicial collapse*. We will also need to allow for infinitely many such moves (or their inverses). Before proceeding to that level of generality, we will spend some time reviewing the classical theory and expanding it to permit infinitely many moves.

3.1. Simplicial complexes: preliminaries. An *abstract simplicial complex* K is a pair $(V, \mathcal{S})$ where V is a set and $\mathcal{S}$ is a collection of nonempty finite subsets of V containing all singletons and satisfying the rule: if $\sigma \in \mathcal{S}$ and $\varnothing \neq \tau \subseteq \sigma$, then $\tau \in \mathcal{S}$. The elements of V are called *vertices* and the elements of $\mathcal{S}$ the *simplices* of K. (Typically, we do not distinguish between a vertex v and the simplex $\{v\}$.) If $\tau \subseteq \sigma \in \mathcal{S}$, we call τ is a *face* of σ; if $\tau \subsetneq \sigma$, it is called a *proper face.* If σ contains exactly $n+1$ vertices, it is called an n *-dimensional simplex* (or simply an n*-simplex*), and we sometimes denote it by σ^n. A *subcomplex* of K is a simplicial complex L, each of whose simplices is a simplex of K; this will be denoted $L \leq K$. As a key example, if σ is a simplex of K then $\partial\sigma \leq K$ is made up of all proper faces of σ. A simplicial complex is *finite* if its vertex set is finite; it is *locally finite* if each vertex belongs to only finitely many simplices.

We define the *geometric realization* $|K|$ of $K = (V, \mathcal{S})$ (also called the *polyhedron corresponding to* K) to be a subset of the real vector space W_V with basis V. By endowing W_V with the standard inner product metric d_V, each subset $F \subseteq V$ containing $k < \infty$ elements, is the basis for a subspace W_F that is isometric to $\mathbb{R}^k$ with the Euclidean metric. For $\sigma = \{v_0, \cdots, v_n\} \in \mathcal{S}$, the *geometric realization* $|\sigma|$ is the convex hull of $\{v_0, \cdots, v_n\}$ in W_σ with the subspace metric. Now define $|K| \subseteq W_v$ to be the union of the geometric realizations of its simplices.

Note that

$$|\sigma| = \left\{ \sum\nolimits_{i=0}^{n} x_i v_i \middle| \, 0 \leq x_i \leq 1 \text{ and } \sum\nolimits_{i=0}^{n} x_i = 1 \right\}$$

If $\mathbf{x} = \sum_{i=0}^{n} x_i v_i \in |\sigma|$, the x_i are called the *barycentric coordinates* of $\mathbf{x}$. The *barycenter* of $|\sigma|$ is the point $\widehat{\sigma} = \sum_{i=0}^{n} \frac{1}{n+1} v_i$. We will refer to any space isometric to $|\sigma|$ as a *standard Euclidean* n*-simplex,* in which case the distance between any pair of distinct vertices is $\sqrt{2}$.[2]

[2]Some authors rescale the metric so that the distance between pairs of vertices in a standard Euclidean n-simplex is 1. For our purposes, that is unnecessary.

There are multiple ways to topologize $|K|$. One can give $|K|$ the initial topology with respect to the individual simplices; in other words, $U \subseteq |K|$ is open if and only if $U \cap |\sigma|$ is open for every $\sigma \in \mathcal{S}$. This is often called the *weak topology* on $|K|$, so we denote it $\mathcal{T}_W$. Alternatively, one can use the inner product metric d_V on W_V to induce the subspace topology on $|K|$. We call this the *barycentric topology* and denote it $\mathcal{T}_B$. It is a classical fact that $\mathcal{T}_W = \mathcal{T}_B$ if and only if K is locally finite; otherwise, $\mathcal{T}_W \supsetneq \mathcal{T}_B$ and the identity map $\mathrm{id}_{|K|} : (|K|, \mathcal{T}_W) \to (|K|, \mathcal{T}_B)$ is a homotopy equivalence. See [**Dow52**].

REMARK 3.1. For the purposes of metric geometry and geometric group theory, the *path length metric on* $|K|$ (also called the *intrinsic metric* in [**BH99**, Chap.I.7]) is often more useful than the barycentic metric. For example, the latter is always bounded and thus ignores the large-scale geometry of $|K|$. Nevertheless, for each $\mathbf{x} \in |K|$, there is an $\varepsilon > 0$ such that balls centered at $\mathbf{x}$ and having radius $< \varepsilon$ are identical under the two metrics. As such, they generate the same topologies.

Abstract simplicial complexes K_1 and K_2 are *isomorphic*, denoted $K_1 \cong K_2$, if there exists a bijection $f : V_1 \to V_2$ between vertex sets that induces a bijection between the sets of simplices. Clearly f induces a corresponding homeomorphism $|f| : |K_1| \to |K_2|$ provided $|K_1|$ and $|K_2|$ are topologized in the same manner. Polyhedra $|L_1|$ and $|L_2|$ are called *piecewise-linearly (PL) homeomorphic* if there exist isomorphic subdivisions L_1' and L_2' of L_1 and L_2.

A *triangulation* of a topological space X is a homeomorphism $h : |K| \to X$ for some simplicial complex K. In that case, we say X is *triangulated* by K.

3.2. Simplicial expansions and collapses. An *elementary combinatorial collapse* $K \overset{e}{\searrow} L$ of K onto a subcomplex L can be performed when K contains an n-simplex σ^n that has a *free* face τ^{n-1} (meaning that τ^{n-1} is a face of no other simplices of K). The subcomplex L is obtained by deleting σ^n and τ^{n-1} from K. This combinatorial move can be realized topologically by a particularly nice type of deformation retraction of $|K|$ onto $|L|$. Due to its importance in this paper, we will carefully describe one such deformation retraction in the following paragraphs. Before that, we note that the reverse of an elementary combinatorial collapse is called an *elementary combinatorial expansion* and denoted $L \overset{e}{\nearrow} K$.

To define the deformation retraction alluded to above, label the vertices of σ^n by $\{v_0, \cdots, v_n\}$ such that $\{v_0, \cdots, v_{n-1}\}$ are the vertices of τ^{n-1}, and (abusing notation slightly) choose a standard Euclidean realization of $|\sigma^n|$ in $\mathbb{R}^n$ with $\left|\tau^{n-1}\right| \subseteq \mathbb{R}^{n-1} \times \{0\}$ having barycenter at the origin and so that $v_n = \left(0, \cdots, 0, \sqrt{\frac{n+1}{n}}\right)$. Let L' be the simplicial complex $\partial\sigma^n - \tau^{n-1}$. For each $\mathbf{x} \in |\sigma^n|$, define $r(\mathbf{x})$ be the unique point where the ray $\overrightarrow{-v_n\mathbf{x}}$ intersects $|L'|$. Then define $J : |\sigma^n| \times [0,1] \longrightarrow |\sigma^n|$ to be the map for which $J|_{\mathbf{x}\times[0,1]}$ is the constant speed linear path from $\mathbf{x}$ to $r(\mathbf{x})$. In particular, J is fixed on $|L'|$, $J_0 = \mathrm{id}_{|\sigma^n|}$, and $J_1 = r$ is a retraction of $|\sigma^n|$ onto $|L'|$. Now define $H : |K| \times [0,1] \longrightarrow |L|$ to be J on $|\sigma^n| \times [0,1]$ while keeping all points of $|L|$ fixed throughout. By a mild abuse of notation, we also let r denote the map $H_1 : |K| \to |L|$ and call it an *elementary simplicial collapse.* The corresponding inclusion map $s : |L| \hookrightarrow |K|$ is called an *elementary simplicial expansion.* It is easy to see that H is continuous under all of the above-mentioned topologies on $|K|$.

Now define a *combinatorial collapse* of K to a subcomplex L_0 to be a finite sequence of elementary combinatorial simplicial collapses $K = L_n \overset{e}{\searrow} L_{n-1} \overset{e}{\searrow} \cdots \overset{e}{\searrow} L_0$, a process denoted loosely by $K \searrow L_0$. The concatenation of the corresponding deformation retractions, gives a deformation retraction of $|K|$ onto $|L_0|$ whose end result $r : |K| \to |L_0|$ is a retraction that we call a *simplicial collapse*. Note that r depends on the specific sequence of elementary collapses used to take K onto L_0 but, regardless of the sequence chosen, it is a homotopy equivalence, with homotopy inverse $|L_0| \hookrightarrow |K|$. Under these circumstances we also write $L_0 \overset{e}{\nearrow} L_1 \overset{e}{\nearrow} \cdots \overset{e}{\nearrow} K$, or simply $L_0 \nearrow K$, and say that L_0 *expands combinatorially* to K and that $|L_0|$ *expands simplicially* to $|K|$.

EXAMPLE 3.2. It is a standard (and useful) fact that, for every finite simplicial complex K, there exists a triangulation of $|K| \times [0,1]$ by a simplicial complex L with subcomplexes $K \times \{i\}$ corresponding to $|K| \times \{i\}$ for $i = 0, 1$, and such that $|L|$ collapses simplicially to $|K \times \{i\}|$ for $i = 0, 1$.

A simplicial complex K is *combinatorially collapsible* if it contains a vertex v such that $K \searrow \{v\}$ or, equivalently, there exists a sequence of elementary combinatorial expansions $\{v\} = L_0 \overset{e}{\nearrow} L_1 \overset{e}{\nearrow} \cdots \overset{e}{\nearrow} L_n = K$ where each $L_i \leq K$. In this case, we say that $|K|$ is *simplicially collapsible.*

REMARK 3.3. The geometric realization of a collapsible simplicial complex is clearly contractible. In fact, collapsibility is often viewed as the combinatorial analog of contractibility. Nevertheless, there are well known examples, such as the *topologist's dunce hat* and *Bing's house with two rooms*, that are contractible but not collapsible. Also clear, from the above definition, is that a collapsible simplicial complex is necessarily finite. We will soon relax key definitions to eliminate that requirement.

REMARK 3.4. The subject of *simple homotopy theory* defines a pair of finite simplicial complexes K and L to be *simple homotopy equivalent* if, beginning with K, there exists a finite sequence of elementary expansions and collapses ending in L. As an example, the dunce hat and Bing's house with two rooms—despite the fact that they are not collapsible—are simple homotopy equivalent to a vertex. By contrast, there exist homotopy equivalent simplicial complexes that are not simple homotopy equivalent. Siebenmann [**Sie70**] has developed a simple homotopy theory for infinite complexes, but it differs substantially from what we will do here. Simple homotopy theory does not play a significant role in this paper.

3.3. Infinite simplicial expansions and collapses. We now extend the classical notions of collapses and expansions to allow for infinitely many moves. We say that a simplicial complex K *collapses combinatorially to* L, or equivalently, that L *expands combinatorially to* K, if there exists a (possibly infinite) sequence of elementary combinatorial simplicial expansions $L = L_0 \overset{e}{\swarrow} L_1 \overset{e}{\swarrow} L_2 \overset{e}{\swarrow} \cdots$ through subcomplexes of K, such that $\cup_{i=0}^{\infty} L_i = K$. In this case, we write $K \searrow L$ or $L \nearrow K$. When extra clarity is desired (and infinitely many moves are used) we will write $K \overset{\infty}{\searrow} L$ or $L \overset{\infty}{\nearrow} K$. We say that K is *combinatorially collapsible* if $\{v\} \nearrow K$, with the option of specifying $\{v\} \overset{\infty}{\nearrow} K$, when infinitely many expansions are required.

Although these new definitions are straightforward and natural, some oddities result. For example, a simplicial complex can be collapsible without having a free face. For example, the standard triangulation of $\mathbb{R}$ is collapsible. Despite such oddities, the generalized definitions are useful. To begin, notice that the assumption $L \nearrow K$ allows one to construct a direct sequence of geometric realizations and inclusion maps

$$|L| = |L_0| \overset{s_1}{\hookrightarrow} |L_1| \overset{s_2}{\hookrightarrow} |L_2| \overset{s_3}{\hookrightarrow} \cdots \tag{3.1}$$

and an inverse sequence of (deformation) retractions.

$$|L| = |L_0| \overset{r_1}{\longleftarrow} |L_1| \overset{r_2}{\longleftarrow} |L_2| \overset{r_3}{\longleftarrow} \cdots \tag{3.2}$$

We say that $|K|$ *collapses simplicially to* $|L|$ and that $|L|$ *expands simplicially to* $|K|$. In this context, we always assume that the retractions used in sequence (3.2) are the simplicial collapses defined earlier. When $|K|$ is endowed with the barycentric or weak topology, a straightforward application of Whitehead's Theorem [**Hat02**] ensures that, $|L| \hookrightarrow |K|$ is a homotopy equivalence—as it was in the finite case. If we opt for the topology $\mathcal{T}_I$ on $|K|$, which is often the case in this paper, proving the inclusion is a homotopy equivalence is more subtle; that will be done in Section 5.2. For now we observe that, in a case of primary interest, everything works out nicely.

PROPOSITION 3.5. *Given the above setup, if K is locally finite, then the topologies $\mathcal{T}_I$, $\mathcal{T}_D$, $\mathcal{T}_W$ and $\mathcal{T}_B$ on $|K|$ are the same.*

PROOF. By definition and [**Dow52**], we know $\mathcal{T}_D = \mathcal{T}_W = \mathcal{T}_B$, so it suffices to show that $\mathcal{T}_D = \mathcal{T}_I$. By Theorem 1.7, we need only show that every $x \in |K|$ is insulated. Toward that end, choose j sufficiently large that every simplex of $|K|$ that contains x is contained in $|L_j|$ and let U be the open star neighborhood of x in $|L_j|$ (and hence in $|K|$). We claim that $r_{j\infty}^{-1}(U) = U$. If not, let $k > j$ be the smallest integer such that $r_{jk}^{-1}(U)$ contains a point not in U. Then $|L_k| = |L_{k-1}| \cup \tau$ and, by definition, r_k maps τ onto a union of its proper faces. By our assumption, one of those faces lies in U, and hence contains x. But then, τ is a simplex of K that contains x and hence lies in U, a contradiction. □

4. Compact expansions and collapses

We are ready to extend the notions of expansion and collapse beyond the simplicial category. That, in itself, is not new. There are well-known extensions to:

- polyhedral complexes, where collapses are performed on convex polyhedra;
- n-manifolds, where topological n-balls are collapsed onto $(n-1)$-balls in their boundary (a process known as *shelling*); and
- CW complexes, where k-cells of various dimensions are collapsed onto potentially crumpled $(k-1)$-cells in their attaching spheres.

The CW version is often viewed as the most general context for defining expansions and collapses. We will argue otherwise, by formulating a version that includes all of the above as special cases. It makes room for new examples and allows some well known constructions to be viewed as special cases of a broader theory.

4.1. Track-faithful deformation retractions. Let $H : C \times [0,1] \to C$ be a self-homotopy of a space C and, for each $t \in [0,1]$, let $H_t : C \to C$ be defined by $H_t(x) = H(x,t)$. For our purposes, we are only interested cases where $H_0 = \mathrm{id}_C$. We call H a *deformation retraction* of C onto A if $A = H_1(C)$ and $H_t|_A = \mathrm{id}_A$ for all t.[3] In that case, we call the retraction $r : C \to A$, defined by $r(x) = H_1(x)$, the *culmination* of the deformation retraction H.

The *track* of a point $x \in C$ under H is the path $\lambda_x : [0,1] \to C$ defined by $\lambda_x(t) = H_t(x)$. We sometimes abuse terminology by conflating λ_x with its image. Call H *track-faithful* if, for every $x \in C$ and $t \in [0,1]$, $H_1(H_t(x)) = H_1(x)$. This condition is equivalent to requiring that, if two tracks intersect, then they end at the same point. We are particularly interested in track-faithful deformation retractions.

EXAMPLE 4.1. If $C = X \times [0,1]$ and $H_t(x,s) = (x,(1-t)s)$ then H is a track-faithful deformation retraction of C onto $X \times \{0\}$.

EXAMPLE 4.2. For any map $f : X \to A$, the deformation retraction described in Example 4.1 induces a track-faithful deformation retraction Map(f) onto A. By contrast, whenever f is a homotopy equivalence, Map(f) deformation retracts onto X, but in some cases there is no track-faithful deformation retraction. An example with $X = A = S^1$ can be obtained by comparing [**Gui01**] with Theorem 5.7 in this paper.

EXAMPLE 4.3. Given a top-dimensional face τ^{n-1} of an n-simplex σ^n and the subcomplex $L' = \partial\sigma^n - \tau^{n-1}$, the deformation retractions of $|\sigma^n|$ onto $|L'|$ and $|K|$ onto $|L|$, described in Section 3.2 and used to define an elementary simplicial collapse, are track-faithful.

EXAMPLE 4.4. Let $\square^n = [-1,1]^n$, $B^{n-1} = \square^{n-1} \times \{1\}$, and $J^{n-1} = \partial\square^n - \operatorname{int} B^{n-1}$. Let $\mathbf{p} = (0, \cdots, 0, 2)$ and for each $\mathbf{x} \in \square^n$, let $\mathbf{q}_x$ be the unique point $\overrightarrow{\mathbf{px}} \cap J$. This gives a continuously varying family of segments $\overline{\mathbf{xq_x}} \subseteq \square^n$ (some of them degenerate) connecting points of $\square^n$ to points of J^{n-1}. By choosing constant speed parametrizations $\lambda_{\mathbf{x}} : [0,1] \to \overline{\mathbf{xq_x}}$ of these segments we obtain a track-faithful deformation retraction of $\square^n$ onto J^{n-1}. We will refer to its culmination as an *elementary cubical collapse.*

EXAMPLE 4.5. A contraction of a space C to a point p is track-faithful if and only if p is fixed by the contraction.

EXAMPLE 4.6. The retractions $r, r' : [-1,1] \to [0,1]$ defined by $r(t) = |t|$ and

$$r'(t) = \begin{cases} 0 & \text{if } t \le 0 \\ t & \text{if } t \ge 0 \end{cases}$$

can each be realized as the culmination of a deformation retraction. But, whereas r' is the culmination of a track-faithful deformation retraction, r is not.

LEMMA 4.7. *Let H_t be a track-faithful deformation retraction of C onto A culminating in $r : C \to A$. Then*

(1) *for each $x \in A$, $r^{-1}(x)$ is contractible, and*
(2) *if $x \in \operatorname{Int} A$, then $r^{-1}(x) = \{x\}$.*

[3]Some authors call this a *strong* deformation retraction.

PROOF. For the first assertion, note that if $y \in r^{-1}(x)$, then $\lambda_y \subseteq r^{-1}(x)$. Therefore, the restriction of H to $r^{-1}(x) \times [0,1]$ is a contraction of $r^{-1}(x)$. For the second assertion, let $x \in \operatorname{Int} A$ and suppose there exists $y \in r^{-1}(x)$ such that $y \neq x$. Then there exists $t_0 \in [0,1]$ such that $\lambda_x(t_0) \in A$ and $\lambda_x(t_0) \neq x$. Let $y_0 = \lambda_x(t_0)$ and note that $\lambda_{y_0} = \{y_0\}$. Our hypothesis ensures that $r(y_0) = H_1(y_0) = H_1(H_{t_0}(y)) = x$, but since $y_0 \in A$, $r(y_0) = y_0$. □

LEMMA 4.8. *If $H : C \times [0,1] \to C$ is a track faithful deformation retraction of C onto A and $J : A \times [0,1] \to A$ is a track faithful deformation retraction of A onto B, then the concatenation of H and J is a track faithful deformation retraction of C onto B with $(H * J)_1 = J_1 \circ H_1$.*

PROOF. By "concatenation of H and J", we mean the homotopy $H * J : C \times [0,1] \to C$ defined by

$$(H * J)(x,t) = \begin{cases} H(x, 2t) & \text{if } 0 \leq t \leq \frac{1}{2} \\ J(x, 2t-1) & \text{if } \frac{1}{2} \leq t \leq 1 \end{cases}$$

Verification of the desired properties is straightforward. □

REMARK 4.9. There are other reasonable candidates for the definition of track-faithful homotopy. For example, one could require that the track of every point lying on λ_x be contained in λ_x. We have chosen the weakest condition that is required in our proofs.

4.2. Compact expansions and collapses. Given a metric space C, a map $\rho : C \to A$ is an *compact collapse* if

a) ρ is the culmination of track-faithful deformation retraction, and
b) $C - A$ has compact closure in C.

In this case, we write $C \overset{\rho}{\searrow} A$ or $A \overset{\rho}{\swarrow} C$; we call the inclusion map $\theta : A \hookrightarrow C$ an *compact expansion* and write $A \overset{\theta}{\nearrow} C$ or $C \overset{\theta}{\nwarrow} A$.

NOTATION 2. As an aid to the reader, we will use the letter "rho" *only* to denote compact collapses and "theta" only to denote compact expansions.

EXAMPLE 4.10. Examples 4.1, 4.2 and 4.5 give compact collapses when the corresponding spaces are compact. The elementary simplicial and cubical collapses in Example 4.3 and 4.4 are also compact collapses.

LEMMA 4.11. *If $C \overset{\rho}{\searrow} A$ and the pair (C, A) is homeomorphic to (C', A'), then $C' \overset{\rho'}{\searrow} A'$ for an appropriately chosen retraction ρ'.*

COROLLARY 4.12. *Let B^n be a topological n-ball and B^{n-1} be an $(n-1)$ ball tamely embedded in ∂B^n. Then $B^n \overset{\rho}{\searrow} B^{n-1}$.*

PROOF. By standard methods in geometric topology $\left(B^n, B^{n-1}\right) \approx \left(\square^n, J^{n-1}\right)$. So the claim follows from Lemma 4.11 and Example 4.4. □

LEMMA 4.13. *Suppose X can be expressed as a union $X = C \cup Y$ of closed subspaces; $A = C \cap Y$; and $C \overset{\rho}{\searrow} A$. Then $X \overset{\rho}{\searrow} Y$. More generally, if $C \overset{\rho}{\searrow} A$ and $f : A \to Y$ is continuous, then $Y \cup_f C \overset{\rho}{\searrow} Y$.*

PROOF. The deformation retraction needed for the initial statement follows from the Pasting Lemma. For the second statement, note that the track-faithful deformation retraction of C onto A induces a track-faithful deformation retraction of the induced quotient space $C/\sim$ onto $f(A) = A/\sim$. Since points in $f(A)$ have constant tracks, this deformation retraction extends via the identity homotopy over Y. □

COROLLARY 4.14. *Every elementary CW collapse (in the sense of* [**Coh73**, p.14]*) of a CW complex K to a subcomplex L can be realized as a compact collapse* $K \overset{\rho}{\searrow} L$.

PROOF. Apply Corollary 4.12 and Lemma 4.13 to the definition. □

The reader may have wondered why we did not include the word "elementary" in our definition of compact collapse. The reason is that, unlike in the more prescriptive simplicial and CW versions, any finite sequence of compact collapses can be combined into a single compact collapse.

LEMMA 4.15. *Suppose*

$$X_n \overset{\rho_n}{\searrow} X_{n-1} \overset{\rho_{n-1}}{\searrow} \cdots \overset{\rho_1}{\searrow} X_0$$

is a finite sequence of compact collapses. Then $X_n \overset{\rho}{\searrow} X_0$*, where* $\rho = \rho_1 \circ \cdots \circ \rho_n$.

REMARK 4.16. Note that every base point preserving contraction of a compactum X is a compact collapse. If we were to allow "noncompact collapses", then all reasonably nice contractible spaces would be collapsible in a single move. That extreme level of flexibility is unlikely to yield a useful theory. By contrast, we will show that a theory built upon compact expansions and collapses can be quite useful.

4.3. Infinite compact expansions and collapses. We are ready for our most general treatment of expansions and collapses—one based on compact expansions and collapses and allowing infinitely many moves.

We say that a metric space X *collapses compactly* to a subspace L, or equivalently, that L *expands compactly to* X, if there exists a (possibly infinite) sequence of compact expansions $L = L_0 \overset{e}{\swarrow} L_1 \overset{e}{\swarrow} L_2 \overset{e}{\swarrow} \cdots$ through subsets of X, such that $\cup_{i=0}^{\infty} L_i = X$. In this case, we write $X \searrow L$ or $L \nearrow X$. For extra clarity we sometimes write $X \overset{\infty}{\searrow} L$ or $L \overset{\infty}{\nearrow} X$. We say that X is *compactly collapsible* if $\{p\} \nearrow X$, for some $p \in X$. Note that, although we can combine finite pieces of an infinite sequence of compact expansions into single compact expansion, an infinite sequence will remain infinite.

REMARK 4.17. It is worth commenting further on the work in [**Fer80**] mentioned in the introduction. Ferry's motivation for studying compact collapses (which he called *contractible retractions*) was an attempt to develop a "simple homotopy theory for compact metric spaces". In analogy with the traditional theory, a pair of compact metric spaces would be *simple homotopy equivalent* if one could be converted to the other by a finite sequence of compact expansions and collapses. That idea was undone by the strength of Ferry's own work. He proved that two compacta satisfy that version of simple homotopy equivalent, if and only if they are homotopy equivalent—a nice theorem, but one which nullified the utility of a new definition. Our work resurrects a portion that program, but with different goals. By utilizing *infinite* sequences of compact expansions and collapses, we will obtain some new results. We are particularly interested in $\mathcal{Z}$-compactifications. For that purpose, Ferry's definition is a perfect fit.

5. Compactifications

A primary goal of this paper is to develop general methods for obtaining nice compactifications. We are especially interested in $\mathcal{Z}$-compactifications of spaces arising in geometric group theory and geometric and algebraic topology; but much our machinery can be applied more widely to obtain other interesting compactifications. To allow for the broadest set of applications, we begin with a flexible set of definitions. The reader is warned that the terminology surrounding this topic—including key definitions, such as $\mathcal{Z}$-set—is not consistent across the literature. Rather than trying to blend with everything written previously, we will choose definitions that work for our purposes and cause as little confusion as possible. In many cases, conflicts in terminology vanish in the presence of certain "niceness" assumptions on the spaces to which that terminology is applied. For a thorough treatment of the various definitions, with proofs of equivalences under appropriate conditions, the reader is directed to [**Anc**].

For this paper, there is no loss in restricting the following definitions to cases where the ambient space Y is compact and metrizable. In that case, we can let d be any metric inducing the correct topology on Y. For an open cover $\mathcal{U}$ of Y, maps $f, g : Y \to Y$ are *$\mathcal{U}$-close* if, for every $x \in Y$, there exists a $U \in \mathcal{U}$ such that $f(x)$ and $g(x)$ are contained in U. These maps are called *$\mathcal{U}$-homotopic* if there exists a homotopy H_t between f and g, each of whose tracks is contained in an element of $\mathcal{U}$. In that case, H_t is called a *$\mathcal{U}$-homotopy.* Clearly $\mathcal{U}$-homotopic maps are $\mathcal{U}$-close.

A subset A of Y is called *mapping negligible* if, for every open cover $\mathcal{U}$ of Y, there is a map $f : Y \to Y$ that is $\mathcal{U}$-close to id_Y and for which $f(Y) \subseteq Y - A$. We say that A is *homotopy negligible* if there exists a homotopy $H : Y \times [0,1] \to Y$ such that $H_0 = \mathrm{id}_Y$ and $H_t(Y) \subseteq Y - A$ for all $t > 0$. For an arbitrary open cover $\mathcal{U}$ of Y, one can arrange—by restricting H to a short interval $[0, \delta]$, then reparametrizing—that this is accomplished by a $\mathcal{U}$-homotopy. If A is closed and mapping negligible, we call it an *Ω-set*; if it is a closed homotopy negligible, we call it a *$\mathcal{Z}$-set.* One of the beauties of homotopy negligible sets (and thus $\mathcal{Z}$-sets) is captured by the following simple observation.

PROPOSITION 5.1. *If A is a homotopy negligible in Y, then $s : Y - A \hookrightarrow Y$ is a homotopy equivalence.*

PROOF. Let $H : Y \times [0,1] \to Y$ be a homotopy with $H_0 = \mathrm{id}_Y$ and $H_t(Y) \subseteq Y - A$ for all $t > 0$. Define $f : Y \to Y - A$ by $f(x) = H_1(x)$. Then $s \circ f = H_1$ which is homotopic to id_Y by assumption. Similarly, $f \circ s = H_1|_{Y-A}$ can be homotoped to id_{Y-A} by applying $H|_{(Y-A)\times[0,1]}$. □

EXAMPLE 5.2. The homotopy negligible subsets of a compact manifold with boundary M^n are precisely the subsets of ∂M^n; the $\mathcal{Z}$-sets are the closed subsets of ∂M^n.

EXAMPLE 5.3. For the space $\overline{X} = X \sqcup R$, described in Example 2.1, R is mapping negligible but not an Ω-set, since it is not closed. It is not homotopy negligible, even though the conclusion of Proposition 5.1 holds.

EXAMPLE 5.4. For the space $\overline{X} = X \sqcup R$, described in Example 2.3(ii), R is an Ω-set but not a $\mathcal{Z}$-set, moreover the conclusion of Proposition 5.1 fails.

A metric compactification $\overline{X} = X \sqcup R$ of a space X is called:

- a *mapping negligible compactification* if R is mapping negligible in $\overline{X}$;
- an *Ω-compactification* if R is an Ω-set in $\overline{X}$;
- a *homotopy negligible compactification* if R is homotopy negligible in $\overline{X}$; and
- a *$\mathcal{Z}$-compactification* if R is a $\mathcal{Z}$-set. in $\overline{X}$.

All compactifications in this paper are obtained via the strategy described in Section 1.1. The question then becomes: When does that procedure yield a compactification satisfying one (or more) of the above definitions?

5.1. Mapping negligible compactifications and Ω-compactifications. Our most general compactification theorem is the following. In the proof, $\simeq$ denotes homotopic maps, and (as previously established) $r_{nm} = r_{n+1} \circ \cdots \circ r_m : C_m \to C_n$.

THEOREM 5.5. *Let $C_0 \xleftarrow{r_1} C_1 \xleftarrow{r_2} C_2 \xleftarrow{r_3} \cdots$ be an inverse sequence of compact metric spaces and retractions and let $X_I = \bigcup_{i=0}^{\infty} C_i$ endowed with the topology $\mathcal{T}_I$. Adopt all other notational conventions established in Section* 1. *Then*

(1) *$\varprojlim \{C_i, r_i\}$ is a mapping negligible compactification of $e(X_I)$,*
(2) *equivalently, $\overline{X_I} = X_I \sqcup R$ is a mapping negligible compactification of X_I,*
(3) *if the conditions in Theorem* 1.7 *hold, then these are Ω-compactifications, and*
(4) *if each r_i is a homotopy equivalence, then the projection maps $\pi_j : \varprojlim \{C_i, r_i\} \to C_j$ and their analogs $\pi'_j : \overline{X_I} \to C_j$ are shape equivalences.*

PROOF. For the first assertion, apply $e \circ \pi_i$ for arbitrarily large i; the second assertion follows immediately. The third is an application of Proposition 1.11.

The final assertion is a routine application of the inverse sequence approach to shape theory, if the C_i are assumed to be finite polyhedra—or even just ANRs. In particular, $C_0 \xleftarrow{r_1} C_1 \xleftarrow{r_2} C_2 \xleftarrow{r_3} \cdots$ then represents the shape of $\varprojlim \{C_i, r_i\}$, and is pro-homotopic to each constant inverse sequence $C_j \xleftarrow{\text{id}} C_j \xleftarrow{\text{id}} \cdots$. For a brief introduction to that approach, see [**Gui16**, §3.7]. If the C_i are permitted to be arbitrary compacta, a more subtle approach is needed. The following argument was communicated to us by Jerzy Dydak. It uses the approach to shapes which considers sets of maps into ANRs.

By [**DS78**, Th.4.1.5] (definition on p.22), $\{C_i, r_i\}$ satisfies the following "continuity condition": Any map $f : \varprojlim \{C_i, r_i\} \to K$, where K is an ANR, factors up to homotopy as $f \simeq g \circ \pi_n$, where $g : C_n \to K$, for some n. And, if $g \circ \pi_n \simeq h \circ \pi_n$, then there is $m > n$ such that $g \circ r_{nm} \simeq h \circ r_{nm}$.

To show that $\pi_0 : \varprojlim \{C_i, r_i\} \to C_0$ is a shape equivalence, we need to show that every map $f : \varprojlim \{C_i, r_i\} \to K$ factors uniquely, up to homotopy, via π_0. It factors as $g \circ \pi_n$ for some $n \geq 0$, so the homotopy inverse $s : C_0 \to C_n$ of r_{0n} satisfies $s \circ \pi_0 \simeq \pi_n$ (as $r_{0n} \circ \pi_n = \pi_0$), and therefore $f \simeq (g \circ s) \circ \pi_0$.

For uniqueness, suppose $g, h : C_0 \to K$ satisfy $g \circ \pi_0 \simeq h \circ \pi_0$. Then for some $n > 1$, $g \circ r_{0n} \simeq h \circ r_{0n}$ and composing with s gives $g \simeq h$. The same argument applies to π_i for $i > 0$. □

REMARK 5.6. Section 2 contains several examples where the projection maps are homotopy equivalences and others where they are shape equivalences but not homotopy equivalences.

5.2. Homotopy negligible compactifications and $\mathcal{Z}$-compactifications. We are ready to prove one of our main theorems.

THEOREM 5.7. *Let $C_0 \overset{\rho_1}{\swarrow} C_1 \overset{\rho_2}{\swarrow} C_2 \overset{\rho_3}{\swarrow} \cdots$ be an inverse sequence of compact metric spaces and compact collapses, and let $X_I = \bigcup_{i=0}^{\infty} C_i$, endowed with the topology $\mathcal{T}_I$. Adopt all other notational conventions established in Section 1. Then there is a homotopy $H : \varprojlim \{C_i, \rho_i\} \times [0,1] \to \varprojlim \{C_i, \rho_i\}$ such that $H_t \left(\varprojlim \{C_i, \rho_i\} \right) \subseteq e(X)$ for all $t > 0$; moreover, for each $i \geq 0$, $H|_{\varprojlim\{C_i,\rho_i\} \times [0, \frac{1}{2^i}]}$ is a compact collapse of $\varprojlim \{C_i, \rho_i\}$ onto $e(C_i)$. By letting $J = \overline{e}^{-1} \circ H \circ (\overline{e}, \mathrm{id})$, it follows that*

1. *$\overline{X_I} = X_I \sqcup R$ is a homotopy negligible compactification of X_I,*
2. *$J|_{\overline{X} \times [0, \frac{1}{2^i}]}$ is a compact collapse of $\overline{X_I}$ onto C_i for each $i \geq 0$,*
3. *X_I deformation retracts onto C_i for each $i \geq 0$, and*
4. *if the conditions in Theorem 1.7 hold, then $\overline{X_I}$ is a $\mathcal{Z}$-compactification,*

PROOF. For each $i \geq 1$, let $\phi_i : C_i \times [0,1] \to C_i$ be a track-faithful deformation retraction of C_i onto C_{i-1} culminating in ρ_i. Then define $H_0 : C_0 \times [0,1] \to C_0$ to be the projection map and, for $i > 0$, inductively define $H_i : C_i \times [0,1] \to C_i$ by

$$H_i(x,t) = \begin{cases} x & \text{if } 0 \leq t \leq 1/2^i \\ \phi_i\left(x, 2^i t - 1\right) & \text{if } 1/2^i \leq t \leq 1/2^{i-1} \\ H_{i-1}\left(\rho_i(x), t\right) & \text{if } 1/2^{i-1} \leq t \leq 1 \end{cases}$$

Using the track-faithful property of the ϕ_i, one can check that the following diagram commutes.

$$\begin{array}{ccccccccc} C_0 \times [0,1] & \overset{\rho_1 \times \mathrm{id}}{\longleftarrow} & C_1 \times [0,1] & \overset{\rho_2 \times \mathrm{id}}{\longleftarrow} & C_2 \times [0,1] & \overset{\rho_3 \times \mathrm{id}}{\longleftarrow} & C_3 \times [0,1] & \overset{\rho_4 \times \mathrm{id}}{\longleftarrow} & \cdots \\ H_0 \downarrow & & H_1 \downarrow & & H_2 \downarrow & & H_3 \downarrow & & \\ C_0 & \overset{\rho_1}{\longleftarrow} & C_1 & \overset{\rho_2}{\longleftarrow} & C_2 & \overset{\rho_3}{\longleftarrow} & C_3 & \overset{\rho_4}{\longleftarrow} & \cdots \end{array}$$

It therefore induces a map H between the inverse limits of the top and bottom rows. Since each bonding map in the top row is a product, the inverse limit of that sequence factors as a product of inverse limits; and since the second factor in each of those product maps is $\mathrm{id}_{[0,1]}$, the second inverse limit is canonically homeomorphic to $[0,1]$. So we have a homotopy

$$H : \varprojlim \{C_i, \rho_i\} \times [0,1] \to \varprojlim \{C_i, \rho_i\}$$

which, by definition, is the restriction of $H_0 \times H_1 \times H_2 \times \cdots : \prod C_i \times [0,1] \to \prod C_i$.

To see that H satisfies the desired properties, recall that, for each $i \geq 0$, $e(C_i)$ is the set of all points $(x_0, x_1, x_2, \cdots)$ in $\varprojlim \{C_i, \rho_i\}$ for which $x_j = x_i$ for all $j \geq i$ (see Theorem 1.9). For $t > 0$, choose j such that $1/2^j \leq t \leq 1/2^{j-1}$ and notice that, for $i \geq j$ and $(x_0, x_1, x_2, \cdots) \in \varprojlim \{C_i, \rho_i\}$

$$H_i(x_i, t) = H_i(\rho_{i+1}(x_{i+1}), t) = H_{i+1}(x_{i+1}, t)$$

so $H((x_0, x_1, x_2, \cdots), t) \in e(C_i) \subseteq e(X)$. Furthermore, if $(x_0, x_1, x_2, \cdots) \in e(C_i)$, and $t \leq 1/2^i$ then $H_i(x_i, t) = x_i$, so by the same formula, $H_j(x_i, t) = x_i$ for all $j > i$. Thus, $H|_{\varprojlim\{C_i,\rho_i\}\times[0,\frac{1}{2^i}]}$ is a deformation retraction onto $e(C_i)$. A similar calculation shows that this homotopy is track faithful. □

REMARK 5.8. The above proof is a generalization and refinement of an argument outlined by Jeremy Brazas in his *Wild Topology* blog. See [**Bra**].

5.3. Compactifications of ANRs. In the context of geometric group theory and algebraic and geometric topology, the spaces Y, X and $\overline{X}$ discussed at the beginning of this section are frequently ANRs. When that is the case, several stronger assertions and equivalences hold. It will be useful to review a few of those.

A metric space W is an *absolute neighborhood retract* (ANR) if, whenever it is embedded as a closed subset of a metric space S, a neighborhood of W in S retracts onto W. We call W an *absolute retract* (AR) if whenever W is embedded as a closed subset of a metric space S, the entire space S retracts onto W. Equivalently, an AR is a contractible ANR. Notice that we are requiring W and S to be metric spaces in these definitions; not all treatments of ANR theory include those assumptions.

Key examples of ANRs include: manifolds; metric CW and simplicial complexes (in particular, all locally finite complexes under the Whitehead topology); and CAT(0) spaces. See [**Hu65**] and [**Ont05**]. The following two well known propositions are useful for identifying larger classes of ANRs.

PROPOSITION 5.9. *A finite-dimensional metric space W is an ANR if and only if it is locally contractible; in other words, for every $x \in W$ and neighborhood U of x, there exists a neighborhood V of x that contracts in U.*

PROPOSITION 5.10. *Being an ANR is a local property. More precisely, every open subset of an ANR is and ANR, and a metric space with the property that every point has an ANR neighborhood is an ANR.*

COROLLARY 5.11. *In order for a locally compact space X to admit a compactification $\overline{X} = X \sqcup R$, in which $\overline{X}$ is an ANR, it is necessary for X to be an ANR.*

Versions of the following are woven into the literature in places where $\mathcal{Z}$-compactifications are used to define and analyze group boundaries. For example, [**Bes96**].

PROPOSITION 5.12. *For a closed subset A of a compact ANR Y, the following are equivalent:*

(1) *A is a $\mathcal{Z}$-set,*
(2) *A is an Ω-set,*
(3) *for every open set $U \subseteq Y$, the inclusion $U - A \hookrightarrow U$ is a homotopy equivalence.*

PROOF. All of these can be obtained from Toruńczyk's Equivalence Theorem from [**Anc**]. See also [**Tor78**]. □

In order to apply Proposition 5.12, the following partial converse to Proposition 5.10 and its corollary can be useful.

PROPOSITION 5.13. *If A is a homotopy negligible subset of a compact metric space Y, and $Y - A$ is an ANR, then Y is an ANR. In other words, every homotopy negligible compactification of an ANR is an ANR.*

PROOF. Apply Hanner's Theorem [**Han51**], as described in [**Tir11**]. □

COROLLARY 5.14. *Suppose $\overline{X} = X \sqcup R$ is an Ω-compactification of an ANR [resp., AR] X. Then $\overline{X}$ is a $\mathcal{Z}$-compactification of X if and only if $\overline{X}$ is an ANR [resp., AR].*

PROOF. The forward and reverse implications follow from Propositions 5.13 and 5.12, respectively. The assertion about ARs relies on Proposition 5.1. □

By using essentially the same argument, one can deduce a strengthening of Proposition 5.13, that will be useful in the next section.

PROPOSITION 5.15. *Let A be a homotopy negligible subset of a compact metric space Y and $H : Y \times [0,1] \to Y$ a corresponding homotopy with $H_0 = \mathrm{id}_Y$ and $H_t(Y) \subseteq Y - A$ for all $t > 0$. Suppose that for each $\varepsilon > 0$, there exist a $t \in [0, \varepsilon]$ and an ANR $B_t \subseteq Y$ such that $H_t(Y) \subseteq B_t$, then Y is an ANR.*

COROLLARY 5.16. *If $\{C_i, \rho_i\}$ is an inverse sequence of compact ANRs and compact collapses, and $\overline{X} = X_I \sqcup R$ is the corresponding homotopy negligible compactification of X_I (as in Theorem 5.7), then $\overline{X}$ is an ANR.*

6. Applications and more examples

We conclude this article with a variety of theorems and examples that can be proved or reinterpreted using the ideas presented in this paper.

6.1. General applications to simplicial and CW complexes and mapping telescopes. The first pair of applications are immediate consequences of Theorem 5.7 Corollary 5.16, together with the general discussion in Sections 1 and 3.

THEOREM 6.1. *Let K be a simplicial complex that collapses combinatorially to a finite subcomplex L, and let $|K|_I$ be the geometric realization of K with the topology $\mathcal{T}_I$ induced by a corresponding inverse sequence of simplicial collapses beginning with $|L|$. Then the compactification $\overline{|K|_I} = |K|_I \sqcup R$ is homotopy negligible, and $\overline{|K|_I}$ is an ANR. Furthermore, there is a compact collapse of $\overline{|K|_I}$ onto $|L|$ and a deformation retraction of $|K|_I$ to $|L|$. If $|L|$ is contractible, then $|K|_I$ and $\overline{|K|_I}$ are ARs.*

COROLLARY 6.2. *Let K be a locally finite simplicial complex that collapses combinatorially to a finite subcomplex L, and let $|K|$ be the geometric realization of K with its standard topology. Then there is a $\mathcal{Z}$-compactification $\overline{|K|} = |K| \sqcup R$ such that $\overline{|K|}$ is an ANR. Furthermore, there is a compact collapse of $\overline{|K|}$ onto $|L|$ and a deformation retraction of $|K|$ to $|L|$. If $|L|$ is contractible, then $|K|$ and $\overline{|K|}$ are ARs.*

PROOF. Along with the previous theorem, apply Propositions 1.11 and 3.5. □

REMARK 6.3. Examples 2.2 and 2.3 show how $\mathcal{T}_I$ can differ from the usual topologies on $|K|$, even when K is locally finite. Examples 2.1 and 2.3 illustrate the importance of having bonding maps that are compact collapses and not just arbitrary deformation retractions.

By combining the above argument with Corollary 4.14, we can obtain a version of Corollary 6.2 for CW complexes.

THEOREM 6.4. *Let X be a locally finite CW complex that collapses, via elementary CW collapses, to a finite subcomplex L. Then there exists a $\mathcal{Z}$-compactification $\overline{X} = X \sqcup R$ such that $\overline{X}$ is an ANR. Furthermore, there is a compact collapse of $\overline{X}$ onto L and a deformation retraction of X to L. If L is contractible, then X and $\overline{X}$ are ARs.*

Versions of the next theorem have played a key role in earlier work on $\mathcal{Z}$-compactifications. See [**CS76**].

THEOREM 6.5. *Let $K_0 \xleftarrow{f_1} K_1 \xleftarrow{f_2} K_2 \xleftarrow{f_3} \cdots$ be an inverse sequence of compact metric spaces. Then there is a $\mathcal{Z}$-compactification*

$$\overline{\mathrm{Tel}\,(K_i, f_i)} = \mathrm{Tel}\,(K_i, f_i) \sqcup R$$

where $R = \varprojlim(\{K_i, f_i\})$. The spaces $\overline{\mathrm{Tel}\,(K_i, f_i)}$ and $\mathrm{Tel}\,(K_i, f_i)$ are ANRs if and only if each K_i is an ANR.

PROOF. The main assertion combines Examples 2.8 and 4.2 and Theorem 5.7. For the follow-up assertion, apply [**Hu65**, IV.1] and Corollary 5.16 to see that $\mathrm{Tel}\,(K_i, f_i)$ and $\overline{\mathrm{Tel}\,(K_i, f_i)}$ are ANRs when all K_i are ANRs. For the converse, note that for each i, $\mathrm{Tel}\,(K_i, f_i)$ contains an open set that retract onto K_i. □

6.2. Applications to CAT(0) spaces. Our first theorem about CAT(0) spaces is well known and obtainable using analytic techniques. The methods developed in this paper provide an alternative approach.

THEOREM 6.6. *Every proper CAT(0) space Y is $\mathcal{Z}$-compactifiable by addition of the visual sphere at infinity. Moreover, $\overline{Y} = Y \sqcup \partial_\infty Y$ is an AR.*

PROOF. The basic properties of this compactification were covered in Example 2.7. The assertions that this is a $\mathcal{Z}$-compactification and that $\overline{Y}$ is an AR follow from 5.7 and Propositions 5.13. □

THEOREM 6.7 ([**Gul23**]). *Every locally finite CAT(0) cube complex is collapsible via a sequence of elementary cubical collapses and therefore admits a $\mathcal{Z}$-compactification $\overline{Y} = Y \sqcup \partial_\square Y$ corresponding to such a sequence.*

REMARK 6.8. The $\mathcal{Z}$-compactification in Theorem 6.7 differs from the one in Theorem 6.6. The proof in [**Gul23**] is combinatorial rather than analytic, and the resulting boundary (referred to as a *cubical boundary*) retains some of the combinatorial structure of Y. Further investigation of that compactification is part of an ongoing project by the authors of this paper.

An interesting sort of partial converse to Theorem 6.7 is contained in [**AF21**]. There it is shown that every collapsible locally finite simplicial complex admits a subdivision that is a CAT(0) cube complex.

6.3. Applications to manifolds. Finite sequences of simplicial and CW expansions and collapses play important roles in the study of compact manifolds. See [**RS82**]. The generalizations developed here can be used to obtain some analogous theorems for noncompact manifolds; we offer a sample in this section. For the sake of brevity, we focus our attention on contractible open manifolds.

A countable simplicial complex K (or its geometric realization $|K|$) is a *combinatorial n-manifold* if the link of each k-simplex of K is PL homeomorphic to either $\left|\partial\sigma^{n-k}\right|$ or $\left|\sigma^{n-k-1}\right|$. The simplices with links of the latter type constitute a subcomplex $\partial K \leq K$. Under these circumstances, $|K|$ satisfies the topological definition of an n-manifold with $\partial|K| = |\partial K|$. A *PL n-manifold* is a space M^n together with a triangulation $h : |K| \to M^n$ by a combinatorial n-manifold; we call this a *PL triangulation* of M^n.

A homeomorphism $h : |K| \to M^n$, where M^n is a topological n-manifold and K is *not* a combinatorial n-manifold, is called a *non-PL triangulation* of M^n. The existence of non-PL triangulations of manifolds is a deep fact first proved by Edwards [**Edw06**]. That work implies that all PL manifolds of dimension ≥ 5 admit non-PL triangulations. By contrast, it is known that non-PL triangulations do not exist in dimensions ≤ 4.

Our first theorem is an easy one.

THEOREM 6.9. *For all n, $\mathbb{R}^n$ admits a simplicially collapsible PL triangulation.*

PROOF. Begin with an n-simplex σ^n, then inductively add PL triangulations of $|\partial\sigma^n| \times [k, k+1]$ of the sort described in Example 3.2. □

The converse of Theorem 6.9 is more striking. It can be viewed as the noncompact analog of an important theorem which asserts that every compact simplicially collapsible combinatorial n-manifold is a PL n-ball [**RS82**, Cor.3.28].

THEOREM 6.10. *If a combinatorial n-manifold $|K|$ is simplicially collapsible, then $|K|$ is PL homeomorphic to $\mathbb{R}^n$.*

PROOF. Let $\{v\} = L_0 \overset{e}{\swarrow} L_1 \overset{e}{\swarrow} L_2 \overset{e}{\swarrow} \cdots$ be a corresponding sequence of elementary combinatorial expansions through subcomplexes such that $\cup L_i = K$. Then each L_i is finite and collapsible, so a regular neighborhood N_i of $|L_i|$ in $|K|$ is a PL n-ball. By choosing wisely (or passing to a subsequence) we can arrange that $N_i \subseteq \operatorname{Int} N_{i+1}$ for all i, so by the Combinatorial Annulus Theorem [**RS82**, Cor.3.19], $N_{i+1} - \operatorname{Int} N_i$ is PL homeomorphic to $S^{n-1} \times [i, i+1]$. Gluing these products together and plugging the hole with N_0 gives a homeomorphism to $\mathbb{R}^n$. □

COROLLARY 6.11. *No triangulation of the Whitehead contractible 3-manifold is simplicially collapsible. More generally, the only open manifolds of dimension ≤ 4 that admit simplicially collapsible triangulations are those PL homeomorphic to $\mathbb{R}^n$.*

PROOF. Every triangulation of a manifold of dimension ≤ 4 is PL, so the assertion follows from Theorem 6.10. □

If we allow non-PL triangulations, a different picture emerges, even for the simplest class of contractible open manifolds.

THEOREM 6.12. *If C^n is a compact contractible PL n-manifold with $n \geq 5$, then* $\operatorname{int} C^n$ *admits a simplicially collapsible non-PL triangulation.*

We begin this proof with a lemma.

LEMMA 6.13. *Let C^n be a compact contractible n-manifold that is not homeomorphic to a ball. Then*

(1) *there does not exist a PL triangulation of C^n that is simplicially collapsible,*
(2) *there exists a non-PL triangulation of C^n that is simplicially collapsible if and only if $n \geq 5$ and ∂C^n admits a PL triangulation. (The latter condition holds whenever $n \geq 6$.)*
(3) *C^n is always compactly collapsible.*

PROOF. Assertion 1) is a consequence of regular neighborhood theory. Suppose such a triangulation exists, and let DC^n be the double of C^n along its boundary. Then, a regular neighborhood of C^n in DC^n is PL homeomorphic to C^n, but a regular neighborhood of a compact collapsible polyhedron is always a PL ball.

Assertion 2) relies on a construction described in [**ADG97**] which triangulates C^n with a simplicial complex of the form

$$T = (\{p_1\} * K_1) \cup_{\{q\}*J} (\{p_2\} * K_2)$$

(the union of two simplicial cones on finite simplicial complexes K_1 and K_2 along a common conical subcomplex $\{q\} * J$ of K_1 and K_2). Since every finite simplicial cone collapses to the subcone over any of its subcomplexes, we can collapse T to the subcomplex $\{p_1, p_2\} * (\{q\} * J)$ which, by associativity of simplicial joins, is the cone $\{q\} * (\{p_1, p_2\} * J)$. From there we can collapse to q.

For the parenthetic remark, note that the boundary of a compact contractible n-manifold has the homology of an $(n-1)$-sphere. The Kirby-Siebenmann obstruction [**KS77**] to a PL triangulation of ∂C^n lies in $H^4(\partial C^n; \mathbb{Z}_2)$, and therefore vanishes when $n \geq 6$.

For assertion 3), choose any contraction of C^n that fixes a point. That contraction is a compact collapse. □

PROOF OF THEOREM 6.12. If we attach an exterior open collar $\partial C^n \times [0, \infty)$ to C^n, the resulting space is homeomorphic to $\operatorname{int} C^n$. Thus,

$$\operatorname{int} C^n \approx C^n \cup (\partial C^n \times [0,1]) \cup (\partial C^n \times [1,2]) \cup (\partial C^n \times [2,3]) \cup \cdots \tag{6.1}$$

By Lemma 6.13, we can choose a simplicially collapsible non-PL triangulation $|K|$ of C^n with a subcomplex $L \leq K$ that triangulates ∂C^n. A sequence of applications of Example 4.10 applied to $|L| \times [i, i+1]$ produces triangulations of $\partial C^n \times [i, i+1]$ which agree on their overlap and collapse to their left-hand boundaries. Assemble these to produce the desired triangulation. □

If we allow arbitrary compact collapses, a still different picture emerges.

THEOREM 6.14. *The interior of every compact contractible n-manifold C^n admits a compact collapse.*

PROOF. Begin with the decomposition of $\operatorname{int} C^n$ provided in (6.1). Since each $\partial C^n \times [i, i+1]$ compactly collapses to $\partial C^n \times \{i\}$ (Example 4.3) and C^n compactly collapses to a point, the theorem follows. □

It is not the case that *every* contractible open manifold is compactly collapsible.

THEOREM 6.15. *The Whitehead contractible 3-manifold W^3 is not compactly collapsible.*

PROOF. By Theorem 5.7, if W^3 were compactly collapsible, it would be $\mathcal{Z}$-compactifiable. Then, by [**GT03**, Th.1.2], W^3 would be semistable at infinity, contradicting [**Geo08**, Prop.16.4.2]. □

REMARK 6.16. The Whitehead manifold is by no means unique. In all dimensions ≥ 3 there exist contractible open manifolds that are not semistable; hence they are not $\mathcal{Z}$-compactifiable; hence they are not compactly collapsible. (Note. For non-manifolds, compact collapsibility does not imply semistability.)

Another interesting collection of contractible open n-manifolds are the exotic universal covers of closed aspherical n-manifolds constructed by Davis in [**Dav83**]. We will take a closer look at the collapsibility properties of those manifolds in [**AGS**]. For now, we mention one particularly nice class.

THEOREM 6.17. *Let K be a flag simplicial complex for which $|K|$ is a closed $(n-1)$-manifold with the same homology as S^{n-1}. Then the right-angled Coxeter group W_K with nerve K acts geometrically on a CAT(0) cube complex X whose underlying space is a contractible open n-manifold that is not homeomorphic to $\mathbb{R}^n$. The space X is cubically collapsible.*

PROOF. The space X is constructed in [**ADG97**]. It is a variation on the Davis complex Σ_K (which is not a manifold). By [**Gul23**], X is cubically collapsible. □

6.4. Applications to 1-dimensional spaces.

THEOREM 6.18. *Every finite valence tree T is simplicially collapsible and admits a $\mathcal{Z}$-compactification $\overline{T} = T \sqcup R$ where $\overline{T}$ is an AR and R is homeomorphic to the space of ends of T.*

PROOF. Endow T with the path length metric; choose a base vertex v_0; and for each $i \in \mathbb{N}$, let K_i be the closed ball of radius i centered at v_0. Then each K_i is a finite subtree and $\overline{K_{i+1} - K_i}$ is a collection of edges, each having one end point in K_{i-1}. Clearly there is a simplicial collapse $\rho_{i+1} : K_{i+1} \to K_i$. Now apply Corollary 6.2. The correspondence between R and the space of ends is almost immediate from the definitions. See, for example, [**Gui16**, §3.3]. □

Provided T is countable, we can prove a weaker result without assuming finite valences.

THEOREM 6.19. *Every countable tree T is a union of finite subtrees $\{v_0\} = T_0 \subseteq T_1 \subseteq T_2 \subseteq \cdots$ with the property that $\overline{T_{i+1} - T_i}$ is a collection of edges, each having one end point in T_{i-1}; thus T is simplicially collapsible. Under the topology induced by the corresponding inverse sequence, T_I admits a homotopy negligible compactification $\overline{T_I} = T_I \sqcup R$ which is an AR.*

PROOF. We leave it to the reader to prove the existence of the collection $\{T_i\}$. From there, the conclusions follow from Theorem 5.7 and Corollary 5.16. □

REMARK 6.20. Example 2.2 is an illustration of Theorem 6.19. Note that T_I is not the usual topology on T. Nevertheless, this theorem provides a method for defining and topologizing the ends of T. For example, the tree in Example 2.2 has no ends.

A *dendrite* is a compact, connected, locally path connected metric space D that is uniquely arc-wise connected. In other words, for any pair of distinct points $p, q \in D$, there is a unique embedded arc with end points p and q. Dendrites are of central importance in the field of continuum theory. They have also played an important role in the study of hyperbolic groups [**Bow99**]. A fundamental theorem from continuum theory provides a link to the previous two theorems.

THEOREM 6.21 ([**Nad92**, Th.10.27]). *Every dendrite D is homeomorphic to the inverse limit of a sequence*

$$\{v_0\} = T_0 \overset{\rho_1}{\swarrow} T_1 \overset{\rho_2}{\swarrow} T_2 \overset{\rho_3}{\swarrow} \cdots$$

of finite trees where $\overline{T_{i+1} - T_i}$ is an arc with one end point in T_i and ρ_{i+1} is the corresponding compact collapse.

In this generality, we cannot assume there are fixed simplicial structures on the T_i making them subcomplexes of T_j for $j \geq i$ and making the ρ_i simplicial collapses. Nevertheless, the proof used for Theorem 6.19 is applicable and gives the following important theorem about dendrites.

COROLLARY 6.22. *Every dendrite D is a compact AR; in particular, D is contractible.*

With additional care, one can say a little more. A point $p \in D$ is an *end point* if it has arbitrarily small neighborhoods whose boundaries are one-point sets. Let $D^{[1]}$ denote the set of end points of D.

THEOREM 6.23. *The set $D^{[1]}$ is homotopy negligible in D.*

PROOF. One strategy for proving Theorem 6.21 is to begin with a countable dense set $X = \{x_0, x_1, x_2, \cdots\} \subseteq D$, then inductively define T_{i+1} to be the union of T_i with the arc A_{i+1} connecting x_{i+1} to x_0. It is easy to see that $T_i \cap A_{i+1}$ is either $\{x_0\}$ or a subarc of A_{i+1}; hence T_{i+1} can be given the structure of a finite tree. (If $T_i = T_{i+1}$, we can omit it from the sequence.) It is a simple fact that $D^{[1]}$ is nowhere dense in D, so we can choose X to be disjoint from D. In that case, an element of $D^{[1]}$ is never a leaf of a T_i. But, for obvious reasons, neither can an element of $D^{[1]}$ be a nonleaf of a T_i. Since D is already compact, our compactification is of the form $D = (\cup T_i) \sqcup R$; hence $D^{[1]} \subseteq R$. □

References

[ADG97] F. D. Ancel, M. W. Davis, and C. R. Guilbault, CAT(0) *reflection manifolds*, Geometric topology (Athens, GA, 1993), AMS/IP Stud. Adv. Math., vol. 2, Amer. Math. Soc., Providence, RI, 1997, pp. 441–445, DOI 10.1090/amsip/002.1/24. MR1470741

[AF21] Karim A. Adiprasito and Louis Funar, *CAT(0) metrics on contractible manifolds*, arXiv:1512.06403v4 [math.GT], 2021.

[AGS] F. D. Ancel, C. R. Guilbault, and P. Sparks, *Infinite boundary connected sums with applications*, Preprint, 2023.

[Anc] Fredric D. Ancel, *Z-sets in ANRs*, Topology at Infinity of Discrete Groups (R. Geoghegan, C. R. Guilbault, and K. Ruane, eds.), Contemporary Mathematics, vol. 812, Amer. Math. Soc., Providence, RI, 2025.

[Bes96] M. Bestvina, *Local homology properties of boundaries of groups*, Michigan Math. J. **43** (1996), no. 1, 123–139, DOI 10.1307/mmj/1029005393. MR1381603

[BH99] M. R. Bridson and A. Haefliger, *Metric spaces of non-positive curvature*, Grundlehren der mathematischen Wissenschaften [Fundamental Principles of Mathematical Sciences], vol. 319, Springer-Verlag, Berlin, 1999, DOI 10.1007/978-3-662-12494-9. MR1744486

[Bow99] B. H. Bowditch, *Treelike structures arising from continua and convergence groups*, Mem. Amer. Math. Soc. **139** (1999), no. 662, viii+86, DOI 10.1090/memo/0662. MR1483830

[Bra] Jeremy Brazas, *Shape injectivity of the earring space (part ii)*, Wild Topology blog (August 6, 2019).

[Coh73] M. M. Cohen, *A course in simple-homotopy theory*, Graduate Texts in Mathematics, Vol. 10, Springer-Verlag, New York-Berlin, 1973. MR362320

[CS76] T. A. Chapman and L. C. Siebenmann, *Finding a boundary for a Hilbert cube manifold*, Acta Math. **137** (1976), no. 3-4, 171–208, DOI 10.1007/BF02392417. MR425973

[Dav83] M. W. Davis, *Groups generated by reflections and aspherical manifolds not covered by Euclidean space*, Ann. of Math. (2) **117** (1983), no. 2, 293–324, DOI 10.2307/2007079. MR690848

[Dow52] C. H. Dowker, *Topology of metric complexes*, Amer. J. Math. **74** (1952), 555–577, DOI 10.2307/2372262. MR48020

[DS78] J. Dydak and J. Segal, *Shape theory: An introduction*, Lecture Notes in Mathematics, vol. 688, Springer, Berlin, 1978. MR520227

[Dug78] J. Dugundji, *Topology*, Allyn and Bacon Series in Advanced Mathematics, Allyn and Bacon, Inc., Boston, Mass.-London-Sydney, 1978. Reprinting of the 1966 original. MR478089

[Edw06] Robert D. Edwards, *Suspensions of homology spheres*, `arXiv:math/0610573` [math.GT], 2006.

[Fer80] S. Ferry, *Homotopy, simple homotopy and compacta*, Topology **19** (1980), no. 2, 101–110, DOI 10.1016/0040-9383(80)90001-4. MR572578

[Geo08] R. Geoghegan, *Topological methods in group theory*, Graduate Texts in Mathematics, vol. 243, Springer, New York, 2008, DOI 10.1007/978-0-387-74614-2. MR2365352

[GT03] C. R. Guilbault and F. C. Tinsley, *Manifolds with non-stable fundamental groups at infinity. II*, Geom. Topol. **7** (2003), 255–286, DOI 10.2140/gt.2003.7.255. MR1988286

[Gui01] C. R. Guilbault, *A non-$\mathcal{Z}$-compactifiable polyhedron whose product with the Hilbert cube is $\mathcal{Z}$-compactifiable*, Fund. Math. **168** (2001), no. 2, 165–197, DOI 10.4064/fm168-2-6. MR1852740

[Gui16] C. R. Guilbault, *Ends, shapes, and boundaries in manifold topology and geometric group theory*, Topology and geometric group theory, Springer Proc. Math. Stat., vol. 184, Springer, [Cham], 2016, pp. 45–125, DOI 10.1007/978-3-319-43674-6_3. MR3598162

[Gul23] D. L. Gulbrandsen, *Collapsibility and Z-Compactifications of Cat(0) Cube Complexes*, ProQuest LLC, Ann Arbor, MI, 2023. Thesis (Ph.D.)–The University of Wisconsin - Milwaukee. MR4675718

[Han51] O. Hanner, *Some theorems on absolute neighborhood retracts*, Ark. Mat. **1** (1951), 389–408, DOI 10.1007/BF02591376. MR43459

[Hat02] A. Hatcher, *Algebraic topology*, Cambridge University Press, Cambridge, 2002. MR1867354

[Hu65] S.-t. Hu, *Theory of retracts*, Wayne State University Press, Detroit, MI, 1965. MR181977

[KS77] R. C. Kirby and L. C. Siebenmann, *Foundational essays on topological manifolds, smoothings, and triangulations*, Annals of Mathematics Studies, No. 88, Princeton University Press, Princeton, NJ; University of Tokyo Press, Tokyo, 1977. MR645390

[MP15] M. M. Marsh and J. R. Prajs, *Internal inverse limits and retractions*, Rocky Mountain J. Math. **45** (2015), no. 4, 1209–1223, DOI 10.1216/RMJ-2015-45-4-1209. MR3418191

[Mun00] J. R. Munkres, *Topology*, Prentice Hall, Inc., Upper Saddle River, NJ, 2000. Second edition of [MR0464128]. MR3728284

[Nad92] S. B. Nadler Jr., *Continuum theory: An introduction*, Monographs and Textbooks in Pure and Applied Mathematics, vol. 158, Marcel Dekker, Inc., New York, 1992. MR1192552

[Ont05] P. Ontaneda, *Cocompact CAT(0) spaces are almost geodesically complete*, Topology **44** (2005), no. 1, 47–62, DOI 10.1016/j.top.2004.01.010. MR2104000

[RS82] C. P. Rourke and B. J. Sanderson, *Introduction to piecewise-linear topology*, Springer Study Edition, Springer-Verlag, Berlin-New York, 1982. Reprint. MR665919

[Sie70] L. C. Siebenmann, *Infinite simple homotopy types*, Indag. Math. **32** (1970), 479–495. Nederl. Akad. Wetensch. Proc. Ser. A **73**. MR287542

[Tir11] C. J. Tirel, *$\mathcal{Z}$-structures on product groups*, Algebr. Geom. Topol. **11** (2011), no. 5, 2587–2625, DOI 10.2140/agt.2011.11.2587. MR2836296

[Tor78] H. Toruńczyk, *Concerning locally homotopy negligible sets and characterization of l_2-manifolds*, Fund. Math. **101** (1978), no. 2, 93–110, DOI 10.4064/fm-101-2-93-110. MR518344

[Whi39] J. H. C. Whitehead, *Simplicial Spaces, Nuclei and m-Groups*, Proc. London Math. Soc. (2) **45** (1939), no. 4, 243–327, DOI 10.1112/plms/s2-45.1.243. MR1576810

[Whi41] J. H. C. Whitehead, *On incidence matrices, nuclei and homotopy types*, Ann. of Math. (2) **42** (1941), 1197–1239, DOI 10.2307/1970465. MR5352

[Whi49] J. H. C. Whitehead, *Combinatorial homotopy. I*, Bull. Amer. Math. Soc. **55** (1949), 213–245, DOI 10.1090/S0002-9904-1949-09175-9. MR30759

[Whi50] J. H. C. Whitehead, *Simple homotopy types*, Amer. J. Math. **72** (1950), 1–57, DOI 10.2307/2372133. MR35437

Department of Mathematical Sciences, University of Wisconsin-Milwaukee, Milwaukee, Wisconsin 53201

Email address: craigg@uwm.edu

Department of Mathematics and Computer Science, Hampden-Sydney College, Hampden-Sydney, Virginia 23943

Current address: Department of Mathematical Sciences, Adams State University, 208 Edgemont Blvd, Alamosa, CO 81101

Email address: dgulbrandsen@adams.edu

Contemporary Mathematics
Volume **812**, 2025
https://doi.org/10.1090/conm/812/16271

Hyperbolic groups and local connectivity

G. Christopher Hruska and Kim Ruane

Dedicated to Mike Mihalik on his 70th birthday.

Abstract. The goal of this paper is to give an exposition of some results of Bestvina–Mess on local connectivity of the boundary of a one-ended word hyperbolic group. We also give elementary proofs that all hyperbolic groups are semistable at infinity and their boundaries are linearly connected in the one-ended case. Geoghegan first observed that semistability at infinity is a consequence of local connectivity using ideas from shape theory, and Bonk–Kleiner proved linear connectivity using analytical methods. The methods in this paper are closely based on the original ideas of Bestvina–Mess.

1. Introduction

In the 1980s, Gromov introduced the notion of a word hyperbolic group and showed that each hyperbolic group can be compactified by attaching a certain boundary at infinity, known as the Gromov boundary [**Gro87**]. Shortly afterward, Mess and Mihalik conjectured that the Gromov boundary of a one-ended hyperbolic group is always locally connected.

In a seminal article, Bestvina–Mess show that the boundary of a one-ended hyperbolic group is locally connected whenever it does not contain a cut point [**BM91**]. Subsequent work of Bowditch, Levitt, and Swarup shows that the boundary can never contain a cut point [**Bow99c, Lev98, Swa96**], which establishes that the conjecture of Mess and Mihalik is true, *i.e.*, the boundary is always locally connected.

Geoghegan observed that local connectivity implies that every hyperbolic group is semistable at infinity, using a deep theorem of shape theory due to Krasinkiewicz (see [**Kra77**]). Semistability is a foundational property central to the study of the fundamental group at infinity (see [**HR**] for more background). It has been conjectured by Geoghegan and Mihalik that every finitely presented group is semistable at infinity.

Additionally, Bonk–Kleiner use analytical methods to show that every hyperbolic group that is not virtually free contains a quasi-isometrically embedded copy of the hyperbolic plane in [**BK05**]. In order to establish this result, Bonk–Kleiner show that the boundary of a one-ended hyperbolic group is linearly connected, a

2020 *Mathematics Subject Classification.* Primary 20F67, 57M07.

Key words and phrases. Hyperbolic group, semistable, locally connected, linearly connected, boundary.

The first author was partially supported by grant #714338 from the Simons Foundation.

stronger quantitative version of local connectivity that is often useful for the analytical study of boundaries.

The goal of this paper is to provide a detailed exposition of the proofs of [**BM91**, §3] and to give simple, elementary proofs of semistability and linear connectivity using the same methods. The essential ingredient in [**BM91**] is a geometric property known as ($\ddagger_M$), which may or may not hold in the Cayley graph of G. If this property holds for some M, one can locally push paths outward away from a basepoint, leading directly to both semistability and linear connectivity. Surprisingly, if ($\ddagger_M$) were to fail for any $M > 0$, then the boundary would contain a cut point, which is known to be false by the Cut Point Theorem of [**Bow99c, Lev98, Swa96**].

In [**Bow99a**], Bowditch has extended some of the ideas related to ($\ddagger_M$) from the hyperbolic setting to the relatively hyperbolic setting. This extension is a key step in his work on local connectivity of boundaries of relatively hyperbolic groups [**Bow99b, Bow01**].

Section 2 is a review of hyperbolic spaces and their boundaries. In Section 3, we introduce the ($\ddagger_M$) condition and use the Cut Point Theorem to show that the Cayley graph of any one-ended hyperbolic group satisfies ($\ddagger_M$) for any $M > 0$. Section 4 contains an elementary proof that all hyperbolic groups are semistable at infinity. In Section 5, we use similar methods to establish the linear connectivity of the boundary of a one-ended hyperbolic group by a short, elementary argument. We conclude by sketching the proof of Bonk–Kleiner's quasi-hyperbolic plane theorem, using linear connectivity.

2. Hyperbolic spaces

In this section, we summarize some well-known background on the structure of δ–hyperbolic spaces and their boundaries. We refer the reader to any of the texts [**GH90, CDP90, BH99, BS07**] for a more detailed introduction.

DEFINITION 2.1 (Hyperbolic spaces). A metric space is a *geodesic space* if each pair of points in X is joined by a geodesic segment. If p, x, y are points of a geodesic space X, the *Gromov product* is the quantity

$$(x|y)_p = \frac{1}{2}\big(d(x,p) + d(y,p) - d(x,y)\big).$$

To understand the geometric significance of the Gromov product, consider a geodesic triangle with vertices x, y, and z, and consider the quantities

$$a = (y|z)_x, \qquad b = (x|z)_y, \qquad \text{and} \qquad c = (x|y)_z.$$

A direct calculation using the definition of the Gromov product reveals that

$$a + b = d(x,y), \qquad a + c = d(x,z), \qquad \text{and} \qquad b + c = d(y,z).$$

A geodesic space X is *δ–hyperbolic* for some $\delta \geq 0$ if for any geodesic triangle with vertices x, y, z in X we have the following. If $y' \in [x,y]$ and $z' \in [x,z]$ are points with $d(x,y') = d(x,z') \leq (y|z)_x$, then $d(y',z') \leq \delta$.

Another way to understand this definition is as follows: if we collapse the triangle onto a tripod with legs of length a, b, and c, then the preimage of any point in the tripod is a set of diameter at most δ. Note that, in a δ–hyperbolic space, each side of any triangle lies in the δ–neighborhood of the union of the other two sides.

Gromov's hyperbolicity inequality states the following.

PROPOSITION 2.2 ([**BS07**], Prop. 2.1.2). *If X is δ–hyperbolic, then*

$$(x|y)_p \geq \min\{(x|z)_p, (z|y)_p\} - \delta$$

for any base point $p \in X$ and any $x, y, z \in X$.

We first introduce the Gromov boundary as a set.

DEFINITION 2.3 (The set $\overline{X} = X \cup \partial X$). Geodesic rays $c, c' \colon [0,\infty) \to X$ are *equivalent* if $d\big(c(t), c'(t)\big)$ remains bounded as $t \to \infty$. The *boundary* ∂X is the set of all equivalence classes of geodesic rays in X, and we let $c(\infty) \in \partial X$ denote the equivalence class of the ray c.

We may identify $\overline{X} = X \cup \partial X$ with the set of all equivalence classes of generalized rays, defined as follows. A *generalized ray* in X is a geodesic $c \colon I \to X$, where I is either an interval $[0, L]$ for some $L > 0$ or $[0, \infty)$. In the case when $I = [0, L]$ we extend c to a map $[0,\infty) \to X$ by setting $c(t) = c(L)$ for all $t \geq L$, and we refer to these as "eventually constant" generalized rays. The point $c(L)$ is the *endpoint* of c and will be denoted by $c(\infty)$. Eventually constant generalized rays are *equivalent* if they have the same endpoint. A geodesic ray and an eventually constant generalized ray are never equivalent.

DEFINITION 2.4 (The topology on $\overline{X}$). Let X be a proper δ–hyperbolic space with basepoint $p \in X$. Let $D > 2\delta$ be a fixed constant. For each geodesic ray c_0 in X with $c_0(0) = p$ we define $V(c_0, r, D)$ to be the set of equivalence classes of generalized rays that have a representative c such that $c(0) = p$ and $d\big(c(r), c_0(r)\big) < D$.

According to [**BH99**, Lem. III.H.3.6], the sets $V(c_0, r, D)$ determine a topology as follows. For each eventually constant generalized ray c_0 based at p, consider the family $\mathcal{V}_{c_0}$ of metric balls about $c_0(\infty)$. For each geodesic ray c_0 based at p and each $D > 2\delta$, consider the family $\mathcal{V}_{c_0}$ of sets $V(c_0, r, D)$ for $r < \infty$. The systems of sets $\mathcal{V}_{c_0}$ determine a topology on $\overline{X}$ in which each $\mathcal{V}_{c_0}$ is a neighborhood base at $c_0(\infty)$. We caution the reader that the basic neighborhoods $V(c_0, r, D)$ are not necessarily open in this topology.

The topology on $\overline{X}$ does not depend on the chosen basepoint or the choice of constant $D > 2\delta$. We refer the reader to [**BH99**, §III.H.3] for more details and proofs of the above assertions.

Each basic neighborhood $V(c_0, r, D)$ in $\overline{X}$ is path connected. Indeed, choose a pair of generalized rays c, c' based at p that intersect the ball of radius D about $c_0(r)$. Then $c(\infty)$ can be joined to $c'(\infty)$ by a piecewise geodesic path

$$c\big([r,\infty]\big) \cup \big[c(r), c_0(r)\big] \cup \big[c_0(r), c'(r)\big] \cup c'\big([r,\infty]\big)$$

that lies inside $V(c_0, r, D)$.

DEFINITION 2.5 (Radial paths). For each neighborhood $V(c_0, r, D)$ the paths described above will be referred to as *radial paths.* We have seen that any two points of $V(c_0, r, D)$ are joined by a radial path.

DEFINITION 2.6 (Extended Gromov product). Let X be a proper δ–hyperbolic space with basepoint $p \in X$. The Gromov product extends to $\overline{X}$ by defining

$$(x|y)_p = \inf \liminf_{i,j\to\infty} (x_i|y_j)_p,$$

where the infimum is taken over all sequences (x_i) and (y_j) converging to x and y respectively. For any sequences $x_i \to x$ and $y_i \to y$, we have

$$(x|y)_p \leq \liminf (x_i|y_i)_p \leq \limsup (x_i|y_i)_p \leq (x|y)_p + 2\delta.$$

See [**BS07**, Lem. 2.2.2].

LEMMA 2.7. *Let X be a proper δ–hyperbolic space with a basepoint p. Choose rays c and c' based at p. For any $s \leq \bigl(c(\infty)\big|c'(\infty)\bigr)_p$ we have $d\bigl(c(s), c'(s)\bigr) \leq \delta$.*

PROOF. For any $\epsilon > 0$, there exists T so that $s \leq \bigl(c(T)\big|c'(T)\bigr)_p + \epsilon$. By the definition of hyperbolicity, setting $s_0 = \bigl(c(T)\big|c'(T)\bigr)_p$ we have

$$d\bigl(c(s_0), c'(s_0)\bigr) \leq \delta.$$

There are two cases: If $s \leq s_0$ then $d\bigl(c(s), c'(s)\bigr) \leq \delta$ by hyperbolicity. Otherwise, $s \in [s_0, s_0 + \epsilon)$, so

$$d\bigl(c(s), c'(s)\bigr) \leq \delta + 2\epsilon.$$

Taking $\epsilon \to 0$, we get $d\bigl(c(s), c'(s)\bigr) \leq \delta$. □

REMARK 2.8 (Slim ideal triangles). A triangle is *ideal* if one or more sides is either a ray or a bi-infinite line. If X is δ–hyperbolic, then each side of an ideal triangle lies in the δ_∞–neighborhood of the union of the other two sides, for some constant δ_∞ depending only on δ. See [**CDP90**, Prop. 2.2.2].

DEFINITION 2.9 (Visual metric). The boundary of a proper δ–hyperbolic space X admits a metric d_∞ with the following property: there exists a parameter $a > 1$ and constants $k_1, k_2 > 0$ such that

$$k_1\, a^{-(\zeta|\xi)_p} \leq d_\infty(\zeta, \xi) \leq k_2\, a^{-(\zeta|\xi)_p}$$

holds for all $\xi, \zeta \in \partial X$. (See [**BS07**, Thm. 2.2.7].) Such a metric d_∞ is called a *visual metric with parameter a* on ∂X.

In the following proposition we relate the distance between certain points on a pair of rays on a given sphere with the distance between their endpoints in the visual metric. This proposition will be used in the proof of Theorem 5.4.

PROPOSITION 2.10. *Let X be a δ–hyperbolic space with basepoint p. Let d_∞ be a visual metric with parameter a on ∂X, and let D be any fixed constant. Then there exists a constant $k_3 = k_3(D)$ such that the following holds. Choose any pair of geodesic rays c, c' based at p. For any $r > 0$,*

$$\textit{if} \quad d\bigl(c(r), c'(r)\bigr) < D \quad \textit{then} \quad d_\infty\bigl(c(\infty), c'(\infty)\bigr) < k_3\, a^{-r}.$$

PROOF. Given a pair of rays c, c' based at p and a constant $r > 0$, let $\xi = c(\infty)$ and $\xi' = c'(\infty)$. The definition of the Gromov product implies that

$$\bigl(c(u)\big|c(v)\bigr)_p = \min\{u, v\}$$

for any $u, v \in [0, \infty)$.

We will show that

$$\bigl(c(r)\big|c'(r)\bigr)_p > r - D/2.$$

Consider the quantities:

$$a = \bigl(p\big|c'(r)\bigr)_{c(r)} \qquad b = \bigl(p\big|c(r)\bigr)_{c'(r)} \qquad s = \bigl(c(r)\big|c'(r)\bigr)_p$$

Definition 2.1 shows that $a + s = r = b + s$, which implies $a = b$. Therefore, $2a = a + b = d\big(c(r), c'(r)\big) < D$. We conclude that

$$s = r - a > r - D/2.$$

For any $\epsilon > 0$ there exists $T > r$ such that

$$\begin{aligned}(\xi|\xi')_p &\geq \limsup_{t\to\infty}\big(c(t)\big|c'(t)\big)_p - 2\delta \geq \big(c(T)\big|c'(T)\big)_p - 2\delta - \epsilon \\ &\geq \min\big\{\big(c(T)\big|c(r)\big)_p, \big(c(r)|c'(r)\big)_p, \big(c'(r)\big|c'(T)\big)_p\big\} - 4\delta - \epsilon \\ &\geq r - (D/2) - 4\delta - \epsilon,\end{aligned}$$

where the third inequality follows by applying Proposition 2.2 twice. Taking $\epsilon \to 0$, we see that

$$(\xi|\xi')_p \geq r - (D/2) - 4\delta.$$

Therefore, we have

$$d_\infty(\xi, \xi') \leq k_2\, a^{-(\xi|\xi')_p} \leq k_2\, a^{(D/2)+4\delta} a^{-r}.$$

□

3. Separating horoballs and global cut points

This section introduces the property $(\ddagger_M)$ and gives an elaboration of the proof of [**BM91**, Prop. 3.3]. In order to create paths in the boundary of a one-ended hyperbolic group G, Bestvina–Mess introduce $(\ddagger_M)$, which may or may not hold in the Cayley graph of G. This property allows one to locally push paths outward away from a basepoint. It turns out that the obstruction to $(\ddagger_M)$ is the existence of a separating horoball in the Cayley graph. Developing this idea, Bestvina–Mess show that if $(\ddagger_M)$ fails to hold for any $M > 0$, then the boundary must contain a cut point. Recall that a *cut point* of a connected topological space Z is a point $p \in Z$ such that $Z - \{p\}$ is disconnected. Subsequently, it was shown that the boundary of a one-ended hyperbolic group can never contain a cut point, which implies that $(\ddagger_M)$ holds for every $M > 0$.

DEFINITION 3.1 (Horoballs). Let X be a proper δ–hyperbolic space, and let δ_∞ be the constant given by Remark 2.8. A function $h\colon X \to \mathbb{R}$ is a *horofunction* about $\eta \in \partial X$ if the following condition holds: There exists a constant $K < \infty$ such that if $a, b \in X$ and $d\big(b, [a,\eta)\big) \leq \delta_\infty$ for some geodesic ray $[a,\eta)$, then $\big|h(a) - h(b) - d(a,b)\big| \leq K$. A *horoball* centered at $\eta \in \partial X$ is a subset $B \subset X$ of the form $h^{-1}(-\infty, 0]$ for some horofunction h about η. Each horoball has a unique center, which is the unique accumulation point of the horoball in ∂X (see [**Bow12**, §5]).

Gromov has considered a different, universal notion of horofunction defined on any proper metric space X via a natural embedding of X into the space of continuous real-valued functions on X (see [**Gro81**]). In the setting of δ–hyperbolic spaces, the "coarse" definition above, due to Bowditch, is more general than Gromov's definition but is well-suited for coarse geometric arguments. The following example illustrates how to use Gromov's construction to produce horoballs in the sense of Definition 3.1.

EXAMPLE 3.2. A closed subset A of a metric space X is the *Kuratowski–Painlevé limit* of a sequence of subsets A_n, denoted $A = \lim_{n\to\infty} A_n$, if

$$A = \big\{\, a \in X \;\big|\; \limsup_{n\to\infty} d(a, A_n) = 0 \,\big\} = \big\{\, a \in X \;\big|\; \liminf_{n\to\infty} d(a, A_n) = 0 \,\big\}.$$

Every sequence of subsets of a separable metric space has a subsequence that is convergent in the above sense by [**Kur66**, §29.VIII].

Let X be a proper δ–hyperbolic space with basepoint y, and let $D \in \mathbb{R}$ be a constant. By [**KMN06**, §2], every sequence (x_n) converging to a point $\eta \in \partial X$ has a subsequence such that the balls $B(x_{n_i}, d(y, x_{n_i}) + D)$ converge to a horoball centered at η.

DEFINITION 3.3. Let X be a geodesic space. We say that *geodesics are almost extendable* in X if there is a constant C such that for all points p and q in X there is a geodesic ray emanating from p that passes within a distance C of q. We typically say that X is *almost extendable* for short. The constant C is an *almost extendability constant* of X. In the literature, almost extendable spaces are sometimes called *visual* spaces (see, for instance, [**BS00**]).

Mihalik observed that infinite word hyperbolic groups are almost extendable (see [**BM91**, §3]). We first review the simple proof of this fact.

PROPOSITION 3.4. *Let X be any proper δ–hyperbolic space with a cocompact group action by a group G. If ∂X has at least two points, then X is almost extendable.*

PROOF. The space X contains a bi-infinite geodesic line. Choose a compact set K whose G–translates cover X, and let $D = \operatorname{diam}(K)$. Choose points $p, q \in X$. By cocompactness, there exists a geodesic line c intersecting the closed ball of radius D centered at q. Consider the ideal geodesic triangle Δ having p as a vertex and c as a bi-infinite side. By Remark 2.8, the triangle Δ is δ_∞–slim. Note that q is at most a distance D from some point $c(t)$ on the line c, and $c(t)$ is within a distance δ_∞ of one of the rays $\bigl[p, c(\infty)\bigr)$ or $\bigl[p, c(-\infty)\bigr)$. Therefore the ray in question comes within a distance $D + \delta_\infty$ of the point q. □

Throughout this section, X is a one-ended δ–hyperbolic space that admits a geometric group action and C is its almost extendability constant.

In [**BM91**], Bestvina–Mess introduce the following property $(\ddagger_M)$, which may or may not be satisfied by X for each constant $M > 0$. The $(\ddagger_M)$ property is key for establishing the connection between local connectivity and separating horoballs.

$(\ddagger_M)$ There is a number $L > 0$ such that for every $R > 0$ and any three points x, y, $z \in X$ with $d(x, y) = d(x, z) = R$ and $d(y, z) \leq M$, there exists a path of length at most L connecting y to z in the complement of the ball of radius $R - C$ centered at x.

The main goal of this section is to prove Corollary 3.7, which states that $(\ddagger_M)$ holds for all $M > 0$.

A horoball B in X is *essentially separating* if B is closed in X and the complement of B has at least two components that each contain a geodesic ray not asymptotic to B.

PROPOSITION 3.5. *Let X be a proper δ–hyperbolic space with a cocompact group action by G. Suppose there exists a constant $M > 0$ such that $(\ddagger_M)$ fails in X. If X is one-ended then it contains an essentially separating horoball.*

The proof of this result depends on the almost extendable property of X, but hyperbolicity plays no essential role in the proof. We note that any proper CAT(0) space with a cocompact group of isometries is almost extendable by [**GO07**].

PROOF. If ($\ddagger_M$) fails, then for each n there is a positive number R_n and a triple of points x_n, y_n, and z_n with $d(x_n, y_n) = d(x_n, z_n) = R_n$ and $d(y_n, z_n) \leq M$ such that any path from y_n to z_n in the complement of the closed ball $B_n = \overline{B}(x_n, R_n - C)$ has length at least n. Since X is almost extendable, we can choose rays c_n and c'_n based at x_n passing within a distance C of y_n and z_n respectively.

After passing to a subsequence, we can assume that either $R_n \to \infty$ or R_n remains bounded as $n \to \infty$. Suppose first that R_n remains bounded. After passing to a further subsequence and applying the cocompact isometry group, we can assume that x_n, y_n, and z_n converge to points $x, y, z \in X$ respectively. We can also assume that c_n converges to a ray from x to some point $\xi \in \partial X$ that passes within C of y. Similarly, c'_n converges to a ray from x to some $\xi' \in \partial X$ that passes within C of z. In this case, the closed balls B_n converge, in the sense of Kuratowski–Painlevé, to a compact set B. The existence of a path α of finite length $< L$ from y to z in the complement of B would imply the existence of a path α' from y_n to z_n in the complement of B_n of length $< L$ for sufficiently large n. However, there is no such path when $n > L$, so B must separate the ends of X containing ξ and ξ'. In particular these ends are distinct, which is impossible since X is one-ended.

Therefore the sequence R_n must limit to infinity. As before we may pass to a subsequence so that y_n, z_n converge to $y, z \in X$ and $x_n \to \eta \in \partial X$ and the balls B_n converge to a closed horoball B centered at η (see Example 3.2). In this case y and z must lie in the complement of B. We may also assume that the rays c_n converge to a geodesic line from η to some $\xi \in \partial X$ that passes within a distance C of y. Similarly the rays c'_n converge to a line from η to some ξ' that passes within C of z. By the same reasoning as above, there is no path of finite length from y to z in the complement of B. As in the previous case, B is an essentially separating horoball. □

The hyperbolicity of the space X plays a key role in the next result, which does not require a cocompact group action.

PROPOSITION 3.6. *Suppose X is a one-ended proper δ–hyperbolic space containing an essentially separating horoball B centered at $\eta \in \partial X$. Then η is a cut point of ∂X.*

PROOF. Fix a basepoint $p \in X$. Let $\eta \in \partial X$ be the center of the horoball B. Choose rays c and c' based at p, such that they have tails in distinct components of $X - B$. Let $\xi = c(\infty)$ and $\xi' = c'(\infty)$. We will show that ξ and ξ' are in distinct components of $\partial X - \{\eta\}$.

Fix a constant $D > 2\delta$, and consider the corresponding neighborhood base for points of ∂X described in Definition 2.4. Since $\overline{B} = B \cup \{\eta\}$ is closed in $\overline{X}$, each point $\zeta \in \partial X - \{\eta\}$ has a basic neighborhood W in $\overline{X} - \overline{B}$. Recall that basic neighborhoods are not necessarily open, but their points may be connected by radial paths, as described in Definition 2.5. Let $\mathcal{U}$ be the set of all interiors of such neighborhoods W. Then $\partial X - \{\eta\}$ is contained in the union of all members of $\mathcal{U}$.

Suppose by way of contradiction that ξ and ξ' are not separated by η in ∂X. Then we may choose a chain $U_1, \dots, U_k$ in $\mathcal{U}$ such that $\xi \in U_1$, $\xi' \in U_k$, and $U_i \cap U_{i+1} \cap \bigl(\partial X - \{\eta\}\bigr)$ is nonempty. Choose a_0 to be a point on the ray c that lies inside U_1. Similarly choose a_k on c' lying inside U_k. For $0 < i < k$, choose

$a_i \in U_i \cap U_{i+1}$. For each i, choose a basic neighborhood $W_i \subseteq \overline{X} - \overline{B}$ such that $U_i \subseteq W_i$. Choose a radial path γ_i from a_{i-1} to a_i lying inside W_i. Since each a_i lies in the interior of a basic neighborhood, we may assume that the paths γ_i and γ_{i+1} use the same generalized ray from p to a_i. The union $\gamma_1 \cup \cdots \cup \gamma_k$ is a connected set K in $\overline{X} - \overline{B}$ containing a_0 and a_k. Inside K is a path of $X - B$ joining a_0 to a_k, contradicting the choice of c and c'. □

We now use the Cut Point Theorem, which asserts the boundary of a one-ended hyperbolic group does not contain a cut point. The Cut Point Theorem combines major contributions of Bowditch, Levitt, and Swarup [**Bow99c**, **Lev98**, **Swa96**].

COROLLARY 3.7. *Suppose X is a one-ended δ–hyperbolic space and G acts geometrically on X. Then $(\ddagger_M)$ holds for all $M > 0$.*

PROOF. By the Cut Point Theorem, the boundary of X does not contain a cut point. Suppose by way of contradiction that $(\ddagger_M)$ fails for some $M > 0$. Then by Proposition 3.5, the space X contains an essentially separating horoball B. Finally, by Proposition 3.6 shows that any such horoball gives rise to a cut point in ∂X, which contradicts the cut point theorem. □

4. Semistability at infinity

The goal of this section is to give a simple, direct proof of semistability for one-ended hyperbolic groups using the condition $(\ddagger_M)$. Indeed, $(\ddagger_M)$ is very close in spirit to a characterization of semistability studied by Mihalik. We also explain how to derive semistability for arbitrary hyperbolic groups from the one-ended case.

Let X be a simply connected space. A *ray* in X is a proper map $c\colon [0,\infty) \to X$. Two rays c and c' in X *converge to the same end* of X if for any compact set $D \subset X$ there exists $N \geq 0$ such that $c([N,\infty)])$ and $c([N,\infty))$ lie in the same component of $X - D$. Convergence to the same end is an equivalence relation whose equivalence classes are called the *ends* of X.

A simply connected space X is *semistable at infinity* if any two rays in X converging to the same end are joined by a proper homotopy.

A theorem of Mihalik ([**Mih83**, Thm. 2.1]) states that semistability is equivalent to the following condition: for some (equivalently any) ray r and each compact set $D \subseteq X$, there is a larger compact set E so that, for every still larger compact set F, each pointed loop α in $X - E$ based on the ray r can be homotoped into $X - F$ via a homotopy in $X - D$ that slides the base point along r.

The following lemma makes precise the idea that $(\ddagger_M)$ allows one to push paths outward in the Cayley graph in a controlled manner.

LEMMA 4.1. *Let G be a one-ended word hyperbolic group acting geometrically on a δ–hyperbolic space X. There is a constant λ such that the following holds. Choose any basepoint $p \in X$. Let c and c' be geodesic rays based at p.*

If there exists a path α of length at most ℓ from c to c' in the complement of the ball $B(p,r)$ then there exists a path β of length at most $\ell\lambda$ from c to c' in the complement of $B(p,r+1)$.

If X is simply connected, there is a further constant $\kappa = \kappa(X)$ such that the following holds. If $r > r_0 + \kappa$ for some positive constant r_0, then for any path α as above, the path β may be chosen homotopic to α by a homotopy consisting of paths from c to c' that lie in the complement of $B(p, r_0)$.

PROOF. Let C be the almost extendability constant of X, and let $M = 6C+3$. Consider the constant $L = L(M)$ given by $(\ddagger_M)$. Choose any radius $r \geq 0$. Consider a path $\alpha\colon [0,\ell] \to X$ in the complement of the ball $B(p,r)$ parameterized so that $d\big(\alpha(j-1),\alpha(j)\big) \leq 1$ for each $j = 1,\dots,\ell$. Since X is almost extendable, for each $j = 0,\dots,\ell$ we may choose a geodesic ray c_j based at p containing a point y_j such that $d\big(y_j,\alpha(j)\big) < C$. We may assume that $c_0 = c$ and $c_\ell = c'$. Then for each $j = 0,\dots,\ell$ the point y_j lies outside the ball $B(y, r-C)$. Choose points $z_0,\dots,z_\ell$ on these rays so that $d(y_j,z_j)$ is at most $2C+1$ and $d(p,z_j)$ is at least $r+C+1$. Then $d(z_{j-1},z_j)$ is at most $6C+3 = M$. By $(\ddagger_M)$, there exists a path β_j of length at most L joining z_{j-1} to z_j in the complement of the ball $B(p,r+1)$. Since the path β formed by concatenating the paths β_j has length at most ℓL, the first conclusion of the lemma follows if we set $\lambda = L$.

Now suppose X is simply connected. By cocompactness, there is a constant H so that any loop γ in X of length at most $M+L$ bounds a disc in the H–neighborhood of γ. Let $\kappa = C+H$. For any fixed radius r_0, suppose now that $r > r_0 + \kappa$. Let η_j be a geodesic from $\alpha(j)$ to y_j. Consider a loop γ_j formed by following α from $\alpha(j-1)$ to $\alpha(j)$, then following η_j to y_j, following c_j from y_j to z_j, following β_j to z_{j-1}, following c_{j-1} to y_{j-1}, and then following η_{j-1} back to $\alpha(j-1)$. Since γ_j is a loop of length at most $M+L$ in the complement of $B(p,r-C)$, it bounds a disc in the complement of the ball $B(p,r-\kappa)$. In particular γ_j bounds a disc disjoint from $B(p,r_0)$. □

PROPOSITION 4.2. *Every one-ended word hyperbolic group is semistable at infinity.*

PROOF. Assume G acts geometrically on a simply connected δ–hyperbolic space X. Choose a basepoint p and a geodesic ray c based at p. We apply Lemma 4.1 with $c = c'$. Using the notation of Lemma 4.1, a loop α outside the ball $B(p,r)$ can be homotoped to a path in the complement of $B(p,r+1)$ by a homotopy in the complement of $B(p,r_0)$. The homotopy involves paths with each of their endpoints on c. At each stage of the homotopy, we insert a subpath of c joining these endpoints so that we now have a homotopy of loops. Iterating this conclusion, we may homotope loops into the complement of an arbitrarily large ball to see that X is semistable at infinity. □

We now discuss how to pass from the one-ended case to the general case to conclude the following result, first observed by Geoghegan using a different argument (see [**BM91**]).

THEOREM 4.3 (Geoghegan). *Every word hyperbolic group G is semistable at infinity.*

This theorem follows from Proposition 4.2 using a result of Mihalik that establishes semistability for graphs of groups over finite edge groups if the vertex groups are known to be semistable [**Mih87**]. In order to keep our explanation relatively self-contained, we sketch a proof of Theorem 4.3 below. In particular, we sketch Mihalik's reduction from the general case to the one-ended case in the setting of hyperbolic groups.

The first ingredient in the proof of this theorem provides a useful graph of groups decomposition by combining several well-known results.

PROPOSITION 4.4. *Let G be a word hyperbolic group. Then G is the fundamental group of a graph of groups whose edge groups are finite and whose vertex groups are either finite or one ended. Furthermore, each vertex group G_v is itself a hyperbolic group, and the inclusion $G_v \hookrightarrow G$ is a quasi-isometric embedding.*

PROOF. Every hyperbolic group G is finitely presented, so Dunwoody's accessibility theorem implies that G splits over finite groups as described [**Dun85**]. It follows that each vertex group G_v is quasiconvex, in the sense that, any geodesic in the Cayley graph of G joining two points of G_v must lie uniformly close to G_v itself. Any quasiconvex subgroup of a hyperbolic group is hyperbolic and quasi-isometrically embedded. See, for instance, Lemma III.Γ.3.5 and Proposition III.Γ.3.7 of [**BH99**]. □

To build a cleaner geometric model space for a multi-ended hyperbolic group, we use the following construction.

PROPOSITION 4.5. *Let G be a word hyperbolic group. There exists a simply connected δ–hyperbolic simplicial complex X on which G acts geometrically such that each finite subgroup of G has a fixed point in X.*

PROOF. Let d_S be the word metric for a finite generating set S. Since a hyperbolic group has finitely many conjugacy classes of finite subgroups [**BH99**, III.Γ.3.2], there exists a constant R so that each finite subgroup has a conjugate of diameter less than R. For sufficiently large R, each finite subgroup stabilizes a simplex in the Rips complex $X = P_R(G, d_S)$ and, thus, the subgroup has a fixed point in X. □

An entirely different proof of Proposition 4.5 is given by Lang, who shows that the injective hull of (G, d_S) is a contractible complex satisfying the conclusion of the preceding result in [**Lan13**].

We are now ready to sketch the proof of the theorem.

PROOF OF THEOREM 4.3. Let the infinite hyperbolic group G be the fundamental group of a graph of groups $\mathcal{G}$ with all edge groups finite and all vertex groups either finite or one-ended. Let T be the Bass–Serre tree for the given splitting. For each one-ended vertex group G_v, let X_v be a simply connected δ–hyperbolic simplicial complex on which G_v acts geometrically such that each finite subgroup of G_v has a fixed point, as in Proposition 4.5. We assume that these spaces and fixed points are chosen equivariantly. For each edge e of T joining vertices v and w, let $p_e \in X_v$ and $q_e \in X_w$ be the chosen points fixed by the edge group G_e. The group G acts geometrically on a δ–hyperbolic space X formed by gluing the spaces $\{X_v\}$ in the pattern of the tree T as follows: Start with the disjoint union of the spaces X_v for $v \in T$, and for each edge of T, attach an edge with endpoints p_e and q_e. We will call these new edges the T–edges of X.

We will see that any pair of rays in X converging to the same end are properly homotopic. Let $\pi \colon X \to T$ be the quotient map that collapses each space X_v to a point, leaving only the T–edges. Recall that a map is *monotone* if the preimage of each point is a connected set. We claim that each proper ray in X is properly homotopic to a ray that projects to a monotone path in T. Indeed, let r be any ray of X that is also an edge path, *i.e.*, traversing each edge linearly. For each point p in a closed T–edge, the complement $X - \{p\}$ has exactly two components that we call *halfspaces*, each of which is simply connected. If H is a halfspace not containing the basepoint $r(0)$, then any subpath of r forming a loop in $\overline{H}$ based at p can be contracted to a point within $\overline{H}$. Thus we may assume that r contains no such loop. But a ray containing no such loop projects to a monotone path $\overline{r} = \pi \circ r$ in T.

Either $\overline{r}$ is eventually constant (so that r has a tail lying inside a single vertex space X_v) or $\overline{r}$ converges to an end of T (so that r travels through an infinite number of vertex spaces, but never visits any of them more than once).

If two rays with monotone projection converge to the same end of X, they must have the same type in T—either their projections to T have tails that lie in the same vertex space X_v or their projections converge to the same end of T. In the first case, they are properly homotopic since, by Proposition 4.2, the space X_v is semistable. In the second case, they are properly homotopic, since each vertex space is simply connected. □

5. Linear connectivity

In this section, we prove two results due to Bonk–Kleiner on linear connectivity of boundaries and the existence of quasi-isometrically embedded hyperbolic planes in hyperbolic groups. The proof of linear connectivity is a slight variation of the elementary proof of [**BM91**, Prop. 3.2].

DEFINITION 5.1. A complete metric space M is *linearly connected* if there exists a constant $\nu < \infty$ such that each pair of points $x, x' \in M$ is contained in a compact, connected set of diameter less than $\nu\, d(x,x')$.

Given two points $x, x' \in M$, a *ρ-chain of length N from x to x'* is a sequence of points $x = b_0, b_1, \ldots, b_N = x'$ such that $d(b_{j-1}, b_j) \leq \rho\, d(x,x')$ for each j.

PROPOSITION 5.2. *Let M be a compact metric space. Suppose there are numbers $\rho < 1$ and $N \in \mathbb{N}$ such that for each $x, x' \in M$ there is a ρ-chain of length N from x to x'. Then M is linearly connected.*

PROOF. Given $x, x' \in M$, let S_1 be a ρ-chain of length N from x to x'. Define S_k recursively by inserting a ρ-chain of length N between each consecutive pair of points of S_{k-1}. Then

$$\operatorname{diam}(S_k) \leq 2N(\rho + \rho^2 + \cdots + \rho^k)\, d(x,x')$$

holds by induction, since

$$\operatorname{diam}(S_1) \leq N\rho\, d(x,x') \quad \text{and} \quad \operatorname{diam}(S_k) \leq \operatorname{diam}(S_{k-1}) + 2N\rho^k\, d(x,x').$$

Therefore, $S = \overline{\bigcup S_k}$ is a compact connected set of diameter at most $\nu\, d(x,x')$ for $\nu = 2N\rho/(1-\rho)$ because $\rho < 1$. □

REMARK 5.3. The proposition above also holds under the weaker hypothesis that M is a complete metric space. Indeed, the inclusion $\bigcup S_k \to M$ determines a map $[0,1] \cap \mathbb{Z}[1/N] \to M$, which is uniformly continuous by a slight variation of the above argument. By completeness, this map extends to a continuous path $I \to M$ from x to x' with image S. Curiously, Hilbert's well-known construction of a space-filling curve may be recovered as a direct consequence: if chains S_k are selected in the square I^2 according to Hilbert's recursive method, the result is a continuous surjection $I \to S = I^2$.

THEOREM 5.4 (Bonk–Kleiner). *If G is any one-ended hyperbolic group, then ∂G is linearly connected with respect to any visual metric.*

COROLLARY 5.5 ([**BM91, Bow99c, Lev98, Swa96**]). *The boundary of a one-ended hyperbolic group is locally connected.*

PROOF OF THEOREM 5.4. Suppose G acts geometrically on a δ–hyperbolic space X. Fix a basepoint $p \in X$, and choose a visual metric d_∞ on ∂X with parameter a. Choose distinct points $x, x' \in \partial X$, and let $s = (x|x')_p$. If c and c' are any rays based at p such that $c(\infty) = x$ and $c'(\infty) = x'$, then $d\big(c(s), c'(s)\big) \leq \delta$ by Lemma 2.7. Let C be the almost extendability constant of X, and let $L = L(\delta)$ be the constant given by $(\ddagger_M)$ using $M = \delta$. Then there exists a path α of length at most L from $c(s)$ to $c'(s)$ in the complement of the ball $B(p, s - C)$.

Let $m > 0$ be a positive integer to be determined later, and let λ be the constant from Lemma 4.1. If we iterate the lemma m times, we see that there exists a path β of length at most $N = \lambda^m L$ from c to c' in the complement of the ball $B(p, s - C + m)$. Fix a parametrization $\beta\colon [0, N] \to X$ so that $d\big(\beta(j-1), \beta(j)\big) \leq 1$ for each $j = 0, \cdots, N$. Since X is almost extendable, for each j we may choose a ray c_j containing a point y_j such that $d\big(y_j, \beta(j)\big) < C$. We will assume that $c_0 = c$ and $c_L = c'$. Then for each j we have

$$d(p, y_j) \geq s - 2C + m \qquad \text{and} \qquad d(y_{j-1}, y_j) \leq 2C + 1.$$

If we let $z_j = c_j(s - 2C + m)$, then we have $d(z_{j-1}, z_j) \leq 2C + 1 + 2\delta$. Let k_1 be given by Definition 2.9, and let $k_3 = k_3(2C + 1 + 2\delta)$ be the constant from Proposition 2.10. Setting $b_j = c_j(\infty) \in \partial X$, we have

$$d_\infty(b_{j-1}, b_j) \leq k_3 a^{-(s-2C+m)} = \frac{k_3}{a^{m-2C}}\, a^{-(x|x')_p} \leq \frac{k_3}{k_1 a^{m-2C}}\, d_\infty(x, x').$$

In particular, there exists a ρ–chain of length N from x to x' for $\rho = k_3/(k_1 a^{m-2C})$. We now fix a value for the parameter m large enough that $a^m > k_3 a^{2C}/k_1$. Then $\rho < 1$, and $N = \lambda^m L$ does not depend on the choice of x and x'. Thus $(\partial X, d_\infty)$ is linearly connected by Proposition 5.2. □

THEOREM 5.6 (Bonk–Kleiner). *Let G be any hyperbolic group that is not virtually free. Then G contains a quasi-isometrically embedded hyperbolic plane.*

The proof of Theorem 5.6 involves the following notions from metric analysis. A metric space M is *doubling* if there is a constant N such that each metric ball can be covered by at most N balls of half the radius. A nonconstant map $f\colon M \to M'$ between metric spaces is *quasisymmetric* if there exists a homeomorphism $\eta\colon [0, \infty) \to [0, \infty)$ such that $d(x, y) \leq t\, d(x, z)$ implies $d\big(f(x), f(y)\big) \leq \eta(t)\, d\big(f(x), f(z)\big)$ for all $x, y, z \in M$ and all $t \geq 0$. A *quasi-arc* in M is a quasisymmetric map $[0, 1] \to M$.

Now that we have established the linear connectivity of the boundary, the proof of Theorem 5.6 is essentially the same as the proof in [**BK05**]. It depends on several results not discussed in detail in this article. The most difficult of these to establish is a delicate result on the existence of quasi-arcs, originally due to Tukia [**Tuk96**], for which a substantially simpler proof was later given by Mackay [**Mac08**]. The version proved by Mackay is more suited for direct use in the proof below.

PROOF OF THEOREM 5.6. By Proposition 4.4 the group G splits as a graph of groups such that each vertex group is a quasi-isometrically embedded hyperbolic group with at most one end. If all vertex groups were finite, then G would be virtually free [**SW79**, Thm. 7.3], so at least one vertex group must be one ended. Therefore, it suffices to prove the theorem when G is one-ended.

The boundary of G is linearly connected by Theorem 5.4. A theorem of Bonk–Schramm states that the boundary of any word hyperbolic group is doubling [**BS00**, Thm. 9.2]. In a doubling, linearly connected, compact metric space, any two points can be joined by a quasi-arc ([**Mac08**]). If S is a sector of the hyperbolic plane, we thus have a quasisymmetric embedding of Gromov boundaries $\partial S \to [0,1] \to \partial G$. Since ∂S is connected, the quasisymmetric embedding extends to a quasi-isometric embedding $S \to G$ as explained in [**BS00**, Thms. 7.4 and 8.2]. Since S contains hyperbolic discs of arbitrarily large radius, a diagonal argument gives a quasi-isometric embedding of the entire hyperbolic plane into G (see [**BH99**, Lem. II.9.34]). $\square$

Acknowledgments

The authors are grateful to Bob Bell and Craig Guilbault for helpful feedback. The authors also thank the referee for many insightful comments that helped to improve the exposition of this paper.

References

[BH99] M. R. Bridson and A. Haefliger, *Metric spaces of non-positive curvature*, Grundlehren der mathematischen Wissenschaften [Fundamental Principles of Mathematical Sciences], vol. 319, Springer-Verlag, Berlin, 1999, DOI 10.1007/978-3-662-12494-9. MR1744486

[BK05] M. Bonk and B. Kleiner, *Quasi-hyperbolic planes in hyperbolic groups*, Proc. Amer. Math. Soc. **133** (2005), no. 9, 2491–2494, DOI 10.1090/S0002-9939-05-07564-7. MR2146190

[BM91] M. Bestvina and G. Mess, *The boundary of negatively curved groups*, J. Amer. Math. Soc. **4** (1991), no. 3, 469–481, DOI 10.2307/2939264. MR1096169

[Bow99a] B. H. Bowditch, *Boundaries of geometrically finite groups*, Math. Z. **230** (1999), no. 3, 509–527, DOI 10.1007/PL00004703. MR1680044

[Bow99b] B. H. Bowditch, *Connectedness properties of limit sets*, Trans. Amer. Math. Soc. **351** (1999), no. 9, 3673–3686, DOI 10.1090/S0002-9947-99-02388-0. MR1624089

[Bow99c] B. H. Bowditch, *Treelike structures arising from continua and convergence groups*, Mem. Amer. Math. Soc. **139** (1999), no. 662, 1–86. MR2000c:20061

[Bow01] B. H. Bowditch, *Peripheral splittings of groups*, Trans. Amer. Math. Soc. **353** (2001), no. 10, 4057–4082, DOI 10.1090/S0002-9947-01-02835-5. MR1837220

[Bow12] B. H. Bowditch, *Relatively hyperbolic groups*, Internat. J. Algebra Comput. **22** (2012), no. 3, 1250016 (66 pages). MR2922380

[BS00] M. Bonk and O. Schramm, *Embeddings of Gromov hyperbolic spaces*, Geom. Funct. Anal. **10** (2000), no. 2, 266–306, DOI 10.1007/s000390050009. MR1771428

[BS07] S. Buyalo and V. Schroeder, *Elements of asymptotic geometry*, EMS Monographs in Mathematics, European Mathematical Society (EMS), Zürich, 2007, DOI 10.4171/036. MR2327160

[CDP90] M. Coornaert, T. Delzant, and A. Papadopoulos, *Géométrie et théorie des groupes* (French), Lecture Notes in Mathematics, vol. 1441, Springer-Verlag, Berlin, 1990. Les groupes hyperboliques de Gromov. [Gromov hyperbolic groups]; With an English summary. MR1075994

[Dun85] M. J. Dunwoody, *The accessibility of finitely presented groups*, Invent. Math. **81** (1985), no. 3, 449–457, DOI 10.1007/BF01388581. MR807066

[GH90] É. Ghys and P. de la Harpe (eds.), *Sur les groupes hyperboliques d'après Mikhael Gromov* (French), Progress in Mathematics, vol. 83, Birkhäuser Boston, Inc., Boston, MA, 1990. Papers from the Swiss Seminar on Hyperbolic Groups held in Bern, 1988, DOI 10.1007/978-1-4684-9167-8. MR1086648

[GO07] R. Geoghegan and P. Ontaneda, *Boundaries of cocompact proper* CAT(0) *spaces*, Topology **46** (2007), no. 2, 129–137, DOI 10.1016/j.top.2006.12.002. MR2313068

[Gro81] M. Gromov, *Hyperbolic manifolds, groups and actions*, Riemann surfaces and related topics: Proceedings of the 1978 Stony Brook Conference (State Univ. New York, Stony Brook, N.Y., 1978), Ann. of Math. Stud., No. 97, Princeton Univ. Press, Princeton, NJ, 1981, pp. 183–213. MR624814

[Gro87] M. Gromov, *Hyperbolic groups*, Essays in group theory, Math. Sci. Res. Inst. Publ., vol. 8, Springer, New York, 1987, pp. 75–263, DOI 10.1007/978-1-4613-9586-7_3. MR919829

[HR] G. Christopher Hruska and Kim Ruane, *An introduction to semistability in geometric group theory.* To appear in L. Paoluzzi and G. Walsh, eds. Connections between geometry, topology, and group theory in low-dimensions (CIRM Jean-Morlet Chair, Spring 2018). `arXiv:1904.12947`.

[KMN06] A. Karlsson, V. Metz, and G. A. Noskov, *Horoballs in simplices and Minkowski spaces*, Int. J. Math. Math. Sci., posted on 2006, Art. ID 23656, 20, DOI 10.1155/IJMMS/2006/23656. MR2268510

[Kra77] J. Krasinkiewicz, *Local connectedness and pointed* 1*-movability* (English, with Russian summary), Bull. Acad. Polon. Sci. Sér. Sci. Math. Astronom. Phys. **25** (1977), no. 12, 1265–1269. MR500986

[Kur66] K. Kuratowski, *Topology. Vol. I*, Academic Press, New York-London; Państwowe Wydawnictwo Naukowe [Polish Scientific Publishers], Warsaw, 1966. New edition, revised and augmented; Translated from the French by J. Jaworowski. MR217751

[Lan13] U. Lang, *Injective hulls of certain discrete metric spaces and groups*, J. Topol. Anal. **5** (2013), no. 3, 297–331, DOI 10.1142/S1793525313500118. MR3096307

[Lev98] G. Levitt, *Non-nesting actions on real trees*, Bull. London Math. Soc. **30** (1998), no. 1, 46–54, DOI 10.1112/S0024609397003561. MR1479035

[Mac08] J. M. Mackay, *Existence of quasi-arcs*, Proc. Amer. Math. Soc. **136** (2008), no. 11, 3975–3981, DOI 10.1090/S0002-9939-08-09444-6. MR2425738

[Mih83] M. L. Mihalik, *Semistability at the end of a group extension*, Trans. Amer. Math. Soc. **277** (1983), no. 1, 307–321, DOI 10.2307/1999358. MR690054

[Mih87] M. L. Mihalik, *Semistability at* ∞*,* ∞*-ended groups and group cohomology*, Trans. Amer. Math. Soc. **303** (1987), no. 2, 479–485, DOI 10.2307/2000678. MR902779

[SW79] P. Scott and T. Wall, *Topological methods in group theory*, Homological group theory (Proc. Sympos., Durham, 1977), London Math. Soc. Lecture Note Ser., vol. 36, Cambridge Univ. Press, Cambridge-New York, 1979, pp. 137–203. MR564422

[Swa96] G. A. Swarup, *On the cut point conjecture*, Electron. Res. Announc. Amer. Math. Soc. **2** (1996), no. 2, 98–100, DOI 10.1090/S1079-6762-96-00013-3. MR1412948

[Tuk96] P. Tukia, *Spaces and arcs of bounded turning*, Michigan Math. J. **43** (1996), no. 3, 559–584, DOI 10.1307/mmj/1029005543. MR1420592

Department of Mathematical Sciences, University of Wisconsin–Milwaukee, PO Box 413, Milwaukee, Wisconsin 53211
Email address: `chruska@uwm.edu`

Department of Mathematics, Tufts University, Medford, Massachusetts 02155
Email address: `kim.ruane@tufts.edu`

Contemporary Mathematics
Volume **812**, 2025
https://doi.org/10.1090/conm/812/16272

Failure of quasi-isometric rigidity for infinite-ended groups

Nir Lazarovich and Emily Stark

ABSTRACT. We prove that an infinite-ended group whose one-ended factors have finite-index subgroups and are in a family of groups with a nonzero multiplicative invariant is not quasi-isometrically rigid. Combining this result with work of the first author proves that a residually-finite multi-ended hyperbolic group is quasi-isometrically rigid if and only if it is virtually free. The proof adapts an argument of Whyte for commensurability of free products of closed hyperbolic surface groups.

1. Introduction

Infinite-ended groups display a strong form of quasi-isometric flexibility: Papasoglu and Whyte [**14**] proved the quasi-isometry type of an infinite-ended group depends only on the quasi-isometry types of its one-ended factors. Commensurability classes, on the other hand, are often finer. For example, while there is one abstract commensurability class of closed hyperbolic surface groups, there are infinitely many abstract commensurability classes of their free products. Indeed, Whyte [**24**, Theorem 1.6] proved that free products of two hyperbolic surface groups are abstractly commensurable if and only if they have equal Euler characteristic.

We adapt Whyte's argument to free products of groups with a nonzero multiplicative invariant, like Euler characteristic. Consequently, we show quasi-isometric rigidity fails for many infinite-ended groups, where a finitely generated group G is *quasi-isometrically rigid* if any group quasi-isometric to G is abstractly commensurable to G, meaning the groups contain isomorphic finite-index subgroups. In particular, by [**10**], our main result gives the following:

THEOREM 1.1. *A residually-finite multi-ended hyperbolic group is quasi-isometrically rigid if and only if it is virtually free.*

Whether all hyperbolic groups are residually finite is an open question of considerable interest in the field. We remark that in the proof of Theorem 1.1 residual-finiteness is only used to ensure that the one-ended factors of the multi-ended hyperbolic group have some proper finite-index subgroup.

Quasi-isometric rigidity is a driving problem in geometric group theory. Many well-studied groups have been shown to be quasi-isometrically rigid, including non-abelian free groups [**7, 20**], fundamental groups of closed hyperbolic surfaces

---

2020 *Mathematics Subject Classification.* 20F65, 20E08, 20E06.

NL was supported by Israel Science Foundation (grant no. 1562/19).
ES was supported by NSF Grant No. DMS-2204339.

[**4**, **8**, **23**], non-uniform lattices in rank-one symmetric spaces [**17**], mapping class groups [**1**], and uniform lattices in certain hyperbolic buildings [**3**, **9**, **25**]. Fundamental groups of closed hyperbolic 3-manifolds and free products of closed hyperbolic surface groups are non-examples, and this paper gives many more. We refer the reader to the book of Drutu–Kapovich [**6**] for additional background.

We show quasi-isometric rigidity for infinite-ended groups fails in the presense of a suitable *multiplicative invariant* (also known as *volume* cf. [**15**]).

DEFINITION 1.2. Let $\mathcal{C}$ be a family of groups closed under taking finite-index subgroups. A *multiplicative invariant on* $\mathcal{C}$ is a function $\rho : \mathcal{C} \to \mathbb{R}$ so that if $G \in \mathcal{C}$ and H is a finite-index subgroup of G, then $\rho(H) = [G : H]\rho(G)$.

EXAMPLE 1.3. The following are examples of multiplicative invariants.

(1) Euler characteristic for the family groups with a finite classifying space; see discussion on generalizations in work of Chiswell [**5**];
(2) ℓ^2-Betti numbers for the family groups with a classifying space with finite skeleta in each dimension; for background, see work of Lück [**11**];
(3) Volume for the family of fundamental groups of closed hyperbolic manifolds of dimension $n \geq 3$ by Mostow Rigidity [**13**]
(4) Volume from an upper or lower volume in the sense of Reznikov [**15**, Section 1]. Examples include deficiency volume and rank volume [**15**] and a volume from topological complexity due to Lazarovich [**10**].

Let $\mathcal{C}$ be a family of one-ended groups and suppose that $\mathcal{C}$ is closed under taking finite-index subgroups. Suppose $\mathcal{C}$ has a multiplicative invariant that is nonzero for every element of $\mathcal{C}$.

THEOREM 1.4. *Let $G_i \in \mathcal{C}$, and suppose $H_i \leq G_i$ is a proper finite-index subgroup for some $n > 1$ and $1 \leq i \leq n$. Then, the groups $G_1 * \ldots * G_n$ and $H_1 * \ldots * H_n$ are quasi-isometric and not abstractly commensurable.*

COROLLARY 1.5. *If G is a finitely presented infinite-ended group that is not virtually free and whose one-ended factors are in $\mathcal{C}$ and contain non-trivial finite-index subgroups, then G is not quasi-isometrically rigid.*

PROOF OF THEOREM 1.1. A multi-ended hyperbolic group which is not virtually free has finitely many one-ended factors. If the group is residually finite, then each factor is a residually-finite hyperbolic group and, in particular, has a proper finite-index subgroup. Every hyperbolic group has finite asymptotic dimension [**16**], and so there exists m such that the asymptotic dimension of all the factors is bounded by m. Let $\mathcal{C}$ be the family of one-ended hyperbolic groups whose asymptotic dimension is bounded by m. By [**10**, Proof of Theorem A], the family $\mathcal{C}$ has a multiplicative invariant, namely $\underline{C}_{2,m}$, that is nonzero for every element of $\mathcal{C}$. The conclusion now follows from Corollary 1.5. $\square$

The free products $G_1 * \ldots * G_n$ and $H_1 * \ldots * H_n$ are quasi-isometric by work of Papasoglu–Whyte [**14**] since G_i and H_i are quasi-isometric for each i. It remains to prove the groups are not abstractly commensurable. To do this, given a multiplicative invariant ρ for a family of groups $\mathcal{C}$, we construct a multiplicative invarant χ_ρ on the family of free products whose non-free free factors are elements of $\mathcal{C}$, and we adapt an argument of Whyte [**24**, Theorem 1.6]. The hypothesis that the multiplicative invariant is nonzero is crucial in the arguments; for example, quasi-isometric free products of finitely generated free abelian groups are commensurable [**2**]. We note that Sykiotis [**22**] defines a related multiplicative invariant for infinite-ended groups through a limiting process; our construction is simpler and more direct.

To prove that two groups are quasi-isometric, often one finds geometric actions of the groups on the same proper metric space - in this case, the groups are said to have a *common model geometry*. Theorem 1.4 can be used to exhibit quasi-isometric groups with no common model geometry. Indeed, such examples arise via finitely generated groups that are not quasi-isometrically rigid, but are *action rigid*, meaning that whenever the group and another finitely generated group act geometrically on the same proper metric space, the groups are abstractly commensurable. Many examples of action rigid free products are given in [**12, 21**], including free products of closed hyperbolic manifold groups.

2. Preliminaries

We will use the notion of a graph of spaces and a graph of groups; we refer the reader to [**18, 19**] for additional background.

DEFINITION 2.1 (Graph notation). A *graph* Γ is a set of vertices $V\Gamma$ together with a set of edges $E\Gamma$. An edge e is oriented with an initial vertex $\iota(e)$ and a terminal vertex $\tau(e)$. For every edge e there is an edge $\bar{e}$ with reversed orientation so that $\iota(e) = \tau(\bar{e})$. Further, $e \neq \bar{e}$ and $e = \bar{\bar{e}}$.

We often suppress the orientation of an edge if it is not relevant to a construction.

DEFINITION 2.2. A *graph of spaces* (X, Γ) consists of the following data:

(1) a graph Γ called the *underlying graph*,
(2) a connected space X_v for each $v \in V\Gamma$ called a *vertex space*,
(3) a connected space X_e for each $e \in E\Gamma$ called an *edge space* such that $X_{\bar{e}} = X_e$, and
(4) a π_1-injective map $\phi_e : X_e \to X_{\tau(e)}$ for each $e \in E\Gamma$.

The *total space* X is the following quotient space:

$$X = \left(\bigsqcup_{v \in V\Gamma} X_v \bigsqcup_{e \in E\Gamma} (X_e \times [0,1]) \right) \Big/ \sim,$$

where $\sim$ is the relation that identifies $X_e \times \{0\}$ with $X_{\bar{e}} \times \{0\}$ via the identification of X_e with $X_{\bar{e}}$, and the point $(x, 1) \in X_e \times [0,1]$ with $\phi_e(x) \in X_{\tau(e)}$.

DEFINITION 2.3. A *graph of groups* $\mathcal{G}$ consists of the following data:

(1) a graph Γ called the *underlying graph*,
(2) a group G_v for each $v \in V\Gamma$ called a *vertex group*,
(3) a group G_e for each $e \in E\Gamma$ called an *edge group*, such that $G_e = G_{\bar{e}}$,
(4) and injective homomorphisms $\Theta_e : G_e \to G_{\tau(e)}$ for each $e \in E\Gamma$ (called *edge maps*).

If $\mathcal{G}$ is a graph of groups with underlying graph Γ, a *graph of spaces associated to* $\mathcal{G}$ consists of a graph of spaces (X, Γ) with points $x_v \in X_v$ and $x_e \in X_e$ so that $\pi_1(X_v, x_v) \cong G_v$ and $\pi_1(X_e, x_e) \cong G_e$ for all $v \in V\Gamma$ and $e \in E\Gamma$, and the maps $\phi_e : X_e \to X_{\tau(e)}$ are such that $(\phi_e)_* = \Theta_e$. The *fundamental group* of the graph of groups $\mathcal{G}$ is $\pi_1(X)$. The fundamental group is independent of the choice of X and is denoted by $\pi_1(\mathcal{G})$.

If $G \cong \pi_1(\mathcal{G})$, where $\mathcal{G}$ is a graph of groups, then we may view each vertex group of $\mathcal{G}$ as a subgroup of G (up to conjugacy). Suppose that $G = \pi_1(\mathcal{G}) = \pi_1(X)$, where $\mathcal{G}$ is a graph of groups and X is the corresponding total space. If $G' \leq G$, then $G' = \pi_1(X')$, where X' covers X. This cover $q : X' \to X$ naturally yields a graph of spaces decomposition of X' with underlying graph Γ', where the edge and vertex spaces are components of the preimages of the edge and vertex spaces in X. Hence, G' is the fundamental group of a corresponding graph of groups $\mathcal{G}'$. Each vertex group of $\mathcal{G}'$ is then conjugate to a subgroup of a vertex group of $\mathcal{G}$.

DEFINITION 2.4. With the notation from the previous paragraph, if G_v is a vertex group of $\mathcal{G}$, we say a vertex group G'_u of $\mathcal{G}'$ *covers* G_v if G'_u is conjugate to a subgroup of G_v via the covering of the corresponding graphs of spaces.

We make use of the following observation.

OBSERVATION 2.5. Suppose $\mathcal{C}$ is a family of groups closed with respect to taking finite-index subgroups and with a multiplicative invariant ρ. Suppose $\mathcal{G}$ is a graph of groups with each vertex group in $\mathcal{C}$ and $G = \pi_1(\mathcal{G})$. Suppose G' is an index-k subgroup of G and $\mathcal{G}'$ is its corresponding graph of groups decomposition. Then for each vertex group $G_v \in \mathcal{G}$, if $K = \{G'_{u_1}, \ldots, G'_{u_n}\}$ is the collection of vertex groups of $\mathcal{G}'$ covering G_v, then

$$\sum_{i=1}^{n} \rho(G'_{u_i}) = k\rho(G_v).$$

3. Multiplicative invariants and commensurability

NOTATION 3.1. For $r \geq 0$, let F_r denote the free group of rank r. Throughout this section let $\mathcal{C}$ be a family of one-ended groups that is closed with respect to taking finite-index subgroups and has a multiplicative invariant ρ. Let $\mathcal{C}'$ denote the family of groups of the form $G = G_1 * \ldots * G_n * F_r$ for some $r \geq 0$, $n \geq 0$, and $G_i \in \mathcal{C}$. Equivalently, $G \in \mathcal{C}'$ if G is the fundamental group of a graph of groups $\mathcal{G}$ with trivial edge groups and each vertex group either trivial or in $\mathcal{C}$.

We will work in the setting of free products, rather than the general infinite-ended case, as the arguments are cleaner. See the proof of Corollary 1.5 regarding the more general setting.

DEFINITION 3.2. Let ρ be a multiplicative invariant on the family of groups $\mathcal{C}$. Define a function $\chi_\rho : \mathcal{C}' \to \mathbb{R}$ by

$$\chi_\rho(G_1 * \ldots * G_n * F_r) = 1 - r - n + \sum_{i=1}^{n} \rho(G_i).$$

REMARK 3.3. The function χ_ρ is a variant of the Euler characteristic. Viewing $G \in \mathcal{C}'$ as the fundamental group of a graph of groups, each vertex in the graph of groups decomposition labeled $G_i \in \mathcal{C}$ is assigned $\rho(G_i)$ rather than 1. In particular, if the multiplicative invariant ρ is Euler characteristic, then χ_ρ is as well.

To prove that χ_ρ is a multiplicative invariant, we will use the following.

DEFINITION 3.4. Let $\mathcal{G}$ be a graph of groups with underlying graph Γ, trivial edge groups, and with nontrivial vertex groups $G_1, \ldots, G_n \in \mathcal{C}$. Define

$$\chi_\rho(\mathcal{G}) = \chi(\Gamma) - n + \sum_{i=1}^{n} \rho(G_i).$$

LEMMA 3.5. *Let $G = G_1 * \ldots G_n * F_r \in \mathcal{C}'$, and suppose $G \cong \pi_1(\mathcal{G})$, where $\mathcal{G}$ is a graph of groups with underlying graph Γ, trivial edge groups, and with nontrivial vertex groups $G_1, \ldots, G_n \in \mathcal{C}$. Then, $\chi_\rho(G) = \chi_\rho(\mathcal{G})$.*

PROOF. By definition,

$$\chi_\rho(G) = 1 - r - n + \sum_{i=1}^{n} \rho(G_i), \text{ and}$$
$$\chi_\rho(\mathcal{G}) = \chi(\Gamma) - n + \sum_{i=1}^{n} \rho(G_i).$$

Thus, we must show $\chi(\Gamma) = 1 - r$. Indeed, it follows from the definition of the fundamental group of a graph of groups that $F_r \cong \pi_1(\Gamma)$. So, $\chi(\Gamma) = \chi(F_r) = 1 - r$. □

PROPOSITION 3.6. *The function $\chi_\rho : \mathcal{C}' \to \mathbb{R}$ is a multiplicative invariant.*

PROOF. Let $G = G_1 * \ldots * G_n * F_r \in \mathcal{C}'$. Realize G as the fundamental group of a graph of groups $\mathcal{G}$ whose underlying graph Γ has vertex set $V\Gamma = \{v_0, \ldots, v_n\}$ and (unoriented) edge set $E\Gamma = \{e_1, \ldots, e_n, a_1, \ldots, a_r\}$, where $e_i = \{v_0, v_i\}$ for $1 \leq i \leq n$, and a_i is a loop based at v_0 for $1 \leq i \leq r$. Suppose each edge group is trivial, the vertex group G_{v_0} is trivial, and the vertex group G_{v_i} is G_i.

Let $G' \leq G$ be an index-k subgroup. As discussed in Section 2, $G' \cong \pi_1(\mathcal{G}')$, where $\mathcal{G}'$ is a graph of groups with:

- k vertices with trivial vertex group that each cover $G_{v_0} = \langle 1 \rangle$,
- k edges with trivial edge group that each cover $G_{a_i} = \langle 1 \rangle$ for $1 \leq i \leq r$,
- k edges with trivial edge group that each cover $G_{e_i} = \langle 1 \rangle$ for $1 \leq i \leq n$, and
- for $i \in \{1, \ldots n\}$, there exists $n_i \in \mathbb{N}_{\geq 1}$ and n_i vertices $v_i^1, \ldots, v_i^{n_i}$ with vertex group $G_{v_i^j}$ that each cover G_{v_i}. Further,

$$\sum_{j=1}^{n_i} \rho(G_{v_i^j}) = k\rho(G_i)$$

  by Observation 2.5 since ρ is a multiplicative invariant.

Thus, $\chi_\rho(\mathcal{G}') = k\chi_\rho(\mathcal{G})$. So, by Lemma 3.5, $\chi_\rho(G') = \chi_\rho(\mathcal{G}') = k\chi_\rho(\mathcal{G}) = k\chi_\rho(G)$. □

As explained in the introduction, Theorem 1.4 follows from the next theorem. The proof is analogous to Whyte's argument in [**24**, Theorem 1.6].

THEOREM 3.7. *Let $\mathcal{C}$ be a family of groups closed with respect to taking finite-index subgroups and with a multiplicative invariant ρ that is nonzero for every element of $\mathcal{C}$. For $m > 1$ and $1 \leq i \leq m$, let $G_i \in \mathcal{C}$, and suppose $H_i \leq G_i$ is a proper finite-index subgroup. Then, the groups $G = G_1 * \ldots * G_m$ and $H = H_1 * \ldots * H_m$ are not abstractly commensurable.*

PROOF. Suppose towards a contradiction that there exist finite-index subgroups $G' \leq G$ and $H' \leq H$ of index k and k', respectively, so that $G' \cong H'$. Suppose that $H' \cong G' \cong G'_1 * \ldots * G'_n * F_r$ where $r \geq 0$ and for each $i \in \{1, \ldots, n\}$, the group G'_i covers G_j and H_ℓ for some $j, \ell \in \{1, \ldots, m\}$. By Proposition 3.6, $\chi_\rho(G') = k\chi_\rho(G) = k'\chi_\rho(H)$. We will show that $k = k'$ by relating these numbers to r, n, and m. This will complete the proof since $\chi_\rho(H) < \chi_\rho(G)$ because ρ is a multiplicative invariant.

CLAIM 3.8. $k = k' = \frac{1-r-n}{1-m}$.

PROOF OF CLAIM. We prove the claim for k; the proof for k' is analogous. First, by Proposition 3.6,

$$\chi_\rho(G') = k\chi_\rho(G) \tag{3.1}$$

$$= k\left(1 - m + \sum_{i=1}^{m} \rho(G_i)\right). \tag{3.2}$$

Next, by the definition of χ_ρ,

$$\chi_\rho(G') = 1 - r - n + \sum_{i=1}^{n} \rho(G'_i). \tag{3.3}$$

For $1 \leq i \leq m$, let $K_i = \{G'_{i_j}\}_{j=1}^{n_i}$ and $i_j \in \{1, \ldots, n\}$ be the collection of free factors of G' so that G'_{i_j} covers G_i. Then, for all $1 \leq i \leq m$,

$$\sum_{j=1}^{n_i} \rho(G'_{i_j}) = k\rho(G_i)$$

by Observation 2.5. Since the K_i partition the set $\{G'_i\}_{i=1}^{n}$, Equation (3) can be rewritten as

$$\chi_\rho(G') = 1 - r - n + \left(\sum_{j=1}^{n_1} \rho(G'_{i_j}) + \ldots + \sum_{j=1}^{n_m} \rho(G'_{i_j})\right) \tag{3.4}$$

$$= 1 - r - n + k\sum_{i=1}^{m} \rho(G_i). \tag{3.5}$$

Combining Equation (2) and Equation (5) yields

$$k\left(1 - m + \sum_{i=1}^{m} \rho(G_i)\right) = 1 - r - n + k\sum_{i=1}^{m} \rho(G_i). \tag{3.6}$$

Thus, $k = \frac{1-r-n}{1-m}$. □

Therefore, $k = k'$, completing the proof of the theorem. □

REMARK 3.9. Note that in the proof of Claim 3.8 we have not used that the multiplicative invariant ρ is non-zero. In fact, the same proof works for the multiplicative invariant $\rho(G) = 1/|G|$ (where $1/\infty = 0$).

REMARK 3.10 (Weakening the hypotheses). If $G = G_1 * \ldots * G_m$ and $H = H_1 * \ldots * H_m$ are in the family $\mathcal{C}'$ and $\chi_\rho(G) > \chi_\rho(H)$, then the proof of the theorem above shows that G and H are not abstractly commensurable. So, for example, one can weaken the hypothesis of the theorem to asking that $H_i \leq G_i$ is a *proper* finite-index subgroup for only one $i \in \{1, \ldots, n\}$.

The general infinite-ended case now follows from Theorem 3.7.

Proof of Corollary 1.5. Suppose G is a finitely presented infinite-ended group. Work of Stallings and Dunwoody [**7, 20**] proves that G is the fundamental group of a finite graph of groups with finite edge groups and vertex groups that have at most one end. If G is not virtually free then there is at least one one-ended factor in this decomposition. Suppose the one-ended vertex groups in this decomposition are $G_1, \ldots, G_n$ for some $n \geq 1$ and belong to a family of groups $\mathcal{C}$ that is closed under taking finite-index subgroups. Suppose $\mathcal{C}$ has a multiplicative invariant that is nonzero for every element of $\mathcal{C}$. Suppose further that each one-ended factor of G contains a non-trivial finite-index subgroup. The group G is quasi-isometric to the free product $G_1 * G_1 * \ldots * G_n$ by work of Papasoglu–Whyte [**14**]. Thus, G is not quasi-isometrically rigid by Theorem 3.7. □

References

[1] J. Behrstock, B. Kleiner, Y. Minsky, and L. Mosher, *Geometry and rigidity of mapping class groups*, Geom. Topol. **16** (2012), no. 2, 781–888, DOI 10.2140/gt.2012.16.781. MR2928983

[2] J. A. Behrstock, T. Januszkiewicz, and W. D. Neumann, *Commensurability and QI classification of free products of finitely generated abelian groups*, Proc. Amer. Math. Soc. **137** (2009), no. 3, 811–813, DOI 10.1090/S0002-9939-08-09559-2. MR2457418

[3] M. Bourdon and H. Pajot, *Rigidity of quasi-isometries for some hyperbolic buildings*, Comment. Math. Helv. **75** (2000), no. 4, 701–736, DOI 10.1007/s000140050146. MR1789183

[4] A. Casson and D. Jungreis, *Convergence groups and Seifert fibered 3-manifolds*, Invent. Math. **118** (1994), no. 3, 441–456, DOI 10.1007/BF01231540. MR1296353

[5] I. M. Chiswell, *Euler characteristics of discrete groups*, Groups: topological, combinatorial and arithmetic aspects, London Math. Soc. Lecture Note Ser., vol. 311, Cambridge Univ. Press, Cambridge, 2004, pp. 106–254, DOI 10.1017/CBO9780511550706.008. MR2073349

[6] C. Druţu and M. Kapovich, *Geometric group theory*, American Mathematical Society Colloquium Publications, vol. 63, American Mathematical Society, Providence, RI, 2018. With an appendix by Bogdan Nica, DOI 10.1090/coll/063. MR3753580

[7] M. J. Dunwoody, *The accessibility of finitely presented groups*, Invent. Math. **81** (1985), no. 3, 449–457, DOI 10.1007/BF01388581. MR807066

[8] D. Gabai, *Convergence groups are Fuchsian groups*, Ann. of Math. (2) **136** (1992), no. 3, 447–510, DOI 10.2307/2946597. MR1189862

[9] F. Haglund, *Commensurability and separability of quasiconvex subgroups*, Algebr. Geom. Topol. **6** (2006), 949–1024, DOI 10.2140/agt.2006.6.949. MR2240922

[10] Nir Lazarovich, *Finite index rigidity of hyperbolic groups*, `arXiv:2302.04484`.

[11] W. Lück, *L^2-invariants: theory and applications to geometry and K-theory*, Ergebnisse der Mathematik und ihrer Grenzgebiete. 3. Folge. A Series of Modern Surveys in Mathematics [Results in Mathematics and Related Areas. 3rd Series. A Series of Modern Surveys in Mathematics], vol. 44, Springer-Verlag, Berlin, 2002, DOI 10.1007/978-3-662-04687-6. MR1926649

[12] Alex Margolis, Sam Shepherd, Emily Stark, and Daniel Woodhouse, *Graphically discrete groups and rigidity*, `arXiv:2303.04843`.

[13] G. D. Mostow, *Strong rigidity of locally symmetric spaces*, Annals of Mathematics Studies, No. 78, Princeton University Press, Princeton, NJ; University of Tokyo Press, Tokyo, 1973. MR385004

[14] P. Papasoglu and K. Whyte, *Quasi-isometries between groups with infinitely many ends*, Comment. Math. Helv. **77** (2002), no. 1, 133–144, DOI 10.1007/s00014-002-8334-2. MR1898396

[15] A. Reznikov, *Volumes of discrete groups and topological complexity of homology spheres*, Math. Ann. **306** (1996), no. 3, 547–554, DOI 10.1007/BF01445265. MR1415078

[16] J. Roe, *Hyperbolic groups have finite asymptotic dimension*, Proc. Amer. Math. Soc. **133** (2005), no. 9, 2489–2490, DOI 10.1090/S0002-9939-05-08138-4. MR2146189

[17] R. E. Schwartz, *The quasi-isometry classification of rank one lattices*, Inst. Hautes Études Sci. Publ. Math. **82** (1995), 133–168 (1996). MR1383215

[18] P. Scott and T. Wall, *Topological methods in group theory*, Homological group theory (Proc. Sympos., Durham, 1977), London Math. Soc. Lecture Note Ser., vol. 36, Cambridge Univ. Press, Cambridge-New York, 1979, pp. 137–203. MR564422

[19] J.-P. Serre, *Arbres, amalgames,* SL_2 (French), Astérisque, No. 46, Société Mathématique de France, Paris, 1977. Avec un sommaire anglais; Rédigé avec la collaboration de Hyman Bass. MR476875

[20] J. R. Stallings, *On torsion-free groups with infinitely many ends*, Ann. of Math. (2) **88** (1968), 312–334, DOI 10.2307/1970577. MR228573

[21] E. R. Stark and D. J. Woodhouse, *Action rigidity for free products of hyperbolic manifold groups*, Ann. Inst. Fourier (Grenoble) **74** (2024), no. 2, 503–544, DOI 10.5802/aif.3585. MR4748179

[22] M. Sykiotis, *Complexity volumes of splittable groups*, J. Algebra **503** (2018), 409–432, DOI 10.1016/j.jalgebra.2018.02.013. MR3780003

[23] P. Tukia, *Homeomorphic conjugates of Fuchsian groups*, J. Reine Angew. Math. **391** (1988), 1–54, DOI 10.1515/crll.1988.391.1. MR961162

[24] K. Whyte, *Amenability, bi-Lipschitz equivalence, and the von Neumann conjecture*, Duke Math. J. **99** (1999), no. 1, 93–112, DOI 10.1215/S0012-7094-99-09904-0. MR1700742

[25] X. Xie, *Quasi-isometric rigidity of Fuchsian buildings*, Topology **45** (2006), no. 1, 101–169, DOI 10.1016/j.top.2005.06.005. MR2170496

Department of Mathematics, Technion-Israel Institute of Technology, Haifa 3200003, Israel

Email address: `lazarovich@technion.ac.il`

Department of Mathematics and Computer Science, Wesleyan University, Middletown, CT 06459

Email address: `estark@wesleyan.edu`

Contemporary Mathematics
Volume **812**, 2025
https://doi.org/10.1090/conm/812/16273

One-endedness of outer automorphism groups of free products of finite and cyclic groups

Rylee Alanza Lyman

ABSTRACT. The main result of this paper is that the outer automorphism group of a free product of finite groups and cyclic groups is semistable at infinity (provided it is one ended) or semistable at each end. In a previous paper, we showed that the group of outer automorphisms of the free product of two nontrivial finite groups with an infinite cyclic group has infinitely many ends, despite being of virtual cohomological dimension two. We also prove that aside from this exception, having virtual cohomological dimension at least two implies the outer automorphism group of a free product of finite and cyclic groups is one ended. As a corollary, the outer automorphism groups of the free product of four finite groups or the free product of a single finite group with a free group of rank two are virtual duality groups of dimension two, in contrast with the above example. Our proof is inspired by methods of Vogtmann, applied to a complex first studied in another guise by Krstić and Vogtmann.

Duality groups. A group G of type FP_∞ is a *duality group of dimension* n in the sense of Bieri–Eckmann [**4**] if there exists a G-module D such that for any G-module A and any integer k, we have an isomorphism

$$H^k(G;A) \cong H_{n-k}(G;D\otimes_{\mathbb{Z}} A),$$

where G acts on $D \otimes_{\mathbb{Z}} A$ diagonally. The dimension, n, turns out to be equal to the cohomological dimension of G, which is thus torsion free, and is therefore determined by G. Similarly, the dualizing module D turns out to be determined by G; it is isomorphic to $H^n(G;\mathbb{Z}G)$.

Many important examples of groups are duality groups, or more generally, *virtual duality groups,* having a duality group as a subgroup of finite index. (Virtual duality groups are thus allowed to have torsion, but they will be virtually torsion free.) Among these are, of course, groups having a closed manifold as a $K(\pi,1)$, but also fundamental groups of knot complements, mapping class groups of closed surfaces, and, most relevant to the current paper, $\mathrm{Out}(F_n)$ [**3**].

This paper is motivated by the following question.

QUESTION A. Let F be a free product of finite and cyclic groups. Is the outer automorphism group of F a virtual duality group?

2020 *Mathematics Subject Classification.* Primary 20F65.

The group $\mathrm{Out}(F)$ is of type FP_∞ and virtually torsion free, so the question makes sense. As we showed, but did not remark, in [**15**], the answer to this question is sometimes "no".

We prove the following theorem.

THEOREM B. *Let* $F = A_1 * \cdots * A_n * F_k$ *be a free product of the nontrivial finite groups* A_i *with a free group of rank* k*. The invariants* n *and* k *determine the number of ends of* $\mathrm{Out}(F)$ *in the following way.*

$$\begin{cases} n \le 1,\ k \le 1 \text{ or } (n,k) = (2,0) & \mathrm{Out}(F) \text{ is finite, i.e. has zero ends.} \\ (n,k) = (3,0),\ (2,1),\ (0,2) & \mathrm{Out}(F) \text{ has infinitely many ends.} \\ \text{otherwise} & \mathrm{Out}(F) \text{ has one end.} \end{cases}$$

Ends of groups. Let us explain. Celebrated results of Freudenthal and Hopf say that a finitely generated group G has 0, 1, 2, or infinitely many *ends.* An *end* of a topological space X is a compatible choice, for every compact subset K of X, of a path component of the complement $X - K$. Put another way, if $K \subset K'$ are compact subsets of X, every path component of $X - K'$ determines a path component of $X - K'$, and in fact we have an inverse system of sets (or spaces, if we'd like to give them the discrete topology). The *space of ends* of X, is the inverse limit

$$\varprojlim_K \pi_0(X - K)$$

and the *number* of ends is its cardinality. What Freudenthal and Hopf prove is that the number of ends of a connected CW-complex, or if you'd like, a Cayley graph, that a finitely generated group G acts on with finite stabilizers and finite (i.e. compact) quotient is in fact an invariant of the group, and that if a group has at least three ends, it has infinitely many.

One way to think about one-endedness, which is in some sense generic, is to say that, supposing we have fixed an exhaustion of X by compact sets $K_0 \subset K_1 \subset \cdots$ with the property that K_n is contained in the interior of K_{n+1}, for any two points "sufficiently far from K_n", say belonging to $X - K_N$ for N sufficiently larger than n, those two points can be connected by a path that avoids K_n. Thus for example the line is not one ended (in fact it has two ends), while the plane is.

End invariants of groups. This reframes one-endedness as 0*-connectivity at infinity,* and naturally leads one to generalize to questions about higher homotopy or homology groups "at infinity". The plane, then, would fail to be simply connected at infinity, since a loop encircling our compact set K_N is not nullhomotopic in $X - K_n$ as soon as K_n is nonempty, while $\mathbb{R}^n$ for larger n *would* be simply connected at infinity.

To ask these questions for groups, one needs to be slightly careful. After all, every finitely generated group has a Cayley graph, and graphs have no higher homotopy groups. In the case of the plane, what is interesting is that every loop *is* nullhomotopic in the plane as a whole, but may *not* remain so after removing a compact set. Thus we put restrictions on the spaces in question, asking them to be n-connected in the ordinary sense, before we ask whether they are n-connected at infinity. This has the effect for groups of restricting our attention to groups with certain *finiteness properties,* which generalize being finitely generated or finitely presented.

Semistability. Even with this restriction, there is a subtlety: the fundamental group—for instance—needs a basepoint, but as we work in the complements of a compact exhaustion no single basepoint will do. Making a choice of basepoint for $\pi_1(X - K_n, \star_n)$ for each n and a path from $\star_n$ to $\star_{n-1}$, we produce a *proper ray* in X. To show that resulting inverse limit

$$\varprojlim_n \pi_1(X - K_n, \star_n)$$

is independent of our choices, we would need to show that any two proper rays (ending in the same end if there is more than one) are properly homotopic. For locally finite complexes X, so in particular for finitely presented groups, this is equivalent by [**11**, Proposition 16.1.2] to the condition that this inverse system of groups is *semistable* or satisfies the *Mittag-Leffler condition:* for each n there exists N such that for $k \geq N$, the image of $\pi_1(X - K_N, \star_n)$ in $\pi_1(X - K_n, \star_n)$ is equal to the image of $\pi_1(X - K_k, \star_k)$ in $\pi_1(X - K_n, \star_n)$.

Unwinding this definition, we see that it is equivalent to the claim that every loop in $X - K_N$ based at $\star_N$ may be pushed off to $X - K_k$ (along our proper ray) by a homotopy that avoids K_n.

The relevance to group cohomology is that if G is *semistable at each end* in the above sense, then $H^2(G, \mathbb{Z}G)$ is a free abelian group [**11**, Theorem 16.5.1]. It turns out that if G is $(n-2)$-connected at infinity, (as the plane and hence $\mathbb{Z}^2$ is 0-connected at infinity, for instance, although merely being $(n-2)$-acyclic will do) then $H^i(G, \mathbb{Z}G) = 0$ for $i < n$, and $H^n(G, \mathbb{Z}G)$ is torsion free [**5, 11**]. (Thus for a one-ended group semistability is somewhere between one-endedness and simple connectivity at infinity.) A group G of type FP_∞ is a duality group of dimension n if and only if $H^i(G, \mathbb{Z}G) = 0$ for $i \neq n$ and $H^n(G, \mathbb{Z}G)$ is nontrivial and torsion free. If additionally G has cohomological dimension n, then $H^i(G, \mathbb{Z}G) = 0$ for $i > n$ and $H^n(G, \mathbb{Z}G)$ is nontrivial.

Thus the combination of $(n-2)$-connectivity at infinity with having virtual cohomological dimension n implies that a group G of type FP_∞ is a virtual duality group.

Additionally, let us remark that it is an open question whether every finitely presented group G satisfies that $H^2(G, \mathbb{Z}G)$ is free abelian, hence whether every finitely presented group is semistable at each end. We are able to answer this in the affirmative for $\mathrm{Out}(F)$ when F is a free product of finite and cyclic groups. The following is our main result.

THEOREM C. *Let $F = A_1 * \cdots * A_n * F_k$ be a free product of the finite groups A_i with a free group of rank k. The group $\mathrm{Out}(F)$ of outer automorphisms of F is semistable at each end.*

COROLLARY D. *For F as above, writing $G = \mathrm{Out}(F)$, we have that $H^2(G, \mathbb{Z}G)$ is free abelian.*

The case of $\mathrm{Out}(F)$. From here we discuss $\mathrm{Out}(F)$, where $F = A_1 * \cdots * A_n * F_k$ is the free product of the finite groups A_i with a free group of rank k, and $\mathrm{Out}(F)$ denotes its *outer automorphism group,* $\mathrm{Aut}(F)/\mathrm{Inn}(F)$. The virtual cohomological dimension of $\mathrm{Out}(F)$ was computed by Krstić and Vogtmann [**14**]; it is equal to the dimension of a certain contractible simplicial complex L that $\mathrm{Out}(F)$ acts on with finite stabilizers and finite quotient. We will call this complex the *spine of (reduced) Outer Space* for F. The groups $\mathrm{Out}(F)$, which are virtually torsion free, are thus virtually of type F, hence certainly of type FP_∞. In any case, the dimension of L can be computed from the invariants n and k by the following rule

$$\dim(L) = \begin{cases} 2k+n-2 & n \geq 2, \\ \max\{2k+n-3, 0\} & \text{otherwise.} \end{cases}$$

Thus, to prove that $\mathrm{Out}(F)$ is a virtual duality group, it suffices to show that it is $(2k+n-4)$-connected at infinity if $n \geq 2$ or $(2k-n-5)$-connected at infinity otherwise. In particular, when L has dimension 2, we need only show that $\mathrm{Out}(F)$ is one ended. There are three such cases: $F = A_1 * A_2 * A_3 * A_4$, $F = A_1 * A_2 * \mathbb{Z}$ and $F = A_1 * F_2$. In the first case, one-endedness follows from a result of Das [**9**], who shows that $\mathrm{Out}(F)$ is *thick* when $F = A_1 * \cdots * A_n$ and $n \geq 4$. The third case is covered by Theorem B, but in the second, Theorem B states that $\mathrm{Out}(F)$ has infinitely many ends: this is part of the main result of [**15**]. This is why the answer to Question A is "no" for some choices of F. We view Theorem B as partial progress towards a full answer to Question A.

In view of Theorem C and the fact that $\mathrm{Out}(A_1 * A_2 * A_3 * A_4)$ and $\mathrm{Out}(A_1 * F_2)$ are one-ended groups of virtual cohomological dimension two, these groups have a well-defined *fundamental group at infinity,* namely the inverse limit of the fundamental groups as discussed in the semistability paragraph above. Because $H^2(G, \mathbb{Z}G)$ is nontrivial, this fundamental group is nontrivial. What is it?

Question E. What is the fundamental group of $\mathrm{Out}(A_1 * A_2 * A_3 * A_4)$ or $\mathrm{Out}(A_1 * F_2)$ at infinity?

Our proof of Theorem B is inspired by Vogtmann's paper [**22**]. There she analyzes the combinatorial structure of links of certain vertices of the Spine of Outer Space (for $F = F_k$) to push paths and homotopies off to infinity. The Spine of Outer Space for general F likewise has a rich combinatorial structure; the present paper is a first contribution towards an understanding of it in general.

Like the papers [**8, 14**], we proceed by putting a kind of "height function" on certain vertices of the complex L of minimal complexity we term "briar patches". (A briar patch may have petals, like a rose, and thorns like a thistle, but may also sprawl somewhat.) Vertices of L correspond to marked graphs of groups (or if you'd rather, actions of F on trees), and the height function records the lengths of immersed loops associated to a predetermined finite set of conjugacy classes (or if you'd rather, the hyperbolic translation lengths of these conjugacy classes).

The proof of contractibility of L in [**8, 14**] goes by showing, using "peak-reduction" techniques, that these height functions behave somewhat like non-singular Morse functions, allowing one to contract the whole complex onto the subcomplex where the function is minimized, and then arguing that by carefully choosing the finite set of conjugacy classes, this minimal subcomplex is manifestly contractible.

Like Vogtmann in [**22**], we'd like to do the opposite: The height function provides an exhaustion of L; for well-chosen finite sets, this is a compact exhaustion, defining for us a sequence of compact subsets of L we call "balls". Analyzing the combinatorics of the spine allows us to understand how our height function changes as we move around the star of a single vertex of L.

Firstly, we restrict our attention to a family of "standard paths", which look like a sequence of one-edge expand–collapse moves (the term is "Whitehead move") between briar patches. In the case $F = F_k$, these moves between roses have the effect of applying Whitehead automorphisms to the fundamental group, hence the name. Supposing that τ_{2k} is a briar patch along such a path which is a kind of local minimum for the height function, the game is to find a new standard path between the adjacent briar patches $\tau_{2(k-1)}$ and $\tau_{2(k+1)}$ that passes through briar patches of height strictly greater than that of τ. By doing so while staying within the star of τ, we have a homotopy between the old path and the new one, and by iteratively applying this method to all local minima, we are able to push our path off to infinity. It turns out that when this strategy of proof works, it suffices to prove semistability as well; we turn to this in Section 3.

Let us remark that our proof of Theorem B is tailored to its application in Theorem C, for which we really need to be able to push paths off to infinity. For a reader interested in an "optimal" proof of Theorem B, we give a sketch of a simpler argument suggested to us by the anonymous referee in an appendix.

1. The Complex L

We work in the *spine of reduced Outer Space for F*, a simplicial complex which we denote as $L = L(F)$. A more detailed definition is given in [**15**]. A vertex of L corresponds to an action of F on a simplicial tree T with finite stabilizers (and in fact trivial edge stabilizers). In this paper it will be more convenient to work in the quotient graph of groups $\mathcal{G} = F\backslash\backslash T$. We assume a certain amount of familiarity with graphs of groups in this paper and refer the reader to [**21**], [**1**], [**20**] and [**17**, Section 1] for background on graphs of groups. We will follow the notation in [**17**], although almost everything we write is standard. The main innovation of [**17**] that we need is the notion of a *homotopy equivalence* between graphs of groups and homotopy of maps of graphs of groups. For us the precise definition is less important than the following consequences: if $\sigma\colon \mathbb{G} \to \mathcal{G}$ is a homotopy equivalence, it determines (up to inner automorphism) an isomorphism $\sigma_\sharp\colon \pi_1(\mathbb{G}) \cong \pi_1(\mathcal{G})$ and under this isomorphism the corresponding Bass–Serre trees S and T have the same subgroups which are *elliptic,* that is, fixing a point of the tree. For a precise definition, see [**17**].

Fix a finite graph of finite groups $\mathbb{G}$, a basepoint $\star \in \mathbb{G}$ and an identification $F \cong \pi_1(\mathbb{G}, \star)$. A *marked graph of groups* $\tau = (\mathcal{G}, \sigma)$ is a graph of finite groups $\mathcal{G}$ together with a homotopy equivalence $\sigma\colon \mathbb{G} \to \mathcal{G}$ called the *marking.* Two marked graphs of groups $(\mathcal{G}, \sigma)$ and $(\mathcal{G}', \sigma')$ are *equivalent* if there is an isomorphism $h\colon \mathcal{G} \to \mathcal{G}'$ such that the following diagram commutes up to homotopy

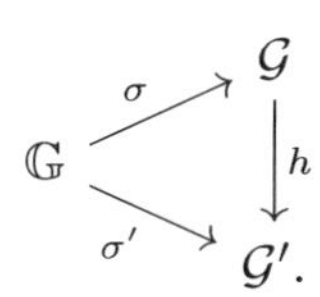

The complex L. A vertex of the complex L is determined by an equivalence class of marked graphs of groups $(\mathcal{G}, \sigma)$ satisfying the following conditions.

(1) Edge groups of $\mathcal{G}$ are trivial.
(2) Each valence-one and valence-two vertex of $\mathcal{G}$ has nontrivial vertex group.
(3) If an (open) edge e separates $\mathcal{G}$, each component of the complement contains a vertex with nontrivial vertex group.

A marked graph of groups $\mathcal{G}$ determining a vertex of L, is *reduced* if collapsing any edge of $\mathcal{G}$ yields a marked graph of groups not homotopy equivalent to $\mathcal{G}$ (because the collapse map cannot be "undone", for example because maybe the resulting graph of groups has an infinite vertex group). The third condition above is equivalent to asking that for each edge e of $\mathcal{G}$, there is a reduced marked graph of groups $\mathcal{G}'$ which may be obtained from $\mathcal{G}$ by collapsing edges in which the edge e is not collapsed. In the language of [**6, 13**], we say the edge e is *surviving.*

Two vertices $(\mathcal{G}, \sigma)$ and $(\mathcal{G}', \sigma')$ are connected by an oriented edge if a marked graph of groups equivalent to $(\mathcal{G}', \sigma')$ can be obtained from $(\mathcal{G}, \sigma)$ by collapsing certain edges in $\mathcal{G}$, and in general L possesses a simplex whenever its 1-skeleton is present. As observed in [**16**, Theorem 3.1], the main result of Krstić–Vogtmann [**14**] implies that L is contractible. The dimension of L is $2k + n - 2$ when $n \geq 2$ and $2k + n - 3$ when $n \leq 1$ and $k > 1$.

We assume, in view of [**22**], throughout this paper that $n \geq 1$. The *edge number* of F is the quantity $\mathrm{en}(F) = 2k + n - 1$. If $\tau = (\mathcal{G}, \sigma)$ is a reduced marked graph of groups with exactly one vertex v of valence at least two, then the edge number of F is the valence of v.

The norm $\|\cdot\|$. Given a finite set W of infinite order elements of F and a reduced marked graph of groups $\tau = (\mathcal{G}, \sigma)$, following Krstić and Vogtmann [**14**] we define a norm on τ as

$$\|\tau\| = \sum_{w \in W} \ell(w),$$

where $\ell(w)$ is the hyperbolic translation length of w on the Bass–Serre tree T of $\mathcal{G}$. The *star* of a reduced marked graph of groups τ is the subcomplex of L spanned by those marked graphs of groups τ' which collapse onto τ. The *ball of radius* r is the union of the stars of reduced marked graphs of groups τ with norm at most r. It is contractible (this is the main technical result of [**14**]). Krstić–Vogtmann prove [**14**, Proof of Proposition 6.3] that for appropriate choice of W, balls of radius r are compact. We therefore choose W such that balls of radius r are compact. This gives us a compact exhaustion of L.

Directions and turns. Let v be a vertex of a marked graph of groups $(\mathcal{G}, \sigma)$. We write $\mathrm{st}(v)$ for the set of oriented edges with initial vertex v. The set of *directions* at v is

$$D_v = \coprod_{e \in \mathrm{st}(v)} \mathcal{G}_v \times \{e\}.$$

(Recall that marked graphs of groups have trivial edge groups in this paper.) There is an obvious action of $\mathcal{G}_v$ on D_v; each orbit is an oriented edge $e \in \mathrm{st}(v)$. A *turn* is the $\mathcal{G}_v$-orbit of a pair $\{(g_1, e_1), (g_2, e_2)\}$ of directions in D_v. A turn is *degenerate* if it is represented by a pair of identical directions and *nondegenerate* otherwise.

The star graph. Let $\tau = (\mathcal{G}, \sigma)$ be a marked graph of groups. For each $w \in W$ thought of as an element of the fundamental group $\pi_1(\mathbb{G}, \star)$, recall that the homotopy equivalence $\sigma\colon \mathbb{G} \to \mathcal{G}$ induces (after choosing basepoints) $\sigma_\sharp\colon \pi_!(\mathbb{G}) \to \pi_1(\mathcal{G})$. Represent the conjugacy class of $\sigma_\sharp(w)$ by a graph-of-groups *edge path*

$$\gamma_w = g_0 e_1 g_1 \ldots e_k g_k$$

which is *cyclically reduced* in the sense that $e\bar{e}$ is not a subpath of any cyclic reordering of γ_w for any oriented edge e of $\mathcal{G}$. We say the path γ_w *takes* the turns $[\{(1, \bar{e}_i), (g_i, e_{i+1})\}]$ for $1 \leq i \leq k-1$ and $[\{(g_k^{-1}, \bar{e}_k), (g_0, e_1)\}]$. Since γ_w is cyclically reduced, these turns are nondegenerate. The *star graph of* τ with respect to W has as vertex set the disjoint union of D_v as v varies over the vertices of $\mathcal{G}$ and an edge connecting a pair of directions (necessarily based at the same vertex v) if the turn they determine is taken by some γ_w. Thus each turn of γ_w based at v corresponds to $|\mathcal{G}_v|$ edges of the star graph. Since turns taken by γ_w are nondegenerate, the star graph has no loop edges. If τ is reduced, then $\|\tau\|$ may be computed from the star graph as

$$\|\tau\| = \frac{1}{2}\sum_{v\in\mathcal{G}}\sum_{d\in D_v}\frac{\text{valence}(d)}{|\mathcal{G}_v|}.$$

Ideal edges. An *ideal edge* based at v in $\mathcal{G}$ is a subset α of D_v with the following properties.

(1) The sets α and $D_v - \alpha$ have at least two elements.
(2) The set α contains at most one element of each $\mathcal{G}_v$-orbit of directions.
(3) There is a direction $(g, e) \in \alpha$ with the property that no direction with underlying edge $\bar{e}$ belongs to α. (We may have that $\bar{e} \notin \text{st}(v)$.)

Write $D(\alpha)$ for the set of directions satisfying item 3 of the above definition. We are mainly interested in the case where τ is reduced and $F = A_1 * \cdots * A_n * F_k$ where $n \geq 1$. In this situation, every vertex of $\mathcal{G}$ has nontrivial vertex group, so item 2 implies that $D_v - \alpha$ has at least two elements if α has at least two elements. There is an ideal edge based at v if and only if v has valence at least two, in which case we say that v is *active*. The action of $\mathcal{G}_v$ on D_v descends to an action on the set of ideal edges based at v, and we say α and α' are *equivalent* if they belong to the same $\mathcal{G}_v$-orbit. An ideal edge α is *contained in* an ideal edge β if both are based at v and α is equivalent to a subset of β. We say α and β are *disjoint* if they are based at different vertices or if they are based at a vertex v and their $\mathcal{G}_v$-orbits are disjoint. We say α and β are *compatible* if one is contained in the other or they are disjoint.

In the case where $\mathcal{G}_v$ is trivial, the set $D_v - \alpha$ is also an ideal edge, and we say α and $D_v - \alpha$ are equivalent, but compatibility remains the same.

Blowing up ideal edges. Given an ideal edge α in a marked graph of groups τ, we describe a new marked graph of groups $\tau^\alpha = (\mathcal{G}^\alpha, \sigma^\alpha)$ obtained from $\tau = (\mathcal{G}, \sigma)$ by *blowing up* the ideal edge α. The graph $\mathcal{G}^\alpha$ is the same as the graph $\mathcal{G}$ with an additional vertex v_α and an additional edge α. The oriented edge α has initial vertex v and terminal vertex v_α. An oriented edge e incident to v in $\mathcal{G}$ is now incident to v_α if there is a direction in the $\mathcal{G}_v$-orbit e in α; all other oriented edges remain incident to the vertices they were incident to in $\mathcal{G}$. All edge groups remain trivial, all vertex groups remain what they were in $\mathcal{G}$, and the new vertex has trivial vertex group. There is a collapse map $\mathcal{G}^\alpha \to \mathcal{G}$ which collapses the edge α and sends an edge e incident to v_α to the edge path ge, where $(g, e) \in \alpha$. This collapse map is a homotopy equivalence; choose a homotopy inverse $f\colon \mathcal{G} \to \mathcal{G}^\alpha$ and define $\sigma^\alpha = f\sigma$.

If τ is reduced, the marked graph of groups τ^α represents a vertex of L: collapsing any edge e in $D(\alpha)$ yields a reduced marked graph of groups τ^α_e in which the edge α is not collapsed. When τ is reduced, we say that τ^α_e is obtained from τ by the *elementary Whitehead move* (α, e). When it is convenient, we will sometimes write (α, d) where $d \in D(\alpha)$ is a direction, rather than an edge. The vertices τ and τ^α_e of L are connected by an edge path of length two in L called an *elementary Whitehead path.* Following [**22**], we denote paths in L by listing their vertices; so the elementary Whitehead path from τ to τ^α_e is $(\tau, \tau^\alpha, \tau^\alpha_e)$.

If α and α' are equivalent, say $\alpha' = g\alpha$, then τ^α and $\tau^{\alpha'}$ are equivalent: one isomorphism $h\colon \mathcal{G}^\alpha \to \mathcal{G}^{\alpha'}$ is the identity on the common edges and vertices (and vertex groups) of $\mathcal{G}^\alpha$ and $\mathcal{G}^{\alpha'}$, sends v_α to v'_α and α to the edge path $g\alpha'$.

If α is contained in the ideal edge β, then α corresponds to an ideal edge α of τ^β based at v_β. If α is based at v and β is based at w and they are disjoint, then α corresponds to an ideal edge α of τ^β based at v and vice versa. In this latter situation we have $(\tau^\alpha)^\beta = (\tau^\beta)^\alpha$.

An *ideal forest* $\Phi = \{\alpha_1, \ldots, \alpha_I\}$ in a reduced marked graph of groups τ is a set of pairwise compatible ideal edges containing at most one element of each equivalence class. By repeatedly blowing up ideal edges in Φ that are maximal with respect to inclusion, we obtain a marked graph of groups $\tau^{\alpha_1,\ldots,\alpha_I}$. The edges $\{\alpha_1, \ldots, \alpha_I\}$ are a *forest* in the sense of [**17**], and collapsing them recovers τ.

The absolute value of an ideal edge. Following Culler and Vogtman [**8**], define the *dot product* of two subsets S and T of the set of vertices of a graph to be the number of (unoriented) edges with one vertex in S and the other in T. The *absolute value* of S is the dot product of S with its complement. An ideal edge corresponds to a subset of the set of vertices of the star graph, so we may compute $|\alpha|$. If $\alpha' = g.\alpha$, then we have $|\alpha| = |\alpha'|$. A direction d likewise has an absolute value $|d| = |\{d\}|$. If $d' = g.d$ for some $g \in \mathcal{G}_v$, then $|d'| = |d|$. Put another way, the absolute value of a direction is actually a property of the underlying oriented edge e, and we will sometimes write $|e|$ for $|d|$.

It follows from [**14**, Lemma 4.7 and Propositions 6.4, 6.5] that if $\Phi = \{\alpha_1, \ldots, \alpha_I\}$ is an ideal forest in a reduced marked graph of groups τ and $\{e_1, \ldots, e_I\}$ is a collection of edges in τ which form a collapsible forest in $\tau^{\alpha_1,\ldots,\alpha_I}$, after possibly reordering and reversing orientation of the e_i, we have directions $d_i \in D(\alpha_i)$ with underlying oriented edges e_i such that collapsing the edges e_i yields a marked graph of groups $\tau^{\alpha_1,\ldots,\alpha_I}_{e_1,\ldots,e_I}$ such that

$$(\star) \qquad \|\tau^{\alpha_1,\ldots,\alpha_I}_{e_1,\ldots,e_I}\| = \|\tau\| + \sum_{i=1}^{I} |\alpha_i| - \sum_{i=1}^{I} |e_i|.$$

A Whitehead move (α, e) is *reductive* if $\|\tau^\alpha_e\| \le \|\tau\|$ and *strictly reductive* if the inequality is strict. By Equation $(\star)$, reductivity is equivalent to the condition that $|\alpha| \le |e|$. An ideal edge α is reductive if every Whitehead move supported by that edge is reductive. The proof of contractibility in [**8, 14**] goes by finding reductive ideal edges compatible with a given collection of ideal edges. In our case, we want to *avoid* reductive ideal edges if possible.

If we have $F = A_1 * \cdots * A_n * F_k$, let $s_1, \ldots, s_k$ be a free basis for F_k, and suppose W contains the elements

$$\{a_i a_j : a_i \in A_i - \{1\},\ a_j \in A_j - \{1\},\ i < j\}$$

assuming $n \ge 2$ along with the elements

$$\{s_i a_j,\ a_j s_i : a_j \in A_j\} \quad \text{and} \quad \{s_i s_j,\ s_i s_j^{-1} : i \neq j\}.$$

Note that W therefore contains the s_i. Throughout this paper, we assume that $\mathrm{Out}(F)$ is infinite, i.e. that $F \neq F_1$, A_1, $A_1 * F_1$ or $A_1 * A_2$.

LEMMA 1.1 (cf. Lemma 2.2 of [**22**]). *For W as above, the absolute value of each edge and each ideal edge in a reduced marked graph of groups τ is nonzero, and the star graph of τ with respect to W has exactly one connected component for each* *vertex of τ. Moreover, if $n \geq 2$, for each vertex v, there exists an oriented edge e with initial vertex v such that there exists an edge in the star graph between every pair of directions in D_v with the same underlying edge e.*

PROOF. Let $\tau = (\mathcal{G}, \sigma)$ be our reduced marked graph of groups. The main idea of the proof is [**21**, 6.5 Corollary 2], which states that if each $w \in W$ is elliptic in (the action of F on the Bass–Serre tree of) some graph of groups with fundamental group identified with F, then actually all of F is elliptic. We produce such splittings by possibly blowing up τ (although strictly speaking we do not always blow up *ideal edges*) and then collapsing all but one edge of the resulting graph of groups. As long as the blown-up graph of groups is minimal, we have a contradiction.

Let us apply this principle. If $|e| = 0$, then each $\sigma_\sharp(w)$ for $w \in W$ does not cross the edge e. Collapse the other edges of $\mathcal{G}$ to obtain a *free splitting* of F, that is, a graph of groups with trivial edge groups and fundamental group identified with F, in which F is elliptic. Since $\mathcal{G}$ was minimal, this is a contradiction, so we conclude that $|e| \neq 0$. If α is an ideal edge of τ with $|\alpha| = 0$, then thinking of α as an edge of τ^α, we can make the same argument to reach a contradiction. Therefore $|\alpha| \neq 0$.

Next we prove that the star graph of τ with respect to W has one component for each vertex v of τ. Write Γ_v for the induced subgraph of the star graph comprising all those edges between directions in D_v. If v has valence one in $\mathcal{G}$, then $n \geq 2$, and letting $\mathcal{G}_v \cong A_i$, the existence of the words $a_i a_j$ (or their inverses) in W implies Γ_v is connected.

So suppose v has valence at least two. If every component of Γ_v has one vertex, then in particular some edge of $\mathcal{G}$ satisfies $|e| = 0$, which we already saw was impossible. So we may suppose some component C of Γ_v has at least two vertices. Since $\mathrm{Out}(F)$ is infinite, assuming Γ_v is disconnected, $\Gamma_v \setminus C$ has at least two vertices. To see this, observe that if $\Gamma_v \setminus C$ is a single direction d, then $|d| = 0$, which implies that the $\mathcal{G}_v$-orbit of d cannot be contained in C, or, if $\mathcal{G}_v$ is trivial, that $|\bar{d}| = 0$, so $\bar{d}$ cannot be contained in C if $\mathcal{G}_v$ is trivial. In other words, if $\Gamma_v \setminus C$ contains exactly one vertex, then it actually contains at least two. Write $\mathcal{G}_C$ for the stabilizer of C in the action of $\mathcal{G}_v$ on Γ_v. Although C does not form an ideal edge in general, we may still apply the procedure in the paragraph "Blowing up ideal edges" to produce a new graph of groups $\tau^C = (\mathcal{G}^C, \sigma^C)$ with one more edge, which we denote as C. The edge C has edge group $\mathcal{G}_C$. The only way $\mathcal{G}^C$ could fail to be minimal is if $\mathcal{G}_C = \mathcal{G}_v$ and either v or the new vertex v_C has valence one in $\mathcal{G}^C$. But together this would imply that either C is empty or all of Γ_v. Collapsing all the edges of $\mathcal{G}^C$ other than C produces a splitting (not necessarily free) of F in which each $w \in W$ is elliptic, a contradiction. Therefore we conclude that Γ_v is connected in this case.

For the final statement, observe that topologically the edge path representing $a_i a_j$ is like a rubber band held at two points: the map factors through a two-to-one branched cover of the circle over the interval. The branch points correspond to turns between a pair of directions with the same underlying edge. As a_i and a_j vary, every vertex of τ (which has nontrivial vertex group because τ is reduced) appears as the image of one of these branch points. In fact, fixing a_j but allowing a_i to vary in A_i, if the branch point corresponding to a_i yields a turn between a pair of directions with underlying oriented edge e, we see that *every* pair of directions with the same underlying oriented edge equal to e are connected by an edge in the star graph. □

For the remainder of the paper, we fix W such that balls are compact and such that the hypotheses and conclusions of the previous lemma hold.

2. Connectivity at Infinity

We fix our free product $F = A_1 * \cdots * A_n * F_k$. Recall from the introduction our dimension formula

$$\dim(L(F)) = \begin{cases} 2k+n-2 & n \geq 2 \\ \max\{2k+n-3, 0\} & \text{otherwise.} \end{cases}$$

Thus it has dimension zero (i.e. is a single point) for $(n,k) = (0,1), (1,0), (1,1)$, and $(2,0)$. It follows that $\mathrm{Out}(F)$ has zero ends (i.e. is finite) in these cases.

The complex L has dimension one, i.e. is a tree, for $(n,k) = (3,0)$ and $(0,2)$. Since it is easy to see that this tree is "bushy", i.e. is not quasi-isometric to the line, it follows that these groups, which have more than one end, actually have infinitely many. In [**15**], we showed that additionally the group $\mathrm{Out}(A_1 * A_2 * \mathbb{Z})$ has infinitely many ends, even though for this choice of (n,k), the complex L has dimension two.

The goal of this section is to show that in all the remaining cases, the complex $L(F)$ is connected at infinity, hence the groups $\mathrm{Out}(F)$ are one ended.

Let B_k denote the ball of radius k. Since we have chosen W such that balls of radius k are compact, it is the union of finitely many stars of reduced marked graphs of groups, where the *star* of a reduced marked graph of groups τ is the subcomplex spanned by the set of all τ' collapsing onto τ. Since L is locally finite and the star of τ is finite, there are only a finite number of reduced marked graphs of groups whose stars have nonempty intersection with B_k. Let C_k denote the union of all of these reduced marked graphs of groups, and let

$$N(k) = \max\{\|\tau\| : \tau \in C_k\}.$$

From here on out, we will say *briar patch* for *reduced marked graph of groups.* To show connectivity at infinity, we will show that any two briar patches with norm at least $N(k)$ can be connected by a path that lies outside of B_k.

To do this, we start with a path between our two briar patches of a given form, and then push it outside of B_k by a homotopy. We say that a path $P = (\gamma_0, \gamma_1, \ldots, \gamma_\ell)$ is *standard* if it has even length $\ell = 2m$, if γ_{2i} is a briar patch for $0 \leq i \leq m$, and if $(\gamma_{2i}, \gamma_{2i+1}, \gamma_{2i+2})$ is an elementary Whitehead path for each i satisfying $0 \leq i \leq m-1$. Since $\mathrm{Out}(F)$ is generated by Whitehead automorphisms in the sense of [**7**] together with the finite group of *factor automorphisms* that fix a certain briar patch, given any two briar patches τ and τ', there is a standard path from τ to τ'.

We will push P toward infinity by eliminating local minima. Consider a briar patch $\tau = \gamma_{2i}$ of minimal norm along P, say $\|\tau\| = m$. The briar patches γ_{2i+2} and γ_{2i-2} are obtained from τ by elementary Whitehead moves (α, e) and (β, f) respectively. By choosing τ carefully, we may assume that $\|\tau^\alpha_e\| > m$, i.e. that $|\alpha| > |e|$ and that $\|\tau^\beta_f\| \geq m$, so that $|\beta| \geq |f|$. We replace the subpath $(\tau^\beta_f, \tau^\beta, \tau, \tau^\alpha, \tau^\alpha_e)$ of P by a new standard path from τ^β_f to τ^α_e that passes through only briar patches of norm strictly greater than m.

LEMMA 2.1. *Suppose* $\dim(L(F)) \geq 2$ *and* F *is not of the form* $A_1 * A_2 * \mathbb{Z}$. *Suppose* τ *is a briar patch of norm* m *and* τ^α_e *and* τ^β_f *are briar patches obtained* *from τ by elementary Whitehead moves satisfying $\|\tau_e^\alpha\| > \|\tau\|$ and $\|\tau_f^\beta\| \geq \|\tau\|$. There is a standard path from τ_e^α to τ_f^β which passes through only briar patches of norm strictly greater than m.*

The bulk of this section is given to proving Lemma 2.1. For now we use it to prove the following piece of Theorem B.

THEOREM 2.2. *If $F = A_1 * \cdots * A_n * F_k$ is such that $\dim(L(F)) \geq 2$ and not of the form $A_1 * A_2 * \mathbb{Z}$, then $L(F)$ is connected at infinity.*

PROOF. In the situation of the statement, suppose ρ and ρ' are briar patches of norm at least $N(k)$, and let P be a standard path from ρ to ρ'. If P does not already lie outside B_k, then there is a briar patch τ of norm $m \leq k$ along P. By assumption on ρ and ρ', we may choose τ such that the adjacent briar patches along P, call them τ_e^α and τ_f^β, which we have already observed are obtained from τ by elementary Whitehead moves, satisfy the assumptions of Lemma 2.1. In fact, we may choose τ to have minimum norm along P and satisfying these assumptions. We may therefore replace the subpath $(\tau_f^\beta, \tau^b, \tau, \tau^\alpha, \tau_e^\alpha)$ by a new standard path from τ_f^β to τ_e^α with only briar patches of norm strictly greater than τ in its interior.

We may repeat this process. At each stage, our new standard path has fewer briar patches of minimum norm m, and since $N(k) > k$, we may even increase this minimum until it is bigger than k, at which point after finitely many repetitions of this process, we have a standard path from ρ to ρ' that avoids B_k, as desired. □

The process in Lemma 2.1 is the opposite of Collins and Zieschang's "Peak Reduction". Like in [**7**], the easy case is when the ideal edges α and β are compatible.

LEMMA 2.3. *Suppose α and β are compatible ideal edges of the briar patch τ, and suppose (α, e) and (β, f) are Whitehead moves such that (α, e) strictly increases and (β, f) does not decrease norm. There is a standard path from τ_e^α to τ_f^β with at most one briar patch in its interior whose norm is strictly greater than the norm of τ.*

PROOF. We follow [**22**, Lemma 3.1]. Consider $(\mathcal{G}, \sigma) = \tau^{\alpha,\beta}$. We think of α and β as edges of $\mathcal{G}$. Our standard path lies in the link of $(\mathcal{G}, \sigma)$ in L and we describe it by listing the edges of $\mathcal{G}$ which must be collapsed. So, for example, we have $\tau_e^\alpha = \{\beta, e\}$.

If $\{e, f\}$ is a (collapsible) forest in $\mathcal{G}$, we may take our path to be

$$(\{\beta, e\}, \{e\}, \{e, f\}, \{f\}, \{\alpha, f\}).$$

If $f \neq e$, the middle vertex of this path is $\tau_{e,f}^{\alpha,\beta}$. Since $|\alpha| > |e|$ and $|\beta| \geq |f|$, Equation ($\star$) shows that this briar patch has norm greater than $\|\tau\|$. If instead $e = f$, this path degenerates to the elementary Whitehead path $(\tau_e^\alpha, \tau_e^{\alpha,\beta}, \tau_e^\beta)$.

Suppose now that $\{e, f\}$ is not a forest in $\mathcal{G}$. One case where this happens is when $\alpha = \beta$. In this case both e and f belong to $D(\alpha)$, so we may take the degenerate path $(\tau_e^\alpha, \tau^\alpha, \tau_f^\alpha)$.

Suppose instead that $\alpha \neq \beta$. Since e and f are individually collapsible in $\mathcal{G}$, they do not form loops and have at least one endpoint with trivial vertex group. If $\{e, f\}$ is not a forest, the subgraph of groups spanned by $\{e, f\}$ must be connected and have fundamental group of the form $A_1 * A_2$, $\mathbb{Z}$ or $A_1 * \mathbb{Z}$. In other words, the subgraph is either a loop on two edges or a path on two edges, and at least one vertex common to both edges has trivial vertex group. In fact, because α and β are distinct, the subgraph must contain both v_α and v_β, both of which have trivial vertex group. It follows that the subgraph is the loop on two edges with fundamental group $\mathbb{Z}$. We see that $\bar{e} \in D(\beta)$ and $\bar{f} \in D(\alpha)$. In other words, $\{\alpha, e\}$ and $\{\beta, f\}$ are forests. If $|e| \geq |f|$, then we may consider the path

$$(\{\beta, e\}\{\beta\}\{\beta, f\}, \{f\}, \{\alpha, f\}).$$

The briar patch τ_f^α corresponding to $\{\beta, f\}$ has norm

$$\|\tau_f^\alpha\| \geq \|\tau_e^\alpha\| > \|\tau\|.$$

if $|f| > |e|$, then we may consider instead the path $(\{\beta, e\}, \{e\}, \{\alpha, e\}, \{\alpha\}, \{\alpha, f\})$. The briar patch τ_e^β corresponding to $\{\alpha, e\}$ has norm

$$\|\tau_e^\beta\| > \|\tau_f^\beta\| \geq \|\tau\|. \qquad \square$$

As in [**22**], the proof of Lemma 2.3 gives a simplicial map of a subdivided square into L which maps the center to $\tau^{\alpha,\beta}$ and the corners to τ, τ_e^α, τ_f^β and a briar patch $\tau_T^{\alpha,\beta}$ for some maximal forest T in $\tau^{\alpha,\beta}$. In the case where $\alpha \neq \beta$ and $e \neq f$, then $\tau_T^{\alpha,\beta}$ is a briar patch with norm strictly greate than τ. The square, see Figure 1, provides a homotopy rel endpoints between the old path and the new one.

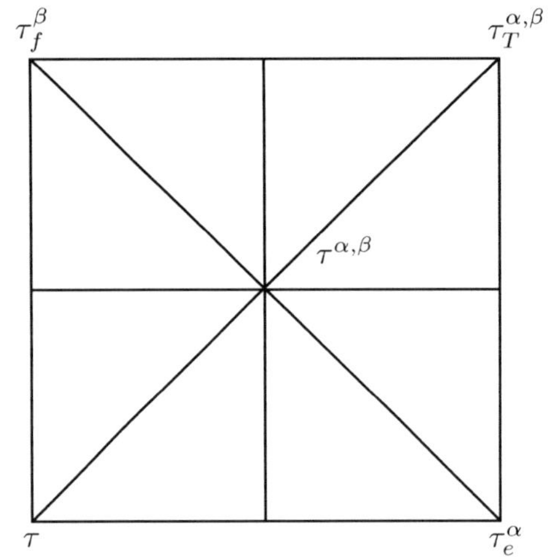

FIGURE 1. The subdivided square in the proof of Lemma 2.3.

Now, in general, we may not assume that α and β are compatible. The strategy for proving Lemma 2.1 is to find a sequence of ideal edges $\alpha = \gamma_0, \gamma_1, \ldots, \gamma_n = \beta$ such that each γ_i is compatible with its neighbors and supports a strictly increasing Whitehead move (γ_i, e_i). By Lemma 2.3, we obtain a standard path from $\tau_{e_i}^{\gamma_i}$ to $\tau_{e_{i+1}}^{\gamma_{i+1}}$, and the concatenation of all of these standard paths will be the standard path we seek. We will describe the outline of the argument in a moment.

Firstly, the *size* of an ideal edge is its cardinality, while the *size* of a Whitehead move is the size of its ideal edge. (Let us remark that in the case where $n = 0$ and an ideal edge α is equivalent to $D_v - \alpha$, we should take the size of α to mean the size of the smaller of the two.) Next, let the *edge number* of F, denoted $\operatorname{en}(F)$, be the quantity $2k + n - 1$. Assuming that $n \geq 1$, if τ is a briar patch in $L(F)$ with one vertex $\star$ with valence greater than one, then the valence of $\star$ is $\mathrm{en}(F)$. Finally, say that a vertex of τ is *active* if it has valence at least two. Active vertices support ideal edges.

In the course of our proof of Lemma 2.1, each ideal edge γ_i aside from α and β will have size two or size three. Our first step is to reduce to the case where τ has one active vertex.

LEMMA 2.4. *Suppose τ has more than one active vertex and that the given ideal edges α and β are not already compatible. Then α and β are compatible with a size-two ideal edge γ (based at a different vertex) which strictly increases norm.*

PROOF. This is a quick corollary of the fact (Lemma 2.8 below) that when $n \geq 2$, which is necessary for τ to have more than one active vertex, each active vertex supports a strictly increasing Whitehead move of size two, and that ideal edges based at different vertices are compatible. □

By Lemma 2.3, Lemma 2.1 follows in this case.

Supposing then instead that τ has only one active vertex, our first step is the following lemma.

LEMMA 2.5. *If τ is a briar patch supporting the Whitehead move (α, e) such that $|\alpha| \geq |e|$, then either α has size two, or α is compatible with a size-two ideal edge which supports a strictly increasing Whitehead move.*

Lemma 2.5 allows us to restrict our attention to the more well-behaved class of size-two Whitehead moves.

Here is our main step. Consider the simplicially subdivided polygons in Figure 2. Further subdivide each one by introducing a new vertex at the midpoint of each edge in the boundary of the polygon. Suppose (α, e) and (β, f) are Whitehead moves of size two based at τ such that $|\alpha| \geq |e|$ and $|\beta| > |f|$. A *good polygon from α to β* in the star of τ is a simplicial map of the further subdivided polygon into L with the following properties.

(1) The center of the polygon is mapped to τ.
(2) Each vertex of the polygon (three for the triangle, four for the rectangle, six for the hexagon) is mapped to a marked graph of groups τ^σ obtained from τ by blowing up an ideal edge σ of size two.
(3) The marked graphs of groups τ^α and τ^β are mapped to vertices of the polygon.
(4) Each midpoint of an edge (three for the triangle, four for the rectangle, six for the hexagon) is mapped to a marked graph of groups τ^σ obtained from τ by blowing up an ideal edge of size three.
(5) Each size-three ideal edge σ is compatible with the size-two ideal edges corresponding to the endpoints of the edge which σ corresponds to the midpoint of.
(6) At most one size-three ideal edge corresponding to a vertex of the polygon is reductive.
(7) At most one size-two ideal edge (specifically α) corresponding to a vertex of the polygon is reductive.

LEMMA 2.6. *Suppose τ has one active vertex supporting Whitehead moves (α, e) and (β, f) of size two such that $|\alpha| \geq |e|$ and $|\beta| > |f|$. Assume that $n \geq 1$,*

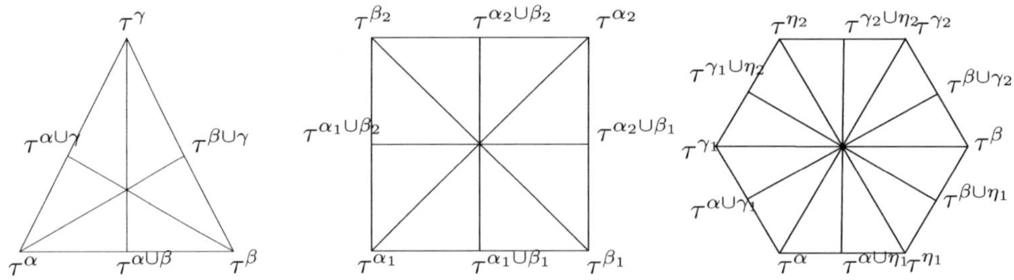

FIGURE 2. The simplicially subdivided polygons in the proof of Theorem 2.2.

en$(F) \geq 3$, *and that* $F \neq A_1 * A_2 * \mathbb{Z}$, *and that* α *and* β *are not compatible. Either (after rechoosing* α *and* β *in their equivalence class) we have that* $\alpha \cup \beta$ *is an ideal edge which is not reductive, or there is a good polygon in the star of* τ *from* α *to* β.

With this lemma in hand, we complete the proof of Lemma 2.1.

PROOF OF LEMMA 2.1. We begin with our given Whitehead moves (α, e) and (β, f) as in the statement. By Lemma 2.4 and Lemma 2.3, if τ has more than one active vertex, then there is a standard path from τ_e^α to τ_f^β with only briar patches of norm strictly greater than $\|\tau\|$ in its interior.

If instead τ has only one active vertex, then by Lemma 2.5, we may replace α and β by ideal edges of size two, call them α' and β' that support Whitehead moves (α', e') and (β', f') satisfying $|\alpha'| \geq |e'|$ and $|\beta'| > |f|$. Then by Lemma 2.6, there is a good polygon in the star of τ. Since there is at most one vertex of the polygon (aside from the vertex τ^α) corresponding to a reductive ideal edge σ, a path around the polygon from $\tau^{\alpha'}$ to $\tau^{\beta'}$ that avoids that vertex τ^σ gives a sequence of ideal edges $\alpha' = \gamma_1, \ldots, \gamma_m = \beta'$ such that each γ_i is compatible with its neighbors and supports a strictly increasing Whitehead move. By Lemma 2.3, we are done. ☐

The remainder of this section is given to proving the statements left unproved above: the main lemma left is Lemma 2.6, but we also need to prove Lemma 2.8, and Lemma 2.5.

Reduction to one active vertex. We turn first to Lemma 2.8, for which our main tool is the following lemma, due to Vogtmann. We give a proof for completeness.

LEMMA 2.7 (Lemma 3.2 of [**22**]). *Let* u, v *and* w *be vertices of a simplicial graph (for example the star graph of* τ*) and* S *the set of other vertices. If* $|\{u, v\}| \leq |v|$ *and* $|\{u, w\}| \leq |w|$, *then* $\{u\} \cdot \{v\} = \{u\} \cdot \{w\}$ *and* $\{u\} \cdot S = 0$.

PROOF. It is clear that if two sets S and T of vertices of a graph are disjoint, then for any set A of vertices, we have

$$A \cdot (S \cup T) = A \cdot S + A \cdot T.$$

Writing S for the set of other vertices as in the statement, we have

$$|\{u, v\}| = \{u\} \cdot S + \{u\} \cdot \{w\} + \{v\} \cdot \{w\} + \{v\} \cdot S$$

and

$$|v| = \{v\} \cdot S + \{v\} \cdot \{u\} + \{v\} \cdot \{w\}.$$

Our first inequality implies that

$$\{u\} \cdot S + \{u\} \cdot \{w\} \leq \{u\} \cdot \{v\}.$$

Calculating with w in place of v, we see that

$$\{u\} \cdot S + \{u\} \cdot \{v\} \leq \{u\} \cdot \{w\}.$$

Therefore we conclude that $\{u\} \cdot S = 0$ and $\{u\} \cdot \{v\} = \{u\} \cdot \{w\}$. □

Using Lemma 2.7, we prove Lemma 2.8.

LEMMA 2.8. *Let τ be a briar patch and suppose that $n \geq 2$. Let v be a vertex of τ of valence at least two. There is a strictly increasing Whitehead move of size two based at v.*

PROOF. Since $n \geq 2$, by Lemma 1.1, the star graph of τ contains an edge betwen a pair of directions d and d' in D_v with the same underlying oriented edge e. Since v has valence at least two, there are a pair of ideal edges $\alpha = \{d, c\}$ and $\alpha' = \{d, c'\}$ where c and c' are distinct and have underlying oriented edges different from e and $\bar{e}$. If both of the Whitehead moves (α, c) and (α', c') are reductive, then by Lemma 2.7, we have $d \cdot S = 0$, where S denotes the set of remaining vertices in the star graph aside from d, c and c'. This contradicts the existence of the edge from d to d'. □

Finding increasing Whitehead moves of size two. We will now turn to the proof of Lemma 2.5. Following Vogtmann, define a *trio* to be a set $\{e, f, g\}$ of three oriented edges with common initial vertex $\star$ of a briar patch τ which does not contain an edge twice, once in each orientation. A trio has the property that $(\{d, d'\}, d)$ is an elementary Whitehead move for any choice of directions d and d' representing edges of the trio. We say the trio *supports* these Whitehead moves.

Vogtmann proves the following lemma using Lemma 2.7.

LEMMA 2.9 (Lemma 3.3 of [**22**]). *For every choice of three directions $\{u, v, w\}$ representing the oriented edges of the trio $\{e, f, g\}$, at least one of the three size-two ideal edges $\{u, v\}$, $\{u, w\}$ and $\{v, w\}$ supports a strictly increasing Whitehead move.*

PROOF. Suppose all Whitehead moves supported by the ideal edges in the statement are reductive. Repeatedly applying Lemma 2.7, we see that the directions u, v and w form an isolated component of the star graph. This contradicts Lemma 1.1. □

LEMMA 2.10. *Suppose (α, e) is a Whitehead move satisfying $|\alpha| \geq |e|$. If the underlying oriented edges of α are of the form $\{e, f, \bar{f}\}$, and we write $\alpha = \{u, v, \bar{v}\}$, then one of the size-two Whitehead moves $\{u, v\}$ or $\{u, \bar{v}\}$ supports a strictly increasing Whitehead move.*

PROOF. Suppose towards a contradiction that the ideal edges $\{u, v\}$ and $\{u, \bar{v}\}$ are reductive. Applying Lemma 2.7, writing S for the set of directions not equal to u, v or $\bar{v}$, we see that $\{u\} \cdot S = 0$ and $\{u\} \cdot \{v\} = \{u\} \cdot \{\bar{v}\}$. Since

$$|v| = \{v\} \cdot \{u\} + \{v\} \cdot \{\bar{v}\} + \{v\} \cdot S = |\bar{v}|,$$

the latter equality implies that $\{v\} \cdot S = \{\bar{v}\} \cdot S$. We have by assumption $|\{u, v, \bar{v}\}| \geq |u|$, which implies that $\{v\} \cdot S \geq \{u\} \cdot \{v\}$. Now,

$$|\{u, v\}| = \{u\} \cdot \{\bar{v}\} + \{v\} \cdot S \quad \text{and} \quad |\{u\}| = \{u\} \cdot \{v\} + \{u\} \cdot \{\bar{v}\},$$

so since we assume $\{u,v\}$ is reductive, we conclude that $\{v\}\cdot\{\bar{v}\}=0$ and that $\{v\}\cdot S=\{u\}\cdot\{v\}$. In particular $|v|=|u|$.

Now, either e, the underlying oriented edge of u, forms a loop, or it does not. If it does, because $\{u\}\cdot S=0$, by $\mathcal{G}_\star$-equivariance of the star graph, every occurence of e (in either orientation) in each $w\in W$ is followed or preceded by an occurence of f, the underlying oriented edge of v. Since $|e|=|f|$, each word has an equal number of occurrences of e and f. Therefore, if we project to the subspace of $H_1(F,\mathbb{Z}/2\mathbb{Z})$ spanned by e and f, we see that each word $w\in W$ has image $(0,0)$ or $(1,1)$. Since these words do not form a basis for $\mathbb{Z}/2\mathbb{Z}\oplus\mathbb{Z}/2\mathbb{Z}$, this contradicts our assumption that the words $w\in W$ satisfy the hypotheses of Lemma 1.1.

Supposing instead that e does not form a loop, then every $w\in W$ contains an even number of occurrences of e and therefore an even number of occurences of f, which is again a contradiction mod 2. □

We are now ready to prove Lemma 2.5.

PROOF OF LEMMA 2.5. Note that if an ideal edge α does not have size two, then it contains a trio unless its underlying edges are of the form $\{e,f,\bar{f}\}$. If it contains a trio, then α is compatible with a size-two ideal edge supporting a strictly increasing Whitehead move by Lemma 2.9, and if not the same holds by Lemma 2.10. □

We need the following elementary observation.

LEMMA 2.11. *Suppose S is a nonempty subset of the vertices of a simplicial graph containing the vertex x. We have that $|S|\le|x|$ if and only if we have that*

$$\{x\}\cdot T\ge\frac{1}{2}|T|,$$

where T is the set $S-\{x\}$. In particular equality holds in one equation if and only if it holds in the other.

PROOF. Write T^c for the complement of T. On the one hand, we have $|x|=\{x\}\cdot T+\{x\}\cdot T^c$, and on the other, we have

$$|S|=\{x\}\cdot T^c+T\cdot(T^c-\{x\}).$$

A bit of algebra finishes the proof. □

COROLLARY 2.12. *Suppose T is a nonempty subset of the vertices of a simplicial graph. There are at most two choices of a vertex $x\notin T$ such that $|T\cup\{x\}|\le|x|$. If there are two choices, say x and x', then we have $|T\cup\{x\}|=|x|$ and $|T\cup\{x'\}|=|x'|$ and in fact elements of T are only connected to each other, x and x'.*

PROOF. By Lemma 2.11, We have that $|T\cup\{x\}|\le|x|$ if and only if $\{x\}\cdot T\ge\frac{1}{2}|T|$. If this holds for another $x'\notin T$, notice that we then have

$$|T|\le\{x\}\cdot T+\{x'\}\cdot T\le|T|.$$

Therefore equality holds throughout and the statement follows. □

For size-two ideal edges, we have the following strengthening.

COROLLARY 2.13. *Suppose u and v are directions based at a vertex $\star$ in a briar patch τ with valence at least three. Suppose that $\star$ has nontrivial vertex group $\mathcal{G}_\star$, and that $\{u,v\}$ is an ideal edge. There is at most one $h\in\mathcal{G}_\star$ such that $\{u,h.v\}\sim\{h^{-1}.u,v\}$ is reductive.*

PROOF. By Corollary 2.12 applied to $T = \{u\}$, we see that there are at most two $h \in \mathcal{G}_\star$ such that $\{u, h.v\}$ is reductive. Supposing that there are two, we may without loss of generality suppose that one is v and the other is $h.v$. Corollary 2.12 allows us to conclude that u is connected only to v and $h.v$ in the star graph. But applying Corollary 2.12 to $T = \{v\}$, we see that v is only connected to u and $h^{-1}.u$. It follows that the $\mathcal{G}_\star$-orbits of u and v together form an isolated component of the star graph. This contradicts Lemma 1.1, since $\star$ has valence at least three. □

We also need the following elaboration of Corollary 2.12.

LEMMA 2.14. *Suppose u, v, w, x, y are five distinct vertices of a simplicial graph. Suppose that $|\{u, v, w\}| \leq \min\{|u|, |v|, |w|\}$, that $|\{u, v\}| \geq \min\{|u|, |v|\}$ and $|\{u, w\}| \geq \min\{|u|, |w|\}$, and that one of the above inequalities involving $\{u, v\}$ and $\{u, w\}$ is strict. If $|\{u, v, x\}| \leq |x|$, then $|\{u, w, y\}| > |y|$.*

PROOF. By two applications of Corollary 2.12, if there exists y such that $|\{u, w, y\}| \leq |y|$, we have $|w| = |\{u, v, w\}| = |v|$, and that u is connected only to v and w; that is, we have that $|u| = \{u\} \cdot \{v, w\}$. But notice that the first chain of equalities together with the inequalities in the statement implies that $|v| = |w| \leq |u|$. So the inequalities in the statement imply by Lemma 2.11 that $\{u\} \cdot \{v\} \leq \frac{1}{2}|u|$ and that $\{u\} \cdot \{w\} \leq \frac{1}{2}|u|$, and that one of the inequalities is strict. This is a contradiction. □

Finding good polygons: the proof of Lemma 2.6. For the remainder of this section, we turn to the proof of Lemma 2.6. Assuming (α, e) and (β, f) are size-two Whitehead moves satisfying $|\alpha| \geq |e|$ and $|\beta| > |f|$ which are not compatible, there are three cases based on the configurations of α and β. We briefly describe the cases.

(1) **Case one:** $\alpha = \{u, v\}$ **and** $\beta = \{u, w\}$**.** In our first case, the directions $\{u, v, w\}$ represent a trio; that is, they are drawn from three distinct *unoriented* edges.
(2) **Case two:** $\alpha = \{u, v\}$ **and** $\beta = \{u, \bar{v}\}$**.** In our second case, we have that $\{u, v, \bar{v}\}$ is an ideal edge, but the three directions u, v and $\bar{v}$ do not form a trio because v and $\bar{v}$ share an underlying *unoriented* edge.
(3) **Case three:** $\alpha = \{u, v\}$ **and** $\beta = \{u, v'\}$, where $v' = g.v$ for some nontrivial $g \in \mathcal{G}_\star$. In our third case, the set of directions $\{u, v, v'\}$ is not an ideal edge.

Case one. In this case $\alpha \cup \beta = \{u, v, w\}$ is an ideal edge. If it is not reductive, we are done.

Supposing $\alpha \cup \beta$ is reductive, the good polygon we will build is the triangle in Figure 2 left. The ideal edge $\gamma = \gamma(g)$ will be of the form $\{v, g.w\} \sim \{g^{-1}.v, w\}$ for some nontrivial element $g \in \mathcal{G}_\star$. Recall that in view of Vogtmann [**22**], we assume that $n \geq 1$. We have the ideal edges $\alpha \cup \gamma(g) = \{u, v, g.w\}$ and $\beta \cup \gamma(g) = \{u, g^{-1}.v, w\}$. By Lemma 2.14, at most one of the size-three ideal edges $\alpha \cup \gamma(g)$ and $\beta \cup \gamma(g)$ is reductive as g varies over the nontrivial elements of $\mathcal{G}_\star$.

First, observe that *no* ideal edge of the form $\gamma(g)$ is reductive when $g \in \mathcal{G}_\star$ is nontrivial. Indeed, if some $\gamma(g)$ were reductive, by Lemma 2.11, we have $\{v\} \cdot \{g.w\} \geq \frac{1}{2}|v|$. Notice that

$$|v| \leq \{v\} \cdot \{g.w\} + \{g^{-1}.v\} \cdot \{w\} \leq |\{u, v, w\}| \leq |v|,$$

so we conclude that equality holds throughout. In fact, by replacing v with w (and viewing $\gamma(g)$ as $\{g^{-1}.v, w\}$), we conclude that $|v| = |w| = |\{u, v, w\}| \leq |u|$, and in particular we have $|u| = \{u\} \cdot \{v, w\}$. As in the proof of Lemma 2.14, this is a contradiction: since at least one of α and β is not reductive, we have by Lemma 2.11 that

$$\{u\} \cdot \{v\} + \{u\} \cdot \{w\} < |u|.$$

Next we show that in fact no ideal edge of the form $\alpha \cup \gamma(g)$ or $\beta \cup \gamma(g)$ is reductive. Recall that by Lemma 2.14, since we assume that $\alpha \cup \beta$ is reductive, there is at most one such edge, suppose it is $\alpha \cup \gamma(g)$. The argument in the case it is of the form $\beta \cup \gamma(g)$ will be identical. By Lemma 2.11, we have that $\frac{1}{2}|\alpha| = \frac{1}{2}|\{u, v\}| = \{w\} \cdot \{u, v\} = \{g.w\} \cdot \{u, v\}$. We will show that $|\alpha \cup \gamma(g)| > |\alpha|$, contradicting our assumption that this ideal edge is reductive. Indeed, we claim that

$$|\{u, v, g.w\}| > 2\{u, v\} \cdot \{w\} = |\alpha|.$$

Indeed, observe that $\{g.w\} \cdot \{g.u, g.v\} = \{w\} \cdot \{u, v\}$ by $\mathcal{G}_\star$-equivariance of the star graph, so it is clear that we have inequality. That the inequality is *strict* follows in the case where $\mathrm{en}(F) > 3$ by Lemma 1.1, since the $\mathcal{G}_\star$-orbits of the directions u, v and w cannot form an isolated component of the star graph. If instead $\mathrm{en}(F) = 3$, so that $n \geq 2$, then again by Lemma 1.1, we must have $\{w\} \cdot \{g.w\} > 0$.

Thus *any* choice of $g \neq 1$ in $\mathcal{G}_\star$ yields a good polygon from α to β.

Case two. Recall that in this case we assume that $\alpha = \{u, v\}$ and $\beta = \{u, \bar{v}\}$, i.e. that $\alpha \cup \beta$ is an ideal edge, but v and $\bar{v}$ share an underlying *unoriented* edge. If $\alpha \cup \beta$ is not reductive, we are done. Supposing this ideal edge is reductive, the good polygon from α to β we will build will be the rectangle in Figure 2 center, where we set $\alpha_1 = \alpha$ and $\beta_1 = \beta$.

To construct this good polygon, we need size-two ideal edges α_2 and β_2 so that $\alpha_1 \cup \beta_2$, $\alpha_2 \cup \beta_1$ and $\alpha_2 \cup \beta_2$ are size-three ideal edges (where we may abuse notation slightly so that these unions may be taken) and so that none of these five ideal edges (including α_2 and β_2) are reductive.

Since we assume that $F \neq A_1 * A_2 * \mathbb{Z}$, there exists a direction w whose $\mathcal{G}_\star$-orbit is disjoint from $\{u, v, \bar{v}\}$; we choose one such. For $h \in \mathcal{G}_\star$, we focus on the ideal edges $\alpha(h) = \{u, h.v\}$, $\beta(h) = \{u, h.\bar{v}\}$, $\gamma(h) = \{\bar{v}, h.w\}$ and $\delta(h) = \{v, h.w\}$. We have $\alpha = \alpha(1)$ and $\beta = \beta(1)$ and will choose α_2 from $\{\alpha(g), \gamma(1), \gamma(g)\}$ and β_2 from $\{\beta(g), \delta(1), \delta(g)\}$ for $g \in \mathcal{G}_\star$ nontrivial. Observe that by abusing notation slightly, for any choice of α_2 and β_2, the size-three ideal edges $\alpha \cup \beta_2$, $\alpha_2 \cup \beta$ and $\alpha_2 \cup \beta_2$ make sense to define.

By Corollary 2.13, at most one of $\gamma(1)$ and $\gamma(g)$ is reductive, and similarly for $\delta(1)$ and $\delta(g)$. Since we assume that $\{u, v, \bar{v}\}$ is reductive, neither $\alpha(g)$ nor $\beta(g)$ can be reductive: if, say, $|\alpha(g)| \leq |v|$, we have that $\{u\} \cdot \{g.v\} \geq \frac{1}{2}|u|$ by Lemma 2.11, but this would that

$$|u| \leq \{u\} \cdot \{g.v\} + \{v\} \cdot \{g^{-1}.u\} < |\{u, v, \bar{v}\}|,$$

contradicting our assumption that this ideal edge is reductive.

Notice that if $\gamma(1)$ is reductive, we have in particular that $\{\bar{v}\} \cdot \{w\} \geq \frac{1}{2}|w|$. Then all four of the ideal edges $\beta(1) \cup \gamma(g)$, $\beta(g) \cup \gamma(g)$, $\delta(1) \cup \gamma(g)$ and $|\delta(g) \cup \gamma(g)$ are greater in absolute value than $\{\bar{v}\} \cdot \{w\} + \{g.\bar{v}\} \cdot \{g.w\} \geq |w|$ and so are not reductive.

From this it follows that if there are two reductive size-two ideal edges under consideration, there is a good polygon from α to β. For instance, if $\gamma(1)$ and $\delta(g)$ are reductive, the relevant path around the good polygon reads

$$\alpha,\ \alpha\cup\delta(1),\ \delta(1),\ \gamma(g)\cup\delta(1),\ \gamma(g),\ \beta\cup\gamma(g),\ \beta.$$

We proceed continuing to assume that $\gamma(1)$ is reductive, but now assuming that neither $\delta(1)$ nor $\delta(g)$ is reductive. If one of $\alpha\cup\beta(g)$, $\alpha\cup\delta(1)$ or $\alpha\cup\delta(g)$ is not reductive, we may use that ideal edge together with ideas above to construct a good polygon from α to β. Explicitly, if $\alpha\cup\delta(1)$ is not reductive, the relevant path reads

$$\alpha,\ \alpha\cup\delta(1),\ \delta(1),\ \delta(1)\cup\gamma(g),\ \gamma(g),\ \beta\cup\gamma(g),\ \beta.$$

We will show that having all three ideal edges reductive leads to a contradiction. In fact, this does not use the assumption that $\gamma(1)$ is reductive. Indeed, by Corollary 2.12, we conclude assuming that $\alpha\cup\beta$ and $\alpha\cup\beta(g)$ are reductive that $|v|\le|u|$. If we had equality, Corollary 2.12 would additionally imply that $\alpha\cup\delta(1)$ and $\alpha\cup\delta(g)$ are greater in norm than $|w|$, in contradiction to our assumption that these ideal edges are reductive. Again by Corollary 2.12, we conclude that $|w|\le|v|$ as well. Therefore in particular by Lemma 2.11, we have $|\alpha|=\{w,g.w\}\cdot\{u,v\}\le|w|$. Since therefore $|w|\ge|\alpha|\ge|v|$, we conclude that equality holds throughout. But it cannot, since this would imply that the $\mathcal{G}_\star$-orbits of the directions u, v and w form an isolated component of the star graph, in contradiction to Lemma 1.1.

Finally, assume that none of the ideal edges $\gamma(1)$, $\gamma(g)$, $\delta(1)$ or $\delta(g)$ are reductive. Now by Corollary 2.12, at most two of the four ideal edges $\alpha\cup\beta_2$, which take the form $\{u,v,x\}$, are less than $|x|$ in norm. If we have $|u|<|v|$, since $\alpha\cup\beta$ is reductive, this is the only ideal edge of that form which is reductive, and similarly for ideal edges of the form $\{u,\bar{v},x\}$. It remains, then, just to show that one of the ideal edges of the form $\alpha_2\cup\beta_2$ is not reductive. Suppose, for instance, then, that $\alpha(g)\cup\beta(g)$ is reductive. This, by Corollary 2.12, implies that $|\{v,\bar{v}\}|=\{u,g^{-1}.u\}\cdot\{v,\bar{v}\}\le|u|$. Since $|u|<|v|$, this implies that $|\{v,\bar{v}\}|<|v|$, so that $\{v\}\cdot\{\bar{v}\}>\frac{1}{2}|v|$, and therefore $\gamma(1)\cup\delta(g)$ and $\gamma(g)\cup\delta(1)$ are both greater in norm than $|v|$, and thus are not reductive unless $|v|<|w|$. But notice that because $|u|<|v|$, reductivity of $\alpha(g)\cup\beta(g)$ implies that $\{u,g.v,g.\bar{v}\}<|g.\bar{v}|=|v|$; by Corollary 2.12 it is the only ideal edge of the form $\{u,g.v,x\}$ to be less than or equal to $|x|$ in norm. In particular $|\alpha(g)\cup\delta(1)|>|w|$ and so is not reductive. This provides a good polygon from α to β in this case.

We therefore assume that $|v|\le|u|$. Now if $|w|<|v|$, for each of the ideal edges $\gamma(1)$, $\gamma(g)$, $\delta(1)$, and $\delta(g)$ at most of the relevant size-three ideal edges containing it can be reductive by Corollary 2.12. The only worry, then, is that both $\gamma(1)\cup\beta$ and $\gamma(g)\cup\beta$ are reductive or, symmetrically, both $\alpha\cup\delta(1)$ and $\alpha\cup\delta(g)$ are reductive. But in either case this implies that $|v|\le|w|$: for instance in the latter case we have $|v|\le|\{u,v\}|=\{w,g.w\}\cdot\{u,v\}\le|w|$. This contradiction provides a good polygon from α to β in this case.

Continuing to assume that $|v|\le|u|$, we now suppose that $|v|\le|w|$. Since the argument in the previous case that not all three ideal edges of the form $\alpha\cup\beta_2$ are reductive did not use the assumption that $\gamma(1)$ was reductive, we conclude that at least one ideal edge of the form $\alpha\cup\beta_2$ is not reductive. Suppose at first that we may choose $\beta_2=\delta(1)$. (By relabelling if necessary, this is equivalent to choosing $\beta_2=\delta(g)$.) Now by symmetry we know that at least one ideal edge of the form

$\alpha_2 \cup \beta$ is not reductive, but we can show more: If both $\delta(1)\cup\beta$ and $\delta(g)\cup\beta$ are reductive, then $|v| = |w| = \{w, g.w\}\cdot\{u,\bar{v}\} = |\{u,\bar{v}\}|$, which is a contradiction to Lemma 1.1, (as well as the assumption that β is not reductive) since it implies that the $\mathcal{G}_\star$-orbits of u, $\bar{v}$ and w form an isolated component of the star graph.

Therefore either $\delta(1)\cup\beta$ or $\delta(g)\cup\beta$ is not reductive. Suppose at first the former is not reductive. If $\delta(1)\cup\gamma(1)$ is not reductive, we are done. Supposing it is, we see by Lemma 2.11 that $\{u,w\}\cdot\{v,\bar{v}\} = |\{v,\bar{v}\}|$. This implies that $|\{v,\bar{v},g.w\}| = |w| + |\{v,\bar{v}\}|$, that $|\{g^{-1}.u, v, g.w\}| \geq |v| + |\{v,\bar{v}\}|$, that $|\{g^{-1}.u,\bar{v},g.w\}| \geq |v| + |\{v,\bar{v}\}|$, and that $|\{g^{-1}.u,v,\bar{v}\}| = |u| + |\{v,\bar{v}\}|$ so these ideal edges are not reductive. If we can show that one of $\alpha\cup\beta(g)$ or $\alpha\cup\delta(g)$ is not reductive and that one of $\alpha(g)\cup\beta$ and $\gamma(g)\cup\beta$ is not reductive, we will have a good polygon from α to β.

Suppose, then, that $\alpha\cup\beta(g)$ and $\alpha\cup\delta(g)$ are reductive. From the former, we conclude that $\{u\}\cdot\{v\} = \{u\}\cdot\{v,g.\bar{v}\} \geq \frac{1}{2}|\{v,g.\bar{v}\}| = |v|$, so equality holds throughout. This implies that $|\bar{v}| = \{w\}\cdot\{\bar{v}\}$. From reductivity of the latter, we conclude using Lemma 2.11 that $|\{u,g.w\}| = \{u\}\cdot\{v\} + \{w\}\cdot\{\bar{v}\}$. In particular $|u| = \{u\}\cdot\{v,g.w\}$ and $|w| = \{w\}\cdot\{g^{-1}.u,\bar{v}\}$. This implies that $|\{u,g.v,\bar{v}\}| = |u| + 2|v|$ and that $|\{g^{-1}.u,v,w\}| = 3|v|$, so neither of these ideal edges are reductive, and we have a good polygon from α to β, the relevant path around which reads

$$\alpha,\ \alpha\cup\delta(1),\ \delta(1),\ \alpha(g)\cup\delta(1),\ \alpha(g)\cup\beta,\ \beta.$$

A symmetric argument dispenses with the case in which both $\alpha(g)\cup\beta$ and $\delta(g)\cup\beta$ are reductive. We have two nested assumptions that need to be dispensed with: the outer one, that $\alpha\cup\delta(1)$ is not reductive, and the inner one, that $\gamma(1)\cup\beta$ is not reductive.

So, still supposing that $\alpha\cup\delta(1)$ is not reductive, suppose that $\gamma(1)\cup\beta$ is reductive. As we noted above, this implies that $\gamma(g)\cup\beta$ is not reductive, and therefore if $\gamma(g)\cup\delta(1)$ is not reductive, we are done. We therefore suppose that it is reductive. We first show that $\gamma(g)\cup\delta(g)$ is not reductive. Indeed, if it were, we see by Corollary 2.12 that $\{u,g.w\}\cdot\{v,\bar{v}\} = |\{v,\bar{v}\}|$. Since $\gamma(g)\cup\delta(1)$ is reductive, we conclude that in fact $\{g.w\}\cdot\{\bar{v}\} = \{g.w\}\cdot\{g.v,\bar{v}\} \geq \frac{1}{2}|\{v,g^{-1}.v\}| = |\bar{v}|$, so equality holds throughout. But this contradicts our assumption that $\gamma(1)\cup\beta$ is reductive, since by Lemma 2.11 that implies that $\{\bar{v}\}\cdot\{u,w\} \geq \frac{1}{2}|\{u,w\}| > 0$.

Thus we are done if $\alpha\cup\delta(g)$ is not reductive. So suppose it is. Applying Lemma 2.11, we see that

$$\begin{aligned}|u| + |w| &\leq \{u\}\cdot\{v,\bar{v}\} + \{w\}\cdot\{g^{-1}.v,\bar{v}\} + \{u\}\cdot\{w\} + \{u\}\cdot\{g.w\}\\ &\leq |u| + |v| - \{u\}\cdot\{w\} - \{g^{-1}.u\}\cdot\{w\},\end{aligned}$$

so we have equality throughout; in particular $\{u\}\cdot\{w\} = \{g^{-1}.u\}\cdot\{w\} = 0$. But this contradicts reductivity of $\gamma(g)\cup\delta(1)$, since we must have $\{w\}\cdot\{v,g^{-1}.\bar{v}\} \geq \frac{1}{2}|\{v,g^{-1}.\bar{v}\}| > 0$. Therefore we again have a good polygon from α to β.

Finally, we supposed that we could choose $\alpha\cup\delta(1)$ to not be reductive. To negate this, in view of the possibility of relabelling, we may assume that $\alpha\cup\delta(1)$ and $\alpha\cup\delta(g)$ are both reductive. This implies that $|w| = |v|$, and thus that $\{w,g.w\}\cdot\{u,v\} = |\{u,v\}|$. As we saw, we must have that $\alpha\cup\beta(g)$ is not reductive. Since this is equal in norm to $|\{u,v\}| + |v|$, we have that these sum to more than $|u|$. By Lemma 2.11, this implies that $\{u\}\cdot\{v\} < |v|$. But reductivity of $\alpha\cup\beta$ actually implies that $\{u\}\cdot\{v\} = \{u\}\cdot\{v,\bar{v}\} \geq \frac{1}{2}|\{v,\bar{v}\}| = |v|$. This contradiction completes the proof in this case.

Case three. Recall that in this case we assume that $\alpha = \{u, v\}$ and $\beta = \{u, g.v\}$, i.e. α and β share both underlying oriented edges. In this case $\alpha \cup \beta$ fails to be an ideal edge. Since $\mathrm{en}(F) \geq 3$, there exists a direction w based at $\star$ whose $\mathcal{G}_\star$-orit is distinct from that of u and v; choose one.

The proof breaks into two cases: either we are able to choose w so that $\{u, v, w\}$ represents a trio, or we are not able to, in which case we may choose w to either be $\bar{u}$ or $\bar{v}$, since we assume that $F \neq A_1 * A_2 * \mathbb{Z}$.

The first subcase. Suppose at first that $\{u, v, w\}$ represents a trio. We consider the ideal edges $\gamma(1) = \{u, w\}$, $\gamma(g) = \{u, g.w\}$, $\delta(1) = \{v, w\}$ and $\delta(g) = \{v, g.w\}$. We will use these ideal edges to build a good polygon (the rectangle in Figure 2) from α to β.

Observe that we cannot have that $\alpha \cup \gamma(1)$ and $\alpha \cup \gamma(g)$ are both reductive. If we did, we conclude by Corollary 2.12 that $|\alpha| = \{w, g.w\} \cdot \{u, v\}$. An argument based on Lemma 1.1 shows that both $\alpha \cup \gamma(1)$ and $\alpha \cup \gamma(g)$ must be strictly greater in norm than this quantity, which in turn is not less than the minimum of $|u|$ and $|v|$, implying neither ideal edge is reductive after all. The same argument applies to all the pairs $\alpha \cup \delta(1)$ and $\alpha \cup \delta(g)$, $\beta \cup \gamma(1)$ and $\beta \cup \gamma(g)$ and $\beta \cup \delta(1)$ and $\beta \cup \delta(g)$.

In fact, provided that, say, $\gamma(1)$ is not reductive, the same argument above proves that $\alpha \cup \gamma(1)$ and $\beta \cup \gamma(1)$ cannot both be reductive. By Corollary 2.13, at most one of $\gamma(1)$ and $\gamma(g)$ is reductive, and similarly for $\delta(1)$ and $\delta(g)$.

So suppose for instance that $\gamma(1)$ and $\delta(g)$ are not reductive. At most one of $\alpha\cup\gamma(1)$ and $\alpha\cup\gamma(g)$ can be reductive; suppose the latter is not. If $\beta\cup\gamma(g) = \beta\cup\delta(1)$ is reductive, then we have that both $\alpha \cup \delta(g) = \alpha \cup \gamma(g)$ and $\beta \cup \delta(g)$ are not reductive. This actually suffices to prove the theorem, but we can work a little harder to complete this to a good polygon. If both $\gamma(g)$ and $\delta(1)$ are reductive, we have $\{w\} \cdot \{g^{-1}.u, v\} = |w|$. This implies that $|\beta \cup \gamma(1)| = |\{u, g.v, w\}| > 2|w|$, so is not reductive. (In fact, with a little more care, it can be shown that the ideal edge $\alpha\cup\gamma(1)$ is also not reductive, so the good polygon with corners α, $\gamma(1)$, β, and $\delta(g)$ has no reductive size-three ideal edges.) If, say, $\delta(1)$ is not reductive, continuing to assume that $\beta \cup \gamma(g) = \beta \cup \delta(1)$ is reductive, we conclude that $\alpha \cup \gamma(1) = \alpha \cup \delta(1)$ is not, and we have a good polygon with corners α, $\delta(1)$, β and $\delta(g)$. A variant of this argument works in all cases.

The second subcase. So suppose instead that $F = A_1 * F_2$: our only choices for w are $\bar{u}$ and $\bar{v}$. We will consider the ideal edges $\gamma(1) = \{u, \bar{v}\}$, $\gamma(g) = \{u, g.\bar{v}\}$, $\delta(1) = \{\bar{u}, v\}$, $\delta(g) = \{\bar{u}, g.v\}$, $\eta(1) = \{\bar{u}, \bar{v}\}$ and $\eta(g) = \{\bar{u}, g.\bar{v}\}$. The good polygon we build will be either the rectangle or the hexagon in Figure 2, right, with α and β seprarated by one vertex of the polygon.

Now, by Corollary 2.13, at most one γ ideal edge, one δ ideal edge and one η ideal edge can be reductive. Suppose that there are three such. For simplicity assume they are $\gamma(1)$, $\delta(1)$ and $\eta(1)$: the other cases are essentially identical. Then by Lemma 2.11, we have that $\{\bar{v}\}\cdot\{u, \bar{u}\} = |\bar{v}|$, while $\{\bar{u}\}\cdot\{v, \bar{v}\} = |\bar{u}|$. In particular $|u| = |v|$. Observe, then, that each of the ideal edges $\alpha \cup \gamma(g)$, $\beta \cup \gamma(g)$, $\alpha \cup \delta(g)$ and $\beta \cup \delta(g)$ are greater in norm than $\frac{3}{2}|v| = \frac{3}{2}|u|$, so none of these ideal edges are reductive and we have a good rectangle from α to β.

Suppose next that there are two reductive size-two ideal edges. Up to swapping u and v, we will consider the cases of $\gamma(1)$ and $\eta(1)$ as well as $\gamma(1)$ and $\delta(1)$; the others are essentially identical. If $\gamma(1)$ and $\eta(1)$ are reductive, then by Lemma 2.11, we have that $|\bar{v}| = \{\bar{v}\}\cdot\{u,\bar{u}\}$. We claim that we have a good rectangle with corners

$$\alpha,\ \gamma(g),\ \beta \text{ and } \delta(1).$$

To see this, suppose that one of the relevant size-three ideal edges is reductive. For example, if $\alpha\cup\gamma(g)$ is reductive, we have

$$|v| \geq \{u\}\cdot\{v\} = \{u\}\cdot\{v, g.\bar{v}\} \geq \frac{1}{2}|\{v, g.\bar{v}\}| = |v|.$$

It follows that $|\beta\cup\gamma(g)| = |u|+|v|+|\bar{v}|$ and $|\beta\cup\delta(1)| = |v|+|\{g^{-1}.u,\bar{u}\}|$, so neither of these ideal edges is reductive. And additionally, by Lemma 2.11, if $\alpha\cup\delta(1)$ is reductive, then $|\bar{u}| = \{\bar{u}\}\cdot\{u,\bar{v}\} < \{u\}\cdot\{u,v,\bar{v}\} = |u|$, which is impossible. The argument if $\beta\cup\gamma(g)$ is reductive is analogous. On the other hand, if $\alpha\cup\delta(1)$ is reductive, it follows from Lemma 2.11 that $|v| = \{v\}\cdot\{u,\bar{u}\} = \frac{1}{2}|\{u,\bar{u}\}|$. Then $|\beta\cup\gamma(g)| = |u| + 2|v|$, so this ideal edge is not reductive. Reductivity of $\beta\cup\delta(1)$ or $\alpha\cup\gamma(g)$ would imply respectively that $\{v\}\cdot\{\bar{u}\} = |v| = |u|$ or $\{v\}\cdot\{u\} = |v|$. But again this would imply in either case that $|u|$ and $|\bar{u}|$ cannot be equal, but they must. The case that $\beta\cup\delta(1)$ is reductive is again analogous. The existence of a good rectangle follows.

If instead $\gamma(1)$ and $\delta(1)$ are reductive, we again have a good rectangle, this time with corners

$$\alpha,\ \gamma(g),\ \beta \text{ and } \delta(g).$$

Indeed, Lemma 2.11 implies that $\{\bar{v}\}\cdot\{u\} \geq \frac{1}{2}|u|$ and $\frac{1}{2}|\bar{v}|$, while $\{v\}\cdot\{\bar{u}\} \geq \frac{1}{2}|\bar{u}|$ and $\frac{1}{2}|v|$. We observe then, that $|\alpha\cup\gamma(g)| \geq \frac{3}{2}|u|$ because we have

$$\{u\}\cdot\{\bar{v}\} + \{v\}\cdot\{\bar{u}\} + \{g.\bar{v}\}\cdot\{g.u\} \leq |\{u, v, g.\bar{v}\}.$$

Similarly we see that $|\alpha\cup\delta(g)| \geq \frac{3}{2}|v|$ and $|\beta\cup\delta(g)| \geq \frac{3}{2}|v|$, so none of these ideal edges are reductive.

If there is only one reductive size-two ideal edge, up to swapping the roles of u and v it is either a γ or an η. Suppose that the reductive ideal edges is $\gamma(1)$, so that by Lemma 2.11, we have $\{u\}\cdot\{\bar{v}\} \geq \frac{1}{2}|v|$ and $\frac{1}{2}|u|$. Since the $\mathcal{G}_\star$-orbits of u, v and $\bar{v}$ cannot form an isolated component of the star graph, we conclude that $|\beta\cup\gamma(g)| > |u|$ and $|\alpha\cup\gamma(g)| > |u|$. The same argument applies to show that $|\alpha\cup\delta(g)| > |v|$ and $|\beta\cup\delta(g)| > |v|$, so we have a good rectangle from α to β. The argument supposing $\gamma(g)$ is reductive is identical.

If instead $\eta(1)$ is reductive, we see by Lemma 2.11 that $\{\bar{u}\}\cdot\{\bar{v}\} \geq \frac{1}{2}|v|$ and $\frac{1}{2}|u|$. This implies that $\alpha\cup\gamma(1)$ and $\beta\cup\gamma(g)$ cannot both be reductive, for if they were, we would have $|\{v,\bar{v}\}| = \{u, g^{-1}.u\}\cdot\{v,\bar{v}\}$ in contradiction to the fact that $\{\bar{v}\}\cdot\{\bar{u}\} > 0$. Similarly at most one of $\alpha\cup\gamma(g)$ and $\beta\cup\gamma(1)$, one of $\alpha\cup\delta(1)$ and $\beta\cup\delta(g)$ and one of $\alpha\cup\delta(g)$ and $\beta\cup\delta(1)$ are reductive.

Now by Corollary 2.12, at most two ideal edges of the form $\alpha\cup\{x\}$ satisfy $|\alpha\cup\{x\}| \leq |x|$, and if there are two, equality holds. If $|u| < |v|$, we see that $\alpha\cup\gamma(1)$ and $\alpha\cup\gamma(g)$ cannot both be reductive. If $|u| > |v|$, the same applies to $\alpha\cup\delta(1)$ and $\alpha\cup\delta(g)$, and if $|u| = |v|$, at most two of these four ideal edges may be reductive. The same holds for size-three ideal edges containing β.

Suppose $|u| < |v|$. If $\alpha\cup\gamma(1)$ and $\beta\cup\gamma(1)$ are reductive, then $\alpha\cup\gamma(g)$ and $\beta\cup\gamma(g)$ are not, and if $\alpha\cup\delta(1)$ is reductive, then $\beta\cup\delta(g)$ is not and regardless of whether $\alpha\cup\delta(g)$ is, we have a good rectangle from α to β. The same proof works if $\alpha\cup\gamma(g)$ and $\beta\cup\gamma(g)$ are reductive. If neither of these occurs, then at most one of the four size-three ideal edges is reductive and we again have a good rectangle.

If we suppose $|v| < |u|$, swapping the roles of u and v completes the proof as above. So suppose $|v| = |u|$. Now at most two of the four size-three ideal edges containing α are reductive, and similarly for β. If there are three or fewer, or more generally, if for some γ or some δ, the relevant size-three ideal edges are both not reductive, we have a good rectangle. If there are four, each involving a different γ or δ, we can build a hexagon. Indeed, because three of the four directions cannot form an isolated component of the star graph, no size-three ideal edge containing $\eta(g)$ is reductive. For example if $\alpha \cup \gamma(1)$ and $\beta \cup \gamma(g)$ are not reductive, one good hexagon has corners

$$\alpha,\ \delta(1),\ \beta,\ \gamma(g),\ \eta(g) \text{ and } \gamma(1).$$

Now suppose no size-two ideal edge is reductive. Again we will distinguish between the case $|u| < |v|$ and the case of equality. In the former case, again we have that not both of $\alpha \cup \gamma(1)$ and $\alpha \cup \gamma(g)$ are reductive and similarly for $\beta \cup \gamma(1)$ and $\beta \cup \gamma(g)$. If only one of these four ideal edges is reductive, we have a good rectangle. If $\alpha \cup \gamma(1)$ and $\beta \cup \gamma(1)$ are both reductive, this would contradict Corollary 2.12 as well. Suppose, then, that $\alpha \cup \gamma(1)$ and $\beta \cup \gamma(g)$ are both reductive. Then $|v| > |u| \geq \{u, g^{-1}.u\} \cdot \{v, \bar{v}\} = |\{v, \bar{v}\}|$, so $\{v\} \cdot \{\bar{v}\} > \frac{1}{2}|v|$. From this it follows that $\delta(1) \cup \eta(1)$, $\delta(g) \cup \eta(1)$, $\delta(1) \cup \eta(g)$ and $\delta(g) \cup \eta(g)$ are all not reductive.

Observe that at most two of the size-three ideal edges $\alpha \cup \delta(1)$, $\beta \cup \delta(g)$ $\gamma(1)\eta(1)$ and $\gamma(g) \cup \eta(g)$ can be reductive by Lemma 2.11. The same holds true for $\alpha \cup \delta(g)$, $\beta \cup \delta(1)$, $\gamma(g) \cup \eta(1)$ and $\gamma(1) \cup \eta(g)$. If at most one of the four size-three ideal edges $\alpha \cup \delta(1)$, $\beta \cup \delta(1)$, $\alpha \cup \delta(g)$ and $\beta \cup \delta(g)$ is reductive, we have a good rectangle. If there are two, say $\alpha \cup \delta(1)$ and $\beta \cup \delta(1)$, we again have a good rectangle by including $\delta(g)$ and either γ ideal edge as corners. A similar statement holds if instead $\alpha \cup \delta(g)$ and $\beta \cup \delta(g)$ are reductive. If $\alpha \cup \delta(1)$ and $\beta \cup \delta(g)$ are reductive but $\alpha \cup \delta(g)$ and $\beta \cup \delta(1)$ are not, then $\gamma(g) \cup \eta(g)$ and $\gamma(1) \cup \eta(1)$ are not reductive and one good hexagon from α to β has corners

$$\alpha,\ \gamma(g),\ \beta,\ \gamma(1),\ \eta(1) \text{ and } \delta(g).$$

A similar statement holds if the two reductive ideal edges containing a δ ideal edge are $\alpha \cup \delta(g)$ and $\beta \cup \delta(1)$.

If three ideal edges containing a δ are reductive, say $\alpha \cup \delta(1)$, $\alpha \cup \delta(g)$ and $\beta \cup \delta(g)$, then $\{v, g.v\} \cdot \{u, \bar{u}\} = |\{u, \bar{u}\}|$, while $\{v\} \cdot \{g.\bar{u}, u\} = \{v\} \cdot \{u\} \geq \frac{1}{2}|\{g.\bar{u}, u\}| = |u|$. But this implies that $|\beta \cup \gamma(g)| = |u| + |\{g.v, g.\bar{v}\}|$, contradicting the assumption that this ideal edge is reductive.

Finally suppose that $|u| = |v|$. Then at most two ideal edges containing any given size-two ideal edge may be reductive by Corollary 2.12. Consider, for example, the γ and δ ideal edges. At most two of these together with α may be reductive and similarly for β. If for some γ or δ neither size-three ideal edge is reductive, we have a good rectangle from α to β, because for at least one of the remaining γ or δ ideal edges, at most one of the relevant size-three ideal edges is reductive.

So suppose that for all γ and δ, at least one of the size-three ideal edges containing it and either α or β is reductive. But then notice that at least one of the size-three ideal edges containing it and either $\eta(1)$ or $\eta(g)$ is not reductive. In fact, we have a good hexagon unless for each size-two ideal edge there are two others for which the corresponding size-three ideal edge is reductive and for, say, $\eta(1)$, either $\eta(1)$ and α have the same set of size-two ideal edges which yield (non)-reductive size-three ideal edges or $\eta(1)$ and β have this property (and therefore similarly for $\eta(g)$.)

Suppose for definiteness that $\alpha \cup \gamma(1)$ and $\alpha \cup \delta(g)$ are reductive, $\beta \cup \gamma(g)$ and $\beta \cup \delta(1)$ are reductive, $\eta(1) \cup \gamma(1)$ and $\eta(1) \cup \delta(g)$ are reductive, $\eta(g) \cup \gamma(g)$ and $\eta(g) \cup \delta(1)$ are reductive. (Notice that swapping for instance $\delta(1)$ and $\delta(g)$ above is not possible by Corollary 2.12.) We claim this is not possible. Indeed, by Lemma 2.11, we have that $\{u, g^{-1}.u\} \cdot \{v, \bar{v}\} = |\{v, \bar{v}\}|$, while $0 = \{\bar{u}\} \cdot \{\bar{v}, g.v\} \geq \frac{1}{2}|\{\bar{v}, g.v\}|$, which is impossible. A similar argument works in all cases; this completes the proof.

3. Semistability at infinity

The goal of this section is to prove Theorem C. Recall that we write B_k for the ball of radius k. We continue to let

$$N(k) = \max\{\|\tau\| : \tau \in C_k\},$$

where C_k denotes the set of reduced marked graphs of groups ("briar patches") whose stars have nonempty intersection with B_k. To prove semistability, we will show that when $L(F)$ is one ended, for a proper ray $\rho\colon \mathbb{R}_{\geq 0} \to L$, with $\rho(k)$ a briar patch such that the sequence $\|\rho(k)\| > N(k)$ is strictly increasing, the inverse sequence

$$\pi_1(L - B_k, \rho(k))$$

satisfies the Mittag-Leffler condition.

THEOREM 3.1. *Suppose $L(F)$ is one ended. If $n \geq N(k)$, then every loop in $L - B_{N(k)}$ based at $\rho(N(k))$ is homotopic to a loop in $L - B_n$ based at $\rho(n)$ by a homotopy that avoids B_k.*

When $L(F)$ is not one ended, we will use our good algebraic understanding of $\mathrm{Out}(F)$ to show that each end of $L(F)$ is semistable.

To prove Theorem 3.1, we need the following proposition.

PROPOSITION 3.2 (cf. Proposition 4.1 of [**22**]). *Every path in $L - B_k$ is homotopic to a standard path in $L - B_k$ by a homotopy in $L - B_k$.*

To prove the proposition, we need a lemma.

LEMMA 3.3. *Let $\mathcal{G}$ be a marked graph of groups representing a vertex of L with two maximal forests F and F'. For each edge $e' \in F' - F$, there exists an edge $e \in F - F'$ such that $F \cup \{e'\} - \{e\}$ is a maximal forest. Put another way, if $\mathcal{G}$ is not a briar patch, the collection of maximal forests in $\mathcal{G}$ forms a* matroid.

Recall that a *matroid* is simply a (nonempty) set, some of whose finite subsets are called *bases.* The bases of a matroid satisfy an *exchange condition* similar to the one in the statement of the lemma. Other examples include the set of spanning trees in a finite graph which is not a rose or the set of bases of a vector space of finite positive dimension.

PROOF. Let F, F' and e' be as in the statement. Since F was maximal, $F \cup \{e'\}$ fails to be a maximal forest, either because as a graph it has rank one or because it connects a pair of vertices a and b with nontrivial vertex group. Let C be either the minimal subgraph supporting the (ordinary) fundamental group of $F \cup \{e'\}$ in the former case or the unique geodesic from a to b in the latter. Now because F' is a maximal forest, C cannot be completely contained in F', so any edge of $F - F'$ contained in C will do as e. □

PROOF OF PROPOSITION 3.2. If γ is a marked graph of groups which lies outside of B_k, then every marked graph of groups it collapses onto is also outside of B_k by definition. Thus if we have a path $P = (\gamma_0, \gamma_1, \ldots, \gamma_\ell)$, by collapsing forests in alternating marked graphs of groups, we may assume that each γ_k with k even is a briar patch. If ℓ is odd, we may extend P by some collapse of γ_ℓ to a path of even length. If γ_{k-1} and γ_{k+1} are briar patches obtained in this way, they are obtained from $\gamma_k = (\mathcal{G}, \sigma)$ by collapsing maximal forests F and F' in $\mathcal{G}$.

By repeatedly applying Lemma 3.3, we see that we may interpolate from F to F' through maximal forests by adding an edge of F' and removing one of F at each step. The path segment $(\gamma_{k-1}, \gamma, \gamma_{k+1})$ is homotopic to the standard path segment

$$(\gamma_{k-1} = \rho_0, \delta_1, \rho_1, \ldots, \delta_s, \rho_s = \gamma_{k+1})$$

where ρ_i is obtained from γ_k by collapsing the ith maximal forest F_i in the interpolation and δ_i is obtained by collapsing $F_{i-1} \cap F_i$. □

With the proposition in hand, we are ready to prove the theorem.

PROOF OF THEOREM 3.1. Let P be a loop in $L - B_{N(k)}$ based at $\rho(N(k))$ and let $n \geq N(k)$. By concatenating at both ends with an arbitrary path in $L - B_{N(k)}$ from $\rho(N(k))$ to $\rho(n)$, we obtain a loop P' based at $\rho(n)$. (Such a path exists by Theorem 2.2.) By Proposition 3.2, P' is homotopic relative to its basepoint to a standard path by a homotopy in $L - B_{N(k)}$. By the proof of Theorem 2.2, since $\|\rho(n)\| > N(n)$, we may push P' outside of B_n by a homotopy. This homotopy strictly increases the norm of briar patches along it. Since P' began in $L - B_{N(k)}$ and $N(k)$ is chosen so that the star of any briar patch in $L - B_{N(k)}$ has empty intersection with the ball B_k, this homotopy avoids B_k. □

To complete the proof of Theorem C, we need only consider the cases where $\mathrm{Out}(F)$ has infinitely many ends. In the case where $\dim L = 1$, we see that $\mathrm{Out}(F)$ is virtually free. Since free groups are semistable at each end (consider the Mittag-Leffler condition in a tree) and by [**11**, Proposition 16.5.3] semistability passes to and is inherited from finite-index subgroups, we see that $\mathrm{Out}(F)$ is semistable at each end in this case.

Thus it remains to consider the case of $F = A_1 * A_2 * \mathbb{Z}$. Observe that A_1 and A_2 are a complete set of representatives of the conjugacy classes of maximal finite subgroups of F, so these conjugacy classes are permuted by $\mathrm{Out}(F)$. There is a normal subgroup of index at most two in $\mathrm{Out}(F)$ comprising the pointwise stabilizer of these conjugacy classes. Call this subgroup G.

If A is a group, the group $\mathrm{Hol}(A) = A \rtimes \mathrm{Aut}(A)$ always admits an automorphism θ (which is an involution when it is nontrivial) sending an element (a, ψ), with $a \in A$ and $\psi \in \mathrm{Aut}(A)$, to the element $(a^{-1}, \iota_{a^{-1}}\psi)$, where $\iota_{a^{-1}}$ is the inner automorphism $x \mapsto axa^{-1}$.

Fix a generator t of a $\mathbb{Z}$ free factor. The group G contains the finite group $(\mathrm{Hol}(A_1) \times \mathrm{Hol}(A_2)) \rtimes C_2$, where the cyclic group of order two acts by θ on each direct factor. Explicitly, the subgroup A_1 acts by left multiplication on t, the subgroup A_2 acts by right multiplication on t, the subgroup $\mathrm{Aut}(A_1)$ acts on the

A_1 free factor, $\mathrm{Aut}(A_2)$ on the A_2 free factor, and C_2 acts by the automorphism

$$\tilde{\theta} = \begin{cases} a_1 & \mapsto a_1 \\ a_2 & \mapsto ta_2t^{-1} \\ t & \mapsto t^{-1}. \end{cases}$$

One checks that the intersection of this group of *automorphisms* with the subgroup of inner automorphisms is trivial, so it injects into $\mathrm{Out}(F)$ and in fact into G.

We will also consider the infinite, one-ended subgroup of automorphisms generated by $\mathrm{Hol}(A_1) \times \mathrm{Hol}(A_2)$ and the automorphism τ which simply inverts t. This subgroup, which we will call H, also injects into G. To see that H is one-ended, observe that it has a map to the finite groups $\mathrm{Aut}(A_1 \times A_2)$ and C_2, defined by first using the universal property to induce an automorphism of $A_1 \times A_2 \times \mathbb{Z}$ and then looking alternately at the action on the torsion subgroup or the action on $\mathbb{Z}$. The intersection H_0 of the kernels, which has finite index in H, comprises those automorphisms which (one checks) act trivially on the A_1 and A_2 free factors and sends t to an element gth, where g and h are elements of $A_1 * A_2$, so H_0 is isomorphic to $(A_1 * A_2) \times (A_1 * A_2)$, a one-ended group which is virtually the direct product of two free groups.

PROPOSITION 3.4. *The group G is isomorphic to the amalgamated free product*

$$((\mathrm{Hol}(A_1) \times \mathrm{Hol}(A_2)) \rtimes C_2) *_{\mathrm{Hol}(A_1)\times\mathrm{Hol}(A_2)} H.$$

It follows from the proposition that G and hence $\mathrm{Out}(F)$ has a finite-index subgroup isomorphic to

$$(F_m \times F_m) * (F_m \times F_m),$$

where F_m is a finite-rank free group. This latter group is a right-angled Artin group, hence is semistable at each end by results of Mihalik [**19**].

PROOF. Firstly, we claim that the subgroup $\mathrm{Hol}(A_1) \times \mathrm{Hol}(A_2)$ together with τ and $\tilde{\theta}$ generate G. Indeed, it is not hard to see that $\mathrm{Aut}(F)$ is generated by these elements together with the family of automorphisms

$$\begin{cases} a_1 \in A_1 & \mapsto a_2a_1a_2^{-1} \\ a_2 \in A_2 & \mapsto a_2 \\ t & \mapsto t \end{cases} \quad \begin{cases} a_1 \in A_1 & \mapsto a_1 \\ a_2 \in A_2 & \mapsto a_1a_2a_1^{-1} \\ t & \mapsto t \end{cases} \quad \begin{cases} a_1 \in A_1 & \mapsto ta_1t^{-1} \\ a_2 \in A_2 & \mapsto a_2 \\ t & \mapsto t, \end{cases}$$

as well as possibly some automorphism permuting A_1 and A_2 in the case they are isomorphic. (See for instance [**12**] for a finite presentation of $\mathrm{Aut}(F)$.) Each of these first three differs from an automorphism in $\langle \mathrm{Hol}(A_1) \times \mathrm{Hol}(A_2), \tilde{\tau}, \tilde{\theta} \rangle$ by an inner automorphism, so the claim follows. To finish the proof, we will use a finite presentation for $\mathrm{Out}(F)$, essentially due to Fouxe-Rabinovitch [**10**], following notation of Marchand [**18**].

Let us fix notation. For ϕ_i an element of $\mathrm{Aut}(A_i)$, we abuse notation and think of ϕ_i as an element of G via its action on the fixed A_i free factor of F. The subgroup of G generated by the ϕ_i as ϕ_i varies in $\mathrm{Aut}(A_i)$ and as i varies over $\{1, 2\}$ is isomorphic to $\mathrm{Aut}(A_1) \times \mathrm{Aut}(A_2)$. Given $i \neq j$ and $\gamma_i \in A_i$, the *partial conjugation* of A_j by γ_i, is the automorphism $\alpha_{ij}^{(\gamma_i)}$ defined as sending $\gamma_j \in A_j$ to $\gamma_i^{-1}\gamma_j\gamma_i$ and fixing A_i and t. We similarly define α_{ti} where we conjugate by t instead of γ_i, so $\gamma_i \in A_i \mapsto t^{-1}\gamma_i t$. Finally we define the *right* and *left transvections* $\rho_i^{(\gamma_i)}$

and $\lambda_i^{(\gamma_i)}$ as acting on t by right (respectively left) multiplication by $\gamma_i \in A_i$. The group generated by the $\lambda_1^{(\gamma_1)}$, the $\rho_2^{(\gamma_2)}$ and the Φ_i is the subgroup $\mathrm{Hol}(A_1) \times \mathrm{Hol}(A_2)$ described above.

As a special case of [**18**, Theorem 5.2], we have that a defining system of relations for G with respect to the generating set

$$\{\mathrm{Aut}(A_1), \mathrm{Aut}(A_2), \alpha_{ij}^{(\gamma_i)}, \alpha_{ti}, \lambda_i^{(\gamma_i)}, \rho_i^{(\gamma_i)}, \tau\}$$

is given by the following. We follow the numbering in Marchand's paper, omitting relations that do not occur. Implicitly one includes relations specifying that, for instance, the subgroup generated by $\lambda_i^{(\gamma_i)}$ for fixed i but variable $\gamma_i \in A_i$ is isomorphic to A_i. (For the worried reader: think of automorphisms as acting on F on the left.)

(1) $\varphi_1\varphi_2 = \varphi_2\varphi_1$, so that $\langle \mathrm{Aut}(A_1), \mathrm{Aut}(A_2)\rangle \cong \mathrm{Aut}(A_1) \times \mathrm{Aut}(A_2)$.
(2) $\varphi_i\alpha_{tj} = \alpha_{tj}\varphi_i$ for $\varphi_i \in \mathrm{Aut}(A_i)$ and $i \neq j$.
(3) $\varphi_i\alpha_{ij}^{(\gamma_i)} = \alpha_{ij}^{(\varphi_i\gamma_i)}\varphi_i$ for $\varphi_i \in \mathrm{Aut}(A_i)$ and $\gamma_i \in A_i$.
(4) $\alpha_{t1}\alpha_{t2} = \alpha_{t2}\alpha_{t1}$.
(5) $\alpha_{ti}\alpha_{ij}^{(\gamma_i)}\alpha_{ti}^{-1} = \alpha_{tj}^{-1}\alpha_{ij}^{(\gamma_i)}\alpha_{tj}$ for $\gamma_i \in A_i$.
(12) $\lambda_i^{(\gamma_i)}\rho_j^{(\gamma_j)} = \rho_j^{(\gamma_j)}\lambda_i^{(\gamma_i)}$ for $\gamma_i \in A_i$ and $\gamma_j \in A_j$, with i and j not necessarily distinct. Also $\tau\lambda_i^{(\gamma_i)} = \rho_i^{(\gamma_i^{-1})}\tau$.
(13) $\alpha_{ij}^{(\gamma_i)}\tau = \tau\alpha_{ij}^{(\gamma_i)}$.
(14) $\alpha_{ti}\tau = \tau\alpha_{ti}^{-1}$.
(15) $\rho_i^{(\gamma_i)}\varphi_j = \varphi_j\rho_i^{(\gamma_i)}$ for $\varphi_j \in \mathrm{Aut}(A_j)$ and $i \neq j$.
(16) $\rho_i^{(\varphi_i\gamma_i)}\varphi_i = \varphi_i\rho_i^{(\gamma_i)}$ for $\gamma_i \in A_i$, $\varphi_i \in \mathrm{Aut}(A_i)$. This, together with some of the relations above implies that the subgroup generated by the $\lambda_1^{(\gamma_1)}$, $\rho_2^{(\gamma_2)}$ and the φ_i is isomorphic to $\mathrm{Hol}(A_1) \times \mathrm{Hol}(A_2)$.
(19) $\rho_i^{(\gamma_i)}\alpha_{ij}^{(\gamma_i)} = \alpha_{ij}^{(\gamma_i)}\rho_i^{(\gamma_i)}$.
(20) $\rho_i^{(\gamma_i)}\alpha_{tj} = \alpha_{tj}\rho_i^{(\gamma_i)}\alpha_{ij}^{(\gamma_i)}$.
(21) $\left(\rho_i^{(\gamma_i)}\right)^{-1}\rho_j^{(\gamma_j)}\rho_i^{(\gamma_i)}\alpha_{ij}^{(\gamma_i)} = \alpha_{ij}^{(\gamma_i)}\rho_j^{(\gamma_j)}$.
(22) $\alpha_{ti}\rho_i^{(\gamma_i)} = \lambda_i^{(\gamma_i)}\alpha_{ti}\varphi_i$, where ϕ_i is the automorphism of A_i defined by $x \mapsto \gamma_i^{-1}x\gamma_i$.
(23) $\tau\varphi_i = \varphi_i\tau$ for $\varphi_i \in \mathrm{Aut}(A_i)$.
(0) $\varphi_i\alpha_{ij}^{(\gamma_i)}\rho_i^{(\gamma_i)}\left(\lambda_i^{(\gamma_i)}\right)^{-1} = 1$, where φ_i is the automorphism of A_i defined by $x \mapsto \gamma_i^{-1}x\gamma_i$. Also $\alpha_{t1}\alpha_{t2} = 1$.

We introduce a new generator $\tilde{\theta}$ and the relation $\tilde{\theta} = \tau\alpha_{t2}$. As remarked already, the relations in items 0 and 12 allow us to discard generators down to the more minimal

$$\{\mathrm{Hol}(A_1) \times \mathrm{Hol}(A_2), \tau, \tilde{\theta}\}.$$

We have already seen that $\langle \mathrm{Hol}(A_1) \times \mathrm{Hol}(A_2), \tilde{\theta}\rangle$ is isomorphic to the semidirect product $(\mathrm{Hol}(A_1) \times \mathrm{Hol}(A_2)) \rtimes C_2$. Easy Tietze transformations yield a defining system of relations for G for the generating set

$$\{\mathrm{Hol}(A_1) \times \mathrm{Hol}(A_2), \tilde{\theta}, \tau\}.$$

Relevant to our purposes, we just need to check that every relation involving $\tilde{\theta}$ takes place in $(\mathrm{Hol}(A_1) \times \mathrm{Hol}(A_2)) \rtimes C_2$ and every relation involving τ takes place in H. Thus we need only examine relations 2, 5, 14, 20, and 22.

(2) The relation $\varphi_1\alpha_{t2} = \alpha_{t2}\varphi_1$ becomes $\varphi_1\tau\tilde{\theta} = \tau\tilde{\theta}\varphi_1$. Because $\mathrm{Hol}(A_1) \times \mathrm{Hol}(A_2)$ is normal in $(\mathrm{Hol}(A_1) \times \mathrm{Hol}(A_2)) \rtimes C_2$, right-multiplying by $\tilde{\theta}^{-1}$ and then using a relation discussed before the proof shows that this relation involves only elements of H. A similar argument works for $\varphi_2\alpha_{t1} = \alpha_{t1}\varphi_2$, which becomes $\varphi_2\tilde{\theta}\tau = \tilde{\theta}\tau\varphi_2$.

(5) This relation, observing that $\alpha_{t2} = \alpha_{t1}^{-1}$, is spurious.

(14) This relation is also spurious: it says either that $\alpha_{t2}\tau\alpha_{t2} = \tau\tilde{\theta}\tau\tau\tilde{\theta} = \tau$, which follows because τ and $\tilde{\theta}$ have order two, or that $\alpha_{t1}\tau\alpha_{t1} = \alpha_{t2}^{-1}\tau\alpha_{t2}^{-1} = \tau$, which follows from the previous relation.

(20) Because $\mathrm{Hol}(A_1) \times \mathrm{Hol}(A_2)$ is normal in the semidirect product, the case of this relation of the form $\rho_2^{(\gamma_2)}\alpha_{t1} = \alpha_{t1}\rho_2^{(\gamma_2)}\alpha_{21}^{(\gamma_2)}$ may be written with only elements of H by writing $\alpha_{t1} = \tilde{\theta}\tau$ and left multiplying by $\tilde{\theta}^{-1}$. The same is true for the relation with the roles of 1 and 2 swapped, where instead we left multiply by $\tilde{\theta}^{-1}\tau^{-1}$ and observe that after conjugating $\rho_1^{(\gamma_1)}$ by τ we obtain an element of $\mathrm{Hol}(A_1) \times \mathrm{Hol}(A_2)$.

(22) The same method of proof as in the previous case works: for example the relation $\alpha_{t1}\rho_1^{(\gamma_1)} = \lambda_1^{(\gamma_1)}\alpha_{t1}\varphi_1$ may be rewritten as $\tau\rho_1^{(\gamma_1)} = \tilde{\theta}^{-1}\lambda_1^{(\gamma_1)}\tilde{\theta}\tau\varphi_1$, which takes place in H.

Thus we have demonstrated that G splits as the amalgamated free product in the statement. □

Appendix: a direct proof of Theorem B

Suppose $F = A_1 * A_2 * \cdots * A_n * F_k$. In view of results of Das [**9**] and Vogtmann [**22**], one-endedness of $\mathrm{Out}(F)$ is only new when both n and k are positive. To illustrate a more elementary attack on the proof of Theorem B, we will give a sketch of an argument supposing that $n \geq 3$ and $k \geq 3$. In view of our proof of Theorem B, it is clear that this argument is not optimal; we leave its improvement as a final question.

As we saw in the previous section, there exists a finite-index subgroup G of $\mathrm{Out}(F)$ comprising those outer automorphisms preserving the conjugacy class of each A_i.

Fix a free basis $\{x_{n+1}, \ldots, x_{n+k}\}$ for F_k. Following notation from the previous section, we have for each nontrivial $\gamma_i \in A_i$, each j satisfying $n+1 \leq j \leq n+k$ and each ℓ satisfying $1 \leq k \leq n+k$, *partial conjugations* $\alpha_{i\ell}^{(\gamma_i)}$ and $\alpha_{j\ell}$ and *right and left transvections,* which when $\ell \in \{n+1, \ldots, n+k\}$ we write as $\rho_{\ell j}$ and $\lambda_{\ell j}$, while when $i \in \{1, \ldots, n\}$, we write as $\rho_{ij}^{(\gamma_i)}$ and $\lambda_{ij}^{(\gamma_i)}$, as they depend on a choice of nontrivial $\gamma_i \in A_i$.

The group G is generated by, one checks, these partial conjugations, transvections, the groups $\mathrm{Aut}(A_i)$ acting on A_i alone, permutations of the x_ℓ for $\ell \in \{n+1, \ldots, n+k\}$ and the *inversions* τ_ℓ which invert x_ℓ and fix the other factors.

Here is the argument we will use; it is due to the anonymous referee: supposing that $\mathrm{Out}(F)$ and hence G had more than one end, it would act on a tree T with finite edge stabilizers. Every one-ended subgroup of G stabilizes a unique vertex of

T. In particular, every subgroup isomorphic to the direct product of two infinite groups stabilizes a unique vertex of T. If $H \leq G$ is an elliptic subgroup in an action of G on a tree T and $g \in G$ normalizes H, observe that g preserves the fixed-point set of H. In particular, if H stabilizes a unique vertex of T, so does its normalizer. If H and H' are elliptic subgroups in the action of G on T, their intersection $H \cap H'$ is also elliptic. If edge stabilizers in T are finite, then if $H \cap H'$ is infinite, we conclude that H and H' fix the same point.

We find a family $\Gamma_1, \ldots, \Gamma_m$ of subgroups of G, each isomorphic to the direct product of two infinite groups, such that $\Gamma_i \cap \Gamma_{i+1}$ is infinite for each i, and such that each element of the above generating set is contained in or normalizes some Γ_i. It follows from the argument above that in any action of G on a tree T with finite edge stabilizers, each Γ_i and their normalizers stabilize a unique vertex of T and moreover these vertices are a single vertex. It follows that G acts with global fixed point on T, from which we conclude that G is one ended.

For each two-element subset of $\{1, \ldots, n\}$, say $\{i, j\}$, it is not hard to see that the partial conjugations $\alpha_{i\ell}^{(\gamma_i)}$ and $\alpha_{j\ell}^{(\gamma_j)}$ as γ_i and γ_j vary in A_i and A_j respectively and ℓ varies in $\{1, \ldots, n\} - \{i, j\}$ together with the left and right transvections $\rho_{i\ell}^{(\gamma_i)}$, $\rho_{j\ell}^{(\gamma_j)}$, $\lambda_{i\ell}^{(\gamma_i)}$ and $\lambda_{j\ell}^{(\gamma_j)}$ as ℓ varies in $\{n+1, \ldots, n+k\}$ generate a group H_{ij} isomorphic to the direct product of $2k + n - 2$ copies of the infinite group $A_i * A_j$. If we fix ℓ satisfying $n + 1 \leq \ell \leq n + k$, the group H_ℓ generated by left and right transvections on the other elements of our fixed free basis with "acting letter" x_ℓ is isomorphic to a free abelian group of rank $2k - 2$. By mixing and matching products of direct factors of the H_{ij} and H_ℓ as i, j and ℓ vary, we obtain a family of subgroups $\Gamma_1, \ldots, \Gamma_m$, each isomorphic to the direct product of two infinite groups, satisfying that $\Gamma_a \cap \Gamma_{a+1}$ is infinite for each a satisfying $1 \leq a \leq m-1$, and such that each partial conjugation or transvection contained in some H_{ij} or H_ℓ is contained in some Γ_a.

It is clear from our construction that because $n \geq 3$ and $k \geq 2$, each partial conjugation or transvection in the generating set for G sketched above normalizes (in fact centralizes) some Γ_a. In fact, the elements of $\mathrm{Aut}(A_i)$ and the inversions also clearly normalize some Γ_a. Since we assume that $k \geq 3$, the same is also true of any transposition of two elements of $\{x_{n+1}, \ldots, x_{n+k}\}$. It follows that the subgroups Γ_a allow us to conclude that G and hence $\mathrm{Out}(F)$ is one ended.

Let us remark that if we knew that each Γ_a was furthermore *undistorted* in G, we could conclude a stronger statement, namely that G and hence $\mathrm{Out}(F)$ is *thick of order one* in the sense of Behrstock, Druţu and Mosher [**2**]. When $k = 0$ and $n \geq 4$, this is the main result of Das's paper [**9**], and for $k \geq 3$ and $n = 0$, this was proven in [**2**].

QUESTION F. Let $F = A_1 * \cdots * A_n * F_k$, and suppose that $\mathrm{Out}(F)$ is one ended. Is it also thick of order one?

Acknowledgments

The author is pleased to thank Kim Ruane for suggesting that Theorem C might be in reach given Theorem B, the anonymous referee for the argument sketched in the appendix, and Mike Mihalik for helpful comments on a preliminary version. This material is based on work supported by the National Science Foundation under Award No. DMS-2202942.

References

[1] H. Bass, *Covering theory for graphs of groups*, J. Pure Appl. Algebra **89** (1993), no. 1-2, 3–47, DOI 10.1016/0022-4049(93)90085-8. MR1239551

[2] J. Behrstock, C. Druţu, and L. Mosher, *Thick metric spaces, relative hyperbolicity, and quasi-isometric rigidity*, Math. Ann. **344** (2009), no. 3, 543–595, DOI 10.1007/s00208-008-0317-1. MR2501302

[3] M. Bestvina and M. Feighn, *The topology at infinity of* $\mathrm{Out}(F_n)$, Invent. Math. **140** (2000), no. 3, 651–692, DOI 10.1007/s002220000068. MR1760754

[4] R. Bieri and B. Eckmann, *Groups with homological duality generalizing Poincaré duality*, Invent. Math. **20** (1973), 103–124, DOI 10.1007/BF01404060. MR340449

[5] N. Brady and J. Meier, *Connectivity at infinity for right angled Artin groups*, Trans. Amer. Math. Soc. **353** (2001), no. 1, 117–132, DOI 10.1090/S0002-9947-00-02506-X. MR1675166

[6] M. Clay, *Deformation spaces of G-trees and automorphisms of Baumslag-Solitar groups*, Groups Geom. Dyn. **3** (2009), no. 1, 39–69, DOI 10.4171/GGD/51. MR2466020

[7] D. J. Collins and H. Zieschang, *Rescuing the Whitehead method for free products. I. Peak reduction*, Math. Z. **185** (1984), no. 4, 487–504, DOI 10.1007/BF01236258. MR733769

[8] M. Culler and K. Vogtmann, *Moduli of graphs and automorphisms of free groups*, Invent. Math. **84** (1986), no. 1, 91–119, DOI 10.1007/BF01388734. MR830040

[9] Saikat Das, *Thickness of* $\mathsf{Out}(A_1 * \cdots * A_n)$, Available at `arXiv:1811.00435` [Math.GR], Nov 2018.

[10] D. I. Fouxe-Rabinovitch, *Über die Automorphismengruppen der freien Produkte. II* (Russian, with German summary), Rec. Math. [Mat. Sbornik] N.S. **9(51)** (1941), 183–220. MR4625

[11] R. Geoghegan, *Topological methods in group theory*, Graduate Texts in Mathematics, vol. 243, Springer, New York, 2008, DOI 10.1007/978-0-387-74614-2. MR2365352

[12] N. D. Gilbert, *Presentations of the automorphism group of a free product*, Proc. London Math. Soc. (3) **54** (1987), no. 1, 115–140, DOI 10.1112/plms/s3-54.1.115. MR872253

[13] V. Guirardel and G. Levitt, *Deformation spaces of trees*, Groups Geom. Dyn. **1** (2007), no. 2, 135–181, DOI 10.4171/GGD/8. MR2319455

[14] S. Krstić and K. Vogtmann, *Equivariant outer space and automorphisms of free-by-finite groups*, Comment. Math. Helv. **68** (1993), no. 2, 216–262, DOI 10.1007/BF02565817. MR1214230

[15] Rylee Alanza Lyman, *A family of CAT(0) outer automorphism groups of free products*, `arXiv:2209.04711` [math.GR], September 2022.

[16] Rylee Alanza Lyman, *Lipschitz metric isometries between Outer Spaces of virtually free groups*, `arXiv:2203.09008` [math.GR], March 2022.

[17] Rylee Alanza Lyman, *Train track maps on graphs of groups*, Available at `arXiv:2102.02848` [math.GR], March 2022.

[18] A. Marchand, *Free representations of outer automorphism groups of free products via characteristic abelian coverings*, J. Group Theory **26** (2023), no. 2, 399–420, DOI 10.1515/jgth-2021-0154. MR4554492

[19] M. L. Mihalik, *Semistability of Artin and Coxeter groups*, J. Pure Appl. Algebra **111** (1996), no. 1-3, 205–211, DOI 10.1016/0022-4049(95)00117-4. MR1394352

[20] P. Scott and T. Wall, *Topological methods in group theory*, Homological group theory (Proc. Sympos., Durham, 1977), London Math. Soc. Lecture Note Ser., vol. 36, Cambridge Univ. Press, Cambridge-New York, 1979, pp. 137–203. MR564422

[21] J.-P. Serre, *Trees*, Springer Monographs in Mathematics, Springer-Verlag, Berlin, 2003. Translated from the French original by John Stillwell; Corrected 2nd printing of the 1980 English translation. MR1954121

[22] K. Vogtmann, *End invariants of the group of outer automorphisms of a free group*, Topology **34** (1995), no. 3, 533–545, DOI 10.1016/0040-9383(94)00042-J. MR1341807

Smith Hall Room 216, 101 Warren St, Newark, New Jersey 07103
Email address: `rylee.lyman@rutgers.edu`

Contemporary Mathematics
Volume **812**, 2025
https://doi.org/10.1090/conm/812/16274

A brief survey of $\mathcal{Z}$-boundaries

Molly A. Moran

ABSTRACT. M. Bestvina defined a $\mathcal{Z}$-boundary on a group G to generalize the theory of boundaries of CAT(0) and hyperbolic groups. This article provides a brief introduction to the theory of $\mathcal{Z}$-boundaries and $\mathcal{EZ}$-boundaries, strategies employed to prove a group admits a $(\mathcal{E})\mathcal{Z}$-boundary, and applications of these group boundaries.

1. Introduction

Given a CAT(0) space X, the visual boundary, $\partial_\infty X$, is defined to be the set of equivalence classes of geodesic rays in X, where two rays are equivalent if they fellow travel. We may think of this boundary as the set of points at infinity for X. If a group G acts geometrically on this CAT(0) space X, we call $\partial_\infty X$ a boundary for G. There are two properties that we wish to observe about the compactification $X \cup \partial_\infty X$. First, we may instantly homotope all points of the visual boundary into the interior of the CAT(0) space by pushing these points in along the geodesic rays. Hence, X and $X \cup \partial_\infty X$ are homotopy equivalent. Secondly, G-translates of compact sets become small near the boundary in $X \cup \partial_\infty X$. CAT(0) groups and their boundaries have been widely studied in part because topological properties of the boundary lead to insights about properties of the group itself.

The significance of these objects motivated Bestvina in [**1**] to extract the relevant properties into one generalized definition of a group boundary, known as $\mathcal{Z}$-boundaries. Roughly speaking, a $\mathcal{Z}$-boundary for a group G is a compactification of a space with the two properties mentioned above, along with a geometric G-action on the space. Shortly after, Farrell and Lafont [**6**] introduced $\mathcal{EZ}$-structures by also requiring that the group action extend to the boundary. The purpose of this survey article is to provide an introduction to $(\mathcal{E})\mathcal{Z}$-boundaries, some examples, and a few applications in hopes of showcasing the value of these boundaries to a wider circle of interest.

2. Background

Since Bestvina introduced the concept of a $\mathcal{Z}$-boundary in [**1**], there have been a number of equivalent versions presented and, in some cases, slightly generalized versions that allow for the study of additional classes of groups. In order to simplify

2020 *Mathematics Subject Classification.* Primary 20F65, 54D35; Secondary 20J06, 20F67.

Written in honor of Mike Mihalik's 70th birthday. Many thanks to Craig Guilbault for his comments and suggestions in the preparation of this manuscript.

and streamline the definition, without changing it, we will use the notion of a model geometry and a model $\mathcal{Z}$-geometry introduced in [**14**].

DEFINITION 2.1. A *model geometry* is a proper, cocompact, metric absolute retract (X,d).

A metric space (X,d) is *proper* if every closed metric ball is compact and it is *cocompact* if there exists a closed metric ball whose translates by elements from the isometry group of X cover X. An *absolute retract*, denoted AR, is a contractible *absolute neighborhood retract*, or ANR. ANR theory provides a common framework to study the types of spaces that are frequently encountered in geometric topology and geometric group theory, such as simplicial complexes, CW complexes, manifolds, and CAT(0) spaces. For the purposes of this survey article, the exact definition and properties of ANRs are not important: when encountered, replace ANR with your favorite example from the list above. For readers interested in a more thorough treatment of ANRs and their place in group boundaries, see [**13**].

We would like to find a compactification of a model geometry (X,d) that, among other things, does not change the homotopy type of the space. One way to achieve this type of compactification is by adding a $\mathcal{Z}$-set to X, known as a *$\mathcal{Z}$-compactification*. A closed subset A of a space Y, is a *$\mathcal{Z}$-set* if there exists a homotopy $H : Y \times [0,1] \to Y$ such that $H_0 = \mathrm{id}_Y$ and $H_t(Y) \subset Y - A$ for every $t > 0$.

A simple example of a Z-set is the boundary of a manifold. The collared neighborhood of the boundary makes it easy to define a homotopy that pushes the boundary into the interior of the manifold along the collar lines. The visual boundary of a CAT(0) space described in Section 1 is also a nice example of a $\mathcal{Z}$-set as the boundary points can be "reeled" in along the geodesic rays used to define the boundary.

The other property of visual boundaries of CAT(0) groups that was highlighted in Section 1 was the size of sets near the boundary when measured in the compactification $X \cup \partial_\infty X$. We add this requirement that sets near the boundary are small in the compactification, known as the *nullity condition*, to obtain a more controlled $\mathcal{Z}$-compactification.

DEFINITION 2.2. A *model $\mathcal{Z}$-geometry* $(\overline{X}, Z, d)$ is a compactification $\overline{X} = X \cup Z$ of a model geometry (X,d) such that

- Z is a $\mathcal{Z}$-set in $\overline{X}$ and
- (nullity condition) for every $R > 0$ and open cover $\mathcal{U}$ of $\overline{X}$, there is a compact set $C \subset X$ so that if $A \subseteq X - C$ and $\mathrm{diam}_d(A) < R$ then $A \subset U$ for some $U \in \mathcal{U}$.

We refer to Z as a *$\mathcal{Z}$-boundary*.

Up to this point, we have been able to define these concepts in the absence of a group. Some results about $\mathcal{Z}$-boundaries can be obtained in the absence of a group action, such as the finite dimensionality of $\mathcal{Z}$-boundaries (see [**12**]). However, the focus for the remainder of the article will be on *group* boundaries. The following definition for $\mathcal{Z}$-boundaries for groups first appeared in [**1**] with the additional requirements that the action be torsion-free and the model geometry be finite-dimensional. These conditions were removed in [**5**]. The equivalent reformulation presented here using model geometries can be found in [**14**].

DEFINITION 2.3. A *$\mathcal{Z}$-structure* on a group G consists of a model $\mathcal{Z}$-geometry $(\overline{X}, Z, d)$ and an action of G on (X, d) that is proper, cocompact, and by isometries (known as a *geometric group action*). In this case, we say Z is a *$\mathcal{Z}$-boundary for G.*

Farrell and Lafont [**6**] introduced an additional condition that the action of G on X also extends to the boundary Z. In this case, we say G admits an *$\mathcal{EZ}$-structure.*

In [**1**], Bestvina asks if all type F groups admit $\mathcal{Z}$-structures. A group has type F if G has a finite $K(G,1)$ complex. Using a deep ANR result from West [**19**], if there is a proper, free, cocompact action of a group on an AR, then the group must have type F (see [**11**] for the full argument). Hence, type F is the appropriate setting for Bestvina's definition of $\mathcal{Z}$-boundaries. In the more generalized setting where actions need not be free, we can instead ask if every group that acts properly and cocompactly on an AR, known as type F^*_{AR}, admits a $\mathcal{Z}$-structure. Alternatively, we could also ask if every group that admits a proper, cocompact, rigid cellular action, known as type F^*, admits a $\mathcal{Z}$-structure. The same questions could also be asked about $\mathcal{EZ}$-structures. All of these questions (including Bestvina's original version) remain open.

In the next section, we provide some known examples and highlight strategies employed to find $(\mathcal{E})\mathcal{Z}$-structures. In Section 4, we give some applications of $(\mathcal{E})\mathcal{Z}$-structures and conclude with a few open questions in Section 5.

3. Examples and Techniques

The first task when showing a group G admits a $\mathcal{Z}$-structure is to identify the "right" model geometry (X, d) that admits a geometric G-action. For example, to show that torsion-free hyperbolic groups G admit $\mathcal{Z}$-structures, in [**2**], Bestvina and Mess need to use a Rips complex $P_d(G)$ with large enough d to ensure $P_d(G)$ is contractible. They then show that the addition of the Gromov boundary to $P_d(G)$ gives a model $\mathcal{Z}$-geometry. Similarly, in [**4**], Dahmani first constructs relative Rips complexes for torsion-free relatively hyperbolic groups G with the bounded coset penetration property and then describes when they are contractible. With this space, Dahmani can define a boundary, topologize the space, and show it satisfies the requirements of a model $\mathcal{Z}$-geometry.

In many cases, the challenge comes after finding the model geometry (X, d). Finding the right boundary could be the obstacle. For example, in [**16**], Osajda and Prztycki have a clear complex to work with for systolic groups: the systolic complex. The challenge arises in defining the boundary. They construct a system of "good" Euclidean geodesics and show this set satisfies the requirements to be a $\mathcal{Z}$-boundary. In some situations, the space and boundary seem clear, but the nullity condition is not satisfied, as in the case of [**18**] for direct products of groups that admit $\mathcal{Z}$-structures. In this case, Tirel defines the right topology on the compactification to ensure that the spherical join of the two boundaries satisfies the nullity condition.

Below, we detail two techniques that have also been successful in showing a class of groups has a $\mathcal{Z}$-boundary in the presence of a good candidate for the model geometry.

3.1. Compression. The Baumslag-Solitar group $BS(m.n)$ has a presentation of the form

$$BS(m,n) = \langle s, t | ts^m t^{-1} = s^n \rangle.$$

In the case that $m \neq n$, this group is neither CAT(0) nor hyperbolic. The universal cover of the presentation complex is homeomorphic to $\mathbb{R} \times T$ where T is the Bass-Serre tree with m incoming edges and n outgoing edges at each vertex. Since $\mathbb{R} \times T$ is a CAT(0) space, compactifying by adding the suspension of the Cantor set to $\mathbb{R} \times T$ gives a nice $\mathcal{Z}$-compactification for this class of groups. However, the compactification is not nice enough to be a $\mathcal{Z}$-structure: certain translates of a fundamental domain, when viewed from a fixed basepoint, grow exponentially, thereby failing the nullity condition. See, for example, Figure 1 below.

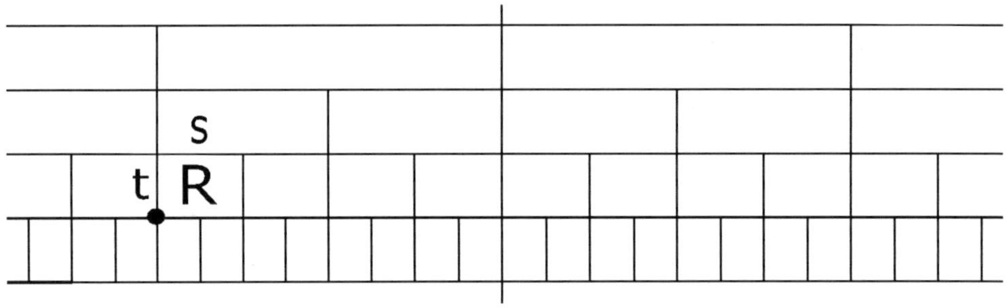

FIGURE 1. A single sheet in BS(1, 2). The vertical translates of the rectangle R grow exponentially

One solution to obtain a $\mathcal{Z}$-structure from this $\mathcal{Z}$-compactification is detailed in [**15**]. Since the uncontrolled growth occurs in the $\mathbb{R}$ direction of the model geometry, the authors conjugate the natural action by a homeomorphism that compresses the $\mathbb{R}$ direction via a log-log function. The modified action on $\mathbb{R} \times T$ ensures that the translates near the boundary are compressed enough to satisfy the nullity condition.

The suspension of the action on $\mathbb{R} \times T$ extends to the boundary, and hence,

THEOREM 3.1 ([**15**]). *All Baumslag-Solitar groups, $BS(m, n)$, admit $\mathcal{EZ}$-structures.*

This trick is generalized in [**10**] to products of CAT(0) spaces where one factor is compressible:

DEFINITION 3.2. A proper metric space Y is *compressible* if for any proper function $\psi : \mathbb{R}^+ \to \mathbb{R}^+$ there is a homeomorphism $h_\psi : Y \to Y$ such that for every compact set $K \subset Y$ with $\text{diam}(K) < \psi(R)$, we have $\text{diam}(h_\psi(K)) < \phi(R)$ where $\phi : \mathbb{R}^+ \to \mathbb{R}^+$ satisfies $\lim_{x\to\infty} \frac{\phi(x)}{x} = 0$.

In [**10**], it is shown that simply connected nonpositively curved Riemannian manifolds are compressible. As in the case of $BS(m, n)$, the fundamental group of a graph of closed nonpositively curved Riemannian n-manifolds has a nice model $\mathcal{Z}$-structure: it is homeomorphic to $T \times \tilde{M}_v$ where $\tilde{M}_v$ is a simply connected nonpositively curved Riemannian n-manifold (same n) and T is the Bass-Serre tree. Conjugating the action to compress the compressible factor, similar to [**15**] gives the following:

THEOREM 3.3 ([**10**]). *Suppose G is the fundamental group of a finite graph of groups where each vertex group is the fundamental group of a closed non-positively curved Riemannian manifold and each edge group is finite index in the corresponding vertex groups. Then G admits a $\mathcal{Z}$-structure.*

Additional technical conditions are required to upgrade the $\mathcal{Z}$-structure to an $\mathcal{EZ}$-structure.

3.2. Boundary Swapping. Now, suppose we know a group G is quasi-isometric to some other group H that admits a $\mathcal{Z}$-structure. Since these groups are quasi-isometric, they are essentially the same from a large-scale perspective. Hence, it seems reasonable that we could just swap the known boundary for H onto G. This technique does work if we start with a model geometry for G.

PROPOSITION 3.4 ([**1**, **13**]). *Suppose G and H are quasi-isometric groups that act geometrically on model geometries (X, d) and (Y, d'), respectively, and Y can be compactified to a $\mathcal{Z}$-structure $(\overline{Y}, Z, d')$ for H. Then X can be compactified by addition of the same boundary to obtain a $\mathcal{Z}$-structure $(\overline{X}, Z, d)$ for G.*

EXAMPLE 3.5. A *generalized Baumslag-Solitar group* is the fundamental group G of a finite graph of groups with all vertex and edge groups $\mathbb{Z}$. These groups also act properly and cocompactly on $\mathbb{R} \times T$ where T is the Bass-Serre tree. In [**20**], Whyte proves that these groups fall into exactly one of the following categories: virtually $\mathbb{Z} \times F_n$, $BS(1, n)$ for some $n > 1$, or quasi-isometric to $BS(2, 3)$. Each of these groups admits a $\mathcal{Z}$-structure by either [**1**] or [**15**]. An application of Proposition 3.4 proves that G also admits a $\mathcal{Z}$-structure.

EXAMPLE 3.6. In [**9**] the authors prove that if G admits a $\mathcal{Z}$-structure $(\overline{X}, Z, d)$, then so does the semi-direct product $G \rtimes \mathbb{Z}$. The semi-direct product acts properly and cocompactly on a mapping telescope $Y = \mathrm{Tel}_f(X)$. This mapping telescope can be shown to be proper homotopy equivalent to the direct product $X \times \mathbb{R}$ and hence the product can be $\mathcal{Z}$-compactified by [**18**]. Unfortunately, this proper homotopy equivalence cannot be upgraded to a quasi-isometry to satisy the conditions of Proposition 3.4. However, a more delicate version of boundary swapping that adds additional control to the size of translates is provided in [**9**]. This version ensures the boundary can be swapped thereby obtaining a $\mathcal{Z}$-structure $(\overline{Y}, S(Z))$ where $S(Z)$ is the suspension of the boundary for G.

From these examples, one may be tempted to conclude that the property of admitting a $\mathcal{Z}$-structure is a quasi-isometry invariant. However, to get started with the technique, we need to know that the group in question has a nice action on an AR. Unfortunately, we cannot guarantee such an action exists, even in the simplest of cases: if H acts geometrically on an AR and H is a finite index subgroup of G, it is unknown if that H-action can be extended to an action by the group G. In other words, it remains an open question if a virtually F^*_{AR} group is F^*_{AR}, or more generally, if type F^*_{AR} is a quasi-isometry invariant.

In [**14**], definitions of coarse (quasi)-$(\mathcal{E})\mathcal{Z}$-structures are provided to avoid these open questions. These generalized versions of Definition 2.3 weaken the geometric G-action to a coarse(quasi)-G-action. In this new setting, the property of admitting a boundary is in fact a quasi-isometry invariant and these coarse(quasi)-$\mathcal{Z}$-boundaries also carry much of the same information as $\mathcal{Z}$-boundaries that we discuss now.

4. Applications of $\mathcal{Z}$-boundaries

It is important to note that a group can admit different non-homeomorphic $\mathcal{Z}$-boundaries. This fact for visual boundaries of CAT(0) groups was first shown in the famous example due to Croke and Kleiner [**3**]. Other simple examples obtained by modifying the visual boundary of Euclidean or hyperbolic n-space are discussed in [**14**, Example 3.7]. However, the boundary of a group is well-defined up to shape equivalence ([**1, 7, 13**]). Additionally, the shape of the boundary of a group is a quasi-isometry invariant.

THEOREM 4.1 ([**13**, Theorem 1.5]). *If quasi-isometric groups G and H admit $\mathcal{Z}$-structures $(\overline{X}, Z_1)$ and $(\overline{Y}, Z_2)$, respectively, then Z_1 and Z_2 are shape equivalent.*

The next set of applications is motivated by results from [**1, 2**] relating properties of the group to topological properties of the $\mathcal{Z}$-boundary. The first result, with versions for groups with and without torsion, relates the cohomological dimension of the group to the topological dimension of its boundary, which is guaranteed to be finite by [**12**].

THEOREM 4.2 ([**1, 2, 5, 14**]). *If a torsion-free group G admits a $\mathcal{Z}$-structure based on a model $\mathcal{Z}$-geometry $(\overline{X}, Z, d)$ then* $\dim Z = \operatorname{cd} G - 1$. *If G has torsion, then* $\dim Z = \operatorname{gcd}_{\mathbb{Z}} Z$.

There is also a beautiful connection between the Cech cohomology of the boundary and the group cohomology.

THEOREM 4.3 ([**1,2,8,14**]). *If group G admits a $\mathcal{Z}$-structure based on a model $\mathcal{Z}$-geometry $(\overline{X}, Z, d)$,then $H^{n+1}(G, RG) \cong \widetilde{\check{H}^n}(Z; R)$ for all integers n and coefficients R.*

One final application requires the existence of an $\mathcal{EZ}$-structure. In particular, for torsion-free groups with $\mathcal{EZ}$-structures, work done by [**6**] shows these groups satisfy the Novikov Conjecture. The Novikov conjecture for groups with torsion is a bit more complicated and has been studied by [**9, 17**]. It remains an open question if *all* groups G that admit an $\mathcal{EZ}$-structure satisfy the Novikov Conjecture.

5. Open Questions

There are many open questions in the study of $(\mathcal{E})\mathcal{Z}$-boundaries. We highlight a few of these questions in this final section.

In Section 2, we mentioned one of the biggest family of open questions related to $\mathcal{Z}$-boundaries:

QUESTION 1. Must every group of type F, F^*, or F^*_{AR} admit a $(\mathcal{E})\mathcal{Z}$-structure?

Many of the techniques described in Section 3 require additional work to determine if the group action extends to the boundary. As this property is desirable for answering questions like the Novikov Conjecture, we propose the following related question:

QUESTION 2. Can every $\mathcal{Z}$-structure be upgraded to an $\mathcal{EZ}$-structure?

Lastly, the discussion surrounding the boundary swapping technique in Section 3.2 and the importance of quasi-isometry invariants in the field of geometric group theory begs the following question.

QUESTION 3. Is the property of admitting a $(\mathcal{E})\mathcal{Z}$-structure a quasi-isometry invariant?

References

[1] M. Bestvina, *Local homology properties of boundaries of groups*, Michigan Math. J. **43** (1996), no. 1, 123–139, DOI 10.1307/mmj/1029005393. MR1381603

[2] M. Bestvina and G. Mess, *The boundary of negatively curved groups*, J. Amer. Math. Soc. **4** (1991), no. 3, 469–481, DOI 10.2307/2939264. MR1096169

[3] C. B. Croke and B. Kleiner, *Spaces with nonpositive curvature and their ideal boundaries*, Topology **39** (2000), no. 3, 549–556, DOI 10.1016/S0040-9383(99)00016-6. MR1746908

[4] F. Dahmani, *Classifying spaces and boundaries for relatively hyperbolic groups*, Proc. London Math. Soc. (3) **86** (2003), no. 3, 666–684, DOI 10.1112/S0024611502013989. MR1974394

[5] A. N. Dranishnikov, *On Bestvina-Mess formula*, Topological and asymptotic aspects of group theory, Contemp. Math., vol. 394, Amer. Math. Soc., Providence, RI, 2006, pp. 77–85, DOI 10.1090/conm/394/07435. MR2216707

[6] F. T. Farrell and J.-F. Lafont, *EZ-structures and topological applications*, Comment. Math. Helv. **80** (2005), no. 1, 103–121, DOI 10.4171/CMH/7. MR2130569

[7] R. Geoghegan, *The shape of a group—connections between shape theory and the homology of groups*, Geometric and algebraic topology, Banach Center Publ., vol. 18, PWN, Warsaw, 1986, pp. 271–280. MR925870

[8] R. Geoghegan and P. Ontaneda, *Boundaries of cocompact proper* CAT(0) *spaces*, Topology **46** (2007), no. 2, 129–137, DOI 10.1016/j.top.2006.12.002. MR2313068

[9] Craig Guilbault, Brendan Burns Healy, and Brian Pietsch, *Group boundaries for semidirect products with Z*, To appear Groups, Geometry, and Dynamics.

[10] C. Guilbault, M. Moran, and K. Schreve, *Compressible spaces and $\mathcal{EZ}$-structures*, Fund. Math. **256** (2022), no. 1, 47–75, DOI 10.4064/fm972-7-2021. MR4361585

[11] C. R. Guilbault, *Weak $\mathcal{Z}$-structures for some classes of groups*, Algebr. Geom. Topol. **14** (2014), no. 2, 1123–1152, DOI 10.2140/agt.2014.14.1123. MR3180829

[12] C. R. Guilbault and M. A. Moran, *A comparison of large scale dimension of a metric space to the dimension of its boundary*, Topology Appl. **199** (2016), 17–22, DOI 10.1016/j.topol.2015.12.001. MR3442590

[13] C. R. Guilbault and M. A. Moran, *Proper homotopy types and $\mathcal{Z}$-boundaries of spaces admitting geometric group actions*, Expo. Math. **37** (2019), no. 3, 292–313, DOI 10.1016/j.exmath.2018.03.004. MR4007577

[14] C. R. Guilbault and M. A. Moran, *Coarse Z-boundaries for groups*, Michigan Math. J. **73** (2023), no. 4, 693–718, DOI 10.1307/mmj/20206001. MR4634977

[15] C. R. Guilbault, M. A. Moran, and C. J. Tirel, *Boundaries of Baumslag-Solitar groups*, Algebr. Geom. Topol. **19** (2019), no. 4, 2077–2097, DOI 10.2140/agt.2019.19.2077. MR3995025

[16] D. Osajda and P. Przytycki, *Boundaries of systolic groups*, Geom. Topol. **13** (2009), no. 5, 2807–2880, DOI 10.2140/gt.2009.13.2807. MR2546621

[17] D. Rosenthal, *Continuous control and the algebraic L-theory assembly map*, Forum Math. **18** (2006), no. 2, 193–209, DOI 10.1515/FORUM.2006.012. MR2218417

[18] C. J. Tirel, *$\mathcal{Z}$-structures on product groups*, Algebr. Geom. Topol. **11** (2011), no. 5, 2587–2625, DOI 10.2140/agt.2011.11.2587. MR2836296

[19] J. E. West, *Mapping Hilbert cube manifolds to ANR's: a solution of a conjecture of Borsuk*, Ann. of Math. (2) **106** (1977), no. 1, 1–18, DOI 10.2307/1971155. MR451247

[20] K. Whyte, *The large scale geometry of the higher Baumslag-Solitar groups*, Geom. Funct. Anal. **11** (2001), no. 6, 1327–1343, DOI 10.1007/s00039-001-8232-6. MR1878322

DEPT. OF MATHEMATICS, THE COLORADO COLLEGE, COLORADO SPRINGS, COLORADO 80903
Email address: mmoran@coloradocollege.edu

SELECTED PUBLISHED TITLES IN THIS SERIES

812 **Ross Geoghegan, Craig R. Guilbault, and Kim Ruane, Editors,** Topology at Infinity of Discrete Groups, 2025

808 **Loring W. Tu, Editor,** Group Actions and Equivariant Cohomology, 2024

807 **Galina Filipuk, Editor,** Recent Progress in Special Functions, 2024

806 **Maximilian Engel, Hildeberto Jardón-Kojakhmetov, and Cinzia Soresina, Editors,** Topics in Multiple Time Scale Dynamics, 2024

805 **Anthony Iarrobino, Pedro Macias Marques, Maria Evelina Rossi, and Jean Vallès, Editors,** Deformation of Artinian Algebras and Jordan Type, 2024

804 **Mahir Bilen Can and Jörg Feldvoss, Editors,** A Glimpse into Geometric Representation Theory, 2024

803 **Peter Gothen, Margarida Melo, and Montserrat Teixidor i Bigas, Editors,** Moduli Spaces and Vector Bundles—New Trends, 2024

802 **Ralph M. Kaufmann, Martin Markl, and Alexander A. Voronov, Editors,** Higher Structures in Topology, Geometry, and Physics, 2024

801 **K. A. Brown, T. J. Hodges, M. Vancliff, and J. J. Zhang, Editors,** Recent Advances in Noncommutative Algebra and Geometry, 2024

800 **Avinoam Mann, Louis H. Rowen, David J. Saltman, Aner Shalev, Lance W. Small, and Uzi Vishne, Editors,** Amitsur Centennial Symposium, 2024

799 **Alberto A. Condori, Elodie Pozzi, William T. Ross, and Alan A. Sola, Editors,** Recent Progress in Function Theory and Operator Theory, 2024

798 **Marat V. Markin, Igor V. Nikolaev, and Carsten Trunk, Editors,** Advances in Functional Analysis and Operator Theory, 2024

797 **Sangita Jha, Mrinal Kanti Roychowdhury, and Saurabh Verma, Editors,** Recent Developments in Fractal Geometry and Dynamical Systems, 2024

796 **John Cremona, John Jones, Jennifer Paulhus, Andrew V. Sutherland, and John Voight, Editors,** LuCaNT: LMFDB, Computation, and Number Theory, 2024

795 **Michael A. Jones, David McCune, and Jennifer M. Wilson, Editors,** Mathematical Analyses of Decisions, Voting and Games, 2024

794 **Jean-Pierre Magnot, Editor,** Recent Advances in Diffeologies and Their Applications, 2024

793 **Abba Gumel, Editor,** Mathematical and Computational Modeling of Phenomena Arising in Population Biology and Nonlinear Oscillations, 2024

792 **Shaoming Guo, Zane Kun Li, and Brian Street, Editors,** Recent Developments in Harmonic Analysis and its Applications, 2024

791 **Mee Seong Im, Bach Nguyen, and Arik Wilbert, Editors,** Algebraic and Topological Aspects of Representation Theory, 2024

790 **Peter Crooks and Alexandru I. Suciu, Editors,** Compactifications, Configurations, and Cohomology, 2023

789 **I. S. Krasil′shchik, A. B. Sossinsky, and A. M. Verbovetsky, Editors,** The Diverse World of PDEs, 2023

788 **I. S. Krasil′shchik, A. B. Sossinsky, and A. M. Verbovetsky, Editors,** The Diverse World of PDEs, 2023

787 **Hemen Dutta, Editor,** Mathematical Modelling, 2023

786 **Hemen Dutta, Editor,** Mathematical Modelling, 2023

785 **A. Leroy and S. K. Jain, Editors,** Algebra and Coding Theory, 2023

784 **Dinh-Liem Nguyen, Loc Hoang Nguyen, and Thi-Phong Nguyen, Editors,** Advances in Inverse Problems for Partial Differential Equations, 2023